APPLICATIONS OF SYSTEMS APPROACHES AT THE FIELD LEVEL

Systems Approaches for Sustainable Agricultural Development

VOLUME 6

Aims and Scope

The book series *Systems Approaches for Sustainable Agricultural Development* is intended for readers ranging from advanced students and research leaders to research scientists in developed and developing countries. It will contribute to the development of sustainable and productive systems in the tropics, subtropics and temperate regions, consistent with changes in population, environment, technology and economic structure.

The series will bring together and integrate disciplines related to systems approaches for sustainable agricultural development, in particular from the technical and the socio-economic sciences, and presents new developments in these areas.

Furthermore, the series will generalize the integrated views, results and experiences to new geographical areas and will present alternative options for sustained agricultural development for specific situations.

The volumes to be published in the series will be, generally, multi-authored and result from multi-disciplinary projects, symposiums, or workshops, or are invited. All books will meet the highest possible scientific quality standards and will be up-to-date. The series aims to publish approximately three books per year, with a maximum of 500 pages each.

The titles published in this series are listed at the end of this volume.

Applications of Systems Approaches at the Field Level

Volume 2

Proceedings of the Second International Symposium on Systems Approaches for Agricultural Development, held at IRRI, Los Baños, Philippines, 6–8 December 1995

Edited by

M. J. KROPFF

Wageningen Agricultural University, Wageningen, The Netherlands

P. S. TENG

International Rice Research Institute, Manila, Philippines

P. K. AGGARWAL

Indian Agricultural Research Institute, New Delhi, India

J. BOUMA

Wageningen Agricultural University, Wageningen, The Netherlands

B. A. M. BOUMAN

Research Institute for Agrobiology and Soil Fertility (AB-DLO), Wageningen, The Netherlands

J. W. JONES

University of Florida, Gainesville, USA

H. H. VAN LAAR

Wageningen Agricultural University, Wageningen, The Netherlands

IRRI

KLUWER ACADEMIC PUBLISHERS

DORDRECHT / BOSTON / LONDON

in cooperation with

International Rice Research Institute

International Consortium for Agricultural Systems Applications

Library of Congress Cataloging-in-Publication Data is available.

ISBN 0-7923-4286-0 (vol. 2)
ISBN 0-7923-4285-2 (vol. 1)
ISBN 0-7923-4287-9 (set)

Published by Kluwer Academic Publishers,
P.O. Box 17, 3300 AA Dordrecht, The Netherlands.

Kluwer Academic Publishers incorporates
the publishing programmes of Martinus Nijhoff,
Dr W. Junk, D. Reidel, and MTP Press.

Sold and distributed in the U.S.A. and Canada
by Kluwer Academic Publishers,
101 Philip Drive, Norwell, MA 02061, U.S.A.

In all other countries, sold and distributed
by Kluwer Academic Publishers Group,
P.O. Box 322, 3300 AH Dordrecht, The Netherlands.

Printed on acid-free paper

Printed and bound in Great Britain by
Hartnolls Limited, Bodmin, Cornwall

Contents

Preface

Agricultural production increased significantly in the second half of this century, mainly as a result of intensification through increasing external inputs per unit of land. Therefore, the enormous increase in food demand as a result of population growth from 1.6 billion in the early 1900s to over 5.7 billion today could be met at a global scale. The supply of food has evolved into a complex system of production activities, land-use needs, industrial processes, trade, market and price mechanisms, and national and international policies. In spite of the past successes, food production must be strongly increased further in the coming decades to meet the continuing demand created by growing populations. This production increase has to be achieved on less land with less labour, less water, and less pesticides and must be sustainable through conserving scarce natural resources. To meet the challenges of increasing food production, ways have to be found to improve the productivity, profitability, and sustainability of the agricultural production systems.

The increased production per unit of land area by increased external inputs – fertilizers and biocides – and often large-scale rural reconstruction, have led to irreversible changes in the landscape, soil quality, diversity of the environment, and the natural resource base by pollution. This has resulted in major questions worldwide in relation to the sustainability of agricultural production systems.

The complexity of the problems and conflicting objectives call for systems approaches in which issues are addressed in an integrated, multidisciplinary way. Systems thinking and systems simulation are indispensable tools for that purpose.

It was towards this end that the theme of this book, and the symposia series, were conceived. Systems approaches for agricultural development are needed if rational strategies for incorporating the role of agriculture into national development are to be made. Mathematical models and computer simulation provide objective tools for applying science to determine options for resource management at field, farm, and regional scales. Yet, these tools will not be very useful unless they incorporate social and economic dimensions. The second SAAD symposium attempted to foster this link between the biophysical and social sciences in the selection of keynote papers and oral presentations dealing with issues at the regional, farm, and field level. This is a further reflection of how systems approaches have definitely moved beyond the research mode into the application mode. The large number and quality of interdisciplinary research teams in different parts of the globe, working to determine land-use options that will meet multiple goals and yet sustain natural resource bases, is a key indicator of this 'coming of age'.

At the field level, the optimization of resource use is a key issue to achieve the different goals with respect to food supply, income, and protection of the environment.

That implies the optimal use of genotype×environment interactions by adjusting genotype and management to local conditions and objectives. Systems approaches are increasingly being used intensively for this purpose to optimize breeding efforts, to determine yield potential in different environments, to optimize water and N use at the field level and to improve crop protection by searching for opportunities for disease prevention and the use of natural enemies to minimize pesticide requirements.

Research on water and nutrient management has conceptually changed in the past decade because of environmental problems with, for example, nitrogen in groundwater and the competition for water resources among various sectors in the society like industry, households, and agriculture. Much research is geared towards matching the seasonal pattern of supply to the demand of the crop at each stage of development to achieve maximum resource-use efficiency and to minimize, for example, nutrient losses to the environment. Optimization of resource use, therefore, often serves different objectives that seem to be contrasting.

Concepts in crop protection have changed in the past decades from exclusion or destruction of pests to pest management. Serious problems with pesticides, such as rapid development of pest resistance, environmental effects of pesticides, and costs, have triggered the development of new approaches and techniques in pest management based on improved knowledge of pest dynamics and their natural enemies, and the interaction between the pest and the crop. An improved understanding of the system will help to identify new control techniques by indicating intervention points and can be useful in decision making in pest management.

This book contains a subset of the papers presented at the Second International Symposium on *Systems Approaches for Agricultural Development (SAAD)*, held at the International Rice Research Institute (IRRI), Los Baños, The Philippines, 6–8 December 1995. The papers in this book address issues at the field level while the companion volume, similarly named, contains papers which address issues at the farm and regional levels. Together, their contents reflect the objectives of SAAD, which were:

- To review the status of applications of systems research and modelling in agricultural research, with special focus on countries where agricultural development is facing major challenges.
- To promote international collaborative activities and to increase awareness of the opportunities for using systems approaches as a tool in research and planning.

We believe these objectives were met by the more than two-hundred scientists from thirty-two countries, when they came together in December 1995.

Production of this volume would not have been possible without the tireless efforts of a support team made up of Ms. J.M.C.A. ‘Peewee’ Cabrera, Ms. R. ‘Say’ Badrina, and Ms. Viring Lantican. Thanks are also due to the Communication and Publication Services of IRRI for editorial assistance.

The second SAAD symposium was guided by an International Steering Committee comprising J.T. Ritchie, J.B. Dent, J.W. Jones, H.F.M. Ten Berge, M.J. Kropff and P.S. Teng, while local organization at Los Baños was done by T.P. Tuong, S. Pandey, K.G. Cassman, G.J.D. Kirk and P.S. Teng. The SAAD symposium was sponsored by IRRI, the International Consortium for Agricultural Systems Applications (ICASA) and by the Simulation and Systems Analysis for Rice Production (SARP) project. We thank IRRI for its financial, logistical and moral support, without which the symposium and this book would not have been possible. The following persons actively contributed to the logistical support during the symposium: Ms. Viring Lantican, Ms. Susan Telosa, Ms. J.M.C.A. 'Peewee' Cabrera, Ms. Anaida Ferrer, Ms. Cecille Lopez, Ms. Perly Villamayor, and Mr. Benjie Nunez.

Lastly, we hope that the two books from this second international symposium on systems approaches will be further testimony of the vision and philosophy provided by the late Professor C.T. De Wit, Wageningen, The Netherlands, which has guided many applications of the approach.

M.J. Kropff
P.S. Teng
P.K. Aggarwal
J. Bouma
B.A.M. Bouman
J.W. Jones
H.H. Van Laar

The challenge of integrating systems approach in plant breeding: opportunities, accomplishments and limitations

P.K. AGGARWAL[1,2], M.J. KROPFF[3], P.S. TENG[1] and G.S. KHUSH[1]

[1]*International Rice Research Institute (IRRI), P.O. Box 933, 1099 Manila, Philippines*
[2]*Division of Environmental Science, Indian Agricultural Research Institute, New Delhi 110012, India*
[3]*Department of Theoretical Production Ecology, Wageningen Agricultural University, P.O. Box 430, 6700 AK Wageningen, The Netherlands*

Key words: plant breeding, simulation, models, plant type, G×E, hybrids, yield potential

Abstract
Plant breeders face enormous challenges in the coming decades to greatly increase food production with less land and higher input use efficiency. Crop breeding has to be more knowledge-intensive and should integrate information from related disciplines to achieve this goal. Systems tools, such as crop simulation models, data bases, geographical information systems (GIS), and optimization and risk analysis techniques, facilitate integration of the effects of climatic, edaphic, management, biotic and socioeconomic factors on crop growth, productivity, and profitability. These tools provide us with an opportunity to accelerate the use of interdisciplinary knowledge in crop breeding. They are useful for evaluating hypotheses and setting breeding goals for different physiological traits in various agroenvironments, using historical weather data and techniques of risk analysis. Physiologically sound models may also assist in extrapolating the performance of a crop population from one site to another, in reducing the number of locations in varietal evaluation programmes, and in describing why a particular variety does not perform well in all locations. They can also play a significant role in determining strategies for minimizing damage to germplasm caused by various pests in different regions. Alternatively, simulation models can assess the relative performance of plant types in scenarios involving various levels of pest infestations in different agro-ecosystems. In this paper, we discuss the expectations of plant breeders from crop physiology and modelling and explore the opportunities offered by the systems approach, review some case studies, assess limitations on the use of models in breeding, and finally comment on future needs.

Introduction

The continued increase in population pressure, urbanization, and income growth is expected to lower food availability in developing countries unless corrective measures are taken (Pinstrup-Anderson and Pandya-Lorch 1995). It is estimated that Asian rice production alone must increase to more than 800 million tons over the next 30 years from the present level of 480 million tons (Zeigler et al. 1996). Increase in area under cultivation does not look feasible in much of the developing world. Further production increase has to come, therefore, from increased productivity per unit area. Zeigler et al. (1996) estimated that irrigated rice yield in Asia must increase by 70–80%, if rainfed yields do not change (Figure 1). The required increase in the irrigated rice productivity will be 50–60%, even if yield in rainfed ecosystems can be increased.

Input use, particularly fertilizer use, is still suboptimal in large parts of Asian countries such as in eastern India. Marginal productivity of input use is high in these

M.J. Kropff et al. *(eds.): Applications of Systems Approaches at the Field Level, 1–23.*

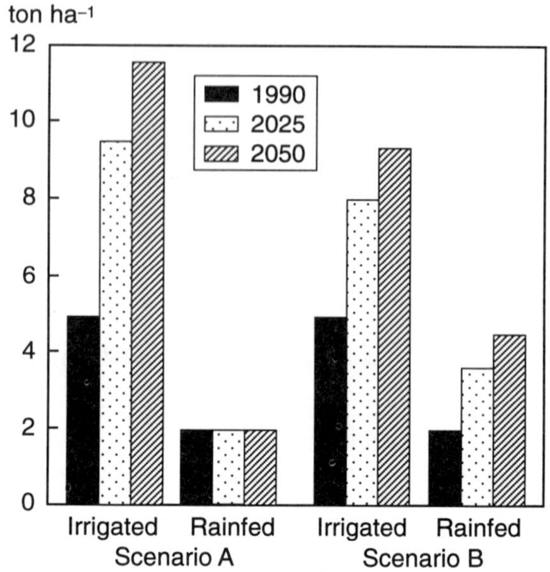

Figure 1. Rice yields needed to meet the projected demands. Scenario A: no change in rainfed rice yields. Scenario B: improvements in average yields in irrigated and rainfed sections (*Source:* Zeigler et al. 1996).

areas, focus must, therefore, be on variety development and improved management. Large yield gaps may be observed in these areas and they can still contribute a lot to food production, provided policy support in terms of infrastructure, credit, and others is put in place to increase and rationalize input use. However, the present food security is dependent on favourable areas that are already high-yielding, such as Punjab in India or Central Luzon in the Philippines. Aggarwal et al. (1995) reported that the economically recoverable gap in such cases may be very small. It is, therefore, essential to develop varieties that will produce more in favourable areas, preferably with a small increase in input use.

At the same time, grain yields in experimental farms are stagnating. The yield potential of rice in tropics has not increased above 10 ton ha^{-1} since IR8 was released 30 years ago, despite significant achievements in attaining yield stability, increasing per day productivity and improving grain quality (Figure 2). The small decline in productivity during the 1980s may be due to decreased soil fertility (Cassman et al. 1995). There is an urgent need to secure past yield gains and further increase the yield potential of rice and other major food crops in favourable areas. From the plant breeding perspective, the principal challenges can be summarized as follows:

- Attaining higher yield potential in favourable environments.
- Increasing stability and yield potential in variable environments.
- Incorporating durable disease and insect resistance.
- Improving quality of economic products.
- Improving efficiency of breeding methods and selection procedures.

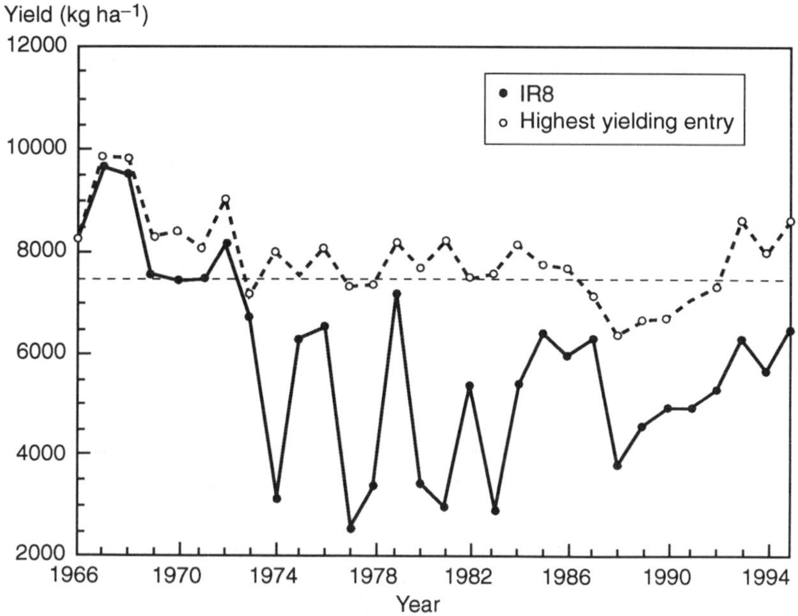

Figure 2. Grain yield of IR8 and the highest yielding single entry in the dry season replicated yield trials with 300–400 elite lines conducted each year at IRRI (G.S. Khush, IRRI, pers. commun.).

In meeting these challenges, plant breeders expect scientific support from disciplines such as genetics, pathology, entomology, physiology, and more recently, tissue culture and molecular biology. Direct use of physiological knowledge in plant breeding has been limited (Rasmusson 1987; Marshall 1991), although Donald (1968), with his concept of plant ideotype, laid the foundation for collaborative research between breeders and plant physiologists. Recent availability of process-based crop models and other systems approaches has provided yet another opportunity to accelerate the use of knowledge from plant physiology and other disciplines in agricultural planning and development. There have been rapid developments in the field of systems approaches in the last two decades (Penning de Vries et al. 1993; Teng and Penning de Vries 1992; Goldsworthy and Penning de Vries 1994). Most of this research was focused on developing a better understanding of crop-weather interactions and on improving improved agronomic production technology. In this paper, our objective is to present a critical review of opportunities offered by systems approaches in plant breeding, summarize recent accomplishments, and discuss future needs.

Higher yield potential: design and evaluation of new plant types

Opportunities

The challenge in ideotype breeding is the identification of critical traits, which in combination with each other will result in higher yield potential. Donald (1968) pioneered research on ideotypes. Today, several ideotypes for major crops have been proposed by different scientific groups. Not many breeders, though, practise ideotype breeding because of lack of proof that a specific trait is important, the large number of traits (as against one trait yield involved), limited genetic variability in suitable donors, and pleiotropic effects (Marshall 1991). The problem gets confounded with the difficulty in finding suitable, efficient screening methods for some of these traits. Molecular approaches may gradually become available for marker-aided selection for yield components and other quantitative traits (Bennett et al. 1994). For example, lack of genetic variability in *indica* rice, which earlier limited the choice of donors to a primary gene pool, is no longer a major constraint. Tissue culture techniques such as embryo rescue and protoplast fusion now allow breeders to exploit genetic variability even from secondary and tertiary gene pools.

Crop simulation models may be useful to determine in a relatively smaller time frame the optimal level of complex traits for increasing yield potential. They may also be useful for examining hypotheses and for setting breeding goals for different traits using historical weather data and techniques of risk analysis. Sensitivity analysis of model parameters is analogous to creation of genetic isolines since only one parameter is changed while keeping the rest of the plant characteristics constant.

A methodological framework for using crop models in the design and evaluation of plant types has been proposed by Aggarwal et al. (1996b). This approach requires a crop simulation model that has been well evaluated in the target environment (Figure 3). The critical crop parameters in the target environment are determined by means of a sensitivity analysis of traits. For these sensitive parameters, the level of genetic variation available in the germplasm is determined from the literature or measured in available germplasm. Alternatively, options to increase variability by genetic means need to be assessed. It is important to determine the range and boundaries of this variation, especially in the parents used in the crossing programmes. Simultaneously, the heritability of such traits and the associated pleiotropic effects must be assessed.

Having determined the extent of variation in critical and heritable traits, hypothetical varieties can be constructed with variation in one or many traits. The Monte Carlo simulation technique, commonly used in quantitative population genetic studies (Crosby 1973), can be used for creating such varieties. In this approach, different traits are considered to be stochastic. This technique allows us to determine the extent of divergence required from the mean value (check variety) of different traits and their consequence on the behaviour of the system (yield potential). The requirement here is to determine the distribution properties of the trait. A large number of hypothetical varieties can be simulated by this approach, each of these varieties is unique in its trait 'make-up'. The varieties 'created' by this approach thus mimic the random segregation behaviour of progeny for different characters when two parents are hybridized. The yield of such hypothetical varieties can then be determined for the target environment. The results also indicate the importance of different traits in a specific environment, as well as set the breeding goals for individual traits.

New techniques like the use of restriction fragment length polymorphism (RFLP)

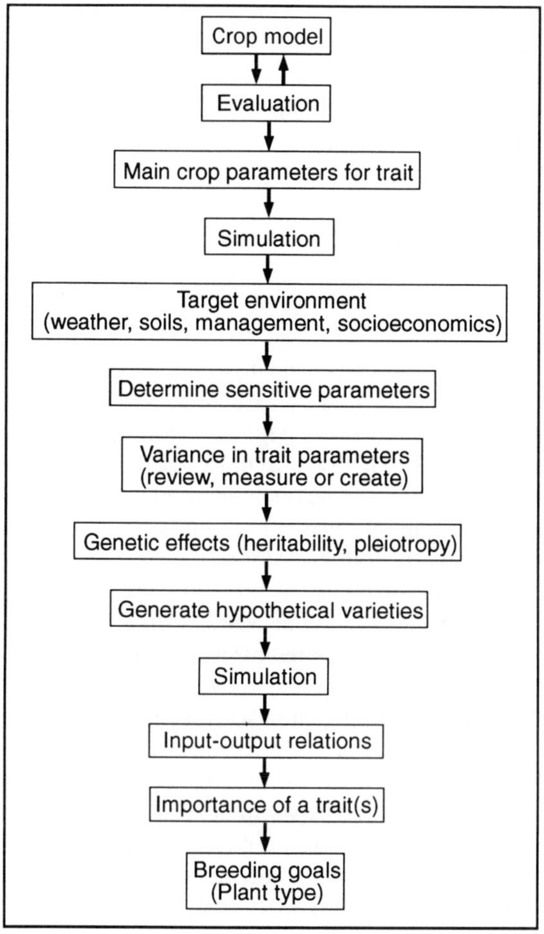

Figure 3. A framework for the use of crop models in design and evaluation of plant types for specific environments (*Source:* Aggarwal et al. 1995).

markers enable tagging of genes. This may lead to marker-aided selection instead of phenotypic selection. By the use of genetically marked chromosomes, it is possible to detect and locate the loci affecting quantitative traits (quantitative trait loci or QTL). Dense genetic linkage maps now exist for many species, including rice (McCouch et al. 1988). The availability of molecular markers and the development of new powerful statistical tools (Jansen 1992) have greatly enhanced the opportunities to unravel complex QTLs into individual genetic factors localized on genetic maps. However, knowledge on the genetic basis of yield-determining traits is strongly complicated by the interactions of the different genotypic traits with environmental factors. Ecophysiological crop models may be the extra step that may help bridge the gap between genetic information on individual traits and the performance of the crop systems in different environments. Although model parameters cannot be seen as true genetic characters (see later in Limitations) in as much as environmental factors may still cause phenotypic variation, these parameters should be easier to link to the genetic component than complex traits like yield ability because the models separate the different processes that determine this yield ability and explicitly account for environmental effects on these processes. It would be challenging to determine if the yield ability can be predicted on the basis of genetic mapping of QTLs in combination with ecophysiological crop models.

Accomplishments

There are numerous examples where crop models have been employed to determine critical factors influencing growth, development and yield of rice, wheat, sorghum, maize, cotton, soybean, peanut, potato, and many other crops (see reviews by Boote and Tollenaar 1994; Kropff et al. 1995). In most cases, the impact of changing traits related to the duration of different phenophases, leaf area development, carbohydrate assimilation and partitioning, N uptake, partitioning and remobilization, and sink capacity was evaluated. It is not our intention to review all of these. We will illustrate this approach with examples from rice.

Dingkuhn et al. (1991) and Penning de Vries (1991) used the crop growth model MACROS to determine the traits required for higher yield potential. They concluded that new rice plant types should possess traits related to enhanced foliar growth during crop establishment in combination with reduced tillering, less foliar growth during later stages, greater export of assimilates from leaves to stems, steeper slopes of the vertical N gradient in the leaf canopy, expanded capacity of the stems to store assimilates, and improved sink capacity with a prolonged grainfilling duration. It is difficult to screen for many of these physiological traits in a large number of genotypes. Some of the morphological traits (low tillering, increased spikelet number for sink capacity) suggested by these simulation approaches also find mention in the present ideotype breeding effort at the International Rice Research Institute, Philippines (Peng et al. 1994).

Kropff et al. (1994b) and Aggarwal et al. (1996b), using a related model ORYZA1 (Kropff et al. 1994a), concluded that any individual trait cannot provide a large advantage in increasing yield potential and that a simultaneous increase in specific leaf nitrogen particularly during grainfilling (stay green), sink capacity, and grainfilling duration is important for increasing rice grain yield potential. Their results also indicated that the relative importance of traits changes with season and year. By using historical weather data of 30 years for the target environment, it was also possible to simulate the probability of higher yields in climatically variable tropical environments and to determine alternative options for similar increase in yield potential (Figure 4).

More recently, Aggarwal et al. (1996a) used a modified version of the ORYZA1 model to determine the impact of simultaneous change in leaf area, nitrogen concentration, duration of different phenophases, and sink capacity on irrigated grain yield

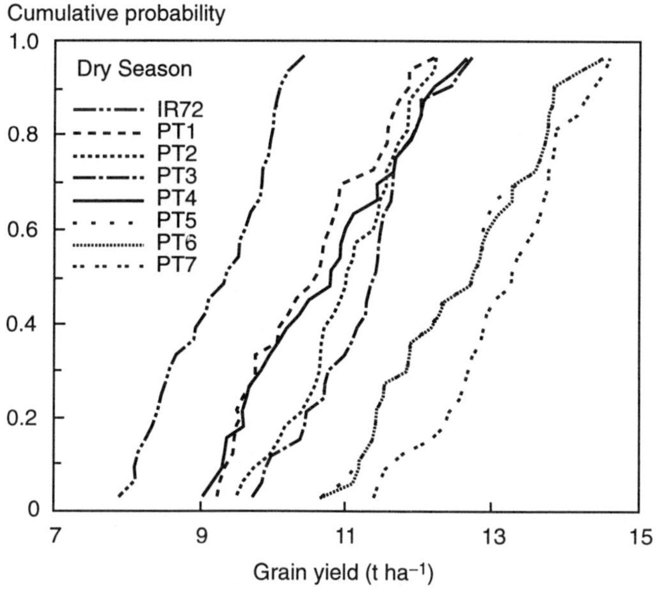

Figure 4. Cumulative probability distributions of simulated grain yield for various rice plant types in the dry season at Los Baños, Philippines (*Source:* Aggarwal et al. 1996b).

at different levels of agronomic management. They used the Monte Carlo simulation techniques to create 500 hypothetical varieties which had random combination of values for different traits. The results showed that IR72, the check variety, has large yield potential that can be realized with better N management. An important conclusion of this simulation study was that at the level of N management practised currently in breeders' plots, the search for higher yield potential may remain elusive, even though the germplasm being screened may possess traits for higher yield potential. These traits need a better N environment in order to express their full potential. Consequently, several lines, although with substantially higher yield potential than IR72, would never be selected since they would yield lower than the check variety under current N management practices.

This example illustrates the fact that significantly higher level of N inputs and their precise management are required for attaining higher yield potential. In such environments, increasing sink capacity is necessary and this can be obtained by increasing the length of the panicle formation phase, or increasing the spikelet growth factor, or achieving higher potential grain weight. Variability in all three factors is well known. In addition, there is a need to maintain high leaf N content and grainfilling duration.

In 1989, IRRI decided to concentrate its efforts on developing rice varieties with 30–50% higher yield potential than the existing semidwarf varieties for tropical environments (IRRI 1989; Khush 1990). This is a rare example of a massive ideotype breeding effort. Based on conceptual, qualitative, physiological models and simulation models (Dingkuhn et al. 1991; Penning de Vries et al. 1993) and genetic studies, a new plant type for direct seeded rice was proposed. The main characteristics of this plant type are: low tillering capacity with 3–4 panicles per plant when direct seeded, no unproductive tillers, 200–250 grains per panicle, 90–100 cm tall, sturdy stems, vigorous root system, multiple disease and insect resistance, growth duration of 110–130 days, harvest index of 0.6, and 13–15 ton ha^{-1} yield potential.

Increased spikelet number establishes greater sink capacity, but as indicated by our simulation results, it is hard to expect increase in yield potential unless agronomic management is optimized; time available for grainfilling duration is extended and leaf N is maintained at a relatively higher level during the filling period. There is also a need to develop a strategy to increase harvest index to 60% in the new plant type. Greater harvest index is generally a result of increased grainfilling duration, higher leaf N concentration during filling (stay green), and sink capacity. Variation in rice grainfilling duration is not well known in rice (Yoshida 1981; Dionora and Kropff 1995), although it is documented in several other crops (Boote and Tollenaar 1994). Greater efforts are being made now to determine variability in this trait in rice as well (G.S. Khush, IRRI 1995, pers. commun.). However, the variation reported in grainfilling duration needs to be carefully examined because correct estimation of variation requires accurate, repetitive, and tedious measurements of number and weight of seeds and definitions of flowering and physiological maturity.

Breeding lines with characteristics similar to those listed above for the new plant type are now available. Insect and disease resistance and suitable grain quality still need to be incorporated into these lines. At present, these lines are being tested in the field. Preliminary agronomic experiments done with these new lines are confounded by pest problems, and it is, therefore, difficult to draw conclusions about the success of the strategy adopted, as well as to test the validity of our simulation results in the field environment.

It can be concluded that simulation models can greatly assist in evaluating the suitability of plant types and in suggesting alternatives. At the same time, they can also help in fine-tuning the required management practices for the best expression of the desired traits, thus increasing the efficiency of crop improvement programmes. However, there are many factors that simulation models ignore. Lodging and pest and diseases increase severally in tropical environments when N application is increased. It is likely that current management practices of moderate N application have been adopted to overcome these constraints. There is a need to examine the 'trade-off trap' between yield loss due to pests and lodging and the increased yield potential due to higher N application to determine if higher yield potential is really possible in field environments.

Exploitation of heterosis

Opportunities

A major constraint to utilizing hybrid seed technology is the need to buy fresh seeds every planting season and the high cost of seed production. Using the available technology of hybrid rice seed production (Virmani and Sharma 1993), yields of

1–2 ton ha^{-1} of hybrid rice seed and 1.5–2.0 ton ha^{-1} of pollen parent seed have been obtained. This limited seed production is largely due to poor outcrossing and asynchronous flowering of male and female parents. Environmental factors such as temperature, humidity, and wind speed at the time of pollination and fertilization play a great role in regulating outcrossing percentage. To synchronize flowering of male and female parents to increase hybrid seed production, trials are conducted at several locations to determine the optimal planting dates. It is a challenge to crop modellers to develop suitable models that simulate the various processes that affect the magnitude of outcrossing and thus seed setting. Such models, if available, would be of great value to plant breeders in developing strategies to increase hybrid seed production. National agricultural research systems and private seed companies are making efforts to identify suitable areas and seasons for hybrid seed production. Systems approaches can very well provide this information in a relatively shorter time frame.

Accomplishments

Crop simulation models that can assist in making decisions about the planting calendar of parents in different environments are now available (Xu 1995). These models simulate and predict the development process of a variety or parental line used in hybrid seed production in a particular environment, as defined by its temperature and daylength. Xu (1995) showed that simulation models predicted optimal planting dates for several parental lines that were very close to the planting dates determined by a large number of experiments. Ideally, dependable weather forecasts (temperature alone) should be available to make accurate predictions. Until good-quality weather forecasts are available, reasonable guidance can also be obtained by using good-quality, historical daily weather data for a few years for the target locations.

Simulating the entire process of outcrossing, which involves flower opening, anthesis, pollen dispersal, pollination, fertilization, and survival of the young seed (embryo), may not be easy to accomplish in view of the limited knowledge available about the mechanisms involved.

Increased productivity in abiotic stress environments

Opportunities

Ecoregional development is the new paradigm of agricultural research. The main guiding principle of this approach for developing countries is increasing crop productivity with efficient use of external inputs such as fertilizers, irrigation, labour, and pesticides while maintaining or increasing the natural resource base. Because different processes may be more important in various production situations, new plant type designs have to be developed for a specific production environment. In situations where there is a considerable yield gap, it might be more efficient to first identify the major constraints and management strategies to improve resource use efficiency for the limiting factors or to reduce pest effects before new crop plant types are designed. Breeding efforts, which is a long-term investment, can then be focused on the major problems that cannot be solved by improved management. For example, recent studies conducted at IRRI have shown that the 'yield decline' in rice observed in experimental fields could be corrected by suitable nutrient management (Cassman et al. 1995). Crop models can greatly facilitate the identification of critical yield-limiting factors. Quantitative understanding of the impact of limiting resource on crop yield on the basis of effects on the different processes through crop models can be instrumental in targeting genotypes to different environments. A large number of traits have been proposed earlier for drought-prone environments (Ludlow and Muchow 1990). Simulation models can assist in determining their real worth in variable environments.

Accomplishments

There are a number of instances where systems approaches have been used to determine critical traits in environments that experience water and nutrient shortages. An interdisciplinary team of scientists had formulated an intermediate plant type of rice that would yield higher in rainfed environments of eastern India. When the hybridization-selection process was initiated, O'Toole and Jones (1987) used regional soil physical characteristics and historical weather data to run a simulation model to test the validity and feasibility of this conceptual plant type. The model was sensitive enough to differentiate between alternative plant types. The simulation results supported the hypothesis that the conceptual intermediate plant type would have higher yield potential than the current traditional type and greater year-to-year stability of yield than the high-yielding types.

Muchow and Carberry (1993) conducted a detailed simulation study for evaluating performance in terms of yield and risk of several crops (sorghum, maize, and kenaf) in the semiarid tropics in Australia. Simulated yield probabilities for different plant types showed that matching crop phenology to the expected water supply is a difficult concept in the climatically variable semiarid tropics and is strongly dependent on attitudes of risk aversion. Crops with higher yield potential were advantageous both in terms of yield and risk in all species despite severe drought stress. The study also showed that the magnitude of crop response to a changed trait varied from year to year, highlighting the limitation of the conventional approach of selecting over a few seasons.

In several rice-growing environments, water may not be a constraint but application of nutrients may be suboptimal. Although current low prices of urea promote N application, policy interventions for environmental control may alter this situation, exposing the crops to N stress. There are few studies where the impact of plant traits on performance in suboptimal N environments has been studied. Aggarwal et al. (1996a) showed that at modest rates of N application, feedbacks among various physiological processes do not allow any plant type to yield significantly larger than the currently available varieties. Increase in grainfilling duration in N-limited environments, however, may increase grain yields by 10%.

Resource-limited environments may have many constraints operating at the same time. There are a number of regions where production system is characterized by subtropical availability of water, nitrogen, phosphorus, and many other nutrients, and weed and pest control. Simulation studies have not been conducted for such multiconstraint environments, largely because models are not yet available to handle all stresses. In the future, the development of such models should be a priority.

Despite this limitation, it is evident that simulation models have a great role to play in determining critical plant traits and setting breeding goals for them in abiotic stress environments. Coupled with soil physical and environmental data bases, they provide powerful tools for hypotheses evaluation that may not be possible in conventional experiments.

Weed control

Opportunities

Direct seeding of rice is catching up in many parts of Asia because of the availability of herbicides and unavailability of labour. Greater use of pesticides may not be environment-friendly. We, therefore, need to develop varieties with greater competitive ability *vis-à-vis* weeds. Competitive ability differs between crops and between cultivars of the same species. Often there is a negative correlation between competitive ability and yield potential (Moody and De Datta 1982). Crop models may be able to suggest plant traits that will provide the rice plant with greater competitive ability against weeds and yet not compromise yield potential.

Accomplishments

An ecophysiological simulation model for interplant competition was used to identify traits that determine the competitive ability of a crop (Kropff and Van Laar 1993). The most important traits were rapid early leaf area development, number of tillers and plant height, and more horizontally oriented leaves in early growth stages (vertical ones later on because of yield potential). In experiments, rice varieties that differed in these traits were evaluated with respect to their competitive ability versus a standard purple variety. The variety with all required traits, Mahsuri, reduced the growth of the purple variety much more than a modern variety such as IR50 (M.J. Kropff and S. Sarkarung 1992, IRRI, pers. commun.).

Increased and more durable resistance to insect pests and diseases

Opportunities

Plant breeding efforts to incorporate insect and disease resistance into crops have been very successful. The availability of molecular techniques has further enhanced the ability to incorporate genes for insect and disease resistance in crops. Such bioengineered plants may have an element of risk when introduced for cultivation.

Breeders would like to incorporate durable resistance to pests in major crops but it is difficult to breed for because durability of resistance is defined *a posteriori*.

Systems approach can facilitate in predicting future performance of a cultivar in multiple and evolving pest scenarios (Teng and Yang 1993). They provide a conceptual framework where crop improvement and pest management are related to other activities of the agroecosystem and enable prediction of potential or actual effect at lower (e.g. gene, tissue) and higher (e.g. community, agroecological zone) levels. Systems models with these linkages are, thus, powerful tools to guide research on improving agricultural systems by breeding more relevant crop cultivars and by deploying more rational pest management practices. These approaches have been used in recent times to investigate the dynamics of pest population and frequency and intensity of disease pressures and to determine opportunities for reducing indiscriminate use of pesticides (Teng 1985). The development and use of pest-crop models will provide a better understanding of the epidemiological effects that various resistance components may have on reducing the rate of pest infection/epidemics. Such information will help in developing better screening methods for evaluating pest-resistant varieties, as well as traits required in a variety under different agronomic and pest environments. Alternatively, simulation models can also be used to assess the relative performance of proposed plant types under scenarios of various levels of pests in different agroecosystems.

Accomplishments

Infection efficiency, latent and infectious period, sporulation intensity, and lesion growth rate are the five major resistance components of epidemiological importance. These resistance components are part of the life cycle of most pathogens, where a single generation is commonly termed a monocycle (Teng et al. 1977). For diseases such as potato late blight and rice blast, epidemic models have proven useful to determine the relative importance of the components of partial resistance – latent period, receptivity, and sporulation. A simulation model was used to determine the critical components of resistance against *Phytophthora* late blight, a major disease of potato (Van Oijen 1992a,b). The disease progress rate was most sensitive to changes in lesion growth rate, followed by infection efficiency and infectious period and latent period. When the available genetic variation was considered, the results showed that both lesion growth rate and infection efficiency are the important characteristics for improving partial resistance to potato late blight. Genetic variation in latent period and infectious period was small. Simulation studies conducted on barley leaf rust also confirmed that resistance components such as infection efficiency and latent period exert more influence on the rate of epidemic development than length of infectious period or sporulation capacity (Teng et al. 1977).

In rice blast disease caused by *Pyricularia grisea*, infection efficiency has been observed to differ significantly between rice cultivars when tested against the same pathotype (Calvero et al. 1992). When these data are used as input to a blast simulation model, they explain the differences in blast epidemics caused on cultivars such as

IR50 and IR72, both of which are known to be susceptible to the disease, but do show differential severity depending on prevailing weather conditions. This points to a useful application of epidemic models, which are based on the individual life cycles of pathogens (monocycles) but allow integration of these monocycles over time, thereby providing a quantification of the polycycle (Teng 1985). Small variation in monocyclic components, such as the infection efficiency and lesion growth rate, can have large effects at the polycyclic level, and because many seedling tests to evaluate resistance use monocyclic data, it can potentially underestimate the value of partial resistance which is a polycyclic trait (Zadoks and Schein 1979).

Systems models that explain the interactions between environment and pathogen adaptation provide guidance on the longevity of resistance. For specific diseases, it is now possible to design deployment strategies that optimize the useful life of major genes in specific environments. These deployment strategies translate into breeding tactics such as 'smart pyramiding' of several genes into the same genotype or selective release of resistance genes in time (Mundt 1994). The spatial deployment of single genes for rice blast and of gene mixtures in the form of multilines or cultivar mixtures has been demonstrated using farmer participatory approaches, but it has been made possible only through information generated with molecular markers (Nelson and Leung 1994).

At a regional level, GIS requires environmental data coupled with epidemic simulation models to further provide geographic delineation of disease and insect pest risk zones, such as those for rice blast (Luo et al. 1996). These zones can also help us in making strategic decisions on deployment strategies for varieties and to determine how long host plant resistance would be expected to last. Historical climate data from sites have been shown to be useful for characterizing the conduciveness of a site to specific wheat diseases (Coakley 1990). Song and Heong (1994) recently provided another illustration where they used a GIS to construct a spatial data base to study pest distributions based on pest surveillance data collected from 152 stations in South Korea. They found that the pattern of insect species changed over time and was related to the cultivation of *japonica* and *indica-japonica* hybrid varieties.

Relative rankings of key and minor pests may also be obtained from combinations of pest profiles and damage matrices, together with statistical techniques like correspondence analysis (Teng and Savary 1992). All of these lead to more locality-suited rice genotypes.

Achieving higher yield potential (see above) in crop plants in the tropics would require a much higher level of input use than is done at present (see earlier sections). Many of the diseases such as blast and bacterial sheath blight are extremely sensitive to microclimate created by luxurious foliar growth in response to nutrients. Perhaps in such situations, our current knowledge and management of many diseases will not be applicable. To address this issue, we need to extrapolate our current information on pathosystem-environment interaction to the targeted systems of higher yield potential and input use. Recent systems analysis and simulation studies have shown that as rice yield increases, incidence of several pests increases much more, whereas incidence of weeds remains the same or decreases (Savary et al. 1996). Therefore,

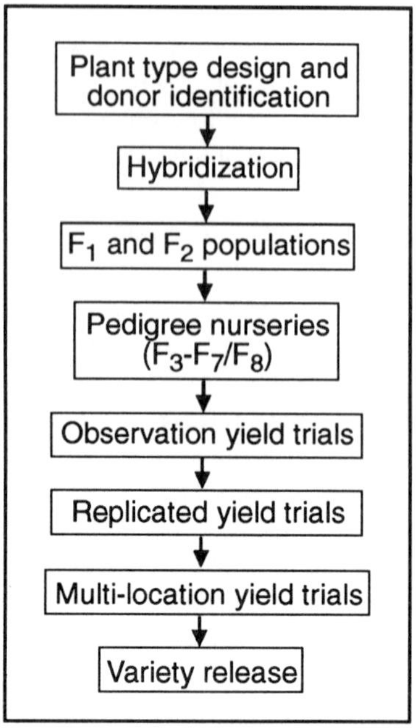

Figure 5. The plant breeding process for variety development.

it is critical that the question of yield potential in the tropics is evaluated in terms of a systems approach which includes strategies for multiple insect and pest resistance, N management, and cost-benefit of input use rather than just achievement of physiological yield potential.

Increased efficiency of plant breeding process

Opportunities

Development and release of a variety is a complex process that may extend over a period of 10–15 years. The entire process of variety development can be divided into several steps (Figure 5). Once the breeding lines have become homozygous, they are bulked and then tested in observational, replicated, and multilocation yield trials. These multisite trials are expensive and need several crop seasons to understand genotype by environment (G×E) interactions. In the state of Tamil Nadu in India, multilocation evaluation of rice is done over a period of 10 years at 100 locations and a variety may be evaluated in as many as 288 experiments before it is released (Palanisamy et al. 1993). Despite extensive testing done with new genotypes using the conventional approach, it is impossible to cover the whole range of environments. It may often happen that a specific environmental challenge may not be available in natural environments in specific years to enable effective discrimination among breeding lines.

Since the systems approach integrates different components of agroecosystems, it can play a great role in increasing the efficiency of plant breeding process, in particular understanding and extrapolating G×E interactions. Crop models together with GIS can facilitate biological characterization of the physical environment (geography, soil, climate, etc.), and thus define key environmental domains for which improved varieties are to be developed. Alternatively, the same methodology can be used to determine the adaptation domains of genotypes. A modelling approach can also provide estimates of yield probability in target environments based on our understanding of G×E interactions. Such studies can help in reducing the number of sites/ seasons required for field evaluation and, thus, increase the efficiency of the whole process of variety development.

Simulation models can be used to assess the relative performance of breeding lines in a relatively short time following the methodology proposed by Aggarwal et al. (1996b). The necessary crop inputs can be measured at a few carefully selected sites and the G×E interactions quantified over other sites. Optimization programs, such as GENCALC (Hunt et al. 1994), are also available to determine the model input parameters from experimental observations. The methodology can be used to assess the performance of various genotypes across a much wider range of management options than would be possible by experimentation. Often there is some uncertainty in model input parameters, which affects the reliability of simulated absolute yields. However, crop models adequately describe the relative trends caused by environmental variations. This is of considerable use in the process of cultivar selection where *ranking* of grain yields is the method used to determine the relative superiority of a genotype, rather than the *absolute* yield values.

Crop simulation can also assist in fine-tuning variety design and selection procedures. Identification of critical traits for increased yield and stability has been discussed earlier. In addition, models can also predict the expected plot yield of spaced plants, as in F_2 generation. Optimal spacing for obtaining the best expression of breeding lines in selection nurseries can also be simulated.

Accomplishments

Environmental characterization. Systems approaches have been used to determine potential and attainable yields for a given level of inputs for various crops (Van Lanen et al. 1991; Aggarwal et al. 1995). Aggarwal (1993) proposed 10 zones that had limited variation in potential and rainfed productivity within a zone. Such a classification is very different from the existing wheat zones recognized by the variety evaluation programme. Chapman and Barreto (1996) recently conducted a detailed study using crop models, weather data bases, and GIS to determine adaptation domains for maize in Latin America and southern Africa. They used phenology and drought resistance to cluster sites. The simulated mid-season drought in southern Africa indicated that there are large areas where drought-tolerant germplasm should be advantageous. These examples illustrate the power of systems approach to better define the crop- and variety-specific environment. There is a need for a critical review

Table 1. Measured grain yields (ton ha^{-1}) and ranks of different genotypes (*Source:* Palanisamy et al. 1995).

Genotype	Measured grain yield	Rank measured	Simulated
AS90043	5.46	1	2
TNAU851979	5.22	2	3
ACM60	5.18	3	1
IET9976	5.12	4	8
ASD18	5.08	5	6
CO37	4.94	6	7
TNAU841434	4.82	7	5
IR64	4.68	8	4

of the agroecological zones used in germplasm evaluation programmes.

Potential yield of cultivars varies with season/year and location. Estimates of such yields for different varieties can establish a reference point for site quality and remove the confounding effects associated with large climatic variation in multi-season evaluation. This index can be used in conventional stability analysis.

Evaluation of new genotypes. Palanisamy et al. (1993) examined the use of the MACROS rice model (Penning de Vries et al. 1989) in assessing the performance (indicated by ranking) of prerelease long-duration rice genotypes in several locations of Tamil Nadu, India. It was found that the model was able to correctly identify two out of the three best genotypes determined in trials. Later, Palanisamy et al. (1995) simulated the relative performance of short-duration genotypes across the state of Tamil Nadu. The average simulated ranking of genotypes across locations showed a reasonable agreement with the measured values (Table 1). The first three lines identified by experimentation were also selected by simulation. Some variability in ranking is possible because the measured grain yields of different genotypes were rather close. For example, the difference in grain yield between the first and third genotype was only 0.28 ton ha^{-1} and that between the first and last genotype was 0.78 ton ha^{-1}. Such small differences between lines commonly observed in many well-conducted advanced multi-environment trials are difficult to simulate with crop models largely because of inaccuracies in measurement of model inputs, and because such differences are generally beyond the crop model's resolution. These differences may not be statistically significant; nevertheless, breeding lines get different ranks. More efforts are needed to achieve a better physiological characterization of the genotypes. Conventional field trials, including the present trial, also encounter stresses such as pests and other management problems which may affect the expression of full yield potential. It is not clear if ranking of genotypes takes care of this problem.

The trials of Palanisamy et al. (1993, 1995) were relatively small with a manageable number of genotypes. However, it is almost impossible to determine crop input parameters for large trials. Programs such as GENCALC (Hunt et al. 1994) can assist but their utility in such applications is still not proven. New statistical tools which do not require this assumption, such as additive main effects and multiplicative interaction models (AMMI) (Gauch 1992) and pattern analysis (Delacy and Cooper

1990), are being used to discriminate among genotypes and to explain G×E interaction. Crop simulation models, together with such statistical tools, can accelerate the evaluation of genotypes (Aggarwal et al. 1996b). This strategy uses limited multi-environment testing (MET) data to estimate genotype interaction scores by AMMI analysis for all test genotypes on one hand, and to identify groups of genotypes with similar interactions via pattern analysis on the other hand. Representative genotypes for each group are identified and their performance simulated over a wider range of target environments. The interaction scores for these new environments are estimated from the simulated responses and combined with the genotype scores from the original MET to extrapolate G×E effects over the wider range of environments. At this stage, however, no multi-environment trials are available where crop input parameters necessary to use crop simulation models are measured.

Relating performance of spaced plants to that of population. Crop models that are sensitive to crop geometry should be able to predict the performance of spaced plants when grown in a community. Several such models are now available. However, the major limitation in using them in the breeding context is that there is very limited plant material available for determining physiological model inputs. The latter often requires destructive plant sampling (such as for estimating partitioning coefficients). The number of segregating plants (or families of plants) may be just too many to parameterize and simulate. In addition, continued genotypic change from F_1 to later generations and reactions to pest profiles may limit the utility of simulation in this selection process.

Pest reaction of new genotypes. Multilocation trials are a common feature in the breeding programmes of most national research systems in Asia, with the implicit goal of determining the expression of resistance in new cultivars to different 'strains' of a pathogen under various environmental conditions. Such trials, however, are not meaningful unless the genetic background of the prevailing pathogen or insect pest population is known, and information is taken at several rice growth stages. The former issue is being addressed with the increased use of molecular markers to characterize pathogen population structures, such as those of bacterial leaf blight and blast of rice, through cooperative networks like the Asian Rice Biotechnology Network. The latter issue is one of scale, since many phenotypic reactions of cultivars or advanced breeding lines are taken at one rice growth stage using monocyclic tests. This limitation may be overcome by knowing how critical resistance components, such as infection efficiency, change with plant age (Calvero et al. 1992), and using this information in a simulation model to produce polycyclic disease progress curves with weather data representative of a location.

Limitations

There are a number of limitations which restrict the scope and quality of applications as indicated above. Key morphological parameters recorded by breeders – for example, early vigour, panicle and tiller number, panicle and plant height, grain size and quality, and pest reaction – are not yet incorporated in most crop physiology based models. Similarly, information on major physiological traits such as compensation, heritability, genetic background, and pleiotropy for the model inputs is generally lacking.

Crop models may not be sensitive to all yield-limiting factors in different environments. For example, in drought-prone environments (particularly uplands), acidity, phosphorus, and other nutrients and weeds can be major constraints. Models used for plant type design in such environments should be sensitive to these factors. Lodging has a high probability in tropical environments, particularly in the wet season and/or in high fertility situations. There is a need to develop models that can integrate the effect of all important factors operating in the field environment, for instance, weather, edaphic conditions, management, incidence and effect of pests, and socio-economics.

In an advanced genotype evaluation trial, differences among the lines are generally small. Most crop models are not sensitive enough to discriminate such small differences. At the same time, due to detailed measurements required for physiological characterization of a genotype, many genotypes such as in a multi-environment trial cannot be simulated.

Simulation modelling-based approaches generally require inputs of crop physiological data to characterize a genotype, soil characteristics, daily weather data, and management information. Crop parameter values could be significantly uncertain due to our imperfect knowledge of their operation, as well as to random errors related with size and number of observations, and systematic errors related with bias in the experimental, measurement, observation, and calibration procedures (Aggarwal 1995; Dionora and Aggarwal, IRRI, 1996, unpubl. data). In addition, many crop input parameters exhibit spatial and temporal variability. Soil and weather inputs required by the crop models also show spatial and temporal variation and may have considerable measurement errors. The stochastic nature of many crop parameters and soil and weather input data may result in uncertainty in the outputs of deterministic crop models (Aggarwal 1995), thereby affecting the precision of simulation results.

The usefulness of crop modelling in agricultural research, including plant breeding, is proven by comparison of simulated data with real field experiments. The progress in evaluating this is often slow because many breeding trials do not have adequate description of site characteristics and field observations. Even simple things such as dates of flowering have different interpretations which are not well documented. Another error generally noted in most trials is lack of documentation and standardization of moisture content of grains.

Synthesis and conclusions

The ideotype concept of Donald (1968) accelerated the applications of interdisciplinary knowledge in plant breeding, particularly that of crop physiology. This research had users (plant breeders), opportunities, short- and long-term objectives, and plenty of advice from other disciplines (in the form of hundreds of papers that appeared since then), indicating the relative importance of different traits for various crops. But perhaps limitations were far too many and the probability (although not determined in that strict sense) of usefulness of research results was low. For example, the importance of physiological traits was not evident in all locations, and even in the same location, with different management practices or seasons. Moreover, it was hard to conclusively demonstrate the importance of a trait because of difficulties in creating isogenic lines/populations needed for such studies. These limitations make it difficult to make conclusions about the usefulness of the trait and, ultimately, of the whole approach (Marshall 1991; Rasmusson 1987). Over time, new tools have become available, understanding of the system has increased, and relevant crop, soil, and weather data have become available. The systems approach, in particular simulation modelling, provides us with an alternative to overcome many of the limitations associated with not being able to see the holistic view. This should facilitate greater integration of interdisciplinary knowledge in plant breeding and, thus, increase its efficiency.

Today, a number of opportunities are available for greater integration of interdisciplinary knowledge in plant breeding programmes. The key areas are design and evaluation of plant types for specific environments and increasing the efficiency of multilocation testing and hybrid seed production programmes. We are now able to explain with simulation models the effects of environmental variability on crop growth and yield. The response of standard cultivars to environment can, thus, be predicted with confidence. The systems approach can assist in optimizing the number of locations and field experiments for multi-environment testing of genotypes. What we still need to do is to develop crop models that are able to explain relatively small genotypic differences among varieties. However, such an endeavour would, while increasing the ability of the models to differentiate small differences among varieties, lead, at the same time, to the requirement of measuring a much larger number of traits. There is a trade-off between simplicity in the model structure and parameters versus the details in the model and the associated number of parameters. Crop physiologists and geneticists need to define and measure a minimum number of efficiently measurable genetic parameters to characterize a genotype.

Crop models may need more refinement for application in other activities of plant breeding such as selection among segregating populations. The power of simulation models to 'create' hypothetical isolines and to answer 'what if' questions for specific agroenvironments is worth exploiting. In particular, the availability of RFLP markers and QTLs for quantitative traits would make it easier to select for agronomic traits. This would help plant breeders in marker-based selection for complex traits which have been difficult to achieve through conventional approaches.

Many of the earlier limitations to greater application of physiology still exist. The issue of physiological methodology needed for precision and useful simulations versus the requirements of an efficient screening methodology for a large number of genotypes still remains unresolved. There is a need to link many of the physiological criteria for yield potential/adaptation with easy-to-measure/score characteristics.

For example, chlorophyll meters can be used to rapidly determine leaf nitrogen content, generally difficult to estimate in a larger number of varieties. Pilot studies are also needed in conventional field trials to verify the results of simulation studies and to determine the extent of genetic variation, heritability, pleiotropy, and the cost of altering a desired trait. It is this lack of complete mechanistic understanding that may restrict progress. Nevertheless, provided this information is made available and the negative linkages among traits and physiological processes are adequately included in crop models, a systems approach can give useful indications as to which characteristics breeders may be able to select for higher yield potential and stability in different agro-environments and for greater efficiency of multilocation trials.

Uncertainties associated with systems tools and their consequence are generally raised as a major limitation to wider application. Efforts are definitely needed to improve precision. However, at the same time, we must remember that our current decision-making process (where the so-called systems approach is not used) is also characterized by a number of limitations, inadequate understanding, and lack of appropriate data. And yet we still have to make decisions. We need to understand and evaluate if the application of systems approach increases, or decreases or does not affect, uncertainty in our decisions. We do not have a perfect understanding of agricultural systems now and perhaps may not ever have. With current emphasis on sustainable ecoregional development and the free market economy, breeding objectives would rapidly be changing. Raising the yield potential may not be the key question; rather, economic and environmental costs associated with different levels of yield potential may be asked. The systems approach, with its well-developed analytical framework, data bases, and powerful simulation models, will be handy to provide answers to many of the queries in a relatively shorter time frame.

There is a need to strengthen research programmes involving plant breeders, physiologists, plant protection scientists, agronomists, and crop modellers to illustrate these possible applications. Simultaneously, user-friendly interface programs should be developed for the use of breeders and other non-modellers to allow greater integration of the systems approach in plant breeding.

References

Aggarwal P K (1993) Agro-ecological zoning using crop growth simulation models: Characterization of wheat environments of India. Pages 97–109 in Penning de Vries F W T, Teng P S, Metselaar K (Eds.) Systems approaches for agricultural development. Kluwer Academic Publishers, Dordrecht, The Netherlands.

Aggarwal P K (1995) Uncertainties in crop, soil and weather inputs used in growth models: Implications for simulated outputs and their applications. Agric. Syst. 48:361–384.

Aggarwal P K, Kalra N, Bandyopadhyay S K, Selvarajan S (1995) A systems approach to analyze production options for wheat in India. Pages 167–186 in Bouma J, Kuyvenhoven A, Bouman B A M, Luyten J C, Zandstra H G (Eds.) Eco-regional approaches for sustainable land use and food production. Kluwer Academic Publishers, Dordrecht, The Netherlands.

Aggarwal P K, Kropff M J, Cassman K G, Ten Berge H F M (1996a) Simulating genotypic strategies to increase irrigated rice yield potential in tropics. Field Crops Res. (in press).

Aggarwal P K, Kropff M J, Matthews R B, McLaren G (1996b) Using simulation models to design new plant types and to analyse genotype and environment interactions. In Cooper M, Hammer G (Eds.) The analysis and exploitation of plant adaptation in agricultural crop improvement programs. CAB International, Wallingford, UK (in press).

Bennett J, Brar D S, Khush G S, Huang N, Setter T L (1994) Molecular approaches. Pages 63–76 in Cassman K G (Ed.) Breaking the yield barrier. International Rice Research Institute, P.O. Box 933, Manila, Philippines.

Boote K J, Tollenaar M (1994) Modeling genetic yield potential. Pages 533–565 in Physiology and determination of crop yield. American Society of Agronomy, Madison, WI, USA.

Calvero S B Jr, Guico E M, Teng P S (1992) Quantifying rice-leaf blast (BI) genetic relationship via the receptivity factor. International Rice Res. Newsl. (IRRI) 17(2):9.

Cassman K G, De Datta S K, Olk D C, Alcantara J C, Samson M I, Descalsota J P, Dizon M A (1995) Yield decline and the nitrogen economy of long-term experiments on continuous, irrigated rice systems in the tropics. Pages 182–222 in Lal R, Stewart B A (Eds.) Soil management: Experimental basis for sustainability and environmental quality. Lewis/CRC Publishers, Boca Raton, FL, USA.

Chapman S C, Barreto H J (1996) Using simulation models and spatial databases to improve the efficiency of plant breeding programs. In Cooper M, Hammer G (Eds.) The analysis and exploitation of plant adaptation in agricultural crop improvement programs. CAB International, Wallingford, UK (in press).

Coakley S M (1990) Using historical weather and pest data for pest zoning. Pages 303–312 in Crop loss assessment in rice. International Rice Research Institute, P.O. Box 933, Manila, Philippines.

Crosby J L (1973) Computer simulation in genetics. John Wiley and Sons, London, UK. 477 p.

Delacy I H, Cooper M (1990) Pattern analysis for the analysis of regional variety trials. Pages 301–334 in Kang M S (Ed.) Genotype by environment interactions and plant breeding. Louisiana State University, Baton Rouge, LA, USA.

Dingkuhn M, Penning de Vries F W T, De Datta S K, Van Laar H H (1991) Concept for a new plant type for direct seeded flooded tropical rice. Pages 17–38 in Direct seeded flooded rice in the tropics. Selected papers from the International Rice Research Conference, 27–31 August 1990, Seoul, Korea. International Rice Research Institute, P.O. Box 933, Manila, Philippines.

Dionora M J A, Kropff M J (1995) Variation in the rate and duration of grain filling in rice genotypes. Pages 123–128 in Aggarwal P K, Kropff M J, Matthews R B, Van Laar H H (Eds.) Applications of systems approaches in plant breeding. SARP Research Proceedings, International Rice Research Institute, P.O. Box 933, Manila, Philippines.

Donald C M (1968) The breeding of crop ideotypes. Euphytica 17:385–403.

Gauch H G (1992) Statistical analysis of regional yield trials. Elsevier Science Publishers, Amsterdam, The Netherlands. 278 p.

Goldsworthy P R, Penning de Vries F W T (Eds.) (1994) Opportunities, use, and transfer of systems research methods in agriculture to developing countries. Kluwer Academic Publishers, Dordrecht, The Netherlands. 366 p.

Hunt L A, Pararajasingham S, Jones J W, Hoogenboom G, Inamura D T, Ogoshi R M (1994) GENCALC: Software to facilitate the use of crop models for analyzing field experiments. Agron. J. 85:1090–1094.

IRRI – International Rice Research Institute (1989) IRRI towards 2000 and beyond. International Rice Research Institute, P.O. Box 933, Manila, Philippines. 66 p.

Jansen R C (1992) A general mixture model for mapping quantitative trait loci by using molecular markers. Theoret. Appl. Genet. 85:252–260.

Khush G S (1990) Varietal needs for environments and breeding strategies. Pages 68–75 in Muralidharan K, Siddiq E A (Eds.) New frontiers in rice research. Directorate of Rice Research, Hyderabad, India.

Kropff M J, Van Laar H H, Matthews R B (Eds.) (1994a) ORYZA1: An ecophysiological model for irrigated rice production. SARP Research Proceedings, International Rice Research Institute, P.O. Box 933, Manila, Philippines. 110 p.

Kropff M J, Peng S, Setter T L, Matthews R B, Cassman K G (1994b) Quantitative understanding of rice yield potential. Pages 5–20 in Cassman K G (Ed.) Breaking the yield barrier. International Rice Research Institute, P.O. Box 933, Manila, Philippines.

Kropff M J, Van Laar H H (Eds.) (1993) Modeling crop-weed interactions. CAB, Wallingford, UK, and International Rice Research Institute, P.O. Box 933, Manila, Philippines. 274p.

Kropff M J, Haverkort A J, Aggarwal P K, Kooman P L (1995) Using systems approaches to design and evaluate ideotypes for specific environments. Pages 417–435 in Bouma J, Kuyvenhoven A, Bouman B A M, Luyten J C, Zandstra H G (Eds.) Eco-regional approaches for sustainable land use and food production. Kluwer Academic Publishers, Dordrecht, The Netherlands.

Ludlow M M, Muchow R C (1990) A critical evaluation of traits for improving crop yields in water-limited environments. Adv. Agron. 43:107–53.

Luo Y, Teng P S, Fabellar N G, TeBeest D O (1996) A rice-leaf blast combination model for simulation of epidemics and yield loss. Agric. Syst. (accepted, under revision).

McCouch S R, Kochert G, Yu Z H, Wang Z Y, Coffman W R, Tanksley S D (1988) Molecular mapping of rice chromosomes. Theor. Appl. Genet. 76:815–829.

Marshall D R (1991) Alternative approaches and perspectives in breeding for higher yields. Field Crops Res. 26:171–190.

Moody K, De Datta S K (1982) Integration of weed control practices for rice in tropical Asia. Pages 37–47 in Soerjani M, Barnes D E, Robertson T O (Eds.) Weed control in small farms. Asian-Pacific Weed Science Society, Biotrop Special Publication No. 15, Bogor, Indonesia.

Muchow R C, Carberry P S (1993) Designing plant types for the semiarid tropics: Agronomists' viewpoint. Pages 37–61 in Penning de Vries F W T, Teng P S, Metselaar K (Eds.) Systems approaches for agricultural development. Kluwer Academic Publishers, Dordrecht, The Netherlands.

Mundt C C (1994) Techniques to manage pathogen co-evolution with host plants to prolong resistance. Pages 193–208 in Teng P S, Heong K L, Moody K (Eds.) Rice pest science and management. International Rice Research Institute, P.O. Box 933, Manila, Philippines.

Nelson R J, Leung H (1994) The use of molecular markers to characterize pathogen populations and sources of disease resistance. Pages 173–192 in Teng P S, Heong K L, Moody K (Eds.) Rice pest science and management. International Rice Research Institute, P.O. Box 933, Manila, Philippines.

O'Toole J C, Jones C A (1987) Crop modelling: Applications in directing and optimizing rainfed rice research. Pages 255–269 in Weather and rice. International Rice Research Institute, P.O. Box 933, Manila, Philippines.

Palanisamy S, Penning de Vries F W T, Mohandass S, Thiyagarajan T M, Kareem A A (1993) Simulation in pre-testing of rice genotypes in Tamil Nadu. Pages 63–75 in Penning de Vries F W T, Teng P S, Metselaar K (Eds.) Systems approaches for agricultural development. Kluwer Academic Publishers, Dordrecht, The Netherlands.

Palanisamy S, Aggarwal P K, Thiyagarajan T M, Ranganathan T B R (1995) Simulating yield and ranking of rice genotypes in multi-location trials. Pages 91–95 in Aggarwal P K, Matthews R B, Kropff M J, Van Laar H H (Eds.) Applications of systems approach in plant breeding. SARP Research Proceedings, International Rice Research Institute, P.O. Box 933, Manila, Philippines.

Peng S, Khush G S, Cassman K G (1994) Evolution of the new plant ideotype for increasing yield potential. Pages 5–20 in Cassman K G (Ed.) Breaking the yield barrier. International Rice Research Institute, P.O. Box 933, Manila, Philippines.

Penning de Vries F W T (1991) Improving yields: Designing and testing VHYVs. Pages 13–19 in Penning de Vries F W T, Kropff M J, Teng P S, Kirk G J D (Eds.) Systems simulation at IRRI. IRRI Res. Pap. Ser. 151, P.O. Box 933, Manila, Philippines.

Penning de Vries F W T, Jansen D M, Ten Berge H F M, Bakema A H (1989) Simulation of ecophysiological processes in several annual crops. Simulation Monographs, Pudoc, Wageningen, The Netherlands. 271 p.

Penning de Vries F W T, Teng P S, Metselaar K (Eds.) (1993) Systems approaches for agricultural development. Kluwer Academic Publishers, Dordrecht/Boston/London and International Rice Research Institute, P.O. Box 933, Manila, Philippines. 542 p.

Pinstrup-Anderson P, Pandya-Lorch R (1995) Prospects for world food security and distribution. Pages 89–111 in Bouma J, Kuyvenhoven A, Bouman B A M, Luyten J C, Zandstra H G (Eds.) Eco-regional approaches for sustainable land use and food production. Kluwer Academic Publishers, Dordrecht, The Netherlands.

Rasmusson D C (1987) An evaluation of ideotype breeding. Crop Sci. 27:1140–1146.

Savary S, Elazegui F A, Teng P S, Pinnschmidt H O (1996) Characterization of rice pest constraints in Asia: An empirical approach. Pages 83–98 in Teng P S, Kropff M J, Ten Berge H F M, Dent J B, Lansigan F P, Van Laar H H (Eds.) Applications of systems approaches at the farm and regional levels. Kluwer Academic Publishers, Dordrecht, The Netherlands.

Song Y H, Heong K L (1994) Use of geographical information systems in analyzing large area distribution and dispersal of rice insects in South Korea. Korean J. Appl. Entomol. 32(3):307–316.

Teng P S (1985) A comparison of simulation approaches to epidemic modeling. Ann. Rev. Phytopathol. 23:351–379.

Teng P S, Blackie M J, Close R C (1977) A simulation analysis of crop yield loss due to rust disease. Agric. Syst. 2:189–198.

Teng P S, Penning de Vries F W T (Eds.) (1992) Systems approaches for agricultural development. Elsevier Applied Sciences, London and New York. 309 p.

Teng P S, Savary S (1992) Implementing the systems approach in pest management. Agric. Syst. 40: 237–264.

Teng P S, Yang X B (1993) Biological impact and risk assessment in plant pathology. Ann. Rev. Phytopathol. 31:495–521.

Van Lanen H A J, Van Diepen C A, Reinds G J, De Koning G H J, Bulens J D, Bregt A K (1991) Physical land evaluation methods and GIS to explore the crop growth potential and its effects within the European Communities. Agric. Syst. 39:307–328.

Van Oijen M (1992a) Evaluation of breeding strategies for resistance and tolerance to late blight in potato by means of simulation modelling. Neth. J. of P. Path. 98:3–11.

Van Oijen M (1992b) Selection and use of a mathematical model to evaluate components of resistance to *Phytophthora infestans* in potato. Neth. J. of P. Path. 98:192–202.

Virmani S S, Sharma H L (1993) Manual for hybrid rice seed production. International Rice Research Institute, P.O. Box 933, Manila, Philippines. 57 p.

Xu Q (1995) A model for simulating hybrid rice seed production. Pages 97–108 in Aggarwal P K, Matthews R B, Kropff M J, Van Laar H H (Eds.) Applications of systems approach in plant breeding. SARP Research Proceedings, International Rice Research Institute, P.O. Box 933, Manila, Philippines.

Yoshida S (1981) Fundamentals of rice crop science. International Rice Research Institute, P.O. Box 933, Manila, Philippines. 269 p.

Zadoks J C, Schein R D (1979) Epidemiology and plant disease management. Oxford University Press, New York, USA. 427 p.

Zeigler R S, Hossain M, Teng P S (1996) Sustainable agricultural development of Asian tropics and subtropics: Emerging trends and potential. In Sustainable agricultural growth in the major ecoregion of the developing world: Prospects to 2020. Paper presented in Virginia, 7–9 November 1994. IFPRI, Washington DC, USA (in press).

Using systems approaches for targeting site-specific management on field level

J. BOUMA[1], J. VERHAGEN[1], J. BROUWER[2] and J.M. POWELL[3]

[1]*Department of Soil Science and Geology, and C.T. de Wit Graduate School for Production Ecology, P.O. Box 37, Wageningen Agricultural University, 6700 AA Wageningen, The Netherlands*

[2]*Department of Agronomy, P.O. Box 341, Wageningen Agricultural University, 6700 AH Wageningen, The Netherlands*

[3]*International Livestock Research Institute, Niamey, Niger; present address: USDA-ARS Dairy Forage Research Center, 1925 Linden Drive, Madison, WI 53706, USA*

Key words: fertilizer management, soil variability, simulation modelling, Site-Specific Management, potatoes, millet, Sahel

Abstract

Site-Specific Management (SSM) implies that management practices are varied within fields based on local soil and crop conditions. Practices include fertilization, tillage, crop protection, seeding and planting. Modern techniques, such as GPS (Global Positioning Systems), on-the-go yield monitoring and remote sensing have shown that variation in crop condition and yield may be considerable within a single field. Such variation, and the potential for SSM, have until recently largely been ignored in agronomic research as emphasis has been on average values. SSM increases the efficiency of the use of natural resources and reduces losses of agrochemicals to the environment. However, variation in time also plays an important role: in SSM assumptions are made about the future when specific weather conditions are unknown. Systems analysis and simulation of crop growth and of solute fluxes are shown to play an important role in defining field-variability patterns in different years. A high-tech example from The Netherlands, dealing with growing potatoes, is compared with a low-tech example from Niger, where a very high spatial variability of millet growth is the basis for site-specific application of animal manure. To allow SSM, field experiments have to be initiated which are based on observed heterogeneity, rather than on assumed homogeneity. This represents a new paradigm for agronomic research.

Introduction

Site-Specific Management (SSM) is developing rapidly internationally in response to an ever increasing volume of data indicating large variation in crop yields within single farmer's fields. Such data are increasingly obtained in the developed world by harvesting equipment with GPS and on-the-go yield monitoring devices. Variations in crop yield may be caused by many factors, such as differences in soil profile characteristics or fertility or local occurrence of compaction and pests and diseases. There appears to be general agreement that finding the causes for these differences and taking appropriate measures in terms of site-specific management will generally be beneficial to the farmer from an economic perspective (Robert et al. 1993, 1994). This also holds for farmers in developing countries, even though here low-tech and low-cost approaches are necessary, while emphasis is on applying scarce available resources as efficiently as possible and not on the need to deal with problems associ-

M.J. Kropff et al. *(eds.): Applications of Systems Approaches at the Field Level, 25–36.*

ated with excess resources (e.g. Bouma et al. 1995a). An added advantage of SSM in developed countries is, of course, the environmental benefit in terms of reduced leaching of agrochemicals beyond the rootzone and possible groundwater pollution.

In this paper, studies will be reviewed in which SSM is being tested under field conditions. However, emphasis will also be placed on implications of the SSM approach for soil and agronomic research in the future.

Classical approaches in soil and agronomy research are focused on defining representative, averaged values for the primary management units of the farmer, which are his separate fields. Chemical fertilization, biocide application, soil tillage, seeding and planting are traditionally based on average conditions in the field. Agronomic and soil-fertility experiments are laid out in randomized blocks in the field and differences among treatments are expressed with statistical techniques which implicitly assume that crop response within a block is homogeneous. If it is not the case, residuals in the statistical procedure are very high! Even though any farmer knows that differences in soil conditions occur within fields, traditionally he has no technical means to vary his management in a systematic, reproducible way. The use of traditional management procedures, therefore, often implies a relatively inefficient use of resources as compared with a condition where SSM is practised. The magnitude of these differences and their implications will vary widely in different countries.

Potential advantages of SSM go, however, considerably beyond narrow financial interests of any particular farmer. They open up the possibility to dynamically manage the soil-plant system not only in space, but also in time. Thus, SSM can contribute significantly to developing sustainable land management systems by increasing the flexibility of land management practices. SSM techniques will be helpful to fine-tune further closing of nutrient cycles which are now out of balance in several developed countries, causing environmental problems. Import of feed grain, for instance, may upset the local nutrient balance by adding more nutrients than are removed by crops and cattle. SSM also will allow, in principle, the implementation of dynamic quality control (DQC) measures during the growing season, as advocated by Larson and Pierce (1992). Such measures are based on defining soil quality parameters with their critical threshold values. Continuous monitoring is followed by specific measures at the very moment that a threshold value is reached. For instance, water is applied by irrigation as soon as the pressure head of the soil water in the rootzone sinks below a critical threshold value. This is being done already widely but not in the SSM mode. A key element in DQC is its proactive character, problems are avoided rather than solved after they have materialized. Both SSM and DQC have to be based on a dynamic characterization of plant growth and fluxes of agrochemicals in the plant-atmosphere-soil-water (PASW) system, which is far from simple because of variability in space and time. To complicate matters even more, variability in time deals with future weather conditions that are unknown. Only very general statistical estimates based on historical information are available. So even if we would be able to continuously monitor water and solute fluxes in the PASW system (a highly improbable proposition because of the enormous cost involved), we still would not be able to predict future conditions because of lack of weather data.

An alternative is, of course, to use process-oriented simulation models for crop growth and solute fluxes which cannot only characterize actual conditions, but potential conditions as well, using historic weather data (e.g. Goldsworthy and Penning de Vries 1994; Bouma et al. 1995b). Use of validated simulation models is a crucial ingredient for realizing SSM because they allow exploratory and rapidly obtained estimates of the effects of different management measures, allowing direct input in the decision-making process. Simulations are made for specific locations in the field and expressions for the entire field are obtained by using interpolation techniques.

Case study: The Netherlands

Introduction

A field of an experimental farm in Wieringermeer, The Netherlands, was thoroughly characterized by soil survey and soil physical techniques, and crop growth was monitored during two growing seasons. Results have been reported elsewhere (Verhagen et al. 1995; Bouma et al. 1995a). Potato yield varied between 30 and 45 ton ha^{-1}, while commercially attractive large potatoes (diameter >5 cm) yielded between 3 and 15 ton ha^{-1}. Our focus here will be on the systems-analysis approach, using modelling, as a tool to ultimately devise an operational decision-support system based on SSM principles.

Systems analysis by modelling

The WAVE model was used to simulate crop growth and nitrate fluxes as a function of different management procedures (Van Clooster et al. 1994). The model is composed of the SWATRE model for water flow, the Johnsson et al. (1987) module for nitrogen transformation and the SUCROS model (Spitters et al. 1989) for plant growth. Water stress is calculated according to Feddes et al. (1978) in which the maximum water uptake is defined as a function of depth. Total water uptake is the integral over the rootzone. Plant growth is reduced at characteristic high and low pressure head values. Nitrate stress is calculated according to Greenwood et al. (1990) in which maximum N-uptake is defined by biomass production. A relation was developed between N concentration (N_c) and crop weight:

$$N_c = 1.35 + 4.05 \exp(-0.26W)$$

in which, N_c is N concentration (%) and W is weight (ton ha^{-1}). Growth is reduced if the actual N concentration (N_a) is below N_c. The reduction of growth is calculated as the ratio of: $(N_a - N_m)/(N_c - N_m)$, in which N_m is the N concentration at which growth stops and is set at 0.5%. The model was used to characterize potato growth and solute fluxes in 1994. Availability of many ground-truth data allowed calibration and validation of the model. Realizing that weather conditions in different years are bound to have a major impact on potato development, runs were also made for seven years ranging from dry to wet. Broad climatic data are presented in Table 1

Table 1. Meteorological data on precipitation (P) and evapotranspiration (ET) for the period before (1 January to 15 May), during (16 May to 6 September) and after (7 September to 31 December) the growing season for seven consecutive years.

	1 Jan. to 15 May		16 May to 6 Sept.		7 Sept. to 31 Dec.	
Year	P (mm)	ET (mm)	P (mm)	ET (mm)	P (mm)	ET (mm)
1988	333	200	360	350	281	82
1989	230	119	258	431	217	87
1990	237	150	215	402	344	75
1991	150	133	216	364	278	75
1992	183	117	363	423	348	81
1993	136	127	277	380	431	69
1994	274	122	236	385	458	67

for the years 1988–94. Runs were made using soil data obtained for 65 point observations and the interpolation technique kriging was used to obtain yield patterns for the entire field (Verhagen et al. 1995).

Pattern recognition

Any SSM system will have to be based on maps indicating areas of the field which show a significantly different behaviour in different years. In any future operational SSM procedure, different types of management are to be defined for each of the areas on the map. Patterns of water-limited yield for the seven years being considered were, therefore, compared by Van Uffelen et al. (1996) to see whether certain general patterns were obtained. Two of the maps, shown in Figure 1, are representative for patterns obtained in relatively dry (1989) and wet (1991) growing seasons respectively. One map for the year 1988 did not show patterns at all. The field being studied consists of more sandy and more clayey spots, and in a dry growing season the sandy spots have lower yields due to water shortage, while in a wet year yields in the clayey spots are lower because of poor aeration associated with wet conditions (e.g. Feddes et al. 1978). In 1988, the weather was such that differences between the soils could not be observed. A systematic analysis for the data of seven years, including application of pattern recognition techniques, resulted in the map in Figure 2, showing two mapping units in four areas to be used as a basis for SSM (Van Uffelen et al. 1996).

Scenario analysis

An analysis of the PASW system implies trade-offs between the need for having adequate production while, at the same time, leaching of agrochemicals to ground or surface waters is avoided, or at least minimized to acceptable levels. An illustration is provided here for 2 of the 65 observation points, one in a sandy, the other in a clayey, soil, corresponding with areas II and I of Figure 2 respectively (Verhagen et al. 1995) (Figure 3). Real field data for 1994 showed a high variability of the nitrogen content in early spring. A recommended fertilizer rate was based on the average

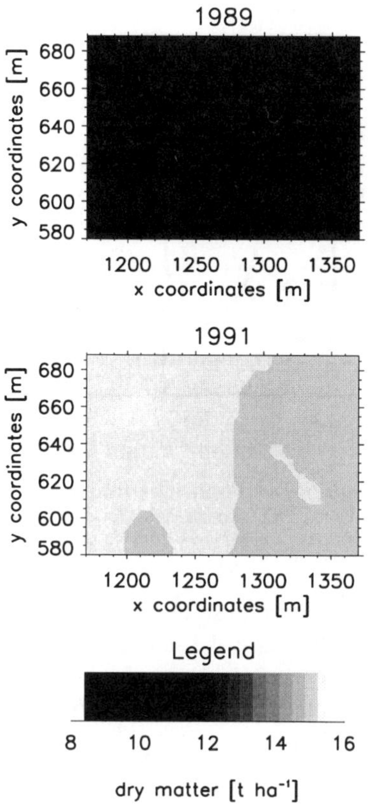

Figure 1. Two maps of potato yields in a Dutch farmer's field obtained by interpolation of simulated point data in the relatively dry growing season 1989 (upper) and the relatively wet growing season 1991.

nitrogen content of the field and on standard tables used by extension agencies. The recommended average rate was 263 kg N ha^{-1}. The WAVE model was applied for 1989 and 1992 (see Table 1) to illustrate the effects of different soil and weather conditions on leaching, assuming the same initial soil-N content in spring (20 kg N ha^{-1}). Leaching is expressed in terms of kg ha^{-1}, where 34 kg ha^{-1} is considered to be a critical threshold value because of Dutch water quality standards. Calculations are made for the entire year, not only for the growing season. Taking the average fertilization rate of 263 kg ha^{-1}, we see leaching of 75 kg ha^{-1} in the sandy spot in the wet year 1992, while leaching was only 10 kg ha^{-1} in the dry year 1989 (Figure 3A). Corresponding values for the clayey site were 30 kg N ha^{-1} and 25 kg N ha^{-1}, both below the critical value of 34 kg N ha^{-1} (Figure 3). Figure 3C shows how a reduction to 135 kg N ha^{-1} leads to acceptable leaching in 1992 on the sandy spot. The model allows calculations of leaching in different years varying the initial N content of the soil in spring and fertilization scenarios including split applications. The latter opens the possibility to arrive at dynamic quality control, as described above.

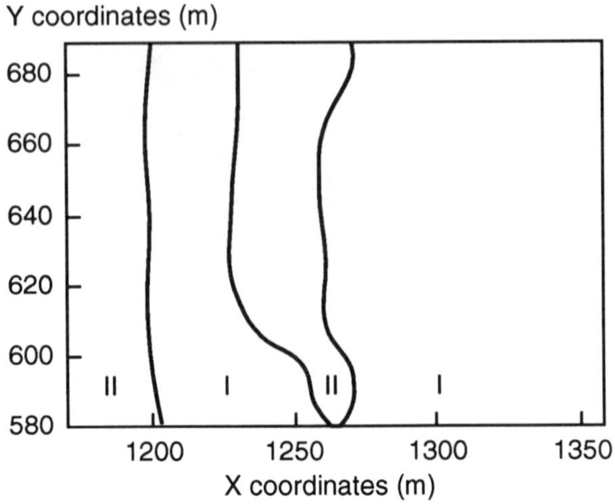

Figure 2. A map for the field, also shown in Figure 1, with two areas of soil which behaved significantly different when considering simulated data for potato yields for eight different years. This map can be the basis for site-specific management practices (*Source:* after Van Uffelen et al. 1996).

The above example is meant to illustrate the potential of modelling for defining optimal fertilization rates. Obviously, the same procedure can be applied to characterize leaching of other agrochemicals. Once simulation values have been obtained for points, interpolation techniques can be applied to obtain expressions for areas of land (e.g. Verhagen et al. 1995).

Future prospects

Having identified subareas within a field that perform significantly different among years, and having a model which allows estimating crop yields and solute fluxes for each of the areas, the tools are available to feed operational SSM decision-support systems. How such systems will have to be organized will be the topic of other papers. The main point to be made here is the use of systems analysis to obtain predictions of crop growth and leaching of agrochemicals for a range of possible management scenarios as a function of different soil and weather conditions. Such ranges will have to reflect economical and environmental threshold values in terms of important indicators such as crop yield, crop quality and soil and water quality. Threshold values are partly to be defined by specialists in other fields, in close mutual interaction.

Case study: the Sahel

Introduction

Sandy soils with a low water holding capacity and a low natural fertility are widespread in the semi-arid Sahelian Zone in Africa (e.g. Brouwer et al. 1993) and are

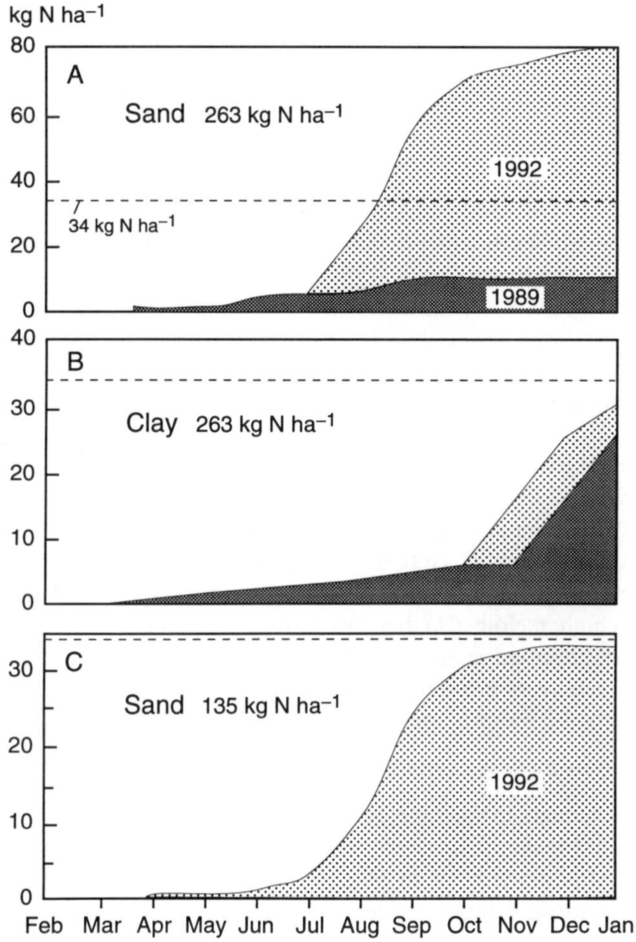

Figure 3. (A) Simulated leaching of nitrate during the years 1989 and 1992 for a sandy spot in the farmer's field. The fertilization rate was 263 kg N ha^{-1}; initial N: 20 kg N ha^{-1}. (B) Same for a clayey spot. (C) A reduced application rate of 135 kg N ha^{-1} resulted in leaching in 1992, just below the threshold value of 34 kg N ha^{-1}.

characterized by a high variability in space on field level and in time due to highly variable weather conditions. Variability in space is due to: (i) small differences in micro-relief and crusting which result in differences in infiltration at short distances varying between 30% and 340% of rainfall (Gaze et al. 1996); (ii) local effects of termites which may enrich the soil; (iii) local effects of trees such as the *Acacia (Faidherbia) albida*; (iv) differences associated with landscape position; and (v) soil fertility gradients around villages.

Chemical fertilizers are too expensive for most farmers. The main options they have for management is to leave crop residues; to try to maximize the impact of animal manure on production by placing the manure exclusively on certain areas within the fields; or to vary management according to the position of the field within the landscape. This is a low-tech approach to site-specific management, but with a potentially greater impact on the production system than the impact of high-tech approaches in the developed world which are coping with excess fertility.

Pattern recognition by on-site research

Spatial variability patterns can be characterized by mapping yield differences within fields as shown by Brouwer et al. (1993) for a field at ICRISAT Sahelian Center in Niger. Using the results of five years of harvesting a 2 ha field as 1325 separate plots of $4\,m \times 3.75\,m$, they showed that the relative yield pattern during a very dry year was by and large the reverse of that during other years. They hypothesized, and supported with other data, that there existed high, relatively fertile but also relatively dry areas in the field, as well as low, relatively moist and leached areas. During years of favourable rainfall, yields would be highest on the fertile high parts of a field, and lower in low parts. During a dry year, yields in the moist low parts would drop a little, but yields in the dry high parts would drop enormously, causing the relative yield pattern to be reversed. Note that comparable differences between dry and wet years were observed in the Dutch experiments, even though absolute yields were, of course, much higher. Note also that variability effects of termites and *Acacia albida* trees were not observed at the site in Niger; such effects would have made comparisons between years more complicated.

Scenario analysis by field experimentation

The main tool farmers in the Sahel have in dealing with yield variability is the application of animal manure. Brouwer and Powell (1995) report on soil aspects of a manure application experiment, in which different quantities of cattle or sheep manure were applied, corresponding to the animals spending one to three nights corralled on the same $4\,m \times 4\,m$ plot in the field. The field was subsequently cropped to millet. One major conclusion was that, at least on the sandy soil as at ICRISAT Sahelian Center, there can be considerable leaching not only of nitrogen, but also of organic carbon and of phosphorus. Calculations showed that, by twelve months after the application of $10\,ton\,ha^{-1}$ of cattle manure (a rate at the high end of what is found in farmers' fields), the nutrient store between 1.5 and 2.0 m depth had increased by $1070\,kg\,ha^{-1}$ of organic carbon; $91\,kg\,ha^{-1}$ of nitrogen, and $19\,kg\,ha^{-1}$ of phosphorus. Similar amounts had probably been leached to beyond 2.0 m depth (Brouwer and Powell 1995).

Clearly, $10\,ton\,ha^{-1}$ of cattle manure need not be applied every year, although that does happen; for instance, around wells in farmers' fields. The results indicate that, to reduce leaching losses, it is more efficient to apply, if possible, smaller amounts of manure fairly frequently, rather than large amounts less frequently (Brouwer and Powell 1995). What is a smaller amount? As further analysis of the data indicates, that depends on the individual plot. Table 2, a modified version of a table in Bouma et al. (1995a), contains yield and nutrient balance data for individual plots from the ILCA manuring experiment. The table shows that, although the average yields depend on the number of nights cattle or sheep were corralled on a plot, the maximum yield is hardly influenced by the number of nights of corralling. In addition, the ratio between highest and lowest yield within a treatment is much greater for the single-night treatments than for the three-nights' treatments. The conclusion must be that one night of manuring each year (about 2.8 ton ha^{-1} of cattle manure and urine, 3.0 ton ha^{-1} of sheep manure and urine; Brouwer and Powell 1995) was enough to attain maximum yields on some plots, not enough on other plots. Similarly, three nights of manuring each year (10 ton ha^{-1} of cattle manure or sheep manure) was most likely too much on some plots, and possibly about right on other plots (Bouma et al. 1995a).

The question remains which plots should have received one night's worth of manure, and which three nights, for most efficient use of the manure available. A first indication of that can also be found in Table 2. The relative wetness (amount of infiltration) for each plot was estimated from water content determinations made on some of the soil samples (column 4). Within each of the four treatments, in the year of observation (1990–91), the most negative net nitrogen and phosphorus balances were found for the wettest plot, i.e. the plot that must have had most leaching (Bouma et al. 1995a). We subsequently related wetness to micro-topographical position. Wetness of individual plots was found to correlate quite well with plot position on the slope, as inferred from the 0.05 m contours on a specially prepared topographic map (see Table 2). In general, within each treatment, plots in concave positions or at the bottom of depressions were found to be wettest, followed by plots on flat crests, then plots on straight slopes; plots on convex slopes were driest.

As plot wetness and position on the slope are correlated, net nutrient balances also correlate well with plot position on the slope. Balances of N as well as P were most negative (or least positive) for the plots in depressions or on concave slopes. Next came the plots on flat crests, then the plots on straight slopes. Nutrient balances were most favourable for plots on convex slopes. All this confirms the logical assumption that wetter plots have more deep drainage and, therefore, more leaching of nutrients. It would, therefore, appear that scarce nutrient resources are best applied on convex slopes, flat crests and possibly straight slopes, in other words, on local high spots and not in depressions.

While such targeted application would no doubt reduce the losses of N, P and C through leaching on these very sandy soils, no account has yet been taken of possible limitation of crop production by lack of water (which is a common enough occurrence in the Sahel!). In dry years or years with a pronounced intra-season drought, low spots may be the only areas where there is enough water available to the crop when that water is most needed (cf. Brouwer et al. 1993; Gaze et al. 1996). This aspect needs to be investigated further.

In summary, emphasis in this case study was only on yield and efficient use of manure. Environmental effects of leaching of nutrients appear to be insignificant. This obviously presents a different focus from the Dutch study case. In this experi-

Table 2. Yields, N and P balances (in kg ha⁻¹ for the top 2.0 m of soil), May 1990–May 1991, using average background values for May 1990. Data from ILCA-ICRISAT-Wageningen Agricultural University manure application experiment, at ICRISAT Sahelian Center, Niamey, Niger (*Source:* Brouwer and Powell 1995; Bouma et al. 1995a).

Plot	Spec.	Nights	Manure kg ha⁻¹	Wetness	Position on slope	Yield in kg ha⁻¹			Ratio highest:lowest yield		Start	In soil total N	Export manure N	End crop N	Net soil N	Start bal. N	In soil total P	Export manure P	End crop P	Net soil total P	bal. P
						Grain	Stover	D.M.	Grain	Stover	D.M.										
79	C	1	2970	3	Concave	356	1428	1783				2486	87	20	2730	177	2008	5.4	5.3	1841	−167
30	C	1	4030	2–3	Straight	1288	4429	5717	3.6	3.1	3.2	2486	105	77	2968	454	2008	6.5	11.6	1846	−157
2	C	1	1530	2	Crest	814	3970	4785				2486	37	70	2713	260	2008	2.3	6.0	1763	−241
9	C	3	8460	3	Crest	974	3914	4888				2486	203	50	2628	−11	2008	12.7	10.9	1943	−67
83	C	3	9830	2	Convex	751	3814	4566	1.4	1.3	1.3	2486	289	*	3076	251	2008	17.7	*	1991	−43
69	C	3	9430	2	Convex	709	3078	3787				2486	260	46	3029	329	2008	16.0	11.6	2257	244
56	S	1	2810	2	Concave	1154	3763	4917	1.6	1.7	1.7	2486	93	51	2842	314	2008	6.0	7.9	1902	−108
44	S	1	3160	1	Convex	717	2181	2898				2486	112	29	3034	464	2008	7.0	4.2	1993	−18
33	S	3	11940	3	Concave	875	3955	4830				2486	425	66	2644	−200	2008	26.2	7.9	1918	−108
88	S	3	9100	2	Straight	626	3966	4593	1.4	1.1	1.1	2486	184	43	2859	236	2008	11.8	12.3	1857	−151
13	S	3	9050	1	Convex	802	3480	4282				2486	349	47	2998	210	2008	21.8	5.8	2082	58
58	cont.	0	0	2–3	Concave	243	855	1098				2486	0	12	2509	35	2008	0.0	1.2	1756	−251
10	cont.	0	0	2–3	Crest	361	1575	1936	2.4	1.9	2.0	2486	0	18	2647	179	2008	0.0	2.1	1843	−163
46	cont.	0	0	2	Straight	153	821	974				2486	0	9	2397	−80	2008	0.0	0.9	1856	−151
93	cont.	0	0	1–2	Straight	357	1373	1730				2486	0	14	2793	321	2008	0.0	1.8	1795	−211

Notes:

Spec.: C = cattle, S = sheep, cont. = control.
Wetness: 1 = low, 2 = average, 3 = high (estimation of infiltration of rainfall).
Start soil: Storage in soil at start of experiment, May 1990 (kg ha⁻¹).
In manure: Nutrient import in manure and urine (kg ha⁻¹) (note: sheep manure data multiplied by 2.0 relative to Brouwer and Powell (1995), because of incomplete recovery of sheep manure in experimental plots).
Export crop: Nutrient export in millet above-ground dry matter (straw, grain and rest of heads) (kg ha⁻¹).
End soil: Storage in soil 12 months into the experiment, May 1991 (kg ha⁻¹).
Net bal. (End soil) − (Start soil) − (In manure) + (Export crop) (kg ha⁻¹), e.g. surplus due to N-fixation or P-desorption, or deficit due to leaching of N or P.

ment, differences in yield were studied as a function of fertilization, while taking into account pre-existing variability patterns. The experiment would have been much less effective if only a standard analysis of variance had been carried out, ignoring the existence of wet and dry plots. Similarly, the use of larger plots could have obscured what was really happening over a distance of only 2–10 m. The method of analysis used was rather laborious, but application of simulation modelling appears rather far-fetched at this time. In future, simulation models will be applied to allow generalization of results obtained. If patterns, as observed here, are representative for large areas in the Sahelian region, results can be extrapolated on the basis of the field experiments only. Before this conclusion can be drawn, however, more on-site experiments will have to be made and they should also extend to the more regional scale. Of major importance is the estimate of the type of weather to be expected. Indications are that weather conditions in early spring provide clues as to the weather to be expected later. Such information would be useful to be able to focus on either a wet or a dry scenario, because this will determine where manure can be placed most efficiently.

Conclusions

Large differences in crop yields of potatoes and millet were reported within a single field for a Dutch and a Sahelian case study. Site-specific management can result in more efficient use of fertilizers in both cases, minimizing leaching of nitrates in the Dutch soil to a threshold value and maximizing yield in the Sahelian soil.

Weather conditions had a major impact on patterns of crop growth as was demonstrated in both case studies. Still, certain patterns can be distinguished that are representative for different years and can, therefore, form a basis for site-specific management.

Simulation modelling was an efficient tool to characterize potato growth and leaching of nitrates, whereby a fertilization rate could be defined corresponding with leaching of nitrates up to a threshold value of 34 kg ha^{-1} being distinguished in Dutch environmental law. Emphasis in the Sahelian case study was on field experiments, using patterns of crop growth which were correlated with the micro-relief. So far, simulation was not applied, but this is being planned for the future.

References

Bouma, J, Brouwer J, Verhagen A, Booltink H W G (1995a) Site specific management on field level: High and low tech approaches. Pages 453–473 in Bouma J, Kuyvenhoven A, Bouman B A M, Luyten J C, Zandstra H G (Eds.) Eco-regional approaches for sustainable land use and food production. Kluwer Academic Publishers, Dordrecht, The Netherlands.

Bouma J, Kuyvenhoven A, Bouman B A M, Luyten L C, Zandstra H G (Eds.) (1995b) Eco-regional approaches for sustainable land use and food production. Kluwer Academic Publishers, Dordrecht, The Netherlands. 505 p.

Brouwer J, Powell J M (1995) Soil aspects of nutrient cycling in a manure application experiment in Niger. Pages 211–226 in Powell J M, Fernández-Rivera S, Williams T O, Renard C (Eds.) Livestock and sustainable nutrient cycling in mixed farming systems of sub-Saharan Africa. Volume II, Technical papers. Proceedings of an International Conference, International Livestock Centre for Africa (ILCA), 22–26 November 1993, Addis Ababa, Ethiopia.

Brouwer J, Fussell L K, Herrmann L (1993) Soil and crop growth variability in the West African semi-arid tropics: A possible risk-reducing factor for subsistence farmers. Agric. Ecosystems and Environment 45:229–238.

Feddes R A, Kowalik P J, Zaradny H (1978) Simulation of field water use and crop yield. Simulation Monographs, Pudoc, Wageningen, The Netherlands. 189 p.

Gaze S R, Simmonds L P, Brouwer J, Bouma J (1996) Measurement of surface redistribution of rainfall and modelling its effect on water balance calculations for a millet field on sandy soil in Niger. J. Hydrology (in press).

Goldsworthy P R, Penning de Vries F W T (Eds.) (1994) Opportunities, use and transfer of systems research methods in agriculture to developing countries. Kluwer Academic Publishers, Dordrecht, The Netherlands. 366 p.

Greenwood P J, Lemair G, Gosse G, Linz P, Drayrot A, Neeteson J J (1990) Decline in percentage N of C3 and C4 crops with increasing plant mass. Annals of Botany 66:425–436.

Johnsson H, Bergstrom L, Jansson P E (1987) Simulated nitrogen dynamics and losses in a layered agricultural soil. Agric. Ecosystems and Environment 18:333–352.

Larson W E, Pierce F J (1992) The dynamics of soil quality as a measure of sustainable management. Paper presented at annual meetings. Soil Sci. Soc. Am., November 1992, Minneapolis, Minn., USA.

Robert P C, Rust RH, Larsen W L (Eds.) (1993) Soil specific crop management. ASA-SSSA-CSSA, Madison, WI, USA. 395 p.

Robert P C, Rust R H, Larsen W L (Eds.) (1994) Site specific management for agricultural systems. ASA-SSSA-CSSA, Madison, WI, USA. 993 p.

Spitters C J T, Van Keulen H, Van Kraalingen D W G (1989) A simple and universal crop growth simulator: SUCROS87. Pages 147–181 in Rabbinge R, Ward S A, Van Laar H H (Eds.) Simulation and systems management in crop protection. Simulation Monographs, Pudoc, Wageningen, The Netherlands.

Van Clooster M, Viaene P, Diels J, Christiaens K (1994) WAVE, a mathematical model for simulating water and agro-chemicals in the soil and environment. Reference and users manual. Inst. Land and Water Management. Catholic University, Leuven, Belgium. 155 p.

Van Uffelen C G R, Verhagen A, Bouma J (1996) Simulation and comparison of land patterns as a basis for site specific management. Agric. Syst. (in press).

Verhagen A, Booltink H W G, Bouma J (1995) Site specific management: Balancing production and environmental requirements at farm level. Agric. Syst. 49:369–384.

New high-yielding, weed competitive rice plant types drawing from *O. sativa* and *O. glaberrima* genepools

M. DINGKUHN, M.P. JONES, B. FOFANA and A. SOW

West Africa Rice Development Association (WARDA), B.P. 2551, Bouaké 01, Ivory Coast

Key words: specific leaf area, assimilate partitioning, growth vigour, leaf area index, leaf nitrogen content

Abstract

Weed competition is the most important yield-reducing factor in upland and hydromorphic rice environments in Africa. Many of the commonly grown tropical *japonica* varieties have a high yield potential, but they compete poorly with weeds. By contrast, the indigenous cultivated *Oryza glaberrima* landraces are highly competitive due to high tillering ability, vigour and leaf area during vegetative growth. But their yield potential is low because of *O. glaberrima*'s specific panicle type and tendency to lodge. Rice breeders at WARDA recently developed stable and fertile progenies from *O. sativa*×*O. glaberrima* crosses that combine the *O. sativa* panicle type with vegetative growth characteristics of *O. glaberrima*. We conducted detailed growth analyses for the two parents (CG14 *O. glaberrima* and WAB56-104 *O. sativa*) and four F7 progenies under moist upland conditions and four levels of N fertilization in Ivory Coast during the 1995 wet season. The objective was to develop crop growth models that help define interspecific plant types combining the weed competitiveness of *O. glaberrima* with the yield potential of *O. sativa*. The superior vegetative vigour and leaf area index (LAI) of the CG14 parent was due to at least five factors, a high specific leaf area (SLA; around $30\,m^2\,kg^{-1}$) which remained constant from seedling stage to maturity, and that was responsible for the pale appearance of the crop; high (>0.6) initial assimilate partitioning to leaves; high leaf N concentration on a dry weight basis; droopy leaves; and extremely high tillering rate. The *glaberrima* parent CG14 had two to three times the LAI, and 1.5 to 2 times the tiller number of WAB56-104, the *O. sativa* parent. The progenies had intermediate SLA, LAI, leaf N content and assimilate partitioning patterns. They combined the superior vigour of CG14 with the panicle structure and sturdy stems of WAB56-104. Preliminary modelling studies using ORYZA1 indicated that early groundcover and high yield potential can be achieved with the same plant type if it has high SLA (*glaberrima* type) during vegetative growth, followed by low SLA (*sativa* type) during the reproductive phase. This was partially achieved in some progenies, resulting in yields up to $5.6\,ton\,ha^{-1}$ on the basis of 25×25 cm hill spacing and $80\,kg\,N\,ha^{-1}$ applied. We will further refine the model and our plant type concepts during 1996 on the basis of weed competition and drought studies, accompanied by a detailed analysis of canopy architecture and leaf gas exchange.

Introduction

Average yields in West Africa's two million hectares of upland rice fields are about one ton grain per hectare (WARDA 1993). Considering that the most commonly grown cultivars, such as the tropical-*japonica* types Moroberekan or OS6, have a yield potential of about four tons, the yield gap is substantial. It reflects the multiple stresses that affect the crop and are largely beyond the control of resource poor farmers. Weed competition is estimated to be the most important yield reducing factor, followed by drought, blast and soil acidity and general soil infertility. Farmers traditionally keep these stresses at the lowest possible level through long periods of

M.J. Kropff et al. *(eds.): Applications of Systems Approaches at the Field Level, 37–52.*

bush fallow. More recently, however, population growth has led to a dramatic reduction of fallow periods in many areas, and thereby aggravated weed pressure and soil infertility (WARDA 1995).

Substantial progress has been made in improving the tolerance of upland rice to soil acidity (e.g. IAC165 from CIAT which is a parent of many of WARDA's acidity tolerant WAB lines (WARDA 1996)), but the weed competitiveness of *O. sativa* upland rices remains poor. By contrast, the African indigenous *Oryza glaberrima* upland cultivars are highly weed competitive due to early vigour and groundcover (Koffi-Goli 1979; WARDA 1996). Many also resist drought, blast and other stresses. But they have been marginalized by the widespread adoption of Asian rice, *O. sativa*, which has a higher yield potential.

O. glaberrima, of which only cultivated forms are known, is now sporadically grown in the upland for its superior grain quality, and in deep water ecosystem where the species' flood tolerant landraces still have a comparative advantage. The species has never been used systematically in rice breeding because of the sterility barrier with *O. sativa* (Bauharmont et al. 1985), its panicle type which has very few secondary branches, its weak stems that easily lodge and its tendency to shatter grain as soon as it matures. Consequently, the marginalization of *O. glaberrima* has resulted in a major, but unquantifiable, loss of biodiversity.

The present study attempted to explore the scope for developing high yielding, strongly weed competitive upland rice plant types drawing from the morpho-physiological characteristics of *O. sativa* and *O. glaberrima*, based on a recent breakthrough in hybridizing the two species (WARDA 1995, 1996). The focus was on growth, dry matter partitioning and yield formation under differential N resources and non-limiting water resources. In conjunction with ongoing experiments on water-limited growth and weed competition, this study will provide the information needed to define environment specific plant type concepts as guidance to breeders. Integration of information and testing of the concepts will be based on systems approaches, in particular crop modelling and environment characterization.

Material and methods

Growth and yield analyses of four *O. sativa* (upland adapted improved tropical *japonica*)×*O. glaberrima* (upland adapted landrace from Senegal) interspecific F7 progenies and their parents were conducted during the 1995 wet season on WARDA's research station at Mbe (7° 52′ N; 5° 6′ W) in Ivory Coast. The male parent material was CG14 (*O. glaberrima*) (V1) and the female parent, WAB56-104 (*O. sativa*) (V2). The progenies were WAB450-1-B-P-160-HB (V3), WAB450-24-3-2-P18-HB (V4), WAB450-24-2-3-P33-HB (V5) and WAB450-1-B-P31-HB (V6). In the following, unless specified otherwise, we refer to V4, the progeny with the most pronounced intermediate characteristics, when discussing the progenies' behaviour. Details of the breeding approach that resulted in the progenies were published elsewhere (WARDA 1994, 1995).

The materials were grown on a comparatively fertile, well-drained alfisol following

Table 1. Weather data recorded at the experimental site. Rs, daily solar radiation; E(pan), pan-A evaporation. Mbe, Ivory Coast, 1995 wet season.

Month	Decade	Temperature (°C)			Rs ($MJ\,m^{-2}\,d^{-1}$)	E(pan) (mm)	Wind ($m\,s^{-1}$)	Rainfall (mm)
		Mean	Min.	Max.				
June	1	25.2	22.3	29.3	18.4	4.95	0.99	49.9
	2	25.7	22.5	29.8	16.9	4.09	1.27	13.8
	3	25.3	22.5	29.2	16.4	4.27	1.25	13.5
July	1	24.5	21.8	28.0	14.0	3.53	0.87	2.6
	2	24.3	22.0	27.7	17.5	2.98	0.83	25.2
	3	24.4	22.3	27.2	15.8	3.48	1.27	10.0
August	1	24.2	22.0	27.3	13.0	2.81	1.00	24.4
	2	24.2	21.0	29.1	15.6	2.59	0.11	128.7
	3	24.6	21.6	29.3	15.2	2.97	0.18	28.0
September	1	24.9	21.7	29.8	16.4	3.42	0.11	102.0
	2	24.5	21.3	29.4	17.7	3.08	0.22	74.2
	3	24.1	21.1	29.3	16.3	3.26	0.25	19.9
October	1	24.6	21.2	31.0	16.1	3.72	0.81	27.0
	2	24.9	20.9	32.0	15.5	3.73	0.65	22.0
	3	24.9	20.9	32.3	16.3	3.40	0.62	23.0

one crop of maize and six years of fallow, using a two-factorial randomized complete block design with three replications. The factors were variety and nitrogen levels (0, 40, 80, and 120 $kg\,ha^{-1}$), half of which was applied basally and the other half top-dressed at 40 days after seeding (DAS). To all plots, 100 $kg\,ha^{-1}$ as triple superphosphate and 50 $kg\,ha^{-1}$ as KCl were applied basally. The soil was well drained and moisture not limiting due to supplementary boom irrigation. Plots measured 3 m×5 m, including a yield area measuring 2 m×2.5 m. Weather data are presented in Table 1.

On 26 June 1995, dry seed was dibbled at a rate of three seeds per hill with a spacing of 0.25 m×0.25 m. Seedlings emerged about 5 days later. Upon seedling establishment, hills were thinned to two plants per hill. Maturity was observed on 11 October (V1), 30 September (V2), 8 October (V3), 12 October (V4), 26 September (V5) and 7 October (V6). The following parameters were measured at 14-day intervals:

- Bulk leaf, stem and panicle dry weight based on four hills (destructive).
- Specific leaf area (SLA) of randomly sampled (exception: flag leaves which were systematically sampled at flowering), fully expanded, healthy leaves (destructive), based on *in situ* area measurement with a LiCor LI-3000 (Lincoln, Nebraska) and subsequent dry weight measurement.
- Tiller number and plant height.
- Areal chlorophyll content of randomly sampled (exception: flag leaves), fully expanded, healthy leaves using a SPAD chlorophyll meter (Minolta). The leaf chlorophyll content was converted into areal and weight-based N content according to Peng et al. (1993).

At maturity, grain yield and yield components were measured as described by Dingkuhn and Le Gal (1996). In addition, we counted the number of primary, secondary and tertiary branches and estimated the degree of shattering by counting the number of attached and missing spikelets for a random sample of 10 panicles per plot at maturity.

In a separate experiment conducted during the 1994 wet season on the same experimental farm, twelve *O. glaberrima* and *O. sativa* lines were studied for their competitiveness with a natural weed flora under upland conditions, based on a two-factorial complete randomized block design with three replications. The factor weed management had two levels: clean weeding throughout, and one single weeding cycle at 43 DAS. Fertilizer inputs were 14 kg N and 20 kg P ha^{-1}. Details of the study were reported by Fofana et al. (1995). Only observations on two lines, CG14 (V1) and WAB56-104 (V2), are reported here.

Results and discussion

Weed competitiveness of the parents

A preliminary study indicated that the *O. glaberrima* parent of the interspecific crosses, CG14, competes much better with weeds than the *O. sativa* parent, WAB56-104 (Figure 1). In clean weeded plots, CG14 showed superior vegetative growth, but had lower yields than WAB56-104. In plots that were weeded once at 42 DAS and then left to competition between rice and a mixed natural population of weeds, however, CG14 produced four times the above-ground biomass and eight times the yield of WAB56-104. Its yield reduction relative to the clean weeded plots was 39%, as compared with 94% for WAB56-104. In this experiment, while the yield reduction due to weeds was particularly dramatic with WAB56-104, a number of improved *O. sativa* cultivars performed poorly in the presence of weeds, compared to the *O. glaberrima* materials (Fofana et al. 1995). Similar observations made at the beginning of WARDA's *O. sativa*×*O. glaberrima* breeding programme formed the rationale for the present growth analyses of parents and interspecific progenies. Evidence has since accumulated indicating that upland adapted *O. glaberrima* landraces form an extraordinarily rich genetic resource for resistance to weeds, drought and a host of diseases.

Grain yields of parents and progenies

The *O. sativa* parent and the progenies showed very similar yields and yield responses to N application, the progenies having consistently (across N levels) but statistically not significantly higher yields (Figure 2). (When comparing pairs of means by *t*-test, however, the progeny V4 significantly outyielded its parents in the 80 kg N treatment.) Yields of the *O. glaberrima* parent did not respond to N application, probably due to lodging which was 100% in all treatments but occurred phenologically earlier at higher N rates. For all N levels, the same grain loss of about one ton per hectare could be attributed to shattering. The *O. sativa* parent and the progenies showed no significant lodging or shattering. When considering both the attached and the shattered

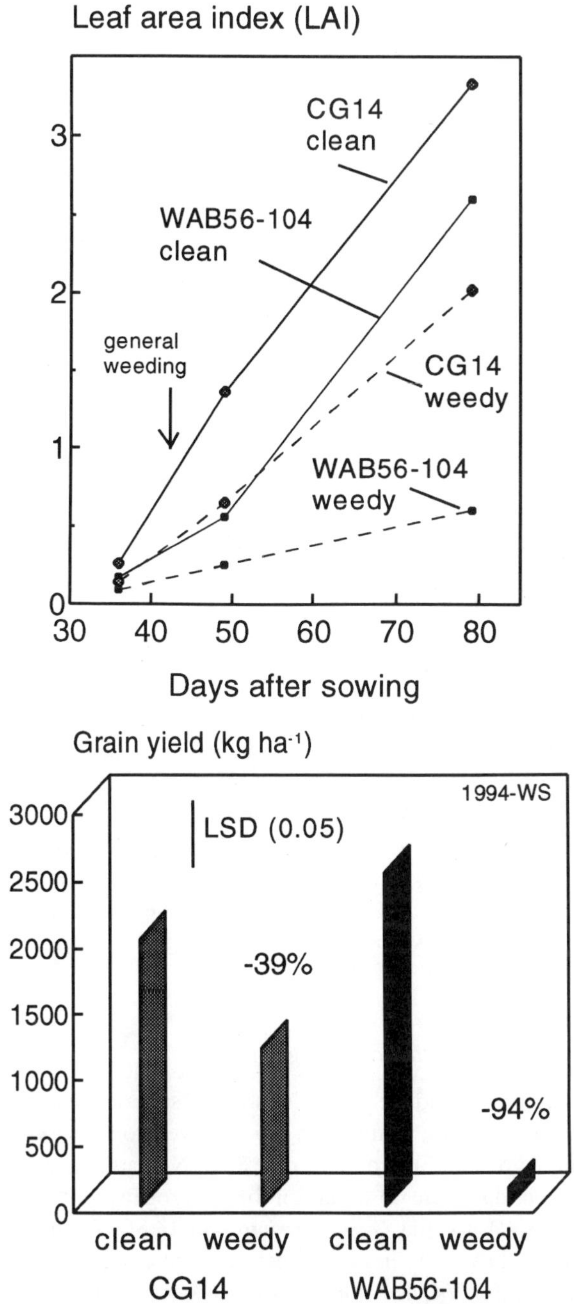

Figure 1. Leaf area index and grain yield of the *O. glaberrima* cultivar CG14 and the *O. sativa* line WAB56-104 grown with differential weed management (clean weeding or one single weeding cycle) under non-water-limited upland field conditions. The LSD (0.05) is based on the factor 'weed management'. WARDA, Mbe, Ivory Coast, 1994 wet season.

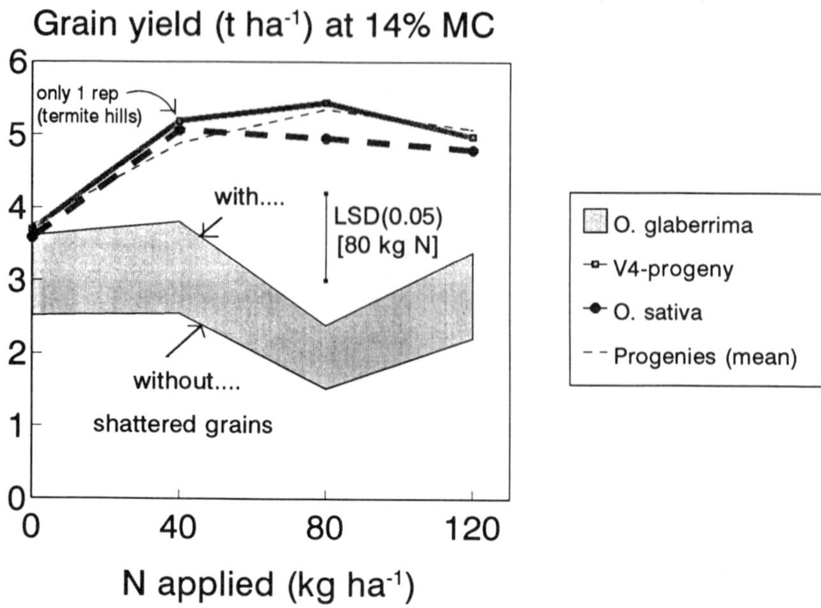

Figure 2. Grain yield as a function of N application rate of the *O. glaberrima* cultivar CG14, the *O. sativa* line WAB56-104 and four interspecific hybrids. WARDA, Mbe, Ivory Coast, 1995 wet season.

grains, all tested genotypes had precisely the same yield at zero N inputs (3.7 ton ha^{-1}), and varietal differences only materialized as N was applied.

The similarity among the yields of the progenies and WAB56-104 was partly due to the high incidence of secondary panicle branches (1.53 secondary branches per primary branch in the *glaberrima* parent, 3.16 in the *sativa* parent and between 3.22 and 3.46 in the progenies). Tertiary branches were generally rare. The progenies had comparatively open panicles that were mostly larger than those of both parents (Figure 3).

Dry matter accumulation and leaf area growth

The *O. glaberrima* parent and the progenies showed superior seedling vigour as compared to the *O. sativa* parent, resulting in rapid above-ground biomass accumulation during exponential growth (Figure 4). For the 80 kg N treatment, relative growth rates (RGR) of 18.7% per day were observed for the V4 progeny between 26 and 39 DAS, as compared to 18.4% for the *O. glaberrima* and 14.8% for the *O. sativa* parent. Between 39 and 53 DAS, the transition period from exponential to linear growth, the RGR was 12.0% for V4, as compared to 11.6% for the *O. glaberrima* and 9.8% for the *O. sativa* parent. Consequently, at any given amount of above-ground biomass, the *O. glaberrima* parent and the progenies grew much faster than the *O. sativa* parent. These differences might have been due to access to resources or resource use

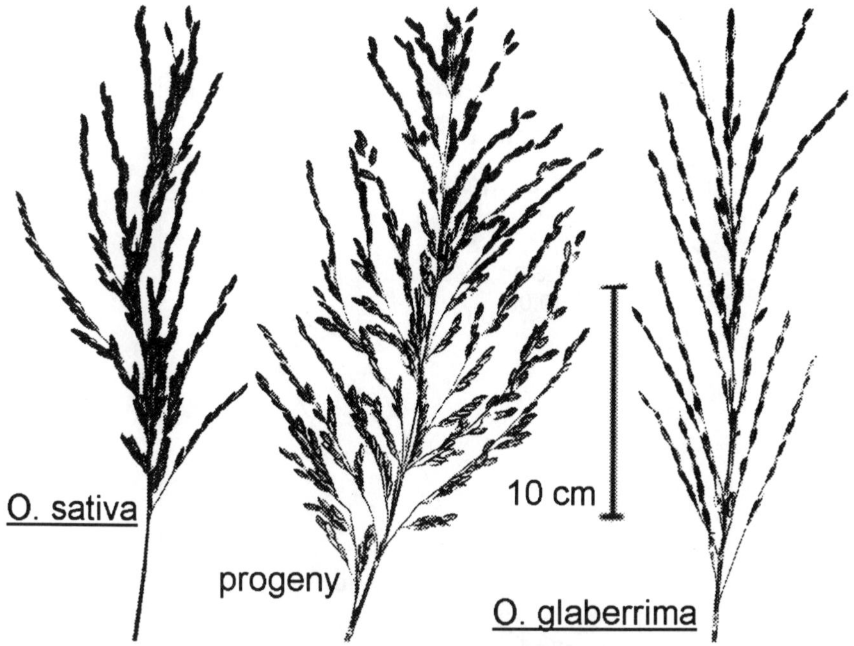

Figure 3. Panicle types observed for the *O. sativa* line WAB56-104, the *O. glaberrima* cultivar CG14 and the interspecific hybrid V4 (for details see text). WARDA, Mbe, Ivory Coast, 1995 wet season.

efficiency within the plant, or both. Preliminary evidence points at a combination of superior N uptake (access) and canopy architecture-related light use efficiency as being major determinants of the observed genotypic differences (data not presented).

The rapid initial growth of the progenies and their *O. glaberrima* parent was associated with faster leaf growth (Figure 5). The *O. glaberrima* parent developed more than twice the leaf area of the *O. sativa* parent under zero-N inputs, and 3.5 times its leaf area in the 80 kg N treatment. The leaf area index (LAI) of the progenies was generally intermediate.

Specific leaf area

Specific leaf area (SLA) was, in part, responsible for the extremely different LAI observed among genotypes. The SLA was strongly affected by genotype and phenological stage (Figure 6). At any given phenological stage and genotype, however, SLA was remarkably stable across replications and N resources. The *O. glaberrima* parent had a high SLA throughout its development. By contrast, the *O. sativa* parent had a much lower initial SLA which even decreased significantly in the course of development, indicating that leaves were getting thicker. The progenies had generally intermediate SLA during early growth stages, followed by a decrease at least as sharp as that of the *O. sativa* parent.

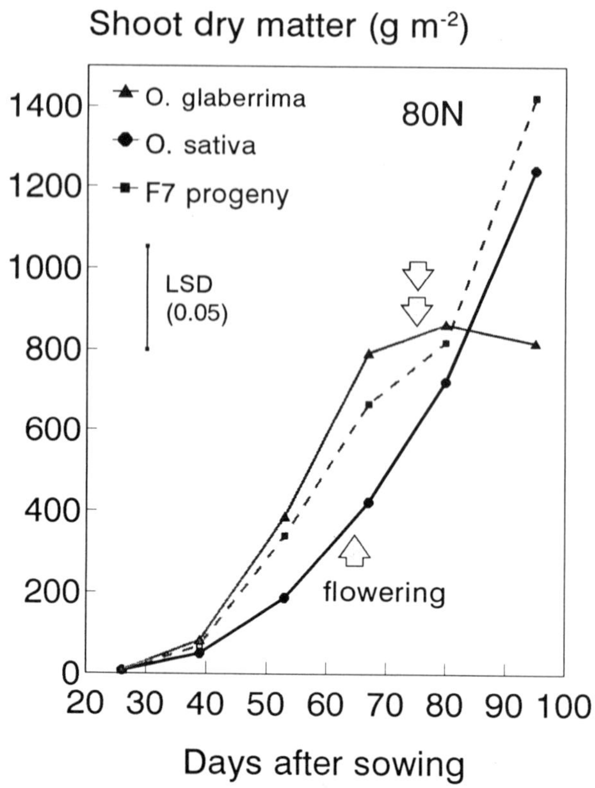

Figure 4. Time courses of above-ground (shoot) dry matter for the *O. glaberrima* cultivar CG14, the *O. sativa* line WAB56-104 and the interspecific F7 progeny V4 (for details see text) in an upland field fertilized with 80 kg N ha^{-1}. WARDA, Mbe, Ivory Coast, 1995 wet season.

We observed linear correlations between LAI and SLA across genotypes at any given sampling date (exemplarily shown for 52 DAS in Figure 6). The slope of these relationships, however, depended on N inputs. The most likely explanation is that LAI was mainly limited by resources (light and N) and the physiological cost of producing leaf area. Genotypes with high SLA achieved a high LAI because less resources were invested per unit area. This, in turn, enabled a higher light harvest and more rapid growth, as confirmed by preliminary modelling studies based on the rice growth model ORYZA1 (Kropff et al. 1994).

Leaf nitrogen content

All test genotypes showed the phenological decrease in dry-weight based leaf N content characteristic of cereals (Figure 7, observations for the zero-N treatment). The *O. glaberrima* parent, however, had a higher N concentration than the *O. sativa* parent throughout the season. The V4 progeny was intermediate during early vegetative growth, followed by a dramatic decrease in N concentration which was probably

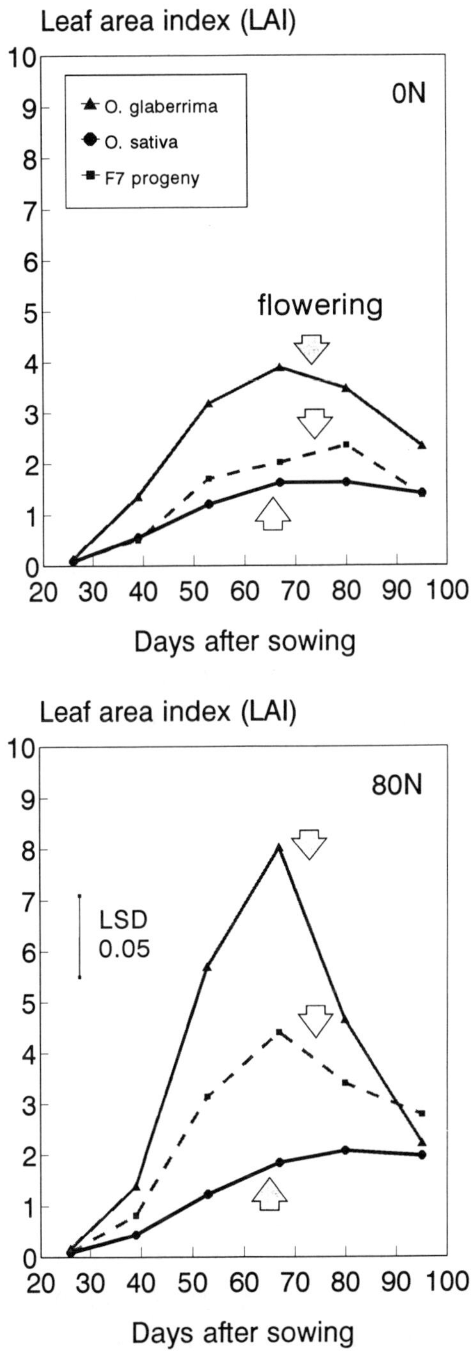

Figure 5. Time courses of leaf area index (LAI) observed in an upland field fertilized with 0 and 80 kg N ha^{-1} for an interspecific rice progeny (WAB 450-24-3-2-P18-HB) and its parents WAB56-104 (*O. sativa*) and CG14 (*O. glaberrima*). WARDA, Mbe, Ivory Coast, 1995 wet season.

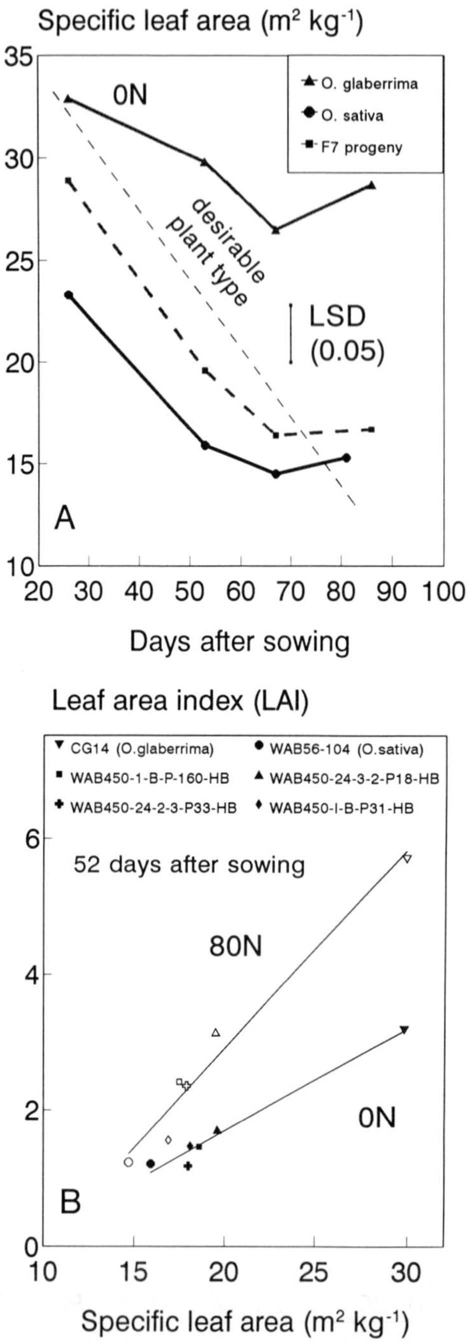

Figure 6. (A) Time courses of specific leaf area (SLA) for an interspecific rice progeny (WAB 450-24-3-2-P18-HB) and its parents WAB56-104 (*O. sativa*) and CG14 (*O. glaberrima*). The broken line indicates the 'ideal' SLA for a high-yielding, weed competitive plant type. (B) Relationship between leaf area index (LAI) and SLA across interspecific progenies and their parents, as observed during late vegetative state in an upland field fertilized with two nitrogen levels. WARDA, Mbe, Ivory Coast, 1995 wet season.

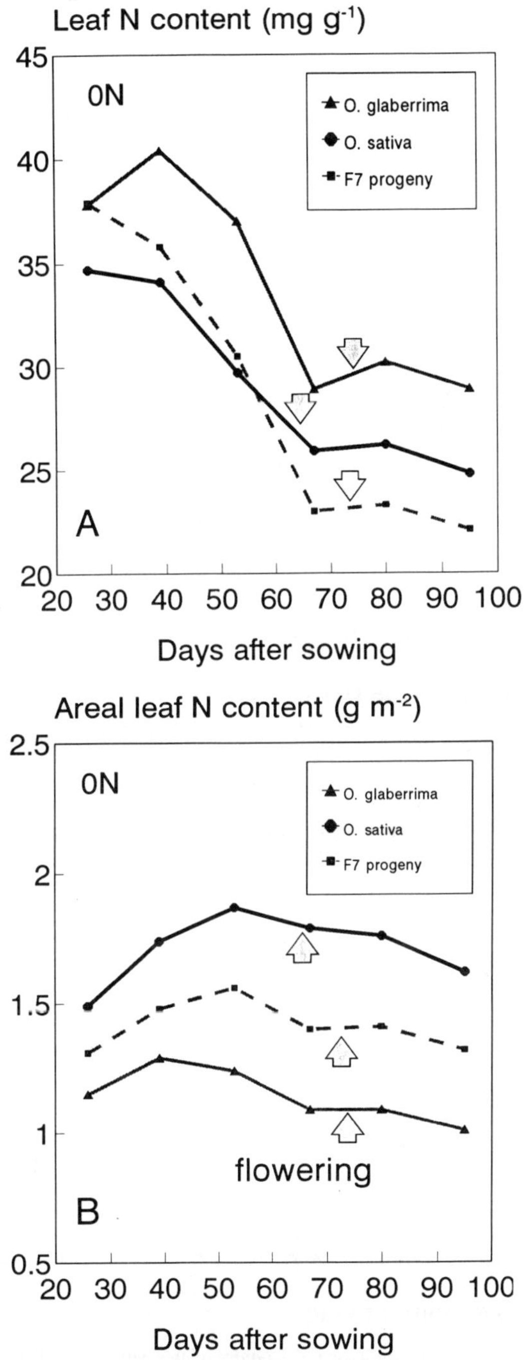

Figure 7. Time courses of (A) weight-based and (B) area-based leaf N content for the *O. glaberrima* cultivar CG14, the *O. sativa* line WAB56-104 and the interspecific hybrid V4 (for details see text) under zero-N inputs. WARDA, Mbe, Ivory Coast, 1995 wet season.

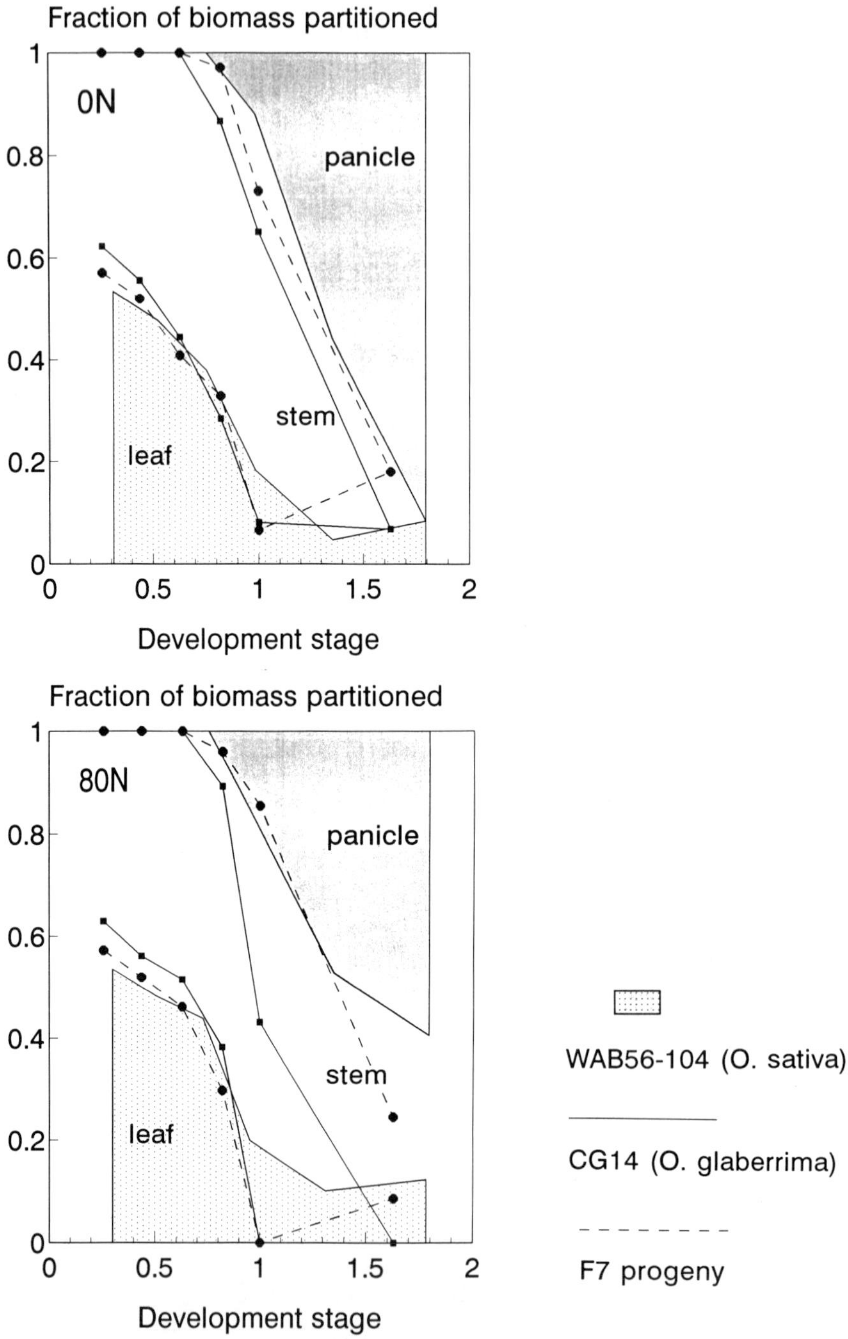

Figure 8. Partitioning patterns for above-ground dry matter between the leaf, stem and panicle for the *O. glaberrima* cultivar CG14, the *O. sativa* line WAB56-104 and the interspecific hybrid V4 (for details see text) under zero and 80 kg N ha^{-1} inputs. Development stages 0, 1 and 2 correspond to germination, flowering and maturity respectively. WARDA, Mbe, Ivory Coast, 1995 wet season.

caused by dilution through rapid growth (Dingkuhn et al. 1990). In the N- fertilized treatments, the drop in N concentration was less pronounced and the inter-specific progenies showed intermediate behaviour throughout the season (data not presented).

When calculated on a per area basis, the lowest leaf N contents were observed in the *O. glaberrima* parent, and the highest in the *O. sativa* parent. This seemingly inverted situation was due to the thinner leaves (higher SLA) of the *O. glaberrima* parent and its progenies. It follows that the generally pale appearance of *O. glaberrima* leaf canopies in this study was not caused by N deficiency (because the weight-based N content was high), but rather by the thin and translucent leaves. It also follows that, when genotypic variability in SLA is high, selection for dark green leaves does not only result in plants with high N content, but also in thicker leaves (low SLA). This may unintentionally favour plants with poor initial vigour and groundcover and, consequently, poor weed competitiveness.

We know little about the relationship between SLA and photosynthetic rates, particularly for genotypes drawing traits from two different species. A study by Furuya et al. (1994) indicated that two *O. glaberrima* cultivars had lower photosynthetic rates than two *O. sativa* cultivars. This observation is consistent with the low area N content measured in our *O. glaberrima* parent, but needs further study.

Dry matter partitioning

Time courses of dry matter partitioning among leaves, stems and panicles were generally similar among the test genotypes (Figure 8) and reflected the patterns usually found in rice (Dingkuhn and Kropff 1996). In contrast to results obtained with high-yielding *indica* rice (Dingkuhn 1996), effects of N nutrition on partitioning were small in this study. Comparing genotypes, the *O. glaberrima* parent appeared to partition more dry matter to leaves than the *O. sativa* parent, with the V4 progeny being intermediate. These trends were consistent across N treatments but have not yet been ascertained statistically. If validated, the partitioning patterns of *O. glaberrima* would indeed be of considerable applied, and even generic, interest. High initial assimilate partitioning rates to leaves, and low rates from mid-season onwards, have been identified as being major components of a high-yielding plant type for direct seeded irrigated conditions (Dingkuhn et al. 1991). The same pattern would improve potential yields for rice varieties with extremely short duration, because it provides for an earlier onset of linear growth. Finally, such partitioning patterns would theoretically increase weed competitiveness in a developmental stage-specific way and, therefore, allow for a high yield potential at the same time.

Tillering

The *O. glaberrima* parent produced twice as many tillers as the *O. sativa* parent in the zero-N treatment (Figure 9). The progenies had slightly less tillers than the *O. sativa* parent and, thus, did not show intermediate behaviour. Both the *O. glaberrima*

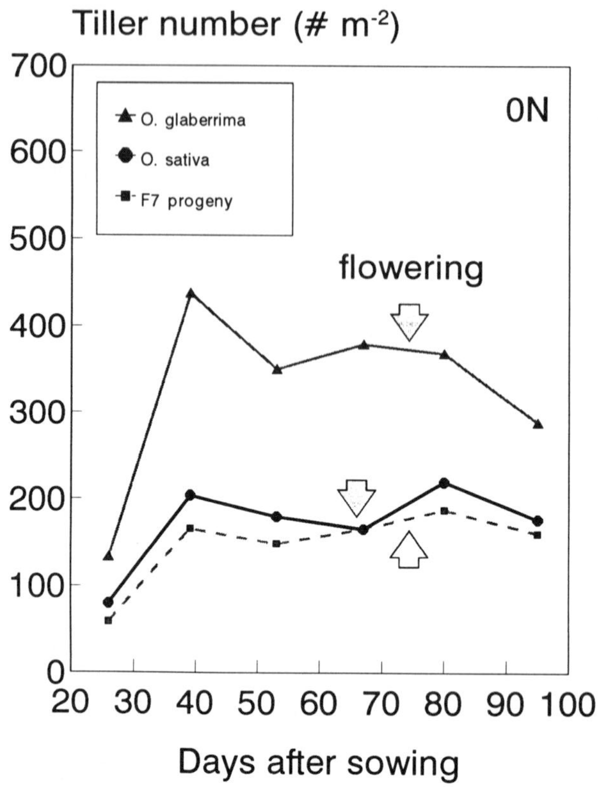

Figure 9. Time courses of tiller number per unit ground area for the *O. glaberrima* cultivar CG14, the *O. sativa* line WAB56-104 and the interspecific hybrid V4 (for details see text) under zero-N inputs. WARDA, Mbe, Ivory Coast, 1995 wet season.

parent and the progenies, however, showed a much stronger response of tillering to N inputs (data not presented).

Previous studies at WARDA showed that tillering is an important determinant of weed competition, along with height and groundcover (WARDA 1995, 1996). New interspecific materials have now been selected which show tillering ability superior to the *O. sativa* parents, a trait inherited from *O. glaberrima.*

Conclusions and outlook

The present study provided morpho-physiological evidence for trait introgressions from *O. glaberrima* into an *O. sativa* background, based on four genetically stable and fully fertile interspecific hybrids. Most of these introgressions are of potentially high adaptive value in a resource-limited production environment, particularly in weed prone upland fields. They include: (i) rapid vegetative growth and leaf area development, at least in part caused by high initial SLA and partitioning of much assimilate to leaves during early growth; and (ii) droopy leaves during early growth

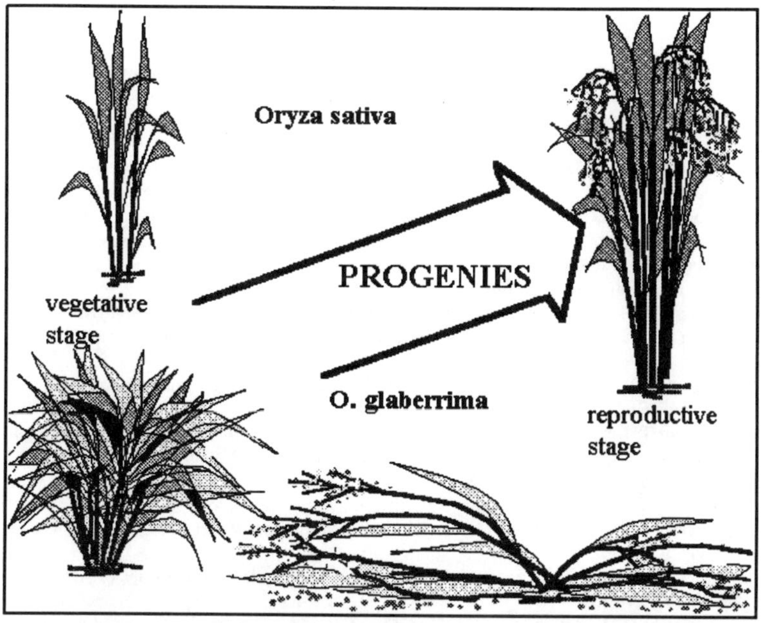

Figure 10. Diagram of the basic concept for the development of a weed competitive, interspecific plant type. The underlying principle is to realize the weed competitive traits of *O. glaberrima* during the vegetative stage, and the agronomically valuable traits of *O. sativa* during the reproductive and ripening stage. This requires a highly dynamic and stage-dependent expression of morphological traits. Preliminary studies indicate a high probability of this being achievable.

stages, theoretically resulting in a high extinction coefficient for solar radiation and, thus, high light-use efficiency and suppression of weeds. Expressed in combination with agronomically useful traits derived from *O. sativa*, such as large panicles, sturdy stems and erect foliage during reproductive stages, these traits are expected to improve yield stability at a high level of potential yield.

Preliminary studies indicate that weed competitiveness is not the only adaptation that new plant types can draw from *O. glaberrima*. Although we know little about the underlying mechanisms, it is already evident that the *O. glaberrima* parent in this study, CG14, is highly drought resistant. Many *O. glaberrima* landraces are also highly resistant to blast, rice yellow mottle virus and the African rice gall midge.

A series of studies has been initiated at WARDA to morpho-physiologically and genetically characterize these traits, develop models to compose and test environment-specific plant type concepts, and identify compatible donors for desirable traits.

The new plant type concepts, as well as the screening tools that will contribute to their realization, will take into account the dynamic nature of morpho-physiological traits. For example, plants will be screened for a droopy, profuse foliage with high SLA during early growth stages, and erect leaves with much lower SLA during reproductive stages. In other words, the new plant types will resemble *O. glaberrima* during early and *O. sativa* during growth stages (Figure 10). Preliminary results indicate that this ambitious objective is indeed feasible.

References

Bauharmont J, Olivier M, Dumont de Chassart M (1985) Cytological observations in some hybrids between the rice species *Oryza sativa* L. and *O. glaberrima* Stend. Euphytica 34:75–81.

Dingkuhn M (1996) Modelling concepts for the phenotypic plasticity of dry matter and nitrogen partitioning in rice. Agric. Syst. (in press).

Dingkuhn M, Schnier H F, De Datta S K, Doerffling K, Javellana C (1990) Response of direct seeded and transplanted flooded rice to nitrogen fertilization. II. Effects of leaf area and nitrogen status on foliage expansion, senescence, canopy photosynthesis and growth. Crop Sci. 30:1284–1292.

Dingkuhn M, Penning de Vries F W T, De Datta S K, Van Laar H H (1991) Concepts for a new plant type for direct seeded flooded tropical rice. Pages 17–39 in Direct seeded flooded rice in the tropics. International Rice Research Institute, P.O. Box 933, Manila, Philippines.

Dingkuhn M, Kropff M J (1996) Rice. Pages 519–547 in Zamski E, Schaffer A A (Eds.) Photoassimilate distribution in plants and crops: Source-sink relationships. Marcel Dekker Inc., New York, Basel, Hong Kong.

Dingkuhn M, Le Gal P-Y (1996) Effect of drainage date on yield and dry matter partitioning in irrigated rice. Field Crops Res. (in press).

Fofana B, Koupeur T, Jones M P, Johnson D E (1995) The development of rice varieties competitive with weeds. Pages 187–192 in Weeds – 1995, Vol. 1, Brighton Crop Protection Conference. 49 Downing Street, Farnham, Surrey GU9 7PH, UK.

Furuya A, Itoh R, Ishii R (1994) Mechanisms of different responses of leaf photosynthesis in African rice (*Oryza glaberrima* Steud.) and rice (*Oryza sativa* L.) to low leaf water potential. Jpn. J. Crop Sci. 63 (4):625–631.

Koffi-Goli (1979) Collection and conservation of existing rice species and varieties of Africa. Agronomie Tropicale 34(3):228–237.

Kropff M J, Van Laar H H, Matthews R B (Eds.) (1994) ORYZA1: An ecophysiological model for irrigated rice production. SARP Research Proceedings, International Rice Research Institute, P.O. Box 933, Manila, Philippines. 110 p.

Peng S, Garcia F, Laza R, Cassman K G (1993) Adjustment for specific leaf weight improves chlorophyll meter's estimate of rice leaf nitrogen concentration. Agron. J. 85:987–990.

WARDA – West Africa Rice Development Association (1993) Medium-Term Plan 1994–1998. West Africa Rice Development Association, BP 2551, Bouaké, Ivory Coast.

WARDA (1994) Annual Report for 1993. West Africa Rice Development Association, BP 2551, Bouaké, Ivory Coast.

WARDA (1995) Annual Report for 1994. West Africa Rice Development Association, BP 2551, Bouaké, Ivory Coast.

WARDA (1996) Annual Report for 1995. West Africa Rice Development Association, BP 2551, Bouaké, Ivory Coast.

Improving rice tolerance to barnyardgrass through early crop vigour: simulations with INTERCOM

J.L. LINDQUIST[1] and M.J. KROPFF[2]

[1]*Department of Agronomy, University of Nebraska, Lincoln, NE 68583-0915, USA*
[2]*Department of Theoretical Production Ecology, Wageningen Agricultural University, P.O. Box 430, 6700 AK Wageningen, The Netherlands*

Key words: crop tolerance, crop-weed competition, rice ecosystem, agricultural development, plant breeding, yield stability

Abstract
Crop yields on-farm rarely approach their production potential, partly as a result of growth reduction due to weed interference. The ability of rice to maintain yield in the presence of barnyardgrass (a measure of tolerance) may be improved if the time between emergence or transplanting and canopy closure is decreased. This may be accomplished through manipulation of crop population density and spatial arrangement, or improvement of early crop vigour. A simulation model of rice-barnyardgrass competition for light (INTERCOM) was used to evaluate the influence of increased early vigour in rice on its tolerance to barnyardgrass. Simulations suggest that increasing early leaf area expansion and height growth rates in rice will reduce yield loss. Correlation between early crop vigour and rice tolerance to barnyardgrass in the field is warranted. Rice cultivars with improved tolerance may improve yield stability in the presence of low barnyardgrass populations.

Introduction

A 67% increase in world rice production may be required to feed the projected population in 2025 (IRRI 1990). Methods of closing the gap between on-farm and potential production must, therefore, be a primary objective for agricultural development research. Since weed interference reduces on-farm rice production by an estimated 25% (Moody 1988), methods to reduce weed interference will aid in this objective.

Herbicides are effective in reducing weed interference. However, use of appropriate compounds may be restricted in some rice production areas due to high costs, limited availability, resistant weed populations, or environmental problems. Alternative weed management strategies must be developed, but the likelihood of developing a neatly packaged, broadly available and applicable, inexpensive, and environmentally benign method of reducing weed interference seems small. Rather, integration of multiple cultural, mechanical, and chemical weed management practices will be necessary to reduce the impact of weeds on crop yields.

The multitude of practices that make integrated weed management desirable also has the unfortunate side-effect of making field research very difficult. There are simply too many factors to include in a comprehensive experiment. Integrated weed management is, therefore, ideally suited for systems research. Simulation models can be used to explore the effects of individual or multiple weed management practices

M.J. Kropff et al. *(eds.): Applications of Systems Approaches at the Field Level, 53–62.*

on crop-weed interactions. Ecophysiological plant growth models may be useful for design-based breeding programmes by predicting performance of hypothetical plant types in specific target environments (Dingkuhn et al. 1993; Hunt 1993).

Improved cultivar competitiveness and tolerance to weeds have been suggested as methods of reducing the negative influence of weeds on crop yield (Callaway 1992; Jordan 1993). A more competitive crop may benefit management by reducing weed seed production. Weed densities in subsequent cropping seasons are, therefore, reduced and long-term yield increased (Lindquist et al. 1995). A crop is more tolerant to weed presence when the impact of each weed on crop yield is reduced. Tolerance improves yield stability in fields where low weed populations are present.

In well-fertilized and irrigated rice production systems, crop-weed competition is expected to be primarily for light. Under these conditions, we assume that a competitive advantage can be conferred on the crop by decreasing the time between crop emergence and canopy closure. This may be accomplished through manipulation of crop population density and spatial arrangement, or an improvement in early crop vigour. Early vigour is commonly measured qualitatively by plant breeders in their breeding trials (e.g. Jennings and Aquino 1968). We quantify early vigour using two measures: early leaf area expansion and early height growth rates.

We use an ecophysiological model for irrigated rice-barnyardgrass competition for light (INTERCOM, Kropff and Van Laar 1993) to explore the influence of early leaf area and height growth rate on simulated rice-barnyardgrass competition. Our objectives were to determine the effects of changing rice early leaf area expansion and height growth rates on rice yield loss. We first examined the effects of leaf and height growth as independent traits, then changed them simultaneously to determine their combined effects. Lindquist and Kropff (1996) further explored the effects of changing rice leaf area expansion rate on simulated barnyardgrass seed production.

The model

INTERCOM, a process-based plant growth and competition simulation model, operates on a daily time step. Beginning at emergence, expansion of leaf area index (LAI) is determined using an exponential function (Kropff 1993):

$$LAI = N \times LA0 \times \exp(RGRL \times GDD) \tag{1}$$

where N is the crop density (plants m^{-2}), $LA0$ the leaf area per plant at seedling emergence (m^2 plant^{-1}), $RGRL$ the relative leaf area growth rate (°C^{-1} d^{-1}), and GDD is growing degree-days (°C d^{-1}) accumulated since emergence. This exponential phase ends when the portion of assimilates partitioned to non-leaf tissue sharply increases, or when shading among individual plants begins. We use a total canopy leaf area index of 1 ($LAI = 1$) as the end of the exponential phase because it marks the beginning of leaf overlap (Kropff 1993).

Following the initial exponential phase, photosynthetic response of individual leaves to absorbed photosynthetically active radiation (PAR) drives daily growth. Carbon gain based upon absorbed PAR at five points within the canopy is integrated over three times during daylight hours to obtain daily biomass increase. INTERCOM for rice simulates competition for light based upon the profile of absorbed photosynthetically active radiation (PAR) in the canopy. The quantity of PAR absorbed by each species is a function of the amount and distribution of photosynthetic area within the canopy, light extinction coefficient, and species height.

In rice, inter-plant competition has a greater influence on tillering than on height growth (Jennings and Aquino 1968). Height growth, therefore, is simulated independent of competitive environment (i.e. competition has no effect on height of either species). Height (*HT*, cm) is simulated for each species using a logistic function (Kropff 1993):

$$HT = HM/(1 + HB \times \exp(-HS \times GDD)) \qquad (2)$$

where *HM* is the maximum height (cm) of the species, *HB* and *HS* ($°C^{-1} d^{-1}$) are coefficients defining the slope of the function. Early height growth is greatly affected by the *HS* coefficient. The parameters *RGRL* and *HS* were used to define early vigour for these simulations.

Since INTERCOM was initially developed for sugarbeet production, a number of other changes were needed to accurately simulate rice-barnyardgrass competition for light in well-fertilized high-yielding irrigated rice ecosystems. The water balance subroutines, for example, were removed because competition for water should not occur under continuously flooded conditions. Two additional improvements were made based upon the rice crop growth model ORYZA1 (Kropff et al. 1994b). Owing to the shock of transplanting, phenological development is delayed following this operation (Kropff et al. 1994c). A species-dependent input parameter is required to define the duration of this delay. INTERCOM uses a saturation function to define the photosynthesis–PAR response curve. In the rice version, the asymptote of this function is determined by the nitrogen content of leaves (Kropff et al. 1994d). Leaf nitrogen content as a function of developmental time is, therefore, required input. Other details of INTERCOM structure and its FORTRAN code are described in Kropff and Van Laar (1993).

Kropff et al. (1993, 1994a) evaluated INTERCOM performance using data from an experiment with irrigated direct seeded rice and barnyardgrass. Dry matter production, leaf area development, and yield were simulated accurately for all treatments. Further tests of model performance were made using eight data sets collected over a wide range of environments. Direct-seeded or transplanted rice yield loss resulting from barnyardgrass interference was predicted accurately by the model (92% of variation accounted for) over this wide range of competition situations (Kropff et al. 1994a).

Material and methods

Early leaf area growth rate

INTERCOM was used to evaluate whether variation in *RGRL* would influence the yield loss–weed density relationship. Field measured *RGRL* values for rice cultivar

IR72 were 0.009 and 0.012 for rice and barnyardgrass, respectively (Kropff and Lotz 1993). In a study of 17 rice cultivars in Indonesia, *RGRL* varied from a low value of 0.005 to a high value of 0.010 $°C^{-1} d^{-1}$ (Sutoro and Makarim 1994). Values reported for other species range from 0.0085 to 0.019 (Kropff and Lotz 1993).

Six rice-barnyardgrass mixture treatments were simulated for each of five *RGRL* values (0.005, 0.007, ..., 0.013, representing hypothetical rice isogenic lines). Direct-seeded rice density was assumed constant at 300 plants m^{-2}. Six barnyardgrass density treatments (0 to 300 plants m^{-2}) were set to emerge simultaneously with the crop. A rectangular hyperbola equation (Cousens 1985) was fit to simulated yield loss on barnyardgrass density obtained for each *RGRL* value.

Early height growth rate

INTERCOM was also used to evaluate whether variation in *HS* influences the rice-barnyardgrass interference relationship. As for *RGRL*, six rice-barnyardgrass mixture treatments were simulated for each of five rice *HS* values (0.0024, 0.0033, 0.0042, 0.0051, 0.006). Field measured *HS* values were 0.0042 and 0.0031 for rice and barnyardgrass, respectively (Kropff and Lotz 1993). Seeding rate, emergence time, and weed density were set to values identical to the simulations for effects of *RGRL*, and the rectangular hyperbola equation was fit to simulated yield loss on barnyardgrass density obtained for each *HS* value.

Combined effects

Qualitative evaluation of early crop vigour includes both early leaf area and height growth (Jennings and Aquino 1968). To examine the combined effects of changing *RGRL* and *HS*, rice-barnyardgrass competition was simulated using a factorial combination of the five *RGRL* and five *HS* values. Barnyardgrass density was held constant at 10 and 80 plants m^{-2} for these simulations.

Simulation results

Early leaf area growth rate

Simulated rice yield loss as a function of barnyardgrass density decreased as rice *RGRL* increased (Figure 1). Estimates of Cousens' (1985) I coefficient (dYL/dN as $N \rightarrow 0$, where YL is % yield loss and N is weed density m^{-2}) obtained from simulated yield loss relationships in Figure 1 are considerably lower when rice *RGRL* is high (Table 1), suggesting that rapid leaf area expansion will improve rice tolerance to barnyardgrass competition.

Early height growth rate

Predicted rice yield loss decreased when early height growth was increased (Figure 2). As with *RGRL*, Cousens' I coefficient decreased as *HS* increased (Table 1), suggesting

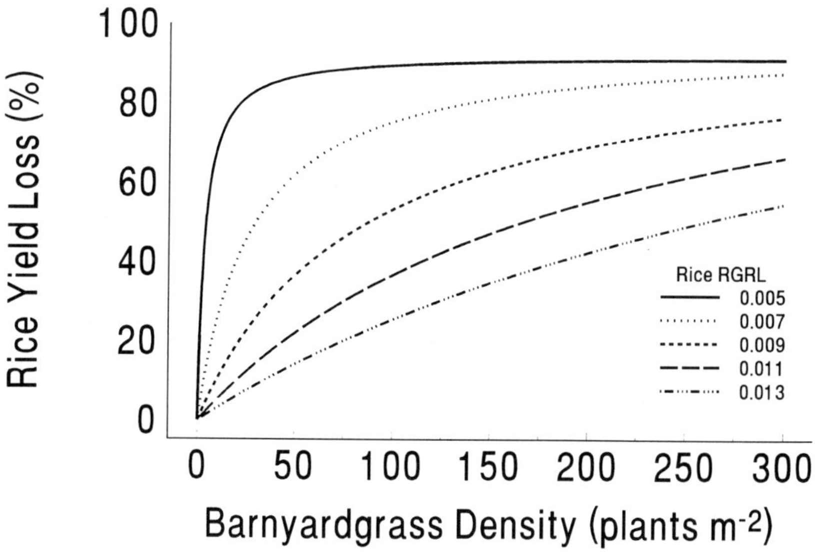

Figure 1. Simulated rice yield loss as a function of barnyardgrass density for five values of rice relative leaf area growth rate (*RGRL*, $°C^{-1} d^{-1}$) values. Lines show best fit of the rectangular hyperbola equation to each simulated data set.

Table 1. Influence of rice early leaf area (*RGRL*) and height (*HS*) growth rates on Cousens' *I* coefficient and simulated weed-free rice yield.

Parameter ($°C^{-1} d^{-1}$)	Value	*I* (% yield loss)	Yield ($kg\,ha^{-1}$)
RGRL	0.005	27.47	6361
	0.007	3.50	6769
	0.009	1.16	6931
	0.011	0.55	7000
	0.013	0.32	7037
HS	0.0024	3.57	6931
	0.0033	1.82	6931
	0.0042	1.16	6931
	0.0051	0.95	6931
	0.0060	0.87	6931

that rapid height growth may improve rice tolerance to barnyardgrass. Difference in simulated rice height growth when *HS* was changed from 0.024 to 0.0033 was greater than when changed from 0.0051 to 0.006 (Figure 3). Likewise, the change in yield loss was greater when *HS* was small.

Combined effects

INTERCOM suggests that simultaneously increasing *RGRL* and *HS* reduces rice yield loss greater than would occur if only one trait were improved, but the benefits

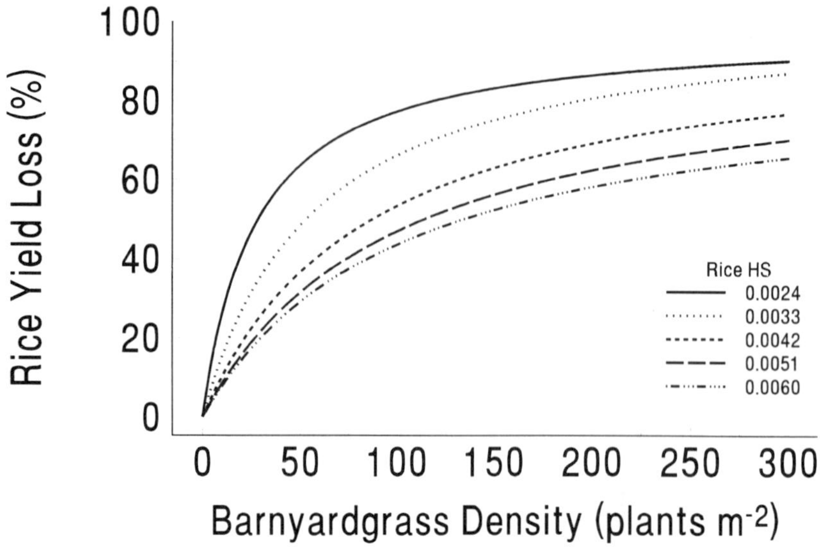

Figure 2. Simulated rice yield loss as a function of barnyardgrass density for five rice early height growth parameter values (*HS*, $°C^{-1}\,d^{-1}$). Lines show best fit of the rectangular hyperbola equation to each simulated data set.

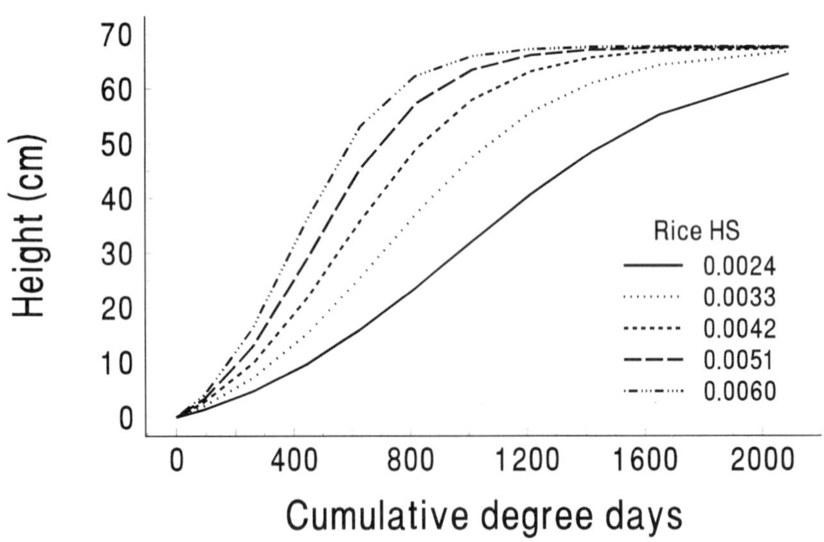

Figure 3. Rice height as a function of growing degree-days accumulated from emergence using equation (2).

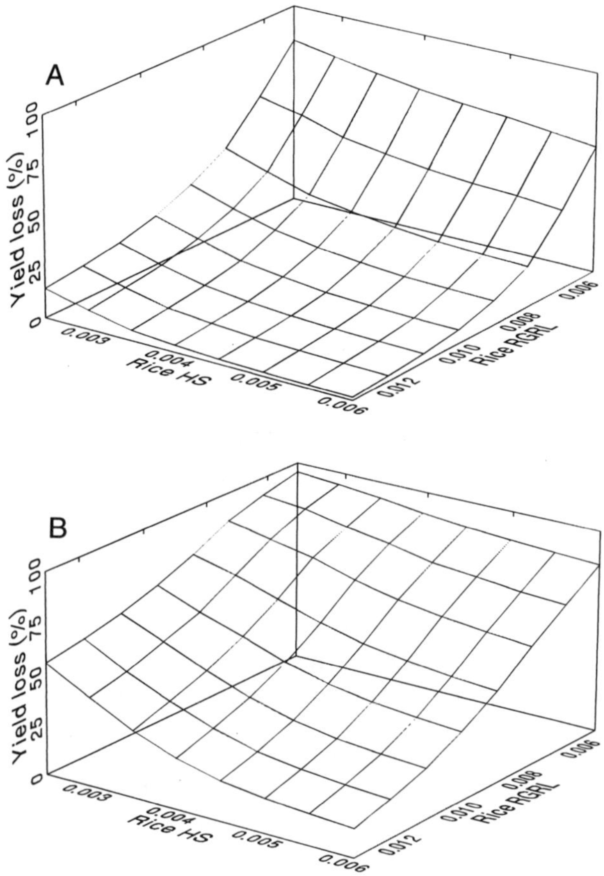

Figure 4. Simulated rice yield loss in mixture with (A) 10 and (B) 300 barnyardgrass plants m^{-2}, as influenced by variation in *RGRL* and *HS*.

were not additive. Improvements in *RGRL* had greater potential benefit than improvements in *HS* (Figure 4).

Discussion

Use of process-based crop growth simulation models to quantitatively identify traits useful for design-driven breeding programmes has received limited attention (Hunt 1993). There are several reasons for this. First, design-driven breeding (e.g. ideotype breeding) has only marginally been incorporated into most applied breeding programmes. Second, traits identified by any simulation model are generally quantitative in nature and, therefore, may be quite difficult to measure, especially when hundreds of genetic lines are being evaluated. Third, while a model may simulate detailed

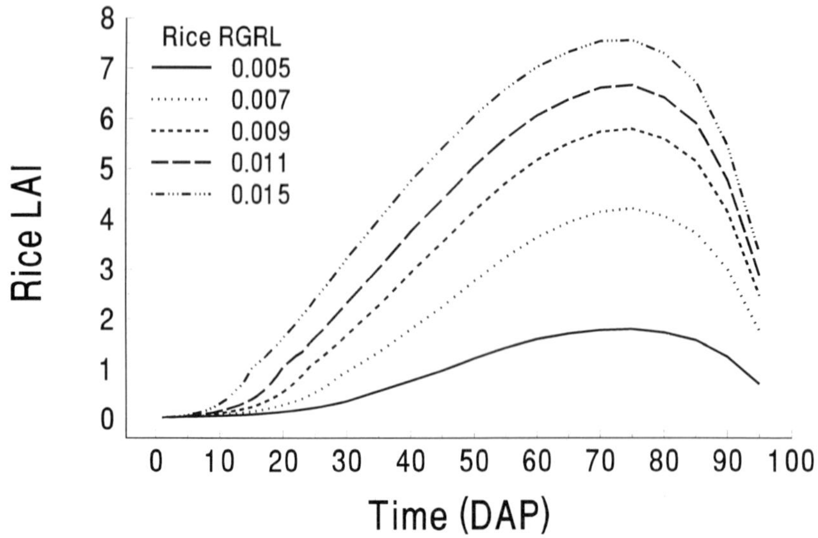

Figure 5. Effects of rice early leaf area growth rate (*RGRL*, °C^{-1} d^{-1}) on simulated rice leaf area index (LAI) grown in mixture with 10 barnyardgrass plants m^{-2}. Time in days after planting.

physiological processes, it is necessarily a simplification of the real system. As a result, feedback relationships that may regulate phenotypic or genotypic correlation among certain traits may not be accounted for.

On-farm yield of a crop cultivar differs from its potential yield as a result of undesirable, sometimes location-specific, conditions. These conditions may be related to nutrient or moisture availability, soil physical or chemical properties, or pests. In underdeveloped agricultural systems where one or more of these conditions may not be alleviated through management, design-driven breeding is a necessary component of yield improvement.

Our simulations suggest that increased early crop vigour correlates well with tolerance to barnyardgrass. However, determination of the relative leaf area and height growth rates requires repeated measurements early in the growing season. This is impractical for a breeder evaluating large numbers of genetic lines. However, recent reports suggest that a single measure of crop canopy area or leaf area index early in the growing season may be a sufficient indicator of crop tolerance (Callaway 1992; Forcella 1987; Jordan 1993). Simulated data shows that LAI does differ among genotypes early in the season, but these differences are relatively small (Figure 5). Owing to the variation likely to occur in a field experiment, it is possible that small differences among genotypes may not be statistically detected if measured early in the season. In a field study of competition among rice genotypes, early vigour resulted in improved competitiveness, but the effects of competition were not detectable until flowering (Jennings and Aquino 1968). INTERCOM suggested that maximum differences in rice LAI among hypothetical genotypes varying in early leaf area growth rate occurred at 70 to 75 days after planting, approximately the time to anthesis

(Figure 5). Weed density maintained during a trial may also have an important influence upon whether significant differences in leaf area index will be detected among genetic lines (Lindquist and Kropff 1996).

INTERCOM predicts that as rice leaf area expansion is increased, rice yield also increases (Table 1). Since differences in biomass partitioning patterns among hypothetical genetic lines were not considered for these simulations, an increase in yield could only occur if total above-ground biomass increased. In practice, some genetic lines of rice are likely to have high early vigour accompanied by a reduction in harvest index due to a trade-off in the fraction of biomass being partitioned to the leaves versus other organs. This negative association between competitiveness and yield might suggest that breeding for improved tolerance to weeds is a step in the wrong direction (see Jennings and Aquino 1968). However, a biomass partitioning pattern that results in rice LAI rapidly approaching a plateau followed by a strong 'stay green' character would optimize rice yield potential (Dingkuhn et al. 1993). Results of these simulations suggest that such a pattern would also optimize rice tolerance to barnyardgrass. Phenotypic and genotypic correlation among desirable 'tolerance' traits and yield may present barriers to breeding for tolerance traits and must be thoroughly explored prior to initiating an ambitious breeding programme. Nevertheless, such a breeding effort may be critical for underdeveloped crop production systems where herbicides are unavailable or costly, or where the probability of water contamination is high.

In developing this version of INTERCOM, we assumed plants compete only for light. Competitive relationships are likely to change under conditions where light is not the most important limiting resource. Traits that confer improved tolerance to weeds in a light-limiting system may be ineffective or even detrimental in a water- or nitrogen-limiting rice production situation. Knowledge of the most limiting resource in a given environment and the response of both crop and weed to that resource in limited supply is extremely important for the identification of traits conferring weed tolerance in other cropping systems.

References

Callaway M B (1992) A compendium of crop varietal tolerance to weeds. Am. J. Alt. Agric. 7:169–180.

Cousens R (1985) An empirical model relating crop yield to weed and crop density and a statistical comparison with other models. J. Agric. Sci. 105:513–521.

Dingkuhn M, Penning de Vries F W T, Miézan K M (1993) Improvement of rice plant type concepts: Systems research enables interaction of physiology and breeding. Pages 19–35 in Penning de Vries F W T, Teng P S, Metselaar K (Eds.) Systems approaches for agricultural development. Kluwer Academic Publishers, Dordrecht, The Netherlands.

Forcella F (1987) Tolerance of weed competition associated with high leaf area expansion rate in tall fescue. Crop Sci. 27:146–147.

Hunt L A (1993) Designing improved plant types: A breeder's viewpoint. Pages 3–17 in Penning de Vries F W T, Teng P S, Metselaar K (Eds.) Systems approaches for agricultural development. Kluwer Academic Publishers, Dordrecht, The Netherlands.

IRRI – International Rice Research Institute (1990) IRRI rice facts. International Rice Research Institute, P.O. Box 933, Manila, Philippines. 5 p.

Jennings P R, Aquino R C (1968) Studies on competition in rice. III. The mechanisms of competition among phenotypes. Evolution 22:529–542.

Jordan N (1993) Prospects for weed control through crop interference. Ecol. Applic. 3:84–91.
Kropff M J (1993) Mechanisms of competition for light. Pages 33–61 in Kropff M J, Van Laar H H (Eds.) Modelling crop-weed interactions. CAB International, Wallingford, UK.
Kropff M J, Lotz L A P (1993) Eco-physiological characterization of the species. Pages 83–104 in Kropff M J, Van Laar H H (Eds.) Modelling crop-weed interactions. CAB International, Wallingford, UK.
Kropff M J, Van Laar H H (Eds.) (1993) Modelling crop-weed interactions. CAB International, Wallingford, UK. 274 p.
Kropff M J, Weaver S E, Lotz L A P, Lindquist J L, Joenje W, Schnieders B J, Van Keulen N C, Migo T R, Fajardo F F (1993) Understanding crop-weed interaction in field situations. Pages 105–136 in Kropff M J, van Laar H H (Eds.) Modelling crop-weed interactions. CAB International, Wallingford, UK.
Kropff M J, Moody K, Lindquist J L, Migo T R, Fajardo F F (1994a) Models to predict yield loss due to weeds in rice ecosystems. Phil. J. Weed Sci. (Special Issue):29–44.
Kropff M J, Van Laar H H, Matthews R B (Eds.) (1994b) ORYZA1: An ecophysiological model for irrigated rice production. SARP Research Proceedings, International Rice Research Institute, P.O. Box 933, Manila, Philippines. 110 p.
Kropff M J, Van Laar H H, Matthews R B, Goudriaan J, Ten Berge H F M (1994c) Description of the model ORYZA1 (Version 1.3). Pages 5–41 in Kropff M J, Van Laar H H, Matthews R B (Eds.) ORYZA1: An ecophysiological model for irrigated rice production. SARP Research Proceedings, International Rice Research Institute, P.O. Box 933, Manila, Philippines.
Kropff M J, Van Laar H H, Matthews R B, Torres R O, Peng S, Cassman K G (1994d) Model parameterization. Pagers 43–55 in Kropff M J, Van Laar H H, Matthews R B (Eds.) ORYZA1: An ecophysiological model for irrigated rice production. SARP Research Proceedings, International Rice Research Institute, P.O. Box 933, Manila, Philippines.
Lindquist J L, Kropff M J (1996) Applications of an ecophysiological model for irrigated rice (*Oryza sativa*) – *Echinochloa* competition. Weed Sci. 44:52–56.
Lindquist J L, Maxwell B D, Buhler D D, Gunsolus J L (1995) Modelling the population dynamics and economics of velvetleaf (*Abutilon theophrasti*) control in a corn (*Zea mays*)–soybean (*Glycine max*) rotation. Weed Sci. 43:269–275.
Moody K (1988) Developing appropriate weed management strategies for small-scale farmers. Pages 319–330 in Altieri M A, Liebmann M (Eds.) Weed management in agroecosystems: Ecological approaches. CRC Press Inc., Boca Raton, FL, USA.
Sutoro, Makarim A K (1994) Variability in crop physiological inputs used in simulation models of rice. Pages 107–112 in Aggarwal P K, Matthews R B, Kropff M J, Van Laar H H (Eds.) Applications of systems approaches in plant breeding. SARP Research Proceedings, International Rice Research Institute, P.O. Box 933, Manila, Philippines.

Recent advances in breeding for drought tolerance in maize

G.O. EDMEADES, M. BÄNZIGER, A. ELINGS, S.C. CHAPMAN and J.-M. RIBAUT

International Maize and Wheat Improvement Centre (CIMMYT), Apdo. Postal 6-641, 06600 Mexico D.F., Mexico

Key words: maize, water stress, breeding, partitioning, anthesis-silking interval, nitrogen

Abstract

Drought is an important cause of instability of national maize grain yields and of the food supply and economy of small-scale maize-based farming systems in the tropics. Water shortages affect maize yields throughout the crop cycle, but most severely at flowering and to a lesser degree at establishment. Maize is often grown in environments thought to be better suited to sorghum and millet, yet because farmers in these areas persist with maize, researchers have an obligation to help stabilize maize production. Recurrent selection, which focused mainly on increasing tolerance to drought during flowering and grainfilling, has successfully increased grain yield over a wide range of moisture regimes. Gains ranged from 80–108 (mean 94) $kg\,ha^{-1}\,yr^{-1}$ when selecting among full-sib (FS) families to 73–144 (mean 111) $kg\,ha^{-1}\,yr^{-1}$ when selecting among S_1 families, and were observed when selections were evaluated at a drought-induced yield level of 1.5–2.4 $ton\,ha^{-1}$. This represents annual gains in severely droughted environments of >5%. Under well-watered conditions where yields ranged from 5.6–8.0 $ton\,ha^{-1}$, gains from the same two selection schemes ranged from 38–108 (mean 73) $kg\,ha^{-1}\,yr^{-1}$ and from 27–89 (mean 59) $kg\,ha^{-1}\,yr^{-1}$, respectively. Gains in yield were mainly due to increased numbers of grains per plant, and were associated with a reduced anthesis-silking interval (ASI) and an increased harvest index. These changes are consistent with increased assimilate partitioning to the ear at flowering, a phenomenon that will need to be addressed by maize models that focus on genotypic responses to stress. Little change was observed in plant water status or foliar senescence rates. Drought-tolerant varieties have shown similar gains when evaluated under low N conditions, suggesting that partitioning of N to the ear has also been improved. The challenge now is to develop drought-tolerant hybrids. Conventionally improved populations have provided a lower frequency of drought-tolerant hybrids than their counterparts with a history of improvement for drought tolerance. Molecular marker-based linkage maps of ASI, an easily observed indicator of drought tolerance at flowering, suggest that this trait could be efficiently transferred to elite inbred lines using marker-assisted backcrossing techniques. Yields of drought-tolerant varieties, lines and hybrids, when grown under well-watered conditions, have shown clearly that drought tolerance does not come at the cost of yield potential. The key to efficient improvement for drought-tolerance lies in the choice of elite adapted germplasm, use of carefully managed drought stress, and selection based on a minimum data set comprising anthesis date, ASI, ears per plant and shelled grain yields. This results in stabilized and improved yield for the maize component of maize-based cropping systems exposed to mid-season and terminal drought in highly variable rainfall environments.

Introduction

Five broad levels of agricultural production can be distinguished (Penning de Vries and Van Laar 1982). At the potential production level, growth occurs under conditions of ample supply of water and nutrients, and is determined by weather conditions (radiation, temperatures) in interaction with the morpho-physiological make-up of

M.J. Kropff et al. *(eds.): Applications of Systems Approaches at the Field Level, 63–78.*

the plant. This circumstance prevails on perhaps 5% of the area planted to maize in the lowland tropics. Low water, soil nitrogen and perhaps phosphorus and zinc availability limit growth for at least part of the growing season, and define production levels 2, 3 and 4 respectively. Together, they define the attainable production level, found on around 95% of the tropical maize growing area. Actual production level may be further reduced if growth is limited by the presence of weeds, pests, diseases or pollutants.

The two largest causes of maize yield loss in the lowland tropics are drought and low fertility. The lowland tropics (< 1200 m elevation) is where the largest area and production of maize in the tropics are found (CIMMYT 1988), and will be the focus of this paper. Losses to drought alone in the lowland tropics may total 15 million tons per year, or around 15% of well-watered production (Edmeades et al. 1992), but can be as high as 60% in severely affected regions such as southern Africa in 1991–92 and again in 1994–95. Severe losses such as these can be expected to lead to a close association between seasonal rainfall and national average maize yields. The normal interseasonal fluctuations in rainfall, however, have also been associated closely with variation in average national maize yields across quite large production regions in eastern and southern Africa (Rötter 1993; Edmeades et al. 1995b), suggesting that drought is a pervasive cause of yield instability in maize-based cropping systems in most years and environments. Localized yield losses can, of course, reach 100% where rainfall falls below 500 mm in the crop season or where soils are sandy or shallow, provided the stress coincides with flowering. Despite the fact that the yields of sorghum and millet are often more stable than maize, they are often lower in good years, and even though it may seem logical to grow sorghum and millet in semi-arid areas, consumer preference often demands that maize be grown instead.

Hunt (1993) noted that the first essential step in breeding is to ensure that crop phenology matches the seasonal cycle of temperature, rainfall and soil water balance. As the maize plant is particularly susceptible to water shortage during flowering (see below), it is important to assess the consequences of stress at a particular time on crop production in a field or region. This should be done either before cultivar development, or later when an elite genotype has been developed and needs to be matched to a suitable growing region (Khush et al. 1995). The latter is especially helpful in unfavourable environments where genotype × environment interactions are large. Simulation models that incorporate the relevant phenological and growth processes can be applied to this process. Examples of such exercises have been reported for tropical maize in southern Africa (Waddington et al. 1995).

Drought at establishment, flowering and grainfilling: tolerance versus escape

Stabilization of maize output in marginal production areas is an exercise in risk management in which genetic improvement for drought tolerance plays a significant part. The probability of drought in maize growing environments in the tropics is highest at the start and end of the rainy season.

Early maturity usually allows cultivars to escape the consequences of terminal drought. If the rainy season is reliable but very short, then earliness is a desirable breeding goal. Evolutionary pressures and farmer selections have resulted in the development of early maturing landrace maizes in dry tropical areas of Indonesia, Kenya, Mexico, Colombia, etc. These escape drought but are usually low yielding when rainfall is not limiting. A more productive strategy would be to develop a later-maturing cultivar with a higher yield under optimal conditions, and a yield under drought which is stabilized by reducing the effects of drought on grain number and size.

A first step in obtaining yield is to have sufficient plant stand. If drought severely reduces stand at the onset of the season, farmers can replant fields with a shorter-duration cultivar or a different species, though this entails additional cost. Preliminary research on improving seedling establishment shows promising improvements in survival under water deficit (M. Bänziger, unpubl. data, 1996). A failure of the rains later in the season when replanting is not possible, however, may lead to a total loss. Monthly rainfall totals in the tropics have a high coefficient of variability, even though mean rainfall appears adequate for maize production. There is always some risk of drought stress at all stages of the season, and although the probability of drought stress is lower at silking, its consequences on yield can be severe since the crop is unusually susceptible to drought at this stage (Grant et al. 1989). Stress from mid- to late-grainfilling also reduces grain yields, but less so than stress at silking (Shaw 1976). Grain number per plant is itself determined by events which occur near flowering, so much emphasis has been placed on this period in our programme in order to stabilize grain yield in marginal environments.

Flowering

The unusual susceptibility of maize to drought stress at flowering is probably because of the spatial separation of male and female flowers on the plant and a reduced proportion of assimilates partitioned to the ear versus the tassel when the plant is stressed at flowering (Bolaños and Edmeades 1993b). Reduction in grain number per plant under drought occurs within the first two weeks after pollination (Grant et al. 1989). A characteristic of maize under drought stress (and most other stresses that reduce photosynthesis per plant) is an increase in the anthesis-silking interval (ASI) and a concomitant increase in the number of barren plants (DuPlessis and Dijkhuis 1967; Bolaños and Edmeades 1993b). For example, a 1-day increase in the ASI is associated with a 28% reduction in kernels per plant, a 3-day increase with 55% and a 5-day increase with 69% (Bolaños and Edmeades 1993b). Pollen shortage is rarely the reason for this type of barrenness (Westgate and Bassetti 1991). Silk growth is very sensitive to plant water potential (Bassetti and Westgate 1993). A short ASI also indicates the diversion of an increased fraction of the plant's current photosynthesis to the ear, since it is associated with rapid ear growth (Edmeades et al. 1993). When late-emerging silks on drought-stressed plants are pollinated, fertilization usually takes place, but grain development is arrested shortly afterwards (Bassetti and

Westgate 1993), giving rise to patchy grain formation, bare ear tips or complete barrenness. In a severely drought-stressed field it is common, therefore, to find a significant proportion of plants, perhaps 30–50%, with no ears at all.

Abortion of fertilized kernels shortly after pollination seems to be due to a reduced flux of assimilates, especially sucrose, to the developing kernel (Schussler and Westgate 1995; Zinselmeier et al. 1995b). The capacity of newly fertilized ovules to attract assimilate reserves in the stem is limited (Schussler and Westgate 1991), though abortion of grains can be prevented by increasing the carbohydrate levels in the stem (Boyle et al. 1991; Zinselmeier et al. 1995b). Schussler and Westgate (1995) and Zinselmeier et al. (1995a) have recently reported that drought at flowering also affects carbohydrate metabolism of the developing ovule, further reducing sucrose flux to the newly formed seed. A long anthesis-silking interval, therefore, is a symptom of reduced assimilate flux to the developing ovule rather than the direct cause of barrenness.

Grainfilling

Stabilization of grain yield over a range of water regimes is best done by selecting for a short ASI under a drought stress that is managed to coincide with flowering, and then proceeding to fill the ear so formed. Provided an ear has been established, the maintenance of a green functional canopy and a capacity to remobilize carbohydrates stored in the stem and husk (Blum 1988) should contribute to high yield under terminal stress. Associations between foliar 'stay green' and yield are often weak (Edmeades et al. 1995b), and reasons for this must be sought in the N balance of the crop at that growth stage. Selection for more grains per plant will likely increase the internal demand for N, and since uptake of N from a dry soil is low, this may result in 'mining' of N from leaves, thus offsetting improvements in 'stay green' resulting from directed selection.

Selection in elite germplasm for tolerance to drought during flowering and grainfilling

Selection procedures

This has been extensively described elsewhere (Bolaños and Edmeades 1993a,b; Bolaños et al. 1993; Byrne et al. 1995). In summary, selection began in the elite lowland tropical white dent population, Tuxpeño Sequía, in 1975. This population underwent eight cycles of recurrent selection among full-sib families in the rain-free winter crop season at Tlaltizapán, Mexico, where timing and intensity of stress can be managed by irrigation. Each of 250 families was grown in single-row plots under three regimes of increasing drought intensity, namely well-watered (WW), intermediate stress (IS; water withdrawn during late flowering and throughout grainfilling) and severe stress (SS; no water applied from 3 weeks before silking onwards). Selection of the best 60–80 families was based on an ideotype with high grain yield under SS and IS, high leaf and stem extension rates under drought, delayed foliar senescence under SS and IS, and reduced canopy temperatures and ASI under IS and SS. Each cycle of full-sib selection required one year to complete.

Improvement commenced in a new and diverse group of five elite maize populations in 1985–86, using a recurrent S_1 selection scheme, and continues in three of these (Edmeades et al. 1995a). Three of these populations (La Posta Sequía; Pool 26 Sequía and TS6) are late maturing, and two (Pool 16; Pool 18 Sequía) are early maturing. Within each selection cycle 500–600 S_1 families were prescreened under drought and heat in the Sonoran Desert at Obregón, NW Mexico. The superior 200-220 families thus identified were grown in Tlaltizapán during winter from remnant seed under the three water regimes described, and the best 50 S_1 families recombined to form the subsequent cycle of selection. The stress level tolerated by S_1 families is less than that for full-sib families, so water was applied once after flowering in the SS treatment. Each of the managed selection environments was used to expose genetic variation for a specific trait. The dry heat of Obregón (temperatures often exceed 37°C) exposed variation for upper leaf senescence, tassel sterility and barrenness. At Tlaltizapán the well-watered environment allowed expression of yield potential, while the IS regime exposed genetic variation for lower leaf senescence and for grain yield. Under the SS regime, variation in grain yield was small, but genetic variability for ears per plant (barrenness) and for ASI was exposed. The heritability for grain yield of S_1 families remained in the range of 0.5–0.6 until yields fell to below 20% of the well-watered control. Selection was principally for increased shelled grain yield, ears per plant and 'stay green' under stress, and for decreased ASI and tassel sterility. Less important traits were upright unrolled leaves, small tassels and lodging resistance. Two years were needed to complete a cycle of two-stage recurrent S_1 selection.

Evaluation of progress

Evaluations of C_1, C_2, C_4, C_6 and C_8 of Tuxpeño Sequía were conducted in two seasons from 1987 to 1989 in large plots at the selection site, using the three water regimes described. Irrigation regimes were grouped over years into droughted and well-watered environments. Selection resulted in a significant increase in grain yield of 108 kg ha^{-1} cycle^{-1} at yield levels ranging from 1 to 8 ton ha^{-1}, and the same rate of gain was observed under well-watered and drought-stressed conditions. At a yield level of 2 ton ha^{-1}, this represented a gain of 6.3% per cycle (Bolaños and Edmeades 1993a). These same genotypes and some conventionally selected checks were grown in 12 environments, seven of which were rainfed and outside Mexico. Gains averaged 90 kg ha^{-1} cycle^{-1} across all sites at a mean yield level of 5.6 ton ha^{-1} (Byrne et al. 1995). This indicated that 83% of the gains from selection in a dry winter environment were observed in a more normal production situation in the summer rainy season.

In populations subjected to S_1 recurrent selection gains were evaluated after 1–3 selection cycles. Gains per cycle were expected to be larger than those under full-sib selection because selection intensity was greater (35% of best families selected

Table 1. Grain yield and other traits of late-maturing maize populations when grown at five droughted sites or at five well-watered locations in Mexico (*Source:* adapted from Edmeades et al. 1995a).

	Yield, droughted (ton ha^{-1})	Yield, well-watered (ton ha^{-1})	Anthesis date droughted (d)	ASI[1] droughted (d)	Ears per plant droughted	Harvest index droughted
Sites	5	5	5	5	5	2
La Posta Sequía C_0	1.91	8.22	83.6	6.7	0.76	0.10
La Posta Sequía C_1	1.95	8.30	83.4	6.1	0.79	0.09
La Posta Sequía C_2	2.38	8.45	82.4	4.2	0.88	0.14
La Posta Sequía C_3	2.53	8.35	82.3	3.4	0.95	0.18
Pop. 43 C_9	1.74	8.00	86.0	7.0	0.79	0.09
Pool 26 Sequía C_0	2.25	7.52	78.0	5.8	0.83	0.16
Pool 26 Sequía C_2	2.59	7.81	77.5	3.7	0.95	0.18
Pool 26 Sequía C_3	2.82	7.87	76.2	2.8	0.98	0.24
Pool 26 C_{23}	2.14	7.97	81.6	5.9	0.81	0.12
Tuxpeño Sequía C_0	1.75	7.48	82.6	6.4	0.73	0.12
Tuxpeño Sequía C_8	2.39	7.78	80.0	2.9	0.93	0.22
TS6 C_1	2.69	8.19	80.8	1.9	0.95	0.24
Mean	2.35	7.96	80.6	4.5	0.88	0.17
LSD	0.35	0.39	0.56	1.34	0.07	0.06
Gain cycle^{-1}						
La Posta Sequía	0.229**[2]	0.053ns	−0.52**	−1.18**	0.07**	0.03**
Pool 26 Sequía	0.288**	0.177*[2]	−0.93**	−1.50**	0.08**	0.04*
Tuxpeño Sequía	0.080**	0.038**	−0.32**	−0.44**	0.02**	0.01**

[1]ASI: Anthesis-silking interval.
[2]Gains significantly different from zero at $P<0.01$ and $P<0.05$ respectively.

Table 2. Evaluation of progress in an early-maturing tropical maize selected for drought tolerance (Pool 18 Sequía) and grown in Mexico (*Source:* adapted from Edmeades et al. 1995b).

Genotype name	Droughted sites mean (ton ha^{-1})	Irrigated sites mean (ton ha^{-1})	Across sites mean (ton ha^{-1})
Sites	4	6	10
Pool 18 Sequía C_0	1.42	5.30	3.54
Pool 18 Sequía C_1	1.51	5.64	3.76
Pool 18 Sequía C_2	1.67	5.47	3.74
Pool 18 Sequía C_3	1.85	5.78	4.00
Pop. 31 C_2	1.55	5.60	3.76
Pop. 31 C_5	1.40	6.00	3.91
Environmental mean	1.46	5.64	3.74
Selection gains			
Pool 18 Sequía (versus C_0)			
ton ha^{-1} cycle^{-1}	0.146**[1]	0.126**	0.134**
Gain % cycle^{-1}	10.3	2.4	3.8
Population 31 (C_5 versus C_2)			
ton ha^{-1} cycle^{-1}	−0.050	0.133**	0.050
Gain % cycle^{-1}	−3.2	2.4	1.3

[1]Gains significantly different from zero, $P<0.01$.

in full-sibs, versus 8% in S_1). Cycles of selection of La Posta Sequía (C_0, C_1, C_2 and C_3) and Pool 26 Sequía (C_1, C_2 and C_3) were compared with the conventionally improved versions of these populations (Pop. 43 C_9 and Pool 23 C_{23}) and with Tuxpeño Sequía C_0 and C_8 and TS6 C_1 (first cycle of S_1 recurrent selection in Tuxpeño Sequía C_6) along with some check cultivars (Edmeades et al. 1995a). Average yields of the environments varied from 1.01 to 10.40 ton ha^{-1}. Five environments were droughted during flowering and grainfilling, and the other five were well-watered. Mean yields were 2.35 ton ha^{-1} and 8.0 ton ha^{-1} respectively. Yield gains across the two populations La Posta Sequía and Pool 26 Sequía averaged 259 kg ha^{-1} (12.4%) cycle^{-1} under drought, and 115 kg ha^{-1} (1.5%) cycle^{-1} under well-watered conditions (Table 1). Yield improvements under drought were paralleled by increases in ears per plant of around 0.075 cycle^{-1} and in harvest index, while ASI declined by 1.3 d cycle^{-1}. Populations became earlier to flower by 0.7 d cycle^{-1}, despite our objective of maintaining constant the time from planting to anthesis. Conventional selection, using international progeny evaluation at mainly well-watered sites did not increase drought tolerance; Pop. 43 C_9 and Pool 26 C_{23} became later by about 1 d cycle^{-1} but improvements in yield, if they occurred at all, were only observed under well-watered conditions. In this evaluation gains of Tuxpeño Sequía C_8 versus C_0 were slightly lower than those reported previously (Bolaños and Edmeades 1993a; Byrne et al. 1995), though TS6 C_1 represented an important improvement over Tuxpeño Sequía C_8 under all environments (Table 1). Principal component analysis of yields in the 10 different environments showed that well-watered and droughted environments were generally orthogonal (Chapman et al. 1996), interpreted by the authors as meaning that selection in only well-watered environments is unlikely to give improvements in yield under drought. They concluded from these and other analyses that selection for drought tolerance has improved broad adaptation, as well as specific adaptation to dry environments.

In a similar set of trials to those described above, rates of progress were determined in Pool 18 Sequía (C_0–C_3) (Table 2). Gains averaged 146 kg ha^{-1} (10.3%) cycle^{-1} under drought, 126 kg ha^{-1} (2.4%) cycle^{-1} under well-watered conditions and 134 kg ha^{-1} (3.8%) cycle^{-1} across sites. This contrasted with a slight loss in yield (−3.2%) under drought and a 2.4% cycle^{-1} yield increase under well-watered conditions in the check entry, Pop. 31, which was subject to conventional selection based on international progeny testing.

As a test of broad adaptation, the most advanced cycle of selection of these populations, other drought-tolerant source populations such as DTP1 and DTP2, and conventional and local check entries were included in International Cultivar tests as part of CIMMYT's Drought Network activities (Table 3). Late entries were part of a 20-entry trial grown at 19 sites, and the early entries were part of a 12-entry trial grown at 21 sites. La Posta Sequía C_3 showed high yields at sites with adequate moisture, as did the conventionally bred elite check, Across 8627 RE. Pool 26 Sequía C_3, a relatively early cultivar, was not stable across sites, performing well in severely stressed sites but poorly in well-watered sites. Stability analyses showed that La Posta Sequía C_3 and several site-specific selections of DTP1 were high yielding and stable.

Table 3. Anthesis date and grain yield (rank in parentheses) of drought-tolerant selections, conventional checks (RE) and local checks included in trials of (A) 20 late-maturing cultivars, and (B) 12 early-maturing cultivars, when grown internationally, 1992–94 (*Source:* adapted from Edmeades et al. 1995b).

	Anthesis (d)	Yield, all sites (ton ha^{-1})	Yield, severe stress (ton ha^{-1})	Yield, moderate stress (ton ha^{-1})	Yield, little stress (ton ha^{-1})
A: Late-maturing trial					
Number of sites	19	19	5	7	7
La Posta Sequía C_3	64.5	3.58 (1)	0.99 (12)	2.95 (6)	6.06 (1)
Pool 26 Sequía C_3	58.4	2.93 (20)	1.37 (1)	2.36 (20)	4.61 (20)
Ngabu (1) DTP1 C_5	61.6	3.36 (4)	1.08 (6)	2.86 (10)	5.48 (9)
DTP2 C_2	60.5	3.35 (6)	1.08 (8)	2.75 (15)	5.57 (4)
TS6 C_1	64.2	3.31 (9)	0.58 (20)	2.99 (4)	5.57 (3)
Across 8627 (RE)	62.8	3.42 (3)	1.17 (2)	2.91 (8)	5.53 (7)
Local check 2	63.6	3.56 (2)	0.88 (18)	3.36 (1)	5.68 (2)
Mean of trial	61.8	3.30	1.01	2.84	5.39
LSD (0.05)	1.36	0.32	0.39	0.55	0.58
B: Early-maturing trial					
Number of sites	21	21	7	7	6
DTP1 C_5 early selection	59.4	2.63 (1)	0.89 (6)	2.70 (1)	4.63 (1)
Pool 18 Sequía C_3	54.3	2.53 (2)	0.99 (3)	2.67 (2)	4.10 (6)
Pool 16 C_{20} Syn. 1	54.2	2.47 (4)	1.03 (1)	2.58 (4)	4.05 (9)
TIWD Drt Tol. Pop. C_0	59.7	2.41 (6)	0.67 (12)	2.48 (5)	4.31 (2)
Santa Rosa 8330 (RE)	58.7	2.29 (9)	1.00 (2)	1.90 (10)	4.18 (4)
Local check 1	58.3	2.50 (3)	0.94 (5)	2.38 (8)	4.18 (3)
Mean of trial	56.8	2.38	0.87	2.36	4.11
LSD (0.05)	1.80	0.28	0.40	0.48	0.51

Grain yield showed a strong dependence on ears per plant (r=0.71**) and on ASI (r=−0.51**). Ears per plant and ASI were also closely associated (r=−0.55**). In the early-maturing trial (Table 3), an early-maturing selection from DTP1 performed well, especially under well-watered conditions. In severely droughted sites earliness had a clear advantage, and Pool 18 Sequía C_3 was the best performer. Pool 18 Sequía C_3 and Pool 16 C_{20} Syn. 1 were stable and high yielding across sites. Grain yield over sites was again heavily dependent on ears per plant (r=0.68**), though not on ASI (r=−0.34ns). In the severely stressed locations, ears per plant and ASI showed a strong interdependence (r=−0.83**) (Edmeades et al. 1995b). Clearly, early-maturing cultivars possessing a high level of drought tolerance have an important role to play in many national programmes.

What changed with selection?

Gains were largely the result of reduced barrenness (i.e. increased ears per plant) under drought, and an associated increase in harvest index, since total biomass production was unaffected by selection. In Tuxpeño Sequía tassel size was reduced, and root biomass in the top 50 cm of soil declined by 35%; there was no change in any trait indicative of plant water status (e.g. predawn or noon water potential; osmotic adjustment; canopy temperature, water extraction profiles) (Bolaños et al. 1993). In all cases, anthesis date became slightly earlier, and ASI became shorter under drought (Table 1). Eight cycles of selection in Tuxpeño Sequía led to significantly faster spikelet and ear growth but also to a 21% reduction in final spikelet number (Edmeades et al. 1993). Thus fewer spikelets were formed, grew more rapidly and were ultimately more successful in forming grain, especially under conditions of drought at flowering. Preliminary data suggests that these changes in ear growth have also occurred in the other late-maturing populations under improvement.

Appropriate secondary traits for selection

A suitable secondary trait must have a high heritability, be easy to measure, be correlated with productivity and show some adaptive value under drought. If it is a direct indicator of yield under stress, then it should be able to be measured as early as possible in the life cycle of the crop. Data from trials of more than 3500 S_1 progenies from several populations suggest that heritabilities of leaf extension rate, canopy temperature and leaf chlorophyll concentration are low, and that those for leaf erectness, tassel size, senescence scores and ASI are relatively high (Bolaños and Edmeades 1996). Of these, only ASI and tassel size appear to have significant adaptive value under drought, though delayed senescence during grainfilling, cool canopy temperatures and upright leaves also form part of our stress-tolerant water-efficient ideotype designed to grow in a competitive situation. Phenotypic and genotypic correlations between secondary traits and grain yield under drought stress are given in Table 4. Variation in yield under stress is almost completely determined by variation in kernel number per plant.

At low yield levels (<20% of yield potential) heritabilities for yield fall. The genetic correlation between grain yield and ASI was −0.65, and that for grain yield and ears per plant increased to 0.90, indicating that these traits become good surrogates for grain yield in these severely stressed environments (Table 4). Selection using ASI and ears per plant, both determined near flowering, can add precision to the routine determination of the yield potential of a genotype under stress. Observations of ASI could be used to eliminate drought-susceptible lines from a breeding nursery without losing a cycle to testing. Changes in ASI due to selection have been related to changes in allelic frequency at genome map positions associated with changes in ASI under drought (Ribaut et al. 1995).

Relationship of drought tolerance to other abiotic stresses

Recent studies have shown that drought tolerant selections also perform well under low N conditions (Table 5). Gains observed under low N averaged 210 $kg^{-1} cycle^{-1}$ for S_1 selection and 86 $kg\,ha^{-1} cycle^{-1}$ for full-sib selection. Annual gains were about the same from both breeding systems, and were very similar in magnitude to those observed under drought. This indicates that selection for improved partitioning of assimilates to the developing ear using drought stress at flowering as the selection

Table 4. Phenotypic and genetic correlations between grain yield and selected traits under severe drought stress for S_1 progenies drawn from several populations. All phenotypic correlations were significant at $P<0.01$. Scores are on a 1 (unrolled, upright or green) to 5 (rolled, lax or dead) scale. Phenotypic correlations are weighted averages of up to 11 progeny trials (with the specified total number of observations); genotypic correlations are unweighted averages for the number of trials specified (*Source:* Bolaños and Edmeades 1996).

	No. of observations	Phenotypic correlation	No. of of trials	Genotypic correlation
Ears per plant	2449	0.77	9	0.90±0.14
Kernels per ear	2227	0.50	7	0.71±0.22
Kernels per plant	2227	0.90	8	0.86±0.15
Kernel weight	2227	0.46	9	0.14±0.17
Days to 50% anthesis	2449	−0.40	9	−0.58±0.12
ASI	2449	−0.53	8	−0.65±0.24
Leaf and stem extension rate	1321	0.10	4	−0.10±0.33
Canopy-air temperature diff.	1089	−0.27	2	−0.20±0.15
Leaf rolling score	2033	−0.18	9	−0.03±0.18
Leaf erectness score	2033	−0.18	1	0.00
Leaf death score	2449	−0.11	9	0.11±0.24
Chlorophyll concentration	905	0.17	3	0.11±0.24
Tassel branch number	1793	−0.16	1	0.15*
Plant height	2449	ns	–	–
Lodging percent	–	–	9	−0.03±0.15

*Calculated from a single evaluation of S_3 lines.

Table 5. Grain yields of original and drought-tolerant cycles of four populations when evaluated under four environments differing in N level at Poza Rica. Gains per cycle of selection and per year were calculated for the means of the two low N experiments only (*Source:* Lafitte and Bänziger 1995).

Sites	Grain yield at				Gain	
	All N levels 5	Low N 2	Medium N 1	High N 2	$cycle^{-1}$	yr^{-1}
Tuxpeño Sequía C_0	4.32	2.26	5.23	5.94		
Tuxpeño Sequía C_8[b]	4.97	2.95	6.03	6.46	86 (3.8)	86 (3.8)
La Posta Sequía C_0	4.81	2.58	5.09	6.91		
La Posta Sequía C_3[c]	5.40	3.28	5.89	7.28	233 (9.0)	117 (4.5)
Pool 26 Sequía C_0	4.72	3.12	4.90	6.23		
Pool 26 Sequía C_3[c]	5.30	3.74	5.36	6.83	207 (6.6)	103 (3.3)
Pool 18 Sequía C_0	3.89	2.51	4.45	4.99		
Pool 18 Sequía C_2 BA[c]	4.17	2.89	4.53	5.29	190 (7.6)	95 (3.8)
Mean	4.70	2.92	5.19	6.24		100 (3.9)
LSD (0.05)	0.49	0.72	1.20	0.70		
P(population)	***[a]	**	*	***		
P(cycle)	***	**	+	*		
P(population×cycle)	ns	ns	ns	ns		

[a]+, *, **, *** indicates significance at $P<0.10$, 0.05, 0.01, and 0.001 respectively.
[b]Full-sib recurrent selection scheme; one cycle yr^{-1}.
[c]S_1 recurrent selection scheme; one cycle needs two years for completion.

screen can simultaneously improve tolerance to drought and to low N. Progeny trials conducted in an early-maturing population have shown a correlation of 0.45 ($P < 0.01$) between ASI observed under low N and under drought (M. Bänziger, unpubl. data, 1995). In a recent study, Uhart and Andrade (1995) noted that under moderate levels of N stress the effects of N on grain number were mediated through effects on carbon metabolism. Under severe levels of N deficiency, however, it seems likely that ASI may also be an indicator of partitioning patterns of N. The relatively low correlation between ASI observed under drought and under low N suggests that each may be under separate genetic control, and this possibility is being investigated at present.

Discussion

Selecting for drought tolerance

Duvick (1995) has proposed that the major goals of tropical maize breeding in the near future should be stabilized yield and broadened adaptation through increased stress tolerance. Results presented here suggest one way of accomplishing this. While it is possible to have stable but low performance in drought-prone environments by selecting for earliness, stable and *improved* yields under drought are only possible by selecting for improved drought tolerance. Present results suggest that selection can increase grain yields by about 35% under conditions where drought during the flowering and grainfilling period has reduced yields by about two-thirds. Gains have averaged about 100 kg ha^{-1} yr^{-1} or about 5% per year. There appear to be no adverse effects in more favoured environments from such a selection strategy: yields under well-watered conditions have actually increased at rates similar to those measured under conventional selection, while gains in drought tolerance from selecting under carefully managed drought stress have been significantly greater. Furthermore, gains observed under drought are also seen under low N, and drought and infertile soils often occur together. The use of drought-tolerant maize will clearly reduce the frequency with which grain yields would fall below some threshold necessary for household food security while, at the same time, reducing the fluctuation between national maize production surpluses and catastrophic deficits. On a note of caution, it should be recalled that even though these gains are significant and important, they may only raise yields under drought from 2 to 2.7 ton ha^{-1}. Careful crop management practices will in all likelihood add at least as much or more. The remainder of the yield gap (often a further 2–5 ton ha^{-1} of grain) can only come from additional water applied to the crop.

A key factor in the success of this selection programme has been the availability of facilities where drought stress can be carefully managed. Even though this led to the use of an atypical winter season selection site, only 17% of the gains were lost when germplasm was transferred to the normal summer production environment. Given the importance of the latter part of the crop life cycle in determining maize grain yields, and the high phenotypic correlation between grains per plant and grain yield under drought stress ($r=0.9^{**}$, Table 4), our focus on factors affecting grains per plant when selecting for drought tolerance in tropical germplasm seems justified. Where resources are scarce, selection should be based on a minimum data set comprising anthesis date, ASI, ears per plant and shelled grain yields. Observation of anthesis date is simply to avoid the unwitting selection of early flowering escapes. Our evidence suggests that selection using these criteria will result in stabilized and improved yields of maize exposed to mid-season and terminal drought in highly variable rainfall environments.

For national programmes the correct choice of germplasm is critical. Landraces normally have a number of agronomic defects that may take years of selection to correct, and for this reason it is recommended that selection begin in *elite* adapted germplasm (Edmeades et al. 1995b). The value of earliness as an escape mechanism cannot be ignored, though very early genotypes are usually low yielding when rains are plentiful. In cases where rainfall intensity and duration are highly variable, it may be appropriate to grow a mixture of an early-maturing cultivar with one which is late-maturing but drought-tolerant. In the event of severe drought, the early component escapes the drought; under moderate stress, the drought-tolerant component provides a stable and increased yield. Should the rains be plentiful, the late-maturing component can exploit these through its higher yield potential. In many tropical areas, maize stands in the field well past the point of physiological maturity, awaiting the availability of labour for hand-harvesting. In these circumstances, a variation of 2–3 weeks in flowering date is unlikely to cause a reduction in harvest quality because of the resulting variation in grain moisture.

Although the research reported here has focused on open-pollinated varieties (OPVs), evidence suggests that these improvements in source populations will carry over to lines and hybrids derived from them. In a current study, a large number of S_2 lines derived at random from three populations improved specifically for drought tolerance, and their conventionally selected equivalents were crossed to two neutral testers (inbred lines) and the hybrids evaluated under drought. The hybrids derived from the drought-tolerant populations outyielded those derived from their conventional equivalents by 20% at a mean yield level of 1.6 ton ha^{-1}. The probability of obtaining a hybrid with a yield of 30–50% greater than this mean value was three to five times greater when lines were derived from drought-tolerant sources, compared with conventionally selected source populations. Population improvement, therefore, has a value in increasing the chance of finding a hybrid with exceptional drought tolerance.

Future challenges

Ideotype for dry environments. Considerable emphasis has been placed on the use of the trait ASI, based on its strong association with grain yield under drought. No other secondary character has shown such consistent association with resistance to barrenness in tropical maize, and it seems likely that it is also an extremely useful trait when selecting temperate maize for drought tolerance (Edmeades et al. 1992).

The trait is easily observed and can be used to estimate yield potential and drought tolerance 6–8 weeks before harvest actually takes place. Genetic variation for ASI is best observed under drought, and it is indicative of genotypic differences in partitioning of assimilates to the ear at flowering. Because of its importance, crop modellers will have to devise ways to incorporate its effects, perhaps as a genetic coefficient, if models are to mimic realistically the effects of stresses which reduce photosynthesis per plant at flowering. The ASI trait has been mapped, and appears to be coded by relatively few important loci. Marker-assisted selection could speed the transfer of regions carrying these loci to elite inbred lines, and thus reduce the time and perhaps the cost associated with breeding procedures for drought tolerance (Ribaut et al. 1995).

It is, however, important to remember that variation in ASI accounts for only 25–35% of the variation in grain yield under drought. There remains ample scope for the identification of additional secondary traits that may help explain the residual variance in grain yield. Such traits might include: relatively slow expansion of green leaf area to conserve soil moisture throughout the crop life cycle; a well-developed *deep* root system for both water and N uptake, resulting in an extended grainfilling period; prolificacy (more than one ear per plant) under low planting densities; remobilization of stored reserves in the stem (Blum 1988). Others which show theoretical benefits, such as small tassel size (reduces shading and direct competition with the ear for assimilates at flowering), osmotic adjustment and increased leaf uprightness (increases the light extinction coefficient, hence efficiency of use of radiation and water) show little association with yield under drought (Table 4). The lack of association may be partly due to our use of small single row plots with inadequate bordering when evaluating families.

Transfer of selection methodology and improved germplasm. In some cases, the elite source germplasm developed by CIMMYT can be the vehicle for technology transfer. Drought-tolerant varieties from Mexico have been released in four countries to date. Most source germplasm, however, will be used in combination with locally adapted germplasm as a varietal cross. Increasingly, CIMMYT is focusing on the development of stress-tolerant hybrids, and it seems likely that drought-tolerant lines, derived from stress-tolerant source populations or by marker-assisted backcrossing methods, will be quite widely used in hybrid combinations in the lowland tropics in the next decade. The utility of the selection methods described above depends on the availability of field facilities at which stress can be carefully controlled and where rainfall does not interfere with stress at flowering. Such sites, located in extremely dry areas or requiring off-season irrigation supply, require a reliable source of water and good experiment station infrastructure. Regrettably, these requirements still remain beyond the present capacity of many national research programmes in the lowland tropics.

Use of models in selection. Models can play a useful role in a drought breeding programme by matching crop phenology, season length and planting dates. Models can also give the breeder a first estimate of the potential value of mixtures of genotypes, and of secondary traits such as rooting depth, leaf angle, osmotic adjustment, etc. Furthermore, models of specific crops must be linked to those which predict the behaviour of the complete farming system and with on-farm trial data, so that a farmer can better optimize output and minimize risk from his whole enterprise. Effective use of models presupposes that the models are reliable, and that complete weather and soils data bases are available. These assumptions are only partly true in the tropics.

Better models for tropical maize. Model development can be divided roughly into the following phases: (1) identification and prioritization of knowledge gaps, followed by targeted research to fill the most important of these; (2) development of modules that accurately describe growth and development of tropical maize; (3) field calibration experiments followed by model calibration; (4) validation of the calibrated model against other experiments; and (5) application of the model to a problem:

Phase 1: The reliable prediction of flowering date and canopy size are basic requirements of models for tropical maize. CERES-Maize, probably the most widely used model for temperate maize (Jones and Kiniry 1986), predicts a final leaf number in lowland tropical maize that is often too large, a flowering date which is too late and has phenology modules that are difficult to calibrate. Also lacking is an explanatory description of the ear-growth rate, and a reasonable prediction of the onset of barrenness under water-limited conditions.

Phase 2: We hypothesize that flowering date can be forecast from the number of leaves under critical daylength (easily obtained in the tropics), plus the genotype's day-length sensitivity. Edmeades et al. (1994) have determined both of these parameters for a wide range of genotypes, and they average respectively 19.2 leaves and 2.3 leaves per hour for lowland tropical maizes. Kernels and ears per plant have been related to plant growth rate during flowering (Tollenaar et al. 1992) and the length of the ASI (Bolaños and Edmeades 1993b). Because quantitative relationships between growth and ASI are lacking, we are attempting to determine the relationship between crop growth rate around flowering and the length of the ASI. Dry matter partitioning, in interaction with environmental conditions, is inadequately understood. This results in poor explanatory descriptions of root versus shoot growth, stem versus leaf and ear growth and tassel versus ear growth.

Phases 3–4: We are currently conducting calibration and validation experiments, and have done so for several years. Validation data are also available for central America and southern and eastern Africa.

Phase 5: Applications that will help set priorities and define selection strategies in drought tolerance research include: causes of yield gaps between attainable and actual yields; timing and intensity of stresses; environmental classification; identification of key test sites and the discarding of duplicate sites; ideotype design; and narrowing of potential management options. A critical component during model development will be the maintenance of effective links with the breeding programme, so that our final recommendations of traits are based firmly on the operational realities faced by field breeders, are accompanied by appropriate information on heritabilities, genetic correlations, etc., and are traits which contribute significantly to yield under a set of jointly agreed-upon environmental conditions.

References

Bassetti P, Westgate M E (1993) Water deficit affects receptivity of maize silks. Crop Sci. 33:279–282.

Blum A (1988) Plant breeding for stress environments. CRC Press, Boca Raton, FL, USA. 223 p.

Bolaños J, Edmeades G O (1993a) Eight cycles of selection for drought tolerance in lowland tropical maize. I. Responses in yield, biomass and radiation utilization. Field Crops Res. 31:233–252.

Bolaños J, Edmeades G O (1993b) Eight cycles of selection for drought tolerance in lowland tropical maize. II. Responses in reproductive behavior. Field Crops Res. 31:253–268.

Bolaños J, Edmeades G O, Martinez L (1993) Eight cycles of selection for drought tolerance in lowland tropical maize. III. Responses in drought-adaptive physiological and morphological traits. Field Crops Res. 31:269–286.

Bolaños J, Edmeades G O (1996) The importance of the anthesis-silking interval in breeding for drought tolerance in tropical maize. Field Crops Res. (in press).

Boyle M G, Boyer J S, Morgan P W (1991) Stem infusion of liquid culture medium prevents reproductive failure of maize at low water potential. Crop Sci. 31:1246–1252.

Byrne P F, Bolaños J, Edmeades G O, Eaton D L (1995) Gains from selection under drought versus multilocation testing in related tropical maize populations. Crop Sci. 35:63–69.

Chapman S C, Crossa J, Edmeades G O (1996) Analysis of genotype by environment interactions across cycles of selection for drought tolerance in tropical maize. I. Pattern and AMMI analysis of discrimination for yield among genotypes. Euphytica (in press).

CIMMYT (1988) Maize production regions in developing countries. CIMMYT, Mexico. 124 p.

DuPlessis D P, Dijkhuis F J (1967) The influence of time lag between pollen shedding and silking on the yield of maize. S. Afric. J. Agric. Sci. 10:667–674.

Duvick D N (1995) Biodiversity, carrying capacity, and people. Paper presented at the 1995 Annual Meeting, American Association for Advancement of Science, 19 February 1995, Atlanta, GA, USA.

Edmeades G O, Bolaños J, Lafitte H R (1992) Progress in breeding for drought tolerance in maize. Pages 93–111 in Wilkinson D (Ed.) Proceedings of the 47th Annual Corn and Sorghum Industry Research Conference 1992. ASTA, Washington DC, USA.

Edmeades G O, Bolaños J, Hernández M, Bello S (1993) Underlying causes for silk delay in lowland tropical maize. Crop Sci. 33:1029–1035.

Edmeades G O, Chapman S C, Lafitte H R (1994) Photoperiod sensitivity of tropical maize cultivars is reduced by cool night temperatures. Presentation at the 86th Annual Meeting of the American Society of Agronomy, 13–18 November 1994, Seattle, Wash., USA.

Edmeades G O, Chapman S C, Bolaños J, Bänziger M, Lafitte H R (1995a) Recent evaluations of progress in selection for drought tolerance in tropical maize. Pages 94–100 in Jewell D C, Waddington S R, Ransom J K, Pixley K V (Eds.) Maize research for stress environments. Proc. Fourth Eastern and Southern African Regional Maize Conference, 28 March–1 April 1994, Harare, Zimbabwe; CIMMYT, Mexico.

Edmeades G O, Bänziger M, Chapman S C, Ribaut J M, Bolaños J (1995b) Recent advances in breeding for drought tolerance in maize. Paper presented at the West and Central Africa Regional Maize and Cassava Workshop, 28 May–2 June 1995, Cotonou, Republic of Benin.

Grant R F, Jackson B S, Kiniry J R, Arkin G F (1989) Water deficit timing effects on yield components in maize. Agron. J. 81:61–65.

Hunt L A (1993) Designing improved plant types: A breeder's viewpoint. Pages 3–17 in Penning de Vries F W T, Teng P S, Metselaar K (Eds.) Systems approaches for agricultural development. Kluwer Academic Publishers, Dordrecht, The Netherlands.

Jones C A, Kiniry J R (1986) CERES-Maize: A simulation model of maize growth and development. Texas A & M University Press, College Station, Tex., USA. 194 p.

Khush G S, Virmani S S, Aggarwal P K (1995) Expectations of plant breeders from crop physiology and modelling. Pages 13–22 in Aggarwal P K, Matthews R B, Kropff M J, Van Laar H H (Eds.) Applications of systems approaches in plant breeding. SARP Research Proceedings, International Rice Research Institute, P.O. Box 933, Manila, Philippines.

Lafitte H R, Bänziger M (1995) Strategies which improve maize varietal performance under low N: Progress due to selection. Pages 57–63 in Jewell D C, Waddington S R, Ransom J K, Pixley KV (Eds.) Maize research for stress environments. Proc. Fourth Eastern and Southern African Regional Maize Conference, 28 March–1 April 1994, Harare, Zimbabwe; CIMMYT, Mexico.

Penning de Vries F W T, Van Laar H H (Eds.) (1982) Simulation of plant growth and crop production. Simulation Monographs, Pudoc, Wageningen, The Netherlands. 308 p.

Ribaut J M, González de León D, Edmeades G O, Huerta E, Hoisington D (1995) Changes in allelic frequencies in a tropical maize population under selection for drought tolerance. Paper presented at the International Congress on Integrated Studies on Drought Tolerance of Higher Plants. 31 August–2 September 1995, Montpellier, France.

Rötter R (1993) Simulation of the biophysical limitations to maize production under rainfed conditions in Kenya. Geographischen Gesellschaft Trier, Germany. 261 p.

Schussler J R, Westgate M E (1991) Maize kernel set at low water potential. II. Sensitivity to reduced assimilates at pollination. Crop Sci. 31:1196–1203.

Schussler J R, Westgate M E (1995) Assimilate flux determines kernel set at lower water potential in maize. Crop Sci. 35:1074–1080.

Shaw R H (1976) Water use and requirements of maize – a review. Pages 119–134 in Agrometeorology of the maize (corn) crop. World Meteorological Organization, Publication 481, Geneva, Switzerland.

Tollenaar M, Dwyer L M, Stewart D W (1992) Ear and kernel formation in maize hybrids representing three decades of grain yield improvement in Ontario. Crop Sci. 32:432–438.

Uhart S A, Andrade F H (1995) Nitrogen deficiency in maize. II. Carbon-nitrogen interaction effects on kernel number and grain yield. Crop Sci. 35:1384–1389.

Waddington S R, Edmeades G O, Chapman S C, Berreto H J (1995) Where to with agricultural research for drought-prone environment? Pages 129–151 in Jewell D C, Waddington S R, Ransom J K, Pixley K V (Eds.) Maize research for stress environments. Proceedings of Fourth Eastern and Southern African Regional Conference, 28 March–1 April 1994, Harare, Zimbabwe; CIMMYT, Mexico.

Westgate M E, Bassetti P (1991) Heat and drought stress in corn: What really happens to the corn plant at pollination? Pages 12–28 in Wilkinson D (Ed.) Proceedings 45th Annual Corn and Sorghum Research Conference, Chicago, 5–6 December 1990. ASTA, Washington DC, USA.

Zinselmeier C, Schussler J R, Westgate M E, Jones R J (1995a) Low water potential disrupts carbohydrate metabolism in maize ovaries. Plant Physiol. 107:385–391.

Zinselmeier C, Lauer M J, Boyer J S (1995b) Reversing drought-induced losses in grain yield: Sucrose maintains embryo growth in maize. Crop Sci. 35:1390–1400.

Potential yield of irrigated rice in African arid environments

M. DINGKUHN and A. SOW
West Africa Rice Development Association (WARDA), B.P. 2551, Bouaké 01, Ivory Coast

Key words: rice, temperature, chilling, arid climate, crop simulation, crop calendars

Abstract
Irrigation schemes for rice in the Sahel, an arid region traversed by several large rivers, were generally designed for rice double cropping. The level and temporal and spatial variability of potential yields in various seasons (wet, cold-dry and hot-dry) is thought to be high because of high solar radiation, but has not been quantified. We conducted a simulation-based study on the spatial, seasonal and year-to-year variability of potential yields. On the basis of growth analyses in rice garden experiments, we developed the rice yield model ORYZA_S which emphasizes the response of rice to extreme temperatures. It combines, in modified form, the phenological model RIDEV and the yield model ORYZA1. Simulation results indicated that potential rice yields in the Sahel are largely driven by temperature, with the solar radiation being abundant most of the time. Among the major physiological determinants of yield were the extremely variable crop duration which largely depends on floodwater temperature, the susceptibility of leaf area expansion to chilling, and spikelet sterility caused by heat or chilling. Temperature also affected kernel weight at maturity and dry matter partitioning to leaves. Potential grain yields vary between 4 ton ha^{-1} in the hot-dry season in continental parts of the Sahel, and 10 tons in the hot-dry season in the Senegal river delta and in the wet season at the southern fringe of the Sahel. Yields dropped close to zero due to sterility in all environments if plants were sown later in the wet season (September–October). On the basis of the spatial and temporal variability of potential yields, and in consideration of results of previous characterization studies, we conclude by proposing environment specific applied research foci for irrigated rice in the Sahel.

Introduction

Irrigated rice in the Sahel is known to have a high physical potential for increased cultivated area and yield. At the same time, it is arguably West Africa's most controversial rice production system. The high yield potential for rice in the Sahel is a combined result of high solar radiation and full water control. These formed the rationale for massive public investments in irrigation schemes, beginning in Mali in the 1920s, and across the whole region in the 1970s and 1980s, when severe droughts devastated rainfed crops and livestock. National research programmes across the region have identified intensified rice production, particularly through double cropping, as one of their foremost objectives.

In many cases, results fell short of expectations: mean grain yields of 3.5 ton ha^{-1} are below anticipated levels; and, in some cases, scheme designs were inadequate for water resources (Diemer and Huibers 1991), or they contributed to soil degradation due to lack of drainage (Bertrand et al. 1993). One of the main discrepancies between projected and actual productivities, however, is in cropping intensity, which

M.J. Kropff et al. *(eds.): Applications of Systems Approaches at the Field Level, 79–99.*

was planned to be between 1.5 and 2. In fact, intensities rarely exceed 1 in the large irrigation schemes of Senegal, Mauritania and Mali.

Recent research has identified a complex of climatic and management problems which render the establishment, management and harvesting of two successive rice crops in one year extremely difficult (Dingkuhn 1995; Le Gal 1995). The risk assessment and planning model RIDEV (Rice Development) was developed to simulate the duration and losses to temperature-induced sterility for any combination of rice genotype, site, planting date and planting method scenario in the Sahel (WARDA 1994; manual in preparation). In applying RIDEV to the spatial and temporal climatic diversity of the Sahel, we identified variety types for cropping in the wet or dry season, or both, and to which agronomically appropriate site-specific crop and crop management calendars can be fitted (WARDA 1994). These tools, however, provided no information on potential rice grain yields.

The present simulation-based study sought to determine the potential grain yield of irrigated rice as a function of planting date and site in the Sahel. Yield potential of irrigated rice in the tropics depends mainly on solar radiation, nitrogen nutrition and the rice variety grown. It is generally believed that in favourable environments and under good management, the dry (sunny) season enables much higher yields than the wet season. The yield ceiling for today's improved rices under favourable tropical dry-season conditions is about 10 ton ha^{-1}. This commonly accepted view, however, is based on observations in the monsoonal humid tropics of Asia.

In the Sahel, in contrast to monsoonal Asia, actual rice grain yields are frequently lower in the dry season than in the wet, and yield stability is generally low in the dry season. Although this seemingly inverted situation might be due in part to management problems in the off-season, the question remains unanswered what the biological potential for rice production actually is in the Sahelian dry season.

The present study sought to determine spatial, seasonal and year-to-year variability of rice potential grain yield in the Sahel using crop growth simulation.

Approach and methodologies

We applied the classical approach of selecting an existing rice model that suited the required level of detail, modified it on the basis of physiological observations in the target environment, validated it with crop observations at key sites and, finally, used it for spatial and temporal extrapolative analyses.

The empirical basis for model development

To develop a rice growth and yield model for the Sahelian environment, on the basis of existing models, we needed the following sets of information:

(1) Evaluation of the general fit of existing models to the Sahelian environment; this required the empirical determination of the genotype-specific morphophysiological parameters used for model parameterization, such as development

stage dependent specific leaf area (SLA) and N content, or the number of spikelets formed per unit of growth during the reproductive phase.

(2) In order to adapt and calibrate the model, additional crop coefficients expressing relationships between critical environmental parameters and crop responses. These included, for example, temperature-dependent upper limits leaf expansion during exponential growth; and the temperature-dependent sink size of kernels.

(3) In order to validate the model, detailed field observations on growth dynamics and yield components for key varieties across the spectrum of climatic conditions encountered in the Sahel.

(4) We needed real-time daily weather records as background for (1) through (3).

The main source of our empirical observations was a series of rice garden trials (Alocilja et al. 1981) conducted at WARDA's experimental stations at Ndiaye (arid-coastal) and Fanaye (arid-continental) in Senegal, based on a continuous monthly planting schedule for key varieties under optimal management (WARDA 1992, 1993; Dingkuhn et al. 1995). Routine crop observations provided growth curves for bulk green and dead leaf blade, stem and panicle dry matter fractions; the dynamics of leaf nitrogen content, based on measurements of leaf chlorophyll content and specific leaf area; tiller number; and yield and yield components. The model was developed using data from 21 planting dates in 1991 and 1992 at Ndiaye (16° 14′N; 16° 14′ W) and Fanaye (16° 33′N; 15° 45′ W) in Senegal. Nitrogen inputs were 150 kg N ha^{-1} as triple split. Details of the experiments were reported by Dingkuhn et al. (1995).

Among the genotypes planted in the rice garden trials, we selected the cultivars Jaya (a medium-duration local check widely grown in Senegal and Mauritania) and IR64 (a short-duration international check) for model development and parameterization. These genotypes have previously been characterized for their photothermal constants and sensitivity to temperature-induced sterility by Dingkuhn and Miézan (1995). Daily minimum and maximum air and water temperatures, air humidity, wind speed and solar radiation were recorded at both experimental sites using LiCor (Lincoln, Nebraska) electronic weather stations.

Characteristics of the adapted model ORYZA_S

We developed the rice model ORYZA_S ('S' for Sahel) from two existing models: the growth and yield model ORYZA1 developed at IRRI (Kropff et al. 1993), and the microclimate-driven rice phenology and spikelet sterility model RIDEV (WARDA 1994; detailed manual in preparation). A more detailed recent version of ORYZA1 (Kropff et al. 1994) which simulates spikelet sterility, grain number and accounts for sink limitations was not available at the time of this study.

ORYZA_S is a deterministic dynamic model which simulates the growth of plant organs at daily increments, taking leaf canopy architecture into account. Emphasis is on the energy, temperature and nitrogen dependency of growth processes, following the modelling techniques described by Penning de Vries et al. (1989). Because the

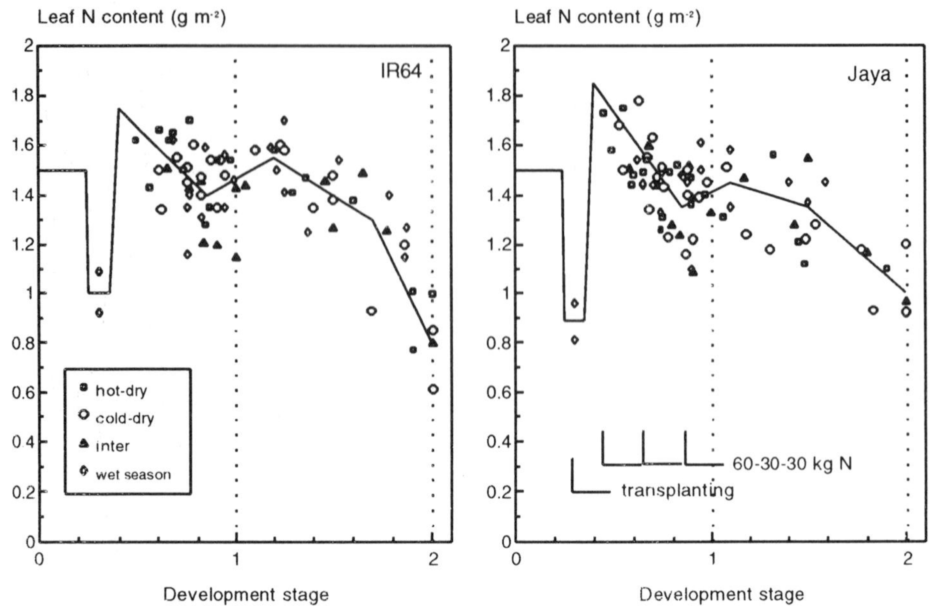

Figure 1. Changes in N content per leaf area in IR64 and Jaya rice as a function of development stage of the crop (phenological time, with 0=germination, 1=flowering and 2=maturity), as observed for 21 crops planted successively at monthly intervals. The hand-fitted line indicates the generalized time course used in the ORYZA_S growth model, Ndiaye, Senegal, 1991 and 1992.

model simulates energy-limited and temperature-limited potential yield, no stresses such as drought or soil toxicities are considered.

Model inputs are solar radiation, minimum and maximum temperatures, geographical latitude (to calculate photoperiod), some properties of the initial plant (seedling) population and a number of genotypic constants. Since the uptake of nitrogen is not simulated, the model operates with forced time courses of leaf nitrogen content, based on empirical observations (Figure 1).

We introduced a number of new functions to improve the model's sensitivity to the variable temperatures in the Sahel. Major modifications, as compared to ORYZA1 and RIDEV, are as follows.

Transplanting shock. Transplanting causes a physiological shock which temporarily affects growth development (Dingkuhn et al. 1990a). To account for this, we simulated a reduction in the maximal photosynthetic rate and the rate of phenological development during a temperature-dependent period. An additional reduction in growth resulted from the low nitrogen content of leaves during the shock period (Figure 1).

Direct effects of temperature on leaf area development. Exponential growth during the early development stages is limited by the leaf area index (LAI) (Dingkuhn et

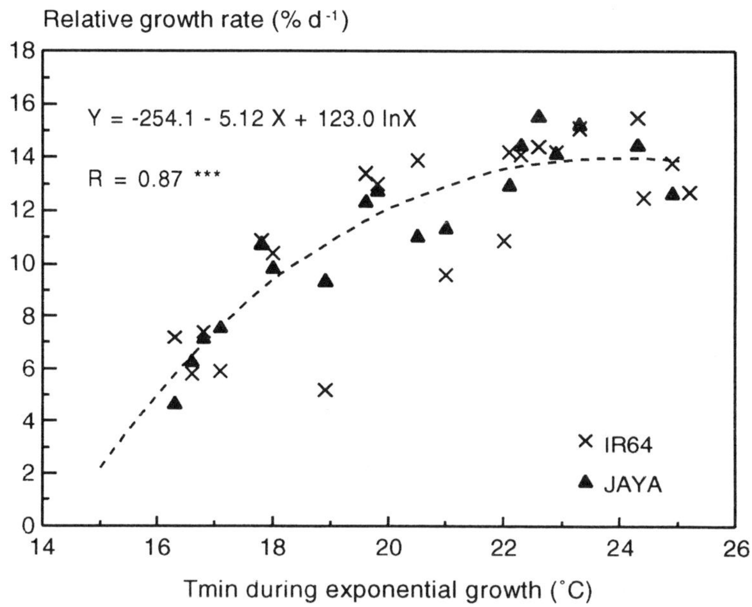

Figure 2. Relationship between the relative growth rate of IR64 and Jaya rice and the minimum air temperature (Tmin) during exponential growth, across 21 sowing dates in 1991 and 1992, Ndiaye, Senegal.

al. 1990b). Leaf area growth, in turn, is mainly driven by temperature and only to a lesser extent by carbon assimilation (Horie et al. 1979). ORYZA1 in its original version simulates early (exponential) growth of LAI as a fixed function of temperature, and switches to photosynthesis-driven leaf expansion once mutual shading of leaves (LAI > 1) becomes a limiting factor for leaf expansion because of reduced carbon assimilation per leaf.

To avoid such a radical (and physiologically improbable) decoupling of leaf weight and area growth, we assumed LAI to be generally driven by assimilation and partitioning, but forced a minimum-temperature dependent upper limit to the relative daily increase of LAI. The forcing mechanism mainly takes effect during exponential growth because the relative daily changes in LAI and dry matter are generally highest during this period. Very low minimum temperatures (e.g. 15°C and below) still affect LAI directly at all stages. A major effect of the forcing mechanism is that leaves not only expand, but also get thicker during cold periods, and they gradually return to their normal, stage-dependent thickness as temperatures increase.

This additional physiological function of the model is crucial in the Sahelian environment because existing models were found to substantially overestimate relative growth rates (RGR) of rice at low temperatures. Figure 2 shows the non-linear relationship between RGR and minimum temperatures observed in the Sahel, which cannot be explained by effects on assimilation alone, because temperatures at midday, when photosynthetic rates are usually highest, were generally in the optimum range for photosynthesis (20–30°C).

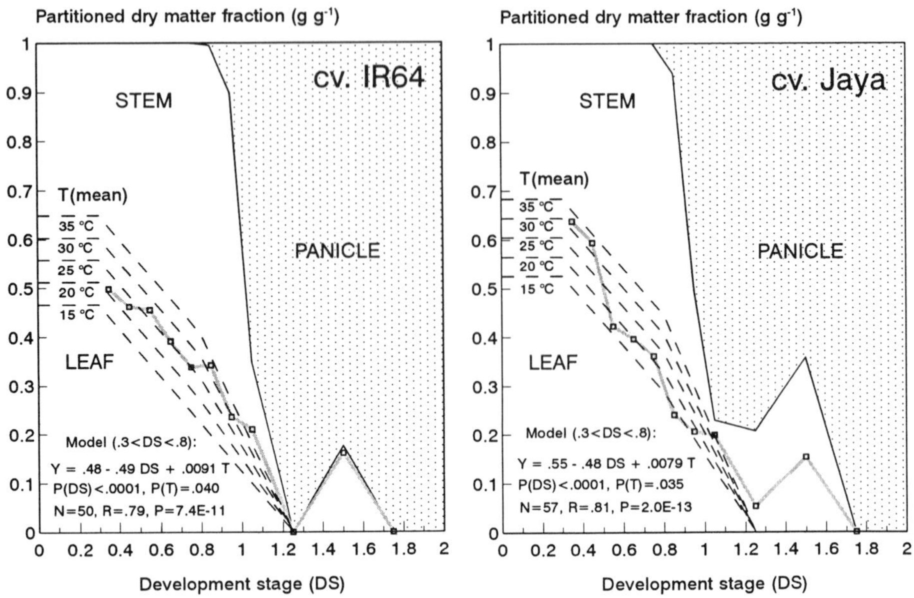

Figure 3. Dry matter partitioning as related to crop development stage (0 = germination, 1 = flowering and 2 = maturity) patterns among shoot organs for IR64 and Jaya rice, based on pooled data from 21 planting dates, Ndiaye, Senegal. Grey lines and individual symbols represent mean observations across thermal environments (planting dates). Superimposed broken lines indicate the dependency of leaf/stem partitioning on the mean air temperature, based on multiple linear regressions of observed partitioning versus T(mean) and development stage, Ndiaye, Senegal, 1991 and 1992.

Temperature effects on assimilate partitioning. The ORYZA1 model programs assimilate partitioning ratios among roots, stems, leaves and the panicle as fixed functions of the developmental stage, thereby assuming no plasticity. Multiple linear regressions of empirically determined leaf partitioning ratios versus air temperature and development stage (DS; $0.3 < DS < 0.8$), however, showed that the fraction of assimilate used for leaf growth is significantly ($P < 0.01$) positively correlated with mean air temperature (Figure 3). Leaf growth, therefore, decreases over-proportionately, compared to other organs, if temperature decreases. We incorporated this relatively weak effect into the model because temperature variations are very high in the Sahel. Independent studies showed that partitioning (and therefore, the morphological plant type) is also affected by N nutrition, necessitating a greater environment-dependent plasticity of partitioning phenomena in future rice models.

Also unlike ORYZA1, partitioning of assimilates to the panicle (storage organ) at pre-flowering was determined by the number of spikelets initiated between the panicle initiation and flowering stages. The number of spikelets initiated and the weight of the developing panicle was considered proportional to the crop growth rate during this period. Temperature affected spikelet number indirectly through assimilation rates.

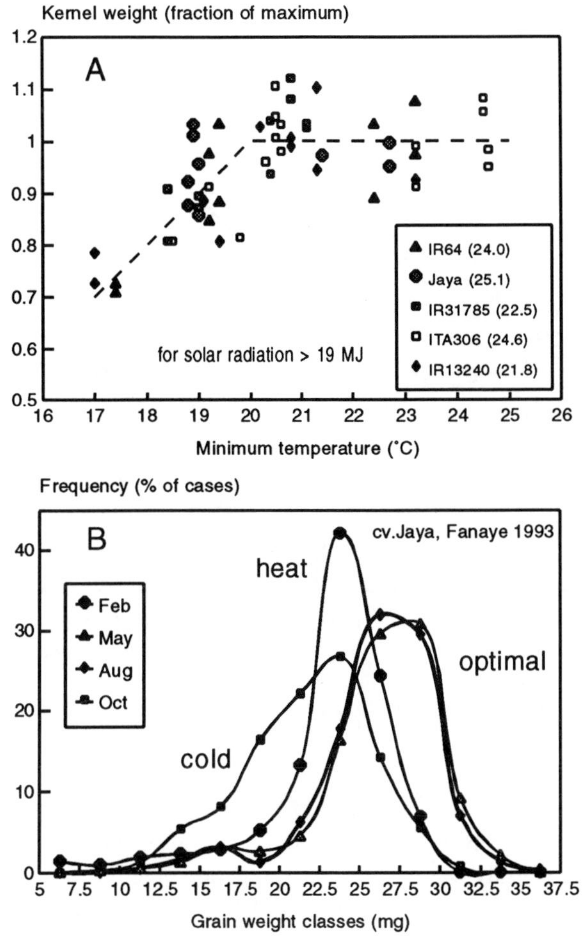

Figure 4. Relationship of rice kernel weight at maturity on the thermal environment. (A) Mean bulk kernel dry weight in relation to the daily minimum air temperature during grainfilling, with low radiation environments disregarded (< 19 MJ). (B) Kernel weight distribution as a function of planting date (*Source:* Cissé 1994).

Temperature effects on grainfilling and yield components. For the arid Sahelian environment, the simulation of a single weight fraction for panicles does not adequately describe yield, because a large fraction of the panicle can consist of unfilled spikelets (Dingkuhn et al. 1995). The two main causes of high numbers of unfilled spikelets are heat- or cold-induced sterility (Nishiyama 1984), or severe source limitations. Previous studies showed that cold-induced spikelet sterility is a major cause of rice yield variability in rice production systems in the Sahel (Le Gal 1995). Three weight fractions of the panicle were simulated: filled fertile spikelets, unfilled fertile spikelets, and sterile spikelets. Fertile spikelets were supposed to be filled

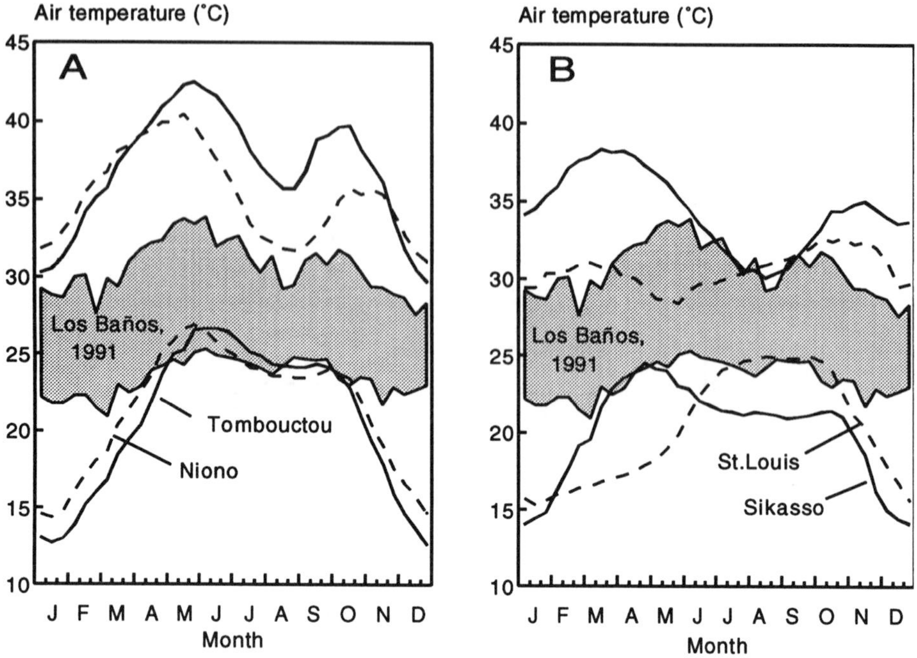

Figure 5. Annual minimum and maximum air temperature profiles for 1950 to 1983 in (A) Niono and Tombouctou (Mali), and (B) in Sikasso (Mali) and St. Louis (Senegal). The shaded zone indicates temperatures in Los Baños, Philippines, for 1991, as a comparison.

sequentially, resulting in a decrease of unfilled and an increase of filled spikelet number. Spikelet sterility was simulated as a function of temperature at critical stages, using the phenological model component RIDEV, based on relationships described by Dingkuhn et al. (1995).

Sink capacity of the panicle was determined by the number of fertile spikelets, multiplied by the potential kernel weight, a genotypic constant. We observed, however, that kernels did not achieve their potential weight when nights were cool (<20°C) or days exceptionally hot (>40°C). At least for the effect of low temperatures, we were able to exclude low solar radiation (e.g. $<19\,MJ\,d^{-1}$) as a potential cause (Figure 4A). Grain weight distribution was normal (Gaussian) under optimal (May – August planting) and hot (February planting) conditions, but showed significant tailing towards lower grain weight classes when nights were cool during ripening (September to December planting) (Figure 4B). In ORYZA_S, we flexibly adjusted potential kernel weight as a function of mean night temperature between development stage 1 (flowering) and 1.5 (dough stage), based on Figure 4B. Heat effects on kernel size were ignored because data were insufficient to establish clear relationships. Detailed observations on the variability of kernel weight distribution under Sahelian conditions were reported by Cissé (1994).

Re-invigorated vegetative growth under sink limitation. The original model ORYZA1 terminates the simulation once all grains are filled or the source is exhausted. In the Sahel, however, sink size can be severely restricted by spikelet sterility which is accompanied by re-invigorated vegetative growth during the later stages of ripening. We simulated this phenomenon, which affects the harvest index, by redirecting unused assimilates to shoot growth.

Approach to extrapolative crop simulation

We used ORYZA_S to simulate potential yields under transplanted conditions for the local check variety Jaya, on the basis of 10 years of historical weather records (1970–79). Simulations were conducted for more than 30 sites in the Sahel and 12 monthly sowing dates. The sites included in the respective data base were described by Dingkuhn (1995).

The present study focuses on four key sites in the Senegal River delta and valley (St. Louis, Rosso, Podor and Bakel), and four key sites in the Niger valley (Niono, Tombouctou, Gao and Niamey). The annual temperature patterns are shown in Figure 5 for a subset of sites. The phenological time courses of leaf N content, as observed in Senegal (Figure 1), were thereby projected onto the other sites because N uptake was not known for these sites.

Simulation results and discussion

Validation of ORYZA_S

Comparison of the models ORYZA_S and ORYZA1 at Ndiaye. Figure 5 shows that the model modifications were crucial. Crop duration for IR64 at Ndiaye was underestimated by 10 days in the wet season (planting in July) and by 30 days in the cold-dry season (November) if the original model ORYZA1 was used (Figure 6A). The phenological model developed by WARDA for the Sahel, RIDEV, came much closer to the observations, largely because of the inclusion of floodwater temperature as a determinant of crop duration (Dingkuhn et al. 1995). The adapted yield model ORYZA_S, which includes components of RIDEV, also performed satisfactorily. Small differences between RIDEV and ORYZA_S were due to the fact that RIDEV uses predetermined patterns of leaf area, whereas ORYZA_S simulates leaf area. Leaf area (or ground cover) affects water temperature.

Yield simulations of ORYZA1 were close to observations for sowing dates between February and August, but they failed to predict the extreme yield reductions observed between September and January, because spikelet sterility was not simulated (Figure 6B). ORYZA_1 simulated accurately the seasonal variations in grain yield but showed a general tendency to overestimate yield. Because these overestimations were partly caused by crop management problems that were not simulated (the objective was to simulate potential, not actual, yields), we made no efforts to further approximate the observations.

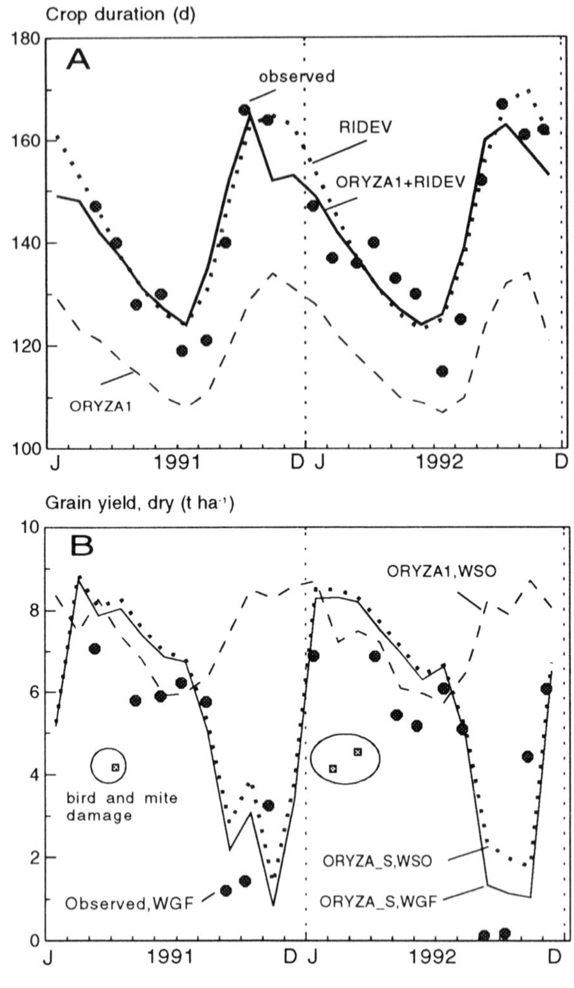

Figure 6. (A) Changes in crop duration for transplanted IR64 rice sown on 21 dates in 1991 and 1992. Field observations (symbols) are compared with simulations using the phenological model RIDEV, the growth model ORYZA1 and the combined model ORYZA_S. (B) Changes in grain yield for the planting dates. WGF, dry weight of filled grains; WSO, weight of the storage organ (whole panicle), Ndiaye, Senegal.

Simulation of yield components. The distinction made in ORYZA_S between the weight of the filled grain (WGF) and the panicle (storage organ, WSO) was of little importance for dates associated with high yields. However, for sowing dates in September to November, when yields were reduced by cold-induced spikelet sterility, the unfilled spikelets accounted for up to 1 ton ha^{-1} of panicle weight (Figure 6B).

Simulations generally matched the observed percentage of unfilled spikelets (Figure 7A). The simulated number of unfilled spikelets consisted of two components, one

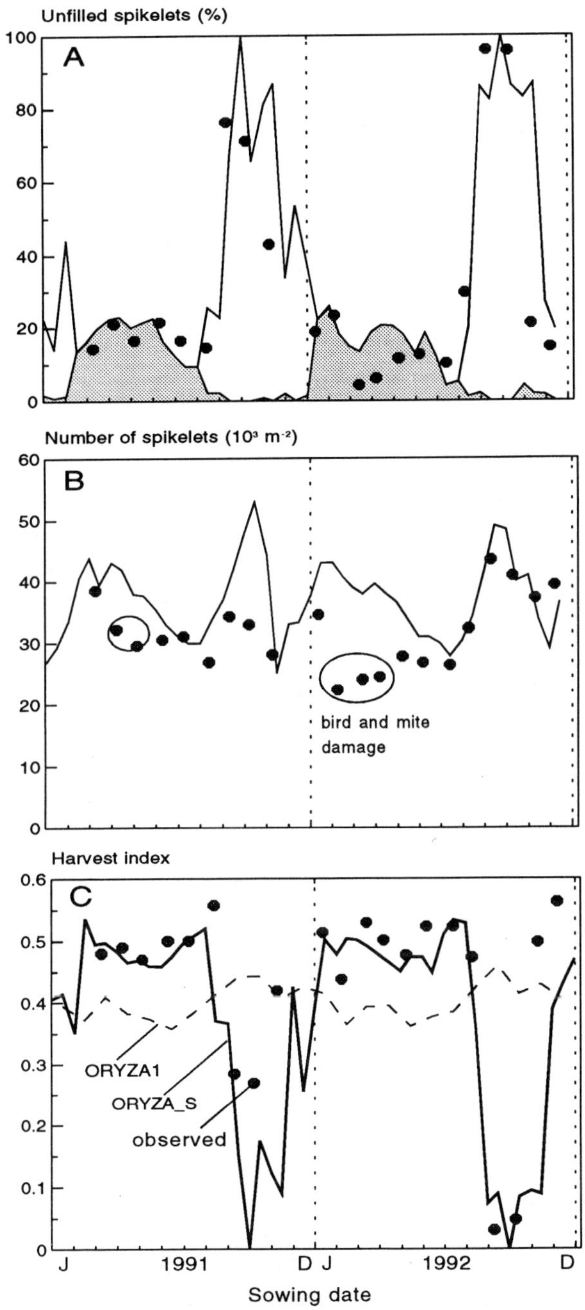

Figure 7. (A) Fraction (%) of unfilled spikelets for IR64 rice sown on 21 planting dates as observed in the field (symbols) and simulated with the model ORYZA_S. Shaded zones represent spikelets that were not filled because of source limitations; the remaining fractions are the result of temperature-induced sterility. (B) Observed (symbols) and simulated total number of spikelets per ground area ($10^3 m^{-2}$). (C) Harvest index of IR64 rice as observed for 21 sowing dates in 1991 and 1992 (symbols), and as simulated with the models ORYZA1 (broken line) and ORYZA_S (solid line).

caused by sterility and one by assimilate source limitations. The two components are difficult to distinguish in the field, but they show very different seasonal patterns (shaded and unshaded zones in Figure 7A). The corresponding seasonal patterns of total numbers of spikelets per unit ground area are shown in Figure 7B.

The most important agronomic indicator of crop performance, apart from yield itself, is the harvest index (HI). The observed HI was around 0.5 for most planting dates but dropped to values close to zero for planting dates in September and October (Figure 7C). ORYZA_S simulated these patterns accurately, whereas the original model ORYZA1 predicted a HI of about 0.4, with little seasonal variation.

Test of yield predictions. The model was validated with independent data sets from two sites (Ndiaye and Fanaye in Senegal) and for two varieties (IR64 and local check Jaya). Observed yields were between 6 and 9 ton ha^{-1} at both sites, except for sowing dates that were associated with cold-induced sterility (September to November). Details of model validation are being published elsewhere (Dingkuhn and Sow 1995).

The validation of the model under Sahelian conditions gave generally acceptable results, but yield was frequently overestimated due to stresses that could not be simulated because they are difficult to measure (e.g. dust accumulated on the leaf surface, spider mite attacks). Furthermore, the dry climate aggravated any crop and water management errors. In environments prone to multiple stresses, it is generally difficult to experimentally validate potential yield models. The regional extrapolative exercises using ORYZA_S as described in the following section should, therefore, be interpreted with appropriate caution.

The model was also validated with a data set for IR64 in the Philippines, based on an N rate of 150 kg ha^{-1} and the leaf N contents actually observed (1988 dry season in Maligaya, Nueva Ecija; authors' data). The model slightly underestimated grain yield (8.3 vs 8.8 ton ha^{-1}) and biological yield (16.1 vs 16.4 ton ha^{-1} dry weight) and, more significantly, underestimated the LAI at the heading stage (8.3 vs 9.9). Considering that ORYZA_S was specifically developed for arid conditions, however, this yield prediction for a site in the humid tropics was considered satisfactory.

A regional survey of simulated potential yields

Thermal characteristics of the test sites. Annual patterns of daily maximum (Tmax) and minimum (Tmin) temperatures are shown for key sites in the Sahel and compared with a humid-tropical site, Los Baños in the Philippines (Figure 5). Two typically 'Sahelian' environments in which irrigated rice is grown are represented by Tombouctou (Mali, 16° 49′ N), at the fringe of the Sahara desert, and Niono (Mali, 14° 18′ N), located at the centre of a large irrigation scheme (Figure 5A). These sites are characterized by annual patterns of Tmin that follow symmetric bell-shaped curves, roughly reflecting (with a one-month lag) changes in day length. Annual Tmin extremes are less than 15°C in January and about 26°C in June. Patterns for Tmax constantly exceed those of Tmin by more than 15°C, except for the short rainy season centred upon August, during which the diurnal variation in temperature is much less.

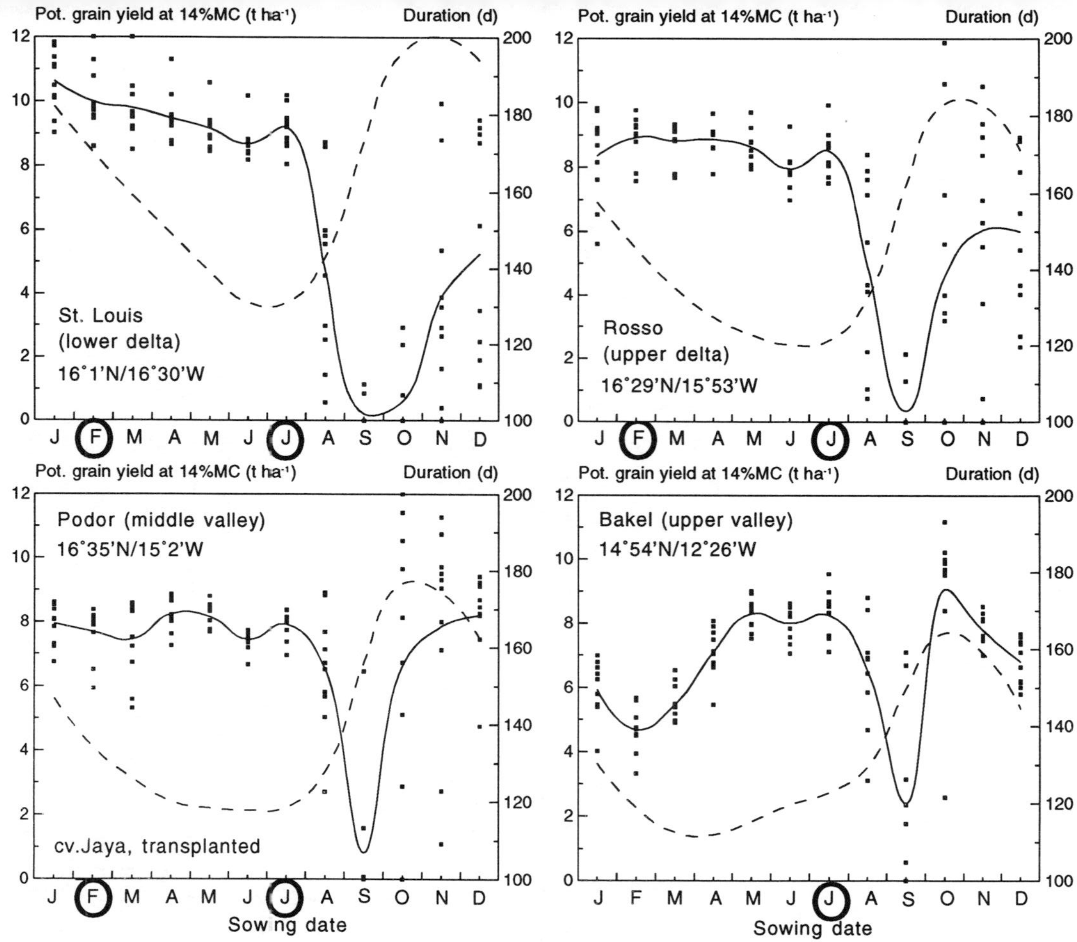

Figure 8. Simulated potential grain yields (solid lines) and crop duration (broken lines) as dependent upon sowing date for transplanted Jaya rice at four sites along the Senegal River (means for 1970–79 weather data; individual years are represented by points). Enhanced months indicate common planting dates for the respective sites.

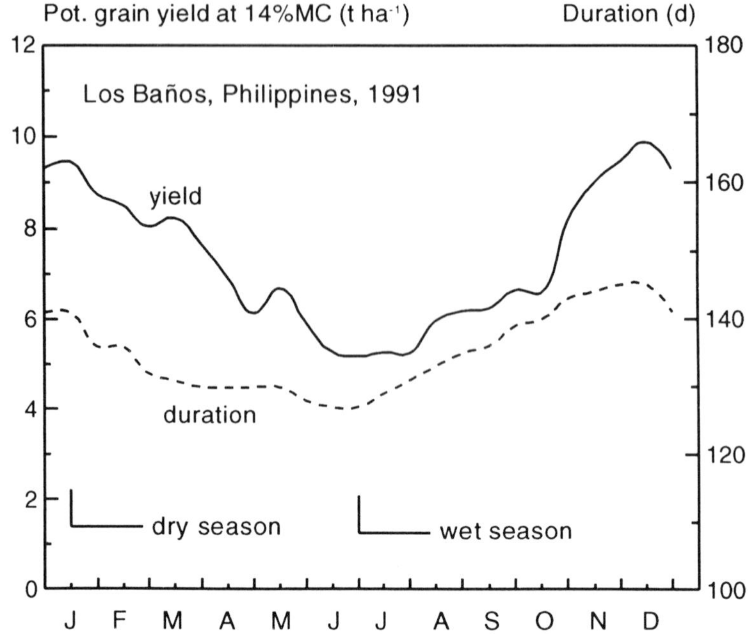

Figure 9. Relationship between grain yield (solid line) and sowing date, as simulated with ORYZA_S for transplanted Jaya rice in Los Baños, Philippines, in 1991. The broken line indicates the respective crop durations.

Three different types of annual thermal patterns occur in the Sahel (Figure 5B). First, the coastal climate (e.g. St. Louis in the Senegal River delta; 16° 1′N) is characterized by a long cool season and moderate Tmax between 28°C and 33°C throughout the year. Second, the great floodplains of the Sahel ranging from 13° N to 17° N have extremely high diurnal temperature fluctuations and Tmax up to 45°C except during a short wet season. Third, the slightly more elevated plateau at the southern fringe of the Sahel has a much cooler and longer wet season between June and September (e.g. Sikasso in Mali; 11° 18′N).

These patterns stand in marked contrast to the climate of the monsoonal tropics (e.g. Los Baños in the Philippines; superimposed in Figure 5), particularly if compared with the critical temperatures for the productivity of tropical rice. The critical Tmin for spikelet sterility of 18°C (WARDA 1994) is virtually never experienced in Los Baños, whereas in the Sahel, Tmin is below 18°C during three to four months of the year. Minimum temperatures below 18°C also limit leaf area growth and potential kernel weight if they occur at critical growth phases.

Potential yields in the Senegal River system. Simulated potential grain yields for the locally grown variety Jaya showed strong seasonal and spatial variations in the Senegal River valley (Figure 8). All sites between the delta (St. Louis) and the upper valley (Bakel) reflected a dramatic yield depression for September sowing dates due to cold-induced spikelet sterility. The planting period associated with this prohibititive yield depression, however, was longer in St. Louis, located on the Atlantic coast, than in the upper valley which has a more continental climate.

Potential yields for the wet season (WS; sowing in July) averaged 9 ton ha^{-1} in the Senegal River delta and 8 ton ha^{-1} in the middle and upper valley. They fluctuated within a relatively narrow range of about two tons among the 10 annual scenarios (1970–79).

Currently recommended rice double-cropping schedules for the Senegal River delta and middle valley call for the second crop to be sown in February (hot-dry season, HDS). Potential yields in the HDS were 10 ton ha^{-1} in St. Louis, 9 in Rosso, 8 in Podor and 5 in Bakel, indicating that the HDS crop is most productive in the coastal zone and least productive in the continental climate of the upper valley. Low yields in the continental (inland) environment were caused by high temperatures through two physiological effects: a short duration of the main (linear) growth phase as it fell into extremely hot periods in the upper valley; and heat-induced spikelet sterility. Within the climate sequence of the river valley, therefore, rice double cropping would have the highest annual yield potential in the delta (up to 17 ton ha^{-1} in Rosso). Actual farmers' yields in this environment, however, average about 9 ton ha^{-1} in double-cropping schemes and 5 ton ha^{-1} where only one wet-season rice crop is grown, indicating a gap between potential and actual yields of 40% to 50%.

We simulated an annual cycle of sowing dates using the same varietal parameters for 1991 in Los Baños, Philippines (Figure 9). Yield potential was highest at about 9 ton ha^{-1} in the dry season (sowing date in January), and lowest at 5.5 ton ha^{-1} in the wet season (sowing in June or July). These results fully agree with results obtained under high inputs for a modern medium-duration variety at IRRI's lowland farm. In contrast to the Sahelian sites, where annual temperature patterns shaped the patterns of yield, potential rice yields in Los Baños were driven by solar radiation.

Potential yields in the Niger River system. The Niger River, along which irrigated rice is grown, is characterized by a continental climate. Rice is usually sown in June, the beginning of the wet season. Rice-rice double cropping on a larger scale is practised only in Niger, represented in this study by Niamey (13° 32′N) (Figure 10). In Niger, the second cold dry-season crop is commonly sown in November or December. Although large public irrigation schemes are found around Niono in Mali (14° 18′N), rice double cropping is relatively uncommon. Between these two important rice-producing areas, the 'knee' of the Niger River reaches as far north as 17°, harbouring small village-based rice irrigation schemes exposed to a desert climate. These locations are represented by Gao and Tombouctou in the present analysis.

At the two sites with more moderate climate, Niamey and Niono, the highest potential yields were simulated for sowing dates in May. Mean potential yields for that month were between 9 and 10 ton ha^{-1}. Yields are only slightly lower for sowing in June and July, followed by a sharp decline caused by cold-induced sterility for sowing in August and September.

Sowing dates in the cold dry season (October to December) were associated with

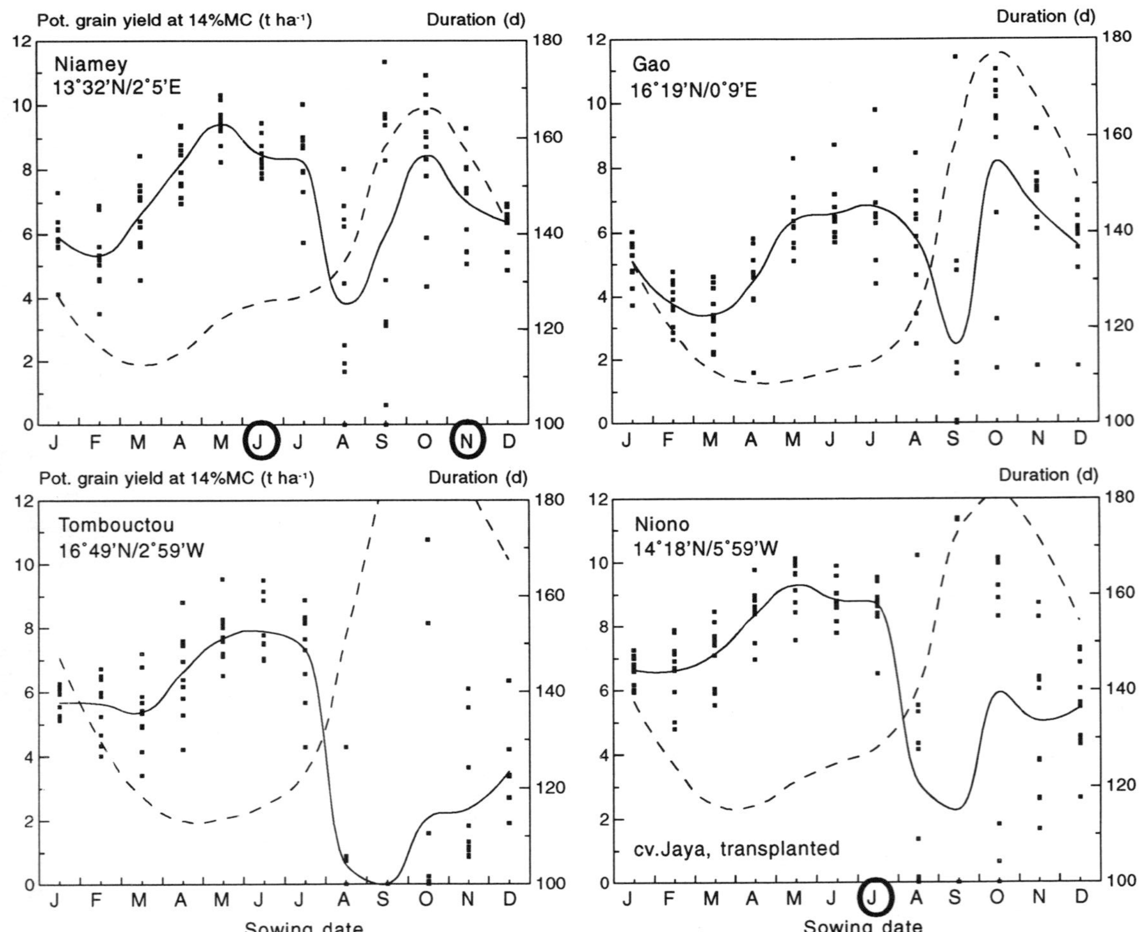

Figure 10. Simulated potential grain yields (solid lines), and crop duration (broken lines) as dependent upon sowing date for transplanted Jaya rice at four sites along the Niger River (means for 1970–79 weather data; individual years are represented by points). Enhanced months indicate common planting dates for the respective sites.

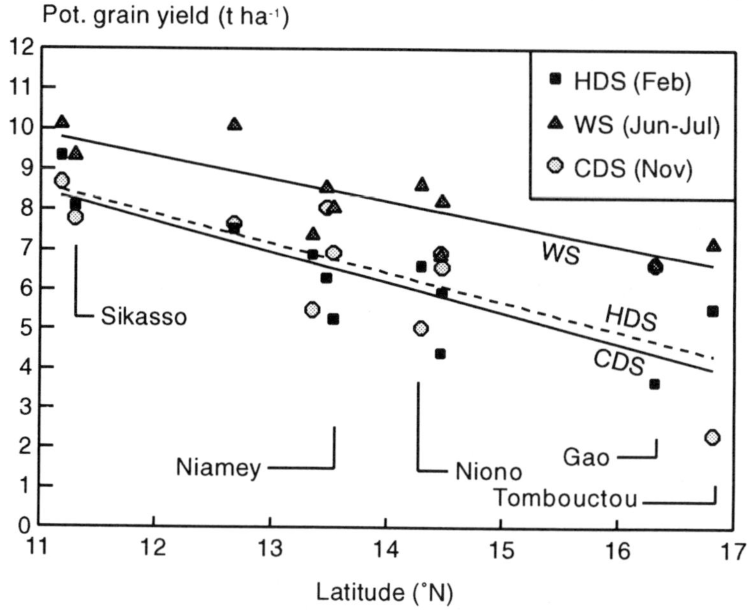

Figure 11. Relationship between simulated potential grain yield of transplanted Jaya rice and geographical latitude in the continental Sahel, separately for the hot dry season (HDS), wet season (WS), and cold dry season (CDS).

high but variable yields. Yield variability during the cold dry season was caused by low temperatures which permitted high biomass production due to long duration, but poor sink capacity caused by sterility and reduced potential grain size. Yield variability in the cold dry season in Niono renders rice production extremely risky, suggesting a delay in sowing until January. Both predictions fully agree with observed cropping calendars at the respective sites (Dingkuhn 1995).

At the near-desert sites Gao and Tombouctou, thermal constraints to potential rice yields were further accentuated. At Tombouctou, the long and intense cold season appears to prohibit rice planting between August and December, whereas a single wet-season crop sown in June may potentially yield about 8 ton ha^{-1}. At Gao, one of the hottest sites in the Sahel, the period during which low temperatures strongly reduced yield potential is comparatively short (sowing dates in September). But potential yields for the other seasons were generally low, such as 4 ton ha^{-1} in the hot dry season and less than 7 ton ha^{-1} in the wet season. Low yields were caused by heat-induced sterility and short crop duration.

North–south potential yield gradients. Simulations for the Niger River valley indicated lower potential yields for the northern locations of Gao and Tombouctou (>16° N), as compared to Niono and Niamey (Figure 10). To further characterize this N–S gradient, we compared 11 sites located in the same river system between 1° N and 17° N (Figure 11).

We found that potential yields for the main cropping seasons (hot dry, wet and cold dry) consistently decrease as latitude increases, between 4 and 10 ton ha^{-1}. The causes for this gradient, however, were not the same for the different seasons. In the main wet season at lower latitudes, such as in Sikasso, a slightly longer crop duration (ca 140 d), comparatively cool nights (about 21°C minimum temperature) which reduce respiration losses, and the absence of sterility caused by extremely high or low temperatures permit high potential yields of around 10 ton ha^{-1}. The lower yields at higher latitudes were due to higher minimum temperatures, which caused a shorter duration and higher respiration rates.

The hot dry and cold dry seasons were found to be potentially less productive than the wet season by 1.5 (11° N) to 2.5 ton ha^{-1} (17° N), because of more severe thermal constraints. In the hot dry season, high maximum temperatures that fall into the reproductive phase (April to May) reduce the duration of the linear growth period and induce sterility, particularly in the north (e.g. Gao). Solar radiation, in such cases, does not limit yields because of intensity, but through a shorter period during which radiation can be intercepted.

In the cold dry season, the duration of the exponential growth phase is extended by low temperatures in December to February, causing slow leaf area growth and low light interception. This is followed by hastened development during the main growth phase, caused by heat in March and April.

The Senegal River delta (ca 16° N), represented in this study by St. Louis and Rosso (Figure 8), deviates markedly from the N–S gradient shown in Figure 10. The hot dry season crop is potentially highly productive because the proximity of the sea moderates air temperatures, even in March to June. At these sites, however, rice double cropping based on a single medium-duration variety is difficult because of extended crop duration in the hot dry season. In practice, therefore, some of the extremely high yield potential in the hot dry season cannot be used in rice double cropping because short-duration varieties are required.

Synopsis of simulation results

At all typically continental-Sahelian test sites (Niamey, Gao, Tombouctou and Niono on the Niger River; Bakel on the Senegal River), potential grain yields are lower in the hot dry season than in the wet season. A timely planted wet-season crop is potentially more productive because spikelet fertility is not threatened by extreme temperatures, and grainfilling falls into a favourable period after the cessation of rains, associated with high solar radiation, moderate day temperatures and cool nights in October and November. To fully utilize the superior yield potential in the wet season, late planting must be avoided because low temperatures at the booting and heading stages can cause total crop failure. The cold dry season offers a superior yield potential in many years, but inter-year variation can be high.

We note that these results stand in contrast to the generally higher productivity of dry-season crops in monsoonal Asia, and even to the roughly equal observed performance of the wet- and dry-season rice crops in Niger. Our results are plausible, however, if we take into account that the wet-season crop in monsoonal Asia generally suffers from low solar radiation which is not the case in the Sahel, and that the wet-season rice crop in Niger and Mali is much more affected by diseases than the dry-season crop.

The situation is again markedly different for the coastal climate in the Senegal River delta, where the cold dry season is not suited to rice cultivation, and where the highest potential yields are observed in the hot dry season (which, as compared to the continental Sahel, is not very hot). Here the main constraint is not yield potential, but a long crop duration in the dry season, which renders rice double crop-ping difficult to manage.

Where rice double cropping is possible, combined annual potential yields vary between 11 (fringe of the Sahara desert, *ca* 16° N) and 18 ton ha^{-1} (transition to the subhumid zone, *ca* 11° N). An economically important exception with very high potential yields in both dry and wet seasons is the Senegal River delta.

Outlook for future research

Our results suggest that in arid environments such as the Sahel, a high yield potential for rice under irrigated conditions should not be taken for granted – despite the abundance of water and solar radiation. The ability of the plant to actually utilize the radiation is limited by several mechanisms. The most important constraints are sink limitations (heat- or cold-induced sterility), a poor cumulative light harvest at mid-season due to hastened development under hot conditions, and long growth lags and poor leaf area development at the seedling stage under cool conditions.

This study, in combination with previous economic studies and analyses of the biophysical feasibility of rice double-cropping calendars in different zones of the Sahel (WARDA 1993, 1994), will guide WARDA's applied research for irrigated rice in the future. The results suggest four foci:

(1) The development of technology packages that would sustainably exploit the superior yield potential of rice double cropping in the climatically moderate southern portions of the Sahel, wherever irrigation water is available during both seasons.

This objective requires placing greater emphasis on the biotic constraints to yield, particularly in the wet season, through integrated pest management and enhanced varietal resistance to diseases, such as blast and rice yellow mottle virus. Constraints in the wet season resemble those in the humid zone, but the dry season requires varieties that are also cold tolerant at the seedling stage. The same medium-duration variety can be grown in both seasons if it combines the necessary adaptations.

(2) The development of technology packages that would exploit the superior yield potential of rice double cropping in zones where the climate is moderated by the sea and irrigation water is not limiting.

This objective requires a continued focus on the photothermal (phenological) responses of genotypes to climate, in order to develop options for either a bivarietal system (one short- and one medium-duration crop) or a monovarietal system (one intermediate variety with stable duration), depending on the socioeconomic context. Substantial progress has been made at WARDA in recent years towards achieving this objective.

(3) The development of cropping systems based on one single, highly productive rice crop per year for zones in the Sahel that have extremely high and low temperatures in the dry season.

This objective requires more conceptual refinement because of little past attention. A number of planting-time calendrial options and varietal duration (or photothermal) types can be considered which might be combined with a different crop grown in the cool season. Simulation can provide valuable assistance to screen options *ex ante*.

(4) The development of cropping systems based on one single, highly productive rice crop per year where not the climate, but water availability and/or socioeconomic considerations stand against rice double cropping.

This objective has been pursued by WARDA's lowland breeding programme in Ibadan, and it will be expanded to the Sahel as appropriate. Vegetable or wheat crops might be considered for before or after rice crop.

We are aware that these preliminary conclusions have been derived from research tools that do not take into account all relevant parameters. Complementary studies on the management of the natural resource base have been initiated (soil conservation and the rehabilitation of degraded soils). But crop simulation will continue to guide us in further refining the applied concepts and in achieving full research coverage of the Sahel environment's diversity.

Acknowledgements

This research was supported by the German Agency for Technical Cooperation (GTZ) and the Federal Ministry for Cooperation (BMZ), Federal Republic of Germany.

References

Alocilja R B, Cervantes E B, Haws L D (1981) A continuous rice production system. The rice garden. International Rice Research Institute, P.O. Box 933, Manila, Philippines. 16 p.

Bertrand R, Keita B, N'Diaye M K (1993) La dégradation des sols des périmètres irrigues des grandes vallées sud-sahariennes (cas de l'Office du Niger au Mali). Cahiers Agricultures 2(5):318–330.

Cissé M (1994) Variabilité du remplissage des grains en riziculture irrigue dans le delta et la moyenne vallée du Fleuve Senegal. Thesis Ingénieur Agronome, November 1994, École Nationale Supérieure d'Agriculture (ENSA), Dakar, Senegal.

Diemer G, Huibers F P (1991) Farmer-managed irrigation in the Senegal river valley: Implications for the current design method. West African Rice Development Association, Bouaké, Ivory Coast. 89 p.

Dingkuhn M, Schnier H F, De Datta S K, Wijangco E, Dörffling K (1990a) Diurnal and developmental changes in canopy gas exchange in relation to growth in transplanted and direct seeded flooded rice. Aust. J. Plant Physiol. 17:119–134.

Dingkuhn M, Schnier H F, De Datta S K, Dörffling K, Javellana C (1990b) Response of direct seeded and transplanted flooded rice to nitrogen fertilization. II. Effects of leaf area and nitrogen status on foliage expansion, senescence, canopy photosynthesis and growth. Crop Sci. 30:1284–1292.

Dingkuhn M, Miézan K M (1995) Climatic determinants of irrigated rice performance in the Sahel. II, Validation of photothermal constants and characterization of genotypes. Agric. Syst. 48:411–434.

Dingkuhn M, Sow A (1995) Simulating potential rice yields for irrigated arid environments. II. Model development. Agric. Syst. (submitted).

Dingkuhn M, Sow A, Samb A, Diack S, Asch F (1995) Climatic determinants of irrigated rice performance in the Sahel. I. Photothermal and microclimatic responses of flowering. Agric. Syst. 48:385–410.

Dingkuhn M (1995) Climatic determinants of irrigated rice performance in the Sahel. III. Characterizing environments by simulating the crop's photothermal responses. Agric. Syst. 48:435–456.

Horie T, De Wit C T, Goudriaan J, Bensink J (1979) A formal template for the development of cucumber in its vegetative stage. Proc. Koninklijke Nederlandse Akademie van Wetenschappen (KNAW, Amsterdam, The Netherlands), Series C, 82(4):433–479.

Kropff M J, Van Laar H H, Matthews R B, Ten Berge H F M (1993) ORYZA1, a basic model for irrigated lowland rice production. SARP Report, International Rice Research Institute, P.O. Box 933, Manila, Philippines. 100 p.

Kropff M J, Van Laar H H, Matthews R B (Eds.) (1994) ORYZA1: An ecophysiological model for irrigated rice production. SARP Research Proceedings. International Rice Research Institute, P.O. Box 933, Manila, Philippines. 110 p.

Le Gal P-Y (1995) Développement de la double culture irrigue au Sahel: Constraintes et perspectives pour les agriculteurs. Contribution to the workshop Irrigated Rice in the Sahel: Perspectives for Sustainable Development, 27–31 March 1995, Ndiaye, Senegal (in press).

Nishiyama I (1984) Climatic influence on pollen formation and fertilization. Pages 153–171 in Tsunoda S, Takahashi N (Eds.) Biology of rice. Elsevier, Amsterdam, Oxford, New York, Tokyo.

Penning de Vries F W T, Jansen D M, Ten Berge H F M, Bakema A (1989) Simulation of ecophysiological processes of growth in several annual crops. Pudoc, Wageningen, The Netherlands. 271 p.

WARDA – West Africa Rice Development Association (1992) Pages 12–22 in Annual Report for 1991. BP 2551, Bouaké, Ivory Coast.

WARDA (1993) Pages 36–39 in Annual Report for 1992. BP 2551, Bouaké, Ivory Coast.

WARDA (1994) Pages 13–28 in Annual Report for 1993. BP 2551, Bouaké, Ivory Coast.

Assessing the potential yield of tropical crops: role of field experimentation and simulation

R.C. MUCHOW[1] and M.J. KROPFF[2]

[1]*CSIRO Division of Tropical Crops and Pastures, 306 Carmody Road, St. Lucia, Qld 4067, Australia*
[2]*Department of Theoretical Production Ecology, Wageningen Agricultural University, P.O. Box 430, 6700 AK Wageningen, The Netherlands*

Key words: solar radiation, temperature, field productivity, simulation

Abstract

This review assesses the role of field experimentation and simulation in better quantifying the potential productivity in different environments, and examines how knowledge on potential productivity can improve the efficiency of the production system. Case studies are presented for the tropical crops maize and rice to highlight the importance of temporal and spatial variation in climatic factors on the yield performance of high-input systems. The key role of field experimentation is to quantify the impacts of temperature and solar radiation on resource capture and resource utilization for a given variety. There is a need to collect minimum data sets of key crop, soil, management and climate variables, and to ensure that climatic data, particularly solar radiation, are regularly calibrated. When knowledge from field experimentation is encapsulated into crop growth simulation models, the gap between actual and potential yield for a given environment can be determined and opportunities for yield improvement can be assessed. Simulation can also be used to quantify temporal and spatial variation in productivity, for use in planning future world food supplies. Given that scope for raising maximum yields is limited, the focus of production research should be on narrowing the gap between actual and potential yields and on the maintenance of maximum yields. By characterizing the frequency of occurrence of constraints in different regions, crop simulation can potentially enhance the efficiency of plant breeding. Given the potential negative impacts of high-input agriculture on the environment, simulation can identify best-practice management options to better align resource supply with crop demand. The development of electronic data bases of climate, soil, management and crop growth data is a priority need, so that the synergy between field experimentation and simulation can be maximized.

Introduction

There is an increasing awareness that food production must be substantially increased in the coming decades to meet the demand created by growing populations (IRRI 1993). This production increase has to be achieved, on less land, with less labour, less water and less pesticides, and it must be sustainable through conserving scarce natural resources. To meet the challenge of increasing food production, ways have to be found to improve the productivity and profitability of agricultural production systems. It has been observed, for example, that annual yield increases for rice, one of the major food crops, are stagnating in many countries (IRRI 1993). A possible reason for the plateauing in yield may be that actual yields are approaching the yield potential. However, good estimates of the yield potential of crops in different environments are necessary and this requires detailed studies in which modelling and experi-

M.J. Kropff et al. *(eds.): Applications of Systems Approaches at the Field Level, 101–112.*

mentation are intertwined (Muchow et al. 1990; Kropff et al. 1995).

Potential productivity, defined as the yield obtained when water or nutrients are non-limiting and in the absence of pests, diseases and weeds, varies across agro-ecologies principally in response to the solar radiation and temperature regime. Such high-input agriculture is important to world food supplies, but increasingly such production systems need to better match resource supply with resource demand, so that resources (e.g. water and nutrients) are not lost from the production system with potential negative impacts on the environment. Both field experimentation and crop growth simulation have a role to play in developing the capability to assess potential productivity. This knowledge on potential productivity can assist in identifying opportunities to improve yield and resource utilization in diverse production systems.

Field experimentation to identify opportunities to improve production efficiency has traditionally been conducted at 'representative' sites and the results are extrapolated to other sites that share similar characteristics (Nix 1984). In climatically variable regions, such experimentation can be costly because of the need to sample a sufficient number of seasons, cultivars, soil types and management scenarios. The generation of probabilistic estimates of yield using crop growth simulation coupled to long-term climatic data has been advocated as a tool to assist decision-making and scenario analysis (Muchow and Bellamy 1991). A major limitation to the approach using crop growth simulation is that model simulations are dependent on current understanding of physiological and soil processes and their interactions. This understanding is by no means complete. Consequently, both field experimentation and crop growth simulation must be viewed as complementary in endeavours to improve the efficiency of the production system.

This paper (i) outlines a framework to assess variation in crop yield, (ii) assesses the role of field experimentation and simulation in better quantifying potential productivity in different environments and (iii) examines how knowledge on potential productivity can be used to improve the efficiency of the production system. Case studies are presented for the tropical crops maize and rice to highlight the importance of temporal and spatial variation in climatic factors on the yield performance of high-input systems.

Framework for assessing variation in crop yield

For a given variety, the two factors which determine crop yield are environment and management. The environment can be considered in terms of two elements: the physical or abiotic environment which includes the attributes of climate (temperature, radiation, rainfall) and soil resource supply (water and nutrients); and the biotic environment which includes the attributes of fauna, weeds and insect pests, and soil and plant diseases. The impact of environment on yield is complex due to the interaction between these elements. For example, climatic conditions control the development and spread of fungal diseases, insects, pests and weeds which can restrict crop growth. Furthermore, climatic elements interact with crop management by setting capricious limits to farm operations.

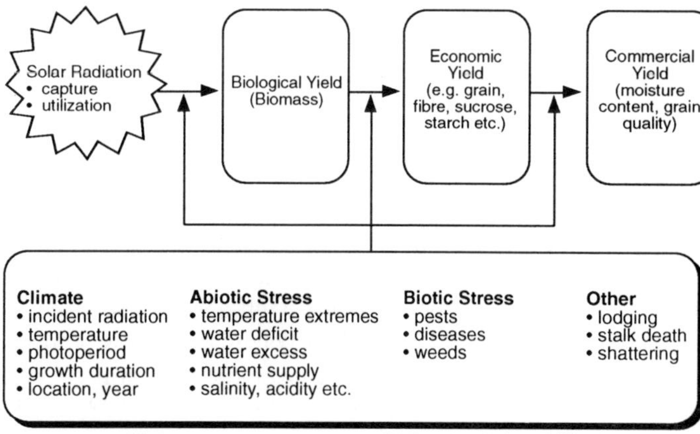

Figure 1. Framework for assessing variation in crop yield.

Quantitatively, these impacts on crop yield can be described in terms of the capture and utilization of solar radiation (Figure 1). For many crop species grown under well-watered conditions and ample nutrition in the absence of pests and diseases, biological yield (biomass production) has been shown to be linearly related to the amount of radiation intercepted (captured) by the crop canopy (Williams et al. 1965; Monteith 1977). This relationship sets a finite limit on yield potential (Sinclair 1994). The slope of the linear relationship, often termed the radiation use efficiency (RUE), is curvilinearly related to light-saturated leaf CO_2 assimilation rate C_A, and at current maximum RUE values, very large increases in C_A would be required to achieve even modest further increases in maximum RUE (Sinclair and Horie 1989). Gifford (1974) reported that canopy biomass accumulation is relatively insensitive to increases in C_A, and in a review, Gifford et al. (1984) concluded that efforts to increase crop growth rates through increasing the rate of net photosynthesis have met with little success for a number of crop species. Similarly, Dwyer and Tollenaar (1989) found no significant differences between new and old maize hybrids in potential leaf photosynthesis. Of more importance in realizing yield potential is the maintenance of high rates of C_A throughout the total growing season. This is particularly so during grain-filling, where a decline in C_A is often related to inadequate nitrogen (N) supply (Kropff et al. 1993; Muchow and Sinclair 1994). Likewise, the prevention of abiotic and biotic stresses (Figure 1) is important in raising yields towards the potential, and the narrowing of this gap between actual and potential yields should be the focus of crop improvement programmes.

That portion of the biological yield that is harvested is referred to as the economic yield and is expressed on a dry weight basis (Figure 1). The proportion of the biological yield that is present as economic yield is referred to as the harvest index. Evans (1993) reviewed the major changes in crops that have been brought about by co-selection for high yield and suitability for modern, often high-input husbandry, and concluded that the increase in harvest index has been one of the most significant advances in crop yield potential. Over the last 30 years, for example, the harvest index of small grain cereals has increased from 0.3 to 0.5. In many crop species, harvest index under potential growth conditions is above 0.5, and an important question is how much more can harvest index be increased. In intensively bred cereals grown under high-input conditions, Austin et al. (1980) estimated a maximum possible harvest index of 0.63, but this is unlikely given that shorter and weaker stems increase the susceptibility to lodging. Whilst harvest index is relatively conservative under high-input conditions, it can be substantially reduced under severe water and nitrogen stress (Muchow 1988, 1989). Sinclair (1988) reported that biomass is more sensitive to stress than is harvest index, with harvest index only decreasing when biomass production is less than 40% of potential biomass production. There is scope under stress conditions to increase harvest index towards the potential, and several studies have shown such improvements (Fischer et al. 1989; Tollenaar et al. 1993).

The commercial yield is that used as a basis of payment to farmers. It is related to the economic yield by parameters such as moisture content and quality. In grain crops, the relationship between economic yield and commercial yield is usually set by a defined grain moisture content and defined quality standard (e.g. grain protein). However, in some crops (e.g. sugarcane), economic return is based on the weight of fresh millable stalks and sucrose concentration on a fresh weight basis. The moisture content of harvested sugarcane is highly variable, and factors such as crop age, climate, water and nitrogen supply affect stalk moisture content (Muchow and Robertson 1994; Robertson and Muchow 1994). To avoid this further complexity of relating economic yield to commercial yield, the focus in this paper will be on biological and economic yield.

Given that there appears to be relatively little scope in increasing the maximum efficiency of radiation utilization in biomass production, nor increasing the partitioning of that biomass to economic yield under potential growth conditions, high yield will be obtained where radiation capture is maximized and radiation utilization is maintained at maximum values. Factors influencing radiation capture and utilization are given in Figure 1.

Radiation capture is strongly dependent on growth duration as determined by crop phenology, and on leaf canopy development. Leaf canopy development as influenced by ambient temperature determines the leaf area index of the crop and, thereby, determines the proportion of the incident radiation which is captured (Muchow and Carberry 1989). The duration of crop growth sets the maximum time that the incident radiation (which varies with location, year and season) can be captured. For many cropping systems, cool and freezing temperature effectively define the growth duration. Similarly, in monsoonal rainfed cropping systems, lack of water at the onset of the dry season effectively defines the length of the growing season. In crops where the reproductive growth comprises the economic yield (e.g. grain crops), the durations of vegetative and reproductive growth as modified by temperature and photoperiod (Muchow and Carberry 1989) also set a limit to the length of the growing season. Where vegetative growth comprises the economic yield (e.g. sugarcane), the duration of crop growth is set by management decisions related to planting, ratooning and harvesting. Radiation capture can also be reduced by stress factors (Figure 1), with N supply having a major influence (Novoa and Loomis 1981; Muchow 1994). N supply determines early leaf area development, the maximum LAI and the rate of senescence (Kropff et al. 1994a).

Radiation utilization under potential growth conditions is depressed at supra-optimal and sub-optimal temperature. For example, in maize maximum RUE occurs at mean daily temperature from 20°C to 40°C, and decreases to zero at a base temperature of 8°C and extreme mean temperature of 50°C (Carberry et al. 1989). Radiation utilization is decreased by water and nitrogen stress (Muchow 1989, 1994). However, even under potential growth conditions, it is difficult to maintain leaf N values sufficiently high for the whole growth duration due to mobilization of leaf N to grain N, with subsequent decrease in radiation utilization during grainfilling (Kropff et al. 1993; Muchow and Sinclair 1994).

Other 'random non-controllable' factors can also affect yield performance under high-input conditions (Figure 1). These factors include waterlogging and lodging, and biomass loss and seed shattering that are difficult to control, even under potential growth conditions. In modern, semi-dwarf rice varieties, lodging is often a constraint to potential yield when N supply is high (Setter et al. 1994). The breeding of semi-dwarf soybeans to overcome the lodging barrier to yield, combined with modified crop management, has led to higher yield potential in soybean in high-yielding environments (Cooper 1985). In high-yielding sugarcane, Muchow et al. (1994) identified an early yield plateau well before scheduled crop harvest when climatic conditions and resource supply were favourable for continued crop growth. Stalk death and subsequent loss of biomass associated with lodging and smothering were significant contributors to the cessation in yield accumulation, with consequent poor efficiency of utilization of radiation, water and nutrients due to poor late growth in these crops. Opportunities for genetic and management manipulations to improve radiation utilization in these circumstances need to be sought.

Role of field experimentation

Field experimentation contributes to understanding of the determinants of potential productivity. Such understanding can be encapsulated into quantitative relationships in crop growth simulation models. As discussed earlier, it is difficult to extrapolate experimental findings both temporally and spatially, where climatic variability is high. Crop growth simulation models when used in conjunction with historical climatic data can take account of temporal and spatial variation in climate, and can be used to assess production risk. However, there is an important role for field experimentation in generating data both for models development and for verification of the results of simulation analysis. The development of data bases of climate, soil, management and crop growth data from field experiments is essential for this purpose.

Muchow et al. (1990) examined the effects of variation in solar radiation and temperature on the potential yield of maize based on field experiments conducted at

Table 1. Effect of temperature and radiation regime on observed and simulated grain yield (at 15.5% moisture) and growth duration from emergence to maturity for maize cultivars at different locations (*Sources:* data from Muchow et al. (1990), except for Lawes (Muchow 1994)).

Location (latitude)	Sowing date and cultivar	Duration (d)	Observed grain yield (ton ha^{-1})	Simulated grain yield (ton ha^{-1})	Mean daily temperature (°C)	Incident radiation (MJ m^{-2} d^{-1})
Katherine, NT, Australia 14° 28′ S	30 Aug. 86 Dekalb XL82	84	9.47	9.60	28.7	25.5
Katherine, NT, Australia 14° 28′ S	6 Feb. 85 Dekalb XL82	95	9.34	9.36	26.3	22.2
Lawes, Q, Australia 27° 34′ S	3 Oct. 90 Dekalb XL82	118	12.2	11.2	24.8	22.1
Champaign, IL, USA 40° 7′ N	4 May 1982 Agway 849X	126	12.7	12.9	21.5	19.9
Grand Junction, CO, USA 39° 4′ N	7 May 84 Funk G4507	138	19.0	19.3	19.8	28.3
Grand Junction, CO, USA 39° 4′ N	28 Apr. 86 Dekalb 656	153	17.1	17.1	18.0	22.4

different locations ranging from latitude 14° S to 40° N. Maize hybrids of similar phenology (18 leaves) were selected, and a subset of this analysis is shown in Table 1. Grain yields were more than double (19 ton ha^{-1}) in the high altitude temperate (Grand Junction) environment compared to the tropical (Katherine) environment (9.4 ton ha^{-1}). In the tropics, the growth duration was much shorter than in higher latitudes, but at a given location potential grain yield was not necessarily related to growth duration. For example, there was little difference in grain yield for the two Katherine sowings shown in Table 1 as the higher incident radiation compensated for the shorter growth duration.

In order to understand the variation in yield across locations, Muchow et al. (1990) developed a simple mechanistic growth model. Crop phenology and leaf growth were calculated from daily mean temperature based on functions derived from the experimental data of Muchow and Carberry (1989). Daily biomass accumulation was calculated by estimating the amount of radiation intercepted and assuming a maximum radiation use efficiency of 1.6 g MJ^{-1}, based on the experimental observations of Muchow and Davis (1988). Grain growth was simulated using a linear increase in harvest index to a maximum value of 0.5 during grainfilling, based on the experimental observations of Muchow (1990). Field experimentation at Katherine provided all the quantitative functions in the model, and when the model was run with inputs of daily maximum and minimum temperature and solar radiation from the experiments at the different locations, there was close agreement between observed and simulated grain yield data (Table 1). Clearly, the role of field experimentation was to provide the quantitative understanding of the determinants of crop yield.

The model analysis showed that both temperature and solar radiation influence the variation in potential maize yield across environments. The primary influence of temperature was on growth duration, with lower temperature increasing the time that the crop can intercept radiation. Biomass accumulation is directly proportional

Table 2. Observed and simulated crop characteristics for field experiments on yield potential in rice in different environments (SE between brackets) (*Source:* Kropff et al. 1994b).

Location	IRRI WS	Japan	IRRI DS	Yanco
Variety	IR72	Nipponbare	IR72	YRL39
Biomass obs. (ton ha^{-1})	11.5 (0.6)	12.8 (0.5)	17.8 (0.6)	28.9 (1.1)
Biomass sim. (ton ha^{-1})	11.5	13.2	15.4	27.3
Yield obs. (ton ha^{-1}, panicle dry wt)	5.7 (0.2)	7.8 (0.5)	9.8 (0.2)	14.7 (0.6)
Yield sim. (ton ha^{-1}, panicle dry wt)	7.26	7.41	9.49	14.3
Maximum LAI obs. ($m^2 m^{-2}$)	4.3 (0.4)	5.3 (0.7)	6.0 (0.5)	14.0 (1.0)
Maximum LAI sim. ($m^2 m^{-2}$)	4.9	8.9	6.4	12.8
Crop duration (d)	109	153	110	159
Grainfilling duration obs. (d)	30	41	30	43
Grainfilling duration sim. (d)	29	32	30	39
Mean temperature (°C)	28.0	23.9	26.5	21.7
Incident radiation (MJ $m^{-2} d^{-1}$)	16.2	15.5	18.6	22.8

to the amount of radiation intercepted, and for a given harvest index grain yield is directly proportional to biomass. Consequently, high maize yield was associated with low temperature and high solar radiation within the range of environments studied (Table 1). Furthermore, this analysis indicates that the solar radiation and temperature regime set a finite limit to potential yield in a given environment.

Kropff et al. (1994b) have observed a similar wide variation in potential rice yields ranging from 6 ton ha^{-1} in tropical environments (wet season, Los Baños, Philippines) to 15 ton ha^{-1} at higher latitudes (Yanco, Australia) (Table 2). Higher yield was associated with lower temperature which resulted in longer duration of all growth phases, particularly the duration of grainfilling. Longer duration of late-maturing rice varieties in the tropics, however, does not translate into higher yield potential, because the duration of grainfilling is shorter than in temperate climates due to higher mean temperature, and grainfilling duration of late, medium and early maturing varieties is the same (Peng et al. 1994; Kropff et al. 1994c). Kropff et al. (1994b) parameterized the ecophysiological rice model ORYZA1 and were able to simulate these effects of solar radiation and temperature on observed potential yields (Table 2). Simulations using this model highlighted the importance of high leaf N for the attainment of potential productivity, and Kropff et al. (1993) concluded that it is essential to match N supply to the requirements of the crop to achieve full yield potential in intensive irrigated rice systems.

This case study also highlights the need to ensure that field experimentation is conducted under 'potential' growth conditions when integrated with potential growth modelling. In the 1980s and early 1990s, maximum yields with current rice varieties were much lower than yields obtained in the early 1970s with the first semi-dwarf variety IR8 (7 vs 9.5 ton ha^{-1}, Kropff et al. 1994c). Rice simulation models like MACROS and CERES-Rice that were calibrated with the available data simulated potential yields of only 8 ton ha^{-1} for the Los Baños environment (Herrera-Reyes and Penning de Vries 1989; U. Singh, IRRI, pers. commun.). Field experiments initiated at IRRI in 1991, with improved agronomic management including N applications higher than the current practice and with late N application at flowering to maintain leaf N during grainfilling, resulted in yields comparable to those reported in the late 1960s for IR8: 6 ton ha^{-1} in the wet season and 10 ton ha^{-1} in the dry season (Cassman et al. 1995; Kropff et al. 1994c). The rice model ORYZA1 has been parameterized using data obtained under these input conditions and is able to simulate performance across different environments (Table 2).

The collection of a minimum data set in field experimentation allows the data to be used in the development and testing of crop growth models. While different models vary in their input specifications for crop and soil data, all models have an essential requirement for climatic data, particularly solar radiation, temperature and rainfall. The importance of climatic data is frequently overlooked, particularly the need for regular calibration of solar radiation, the primary driving variable of crop growth and a key determinant of the reliability of simulation estimates of crop yield. The collection of climate, crop, soil and management data for model development and testing is expensive. There is a need to ensure that data collected in field experimentation are sufficiently complete to maximize their value in overcoming production constraints. Nix (1984) introduced the concept of a *minimum* data set and suggested a hierarchy of levels appropriate to different applications. There is also a need for electronic data bases to allow systematic filing of all the necessary data for model analysis. Examples of data bases for this purpose include DSSAT (Tsuji et al. 1994) and SUGARBAG (Prestwidge et al. 1994), and these publications also specify the minimum data set required for different growth conditions (e.g. potential growth, and water- and nitrogen-limiting conditions). Even for field experimentation alone, electronic data bases can function as an experimental diary for use by individual researchers, or as an electronic filing cabinet for systematic storage of data from an experiment.

In many situations, the data collected in data bases do not directly provide the input data required for specific simulation models, as is often the case with data bases on soil characteristics. In such situations, the so-called transfer functions which relate available data to parameters needed by the model have to be developed. Bouma et al. (1993) developed soil transfer functions to relate data from standard soil surveys (i.e. texture) to soil physical properties that are required for mechanistic hydrological models.

Role of crop growth simulation

Crop simulation has a role in the interpretation of data from field experiments and in explaining yield differences in a range of environments. Knowledge of the benchmark yield for a given cultivar, location and season allows assessment of how far current yields are below the potential yield and what is the scope for yield improvement. Simulation also has a role, in conjunction with historical climatic data, to quantify the temporal and spatial variation in potential productivity across agroecologies, and when combined with spatial analysis provides a powerful tool in planning future world food supplies (Penning de Vries et al. 1995). By characterizing the types of environmental challenges and their frequencies of occurrence, crop growth simulation can potentially enhance the efficiency of plant breeding and selection programmes. Given the potential negative impacts of high-input agriculture on the environment, simulation also has an important role to play in better quantifying the resource requirements to attain potential yield in different environments, so that crop management can better align resource supply with crop demand (Kropff et al. 1994b; Cassman et al. 1993).

The data on potential maize and rice yields presented in Tables 1 and 2 show that the solar radiation and temperature regime results in considerable variation in yield across seasons and locations. The simulated benchmark potential yield can be compared with the actual yield for a given season to assess the scope for further yield improvement. The power of crop growth simulation is that it can account for some of the factors affecting radiation capture and utilization outlined in Figure 1, and it can identify other factors potentially constraining yield. Park and Sinclair (1993) combined a simulation approach using actual climatic data with reported farm maize yield data in Champaign County, Illinois, USA, to show that the gap between actual and simulated maize yield diminished between 1965 and 1988. They concluded that as the gap continues to narrow, the trend of increasing yield is likely to diminish as maize yield approaches the limit set by climatic factors. They also observed that the skewness of the distribution of individual farm yield varied across years, indicating variable climatic constraint (water supply) to productivity. This approach can be used to identify opportunities for overcoming production constraints and enhancing stability of food supplies.

The temporal and spatial variation in potential yield can be examined using a simulation approach. Kropff et al. (1993) used ORYZA1 with long-term climatic data from two sites in the Philippines (Los Baños and Nueva Ecija), and with simulated monthly sowings, showed that there were large within-year and year-to-year variation in rice yield potential, which reflected yearly differences in climate at both sites. At Nueva Ecija, yield differences between the wet and dry season were not as distinct as, and maximum yield potential was higher than, at Los Baños. When expanded for a range of sites, and potentially linked to a Geographic Information System, this is a powerful tool to define crop resource requirements and management inputs for the attainment of maximum productivity, and to predict variation in food supplies.

In addition to assisting the optimization of crop management, simulation also has a role to play in assisting genetic improvement in the cropping system. The types of environments encountered within a target population of environments play a dominant role in determining crop performance, genetic variation for quantitative traits and, therefore, the relative performance of genotypes. Muchow et al. (1991) considered some limitations to effective sampling of the target population of environments with traditional multi-environment trials. Better characterization of environments to define the spectrum of environmental challenges offers the potential to improve the efficiency of selection within multi-environment trials, and this may lead to better choice of environments for field trials, thereby increasing gains from selection. Muchow et al. (1996) outlined a possible approach of using crop simulation with historical climatic data to describe environments and then using pattern analysis to characterize the types and frequency of occurrence of different types of environments. This approach can be linked to spatial data bases to identify suitable testing locations, either those that best represent target environments or those that present specific challenges to germplasm under test (Chapman and Barreto 1996). Crop simulation can also assist in the evaluation of potentially useful crop traits (Hammer et al. 1996). Kropff et al. (1994c) used different models to determine the physiological traits that would be required by a new rice plant type with a higher yield potential in the tropical dry season than the current modern high-yielding varieties. The potential to combine statistical methodologies with crop simulation approaches to better define genotype × environment interactions and improve breeding efficiency is further discussed by Cooper and Hammer (1996).

The attainment of potential productivity requires high resource inputs. As discussed above, high leaf N is required to maximize productivity especially during grainfilling (Kropff et al. 1993). It is essential to match the seasonal pattern of N supply to the N demand of the crop at each stage of development to achieve full yield potential (Cassman et al. 1993; Ten Berge et al. 1994). However, N is mobile in the soil system, and with the required high rates of N application the possibility exists for losses (e.g. leaching), and negative environmental impact. Simulation has a role to play in defining both the crop demand and the potential losses from the system under different climatic conditions. Keating et al. (1991) have reviewed the range of models available to describe the interaction of N supply with climatic factors on yield. However, further development of models is required to simulate the crop-soil system, and to assist in the development of strategies that best match resource supply with resource demand with minimal loss of resources to the environment. Perhaps this remains as the biggest challenge to sustaining potential productivity in the cropping system.

References

Austin R B, Bingham J, Blackwell R D, Evans L T, Ford M A, Morgan C L, Taylor M (1980) Genetic improvements in winter wheat yields since 1990 and associated physiological changes. J. Agric. Sci. (Camb.) 94:675–689.

Bouma J, Wopereis M C S, Wösten J H M, Stein A (1993) Soil data for crop-soil models. Pages 207–220 in Penning de Vries F W T, Teng P S, Metselaar K (Eds.) Systems approaches for agricultural development. Kluwer Academic Publishers, Dordrecht, The Netherlands.

Carberry P S, Muchow R C, McCown R L (1989) Testing the CERES-Maize simulation model in a semi-arid tropical environment. Field Crops Res. 20:297–315.

Cassman K G, Kropff M J, Gaunt J L, Peng S (1993) Nitrogen use efficiency of rice reconsidered: What are the key constraints? Plant and Soil 155/156:359–362.

Cassman K G, De Datta S K, Olk D C, Alcantara J M, Samson M I, Descalsota J P, Dizon M A (1995) Yield decline and the nitrogen economy of long-term experiments on continuous, irrigated rice systems in the tropics. Advances in Soil Science (in press).

Chapman S C, Barreto H J (1996) Using simulations models and spatial databases to improve the efficiency of plant breeding programs. In Cooper M and Hammer G L (Eds.) Plant adaptation and crop improvement. ICRISAT and IRRI, CAB International, Wallingford, UK (in press).

Cooper M, Hammer G L (Eds.) (1996) Plant adaptation and crop improvement. ICRISAT and IRRI, CAB International, Wallingford, UK (in press).

Cooper R L (1985) Breeding semi-dwarf soybeans. Plant Breeding Reviews 3:289–311.

Dwyer L M, Tollenaar M (1989) Genetic improvement in photosynthetic response of hybrid maize cultivars, 1959 to 1988. Can. J. Plant Sci. 69:81–91.

Evans L T (1993) Processes, genes, and yield potential. Pages 687–701 in Buxton D R, Shibles R, Forsberg R A, Blad B L, Asay K H, Paulsen G M, Wilson R F (Eds.) International crop science I. CSSA Madison, WI, USA.

Fischer K S, Edmeades G O, Johnson E C (1989) Selection for the improvement of maize yield under moisture-deficits. Field Crops Res. 22:227–243.

Gifford R M (1974) A comparison of potential photosynthesis, productivity and yield of plant species with differing photosynthetic metabolism. Aust. J. Plant Physiol. 1:107–117.

Gifford R M, Thorne J H, Hitz W D, Gianquinta R T (1984) Crop productivity and photoassimilate partitioning. Science 225:801–808.

Hammer G L, Butler D, Muchow R C, Meinke H (1996) Integrating physiological understanding and plant breeding via crop modelling and optimisation. In Cooper M, Hammer G L (Eds.) Plant adaptation and crop improvement. ICRISAT and IRRI, CAB International, Wallingford, UK (in press).

Herrera-Reyes C G, Penning de Vries F W T (1989) Evaluation of a model for simulating the potential production. Philipp. J. Crop Sci. 14(1):21–32.

IRRI – International Rice Research Institute (1993) Rice research in a time of change. International Rice Research Institute, P.O. Box 933, Manila, Philippines. 79 p.

Keating B A, Godwin D C, Watiki J M (1991) Optimizing nitrogen inputs in response to climatic risk. Pages 329–358 in Muchow R C, Bellamy J A (Eds.) Climatic risk in crop production: Models and management for the semiarid tropics and subtropics. CAB International, Wallingford, UK.

Kropff M J, Cassman K G, Van Laar H H, Peng S (1993) Nitrogen and yield potential of irrigated rice. Plant and Soil 155/156:391–394.

Kropff M J, Van Laar H H, Matthews R B (Eds.) (1994a) ORYZA1: An ecophysiological model for irrigated rice production. SARP Research Proceedings, International Rice Research Institute, P.O. Box 933, Manila, Philippines. 110 p.

Kropff M J, Cassman K G, Van Laar H H, Torres R O, Liboon S P, Williams R L, Horie T (1994b) Evaluation of the model ORYZA1. Pages 57–65 in Kropff M J, Van Laar H H, Matthews R B (Eds.) ORYZA1: An ecophysiological model for irrigated rice production. SARP Research Proceedings, International Rice Research Institute, P.O. Box 933, Manila, Philippines.

Kropff M J, Cassman K G, Peng S, Matthews R B, Setter T L (1994c) Quantitative understanding of yield potential. Pages 21–38 in Cassman K G (Ed.) Breaking the yield barrier. International Rice Research Institute, P.O. Box 933, Manila, Philippines.

Kropff M J, Williams R L, Horie T, Angus J F, Singh U, Centeno H G, Cassman K G (1995) Predicting the yield potential of rice in different environments. Pages 657–664 in Humphreys E, Murray E A, Clampett W S, Lewinn L G (Eds.) Temperate rice: Achievements and potential. Proceedings of the Temperate Rice Conference, February 1994, Yanco, Australia.

Monteith J L (1977) Climate and the efficiency of crop production in Britain. Phil. Trans. R. Soc. London B 281:277–294.

Muchow R C (1988) Effect of nitrogen supply on the comparative productivity of maize and sorghum in a semiarid tropical environment. III. Grain yield and nitrogen accumulation. Field Crops Res. 18:31–44.

Muchow R C (1989) Comparative productivity of maize, sorghum and pearl millet in a semi-arid tropical environment. II. Effect of water deficits. Field Crops Res. 20:207–219.

Muchow R C (1990) Effect of high temperature on grain growth in field grown maize. Field Crops Res. 23:146–158.

Muchow R C (1994) Effect of nitrogen on yield determination in irrigated maize in tropical and subtropical environments. Field Crops Res. 38:1–13.

Muchow R C, Bellamy J A (Eds.) (1991) Climatic risk in crop production: Models and management for the semiarid tropics and subtropics. CAB International, Wallingford, UK. 548 p.

Muchow R C, Carberry P S (1989) Environmental control of phenology and leaf growth in a tropically adapted maize. Field Crops Res. 20:221–236.

Muchow R C, Davis R (1988) Effect of nitrogen supply on the comparative productivity of maize and sorghum in a semi-arid environment. II. Radiation interception and biomass accumulation. Field Crops Res. 18:17–30.

Muchow R C, Robertson M J (1994) Relating crop nitrogen uptake to sugarcane yield. Pages 122–130 in Proceedings of the 16th Conference, Australian Society of Sugarcane Technologists, 26–29 April 1994. Townsville, Queensland, Australia.

Muchow R C, Sinclair T R (1994) Nitrogen response of leaf photosynthesis and canopy radiation use efficiency in field-grown maize and sorghum. Crop Sci. 34:721–727.

Muchow R C, Sinclair T R, Bennett J M (1990) Temperature and solar radiation effects on potential maize yield across locations. Agron. J. 82:338–343.

Muchow R C, Hammer G L, Carberry P S (1991) Optimising crop and cultivar selection in response to climatic risk. Pages 235–262 in Muchow R C, Bellamy J A (Eds.) Climatic risk in crop production: Models and management for the semiarid tropics and subtropics. CAB International, Wallingford, UK.

Muchow R C, Spillman M F, Wood A W, Thomas M R (1994) Radiation interception and biomass accumulation in a sugarcane crop grown under irrigated tropical conditions. Aust. J. Agric. Res. 45:37–49.

Muchow R C, Cooper M, Hammer G L (1996) Characterising environmental challenges using models. In Cooper M, Hammer G L (Eds.) Plant adaptation and crop improvement. ICRISAT and IRRI, CAB International, Wallingford, UK (in press).

Nix H A (1984) Strategies for crop research. Proceedings of the Agronomy Society of New Zealand 10:107–110.

Novoa R, Loomis R S (1981) Nitrogen and plant production. Plant Soil 58:177–204.

Park W I, Sinclair T R (1993) Consequences of climate and crop yield limits on the distribution of corn yields. Rev. Agric. Econ. 15:483–493.

Peng S, Khush G S, Cassman K G (1994) Evolution of the new plant ideotype for increased yield potential. Pages 5–20 in Cassman K G (Ed.) Breaking the yield barrier. International Rice Research Institute, P.O. Box 933, Manila, Philippines.

Penning de Vries F W T, Van Keulen H, Rabbinge R (1995) Natural resources and limits of food production in 2040. Pages 65–87 in Bouma J, Kuyvenhoven A, Bouman B A M, Luyten J C, Zandstra H G (Eds.) Ecoregional approaches for sustainable land use and food production. Kluwer Academic Publishers, Dordrecht, The Netherlands.

Prestwidge D B, Robertson M J, Muchow R C, Hammer G L, Hughes R M, Keating B A, Kingston G, Liu D L, Wood A W (1994) SUGARBAG: A database system for sugarcane crop growth, climate, soils and management data. CSIRO Division of Trop. Crops and Pastures, Trop. Agron. Tech. Mem. No. 84, St. Lucia, Queensland, Australia.

Robertson M J, Muchow R C (1994) Future research challenges for efficient crop water use in sugarcane production. Pages 193–200 in Proceedings of the 16th Conference, Australian Society of Sugarcane Technologists, 26–29 April 1994. Townsville, Queensland, Australia.

Sinclair T R (1988) Selecting crops and cropping systems for water-limited environments. Pages 87–94 in Bidinger F R, Johansen C (Eds.) Drought research priorities for the dryland tropics. ICRISAT, Patancheru, India.

Sinclair T R (1994) Limits to crop yield? Pages 509–532 in Boote K J, Bennett J M, Sinclair T R, Paulsen G M (Eds.) Physiology and determination of crop yield. ASA, CSSA, and SSSA, Madison WI, USA.

Sinclair T R, Horie T (1989) Leaf nitrogen, photosynthesis, and crop radiation use efficiency: A review. Crop Sci. 29:90–98.

Setter T L, Peng S, Kirk G J D, Virmani S S, Kropff M J, Cassman KG (1994) Physiological considerations and hybrid rice. Pages 39–62 in Cassman K G (Ed.) Breaking the yield barrier. International Rice Research Institute, P.O. Box 933, Manila, Philippines.

Ten Berge H F M, Wopereis M C S, Shin J C (Eds.) (1994) Nitrogen economy of irrigated rice: Field and simulation studies. SARP Research Proceedings, International Rice Research Institute, P.O. Box 933, Manila, Philippines. 306 p.

Tollenaar M, McCullough D E, Dwyer L M (1993) Physiological basis of the genetic improvement of corn. Pages 183–236 in Slafer G A (Ed.) Genetic improvements in field crops: Current status and development. Marcel Dekker Inc., New York, USA.

Tsuji G Y, Uehara G, Balas S (1994) DSSAT, Version 3: A decision support system for agrotechnology transfer. University of Hawaii, Honolulu, HI, USA. 164 p.

Williams W A, Loomis R S, Lepley C R (1965) Vegetative growth of corn as affected by population density. I. Productivity in relation to interception of solar radiation. Crop Sci. 5:211–215.

Evaluation of the CROPGRO-Soybean model over a wide range of experiments

K.J. BOOTE[1], J.W. JONES[2], G. HOOGENBOOM[3] and G.G. WILKERSON[4]

[1]*Department of Agronomy, University of Florida, Gainesville, FL 32611, USA*

[2]*Department of Agricultural Engineering, University of Florida, Gainesville, FL 32611, USA*

[3]*Department of Agricultural Engineering, Georgia Station, University of Georgia, Griffin, GA 30223, USA*

[4]*Crop Science Department, North Carolina State University, Raleigh, NC 27695, USA*

Key words: soybean modelling, weather effects, growth, seed yield, life cycle prediction

Abstract

Crop simulation models are increasingly being used to predict yield responses to soil, weather, and management conditions. This requires that the models be evaluated for their abilities to accurately respond to those factors. Our objective in this paper was to evaluate the recently released CROPGRO-Soybean model for its ability to simulate soybean growth, seed yield, flowering dates, and season lengths over a wide range of conditions. Inputs (weather, soil characteristics, management practices) and data on growth and yield were assembled for several cultivars from various locations in the USA. The Bragg cultivar was evaluated in multiple years at Gainesville under varying water supply and showed the model ability to predict water limitations on growth. The Williams cultivar was evaluated in multiple years at sites in Iowa, Ohio, and Florida and illustrated CROPGRO ability to predict in different locations and climatic environments. Planting date studies were simulated for three cultivars in North Carolina to evaluate model ability to predict yield response to planting date. Model ability to predict growth and yield in two 'on-farm' soybean trials was evaluated. Three of the experiments (Williams cultivar in Florida, planting date trials, and on-farm trials) represent independent data never used in model calibration and illustrate the ability of CROPGRO to predict growth and yield in new locations and environments. We conclude that CROPGRO-Soybean gives reasonable predictions under a wide range of environmental conditions.

Introduction

Crop simulation models are increasingly being used to predict yield response to soil, weather, and management conditions. This requires that models released for the purpose of crop management under such conditions be evaluated for their abilities to accurately respond to those factors. The CROPGRO-Soybean model, V3.0 released in 1994 and V3.1 in 1996, is a dynamic crop growth model developed for such management purposes. It is substantially modified from the SOYGRO V5.42 growth model (Wilkerson et al. 1983; Jones et al. 1989) and our present objective is to evaluate the CROPGRO model for ability to accurately simulate soybean growth responses under a wide range of conditions.

The CROPGRO model is process-oriented and simulates crop carbon balance, crop and soil N balance, and soil water balance. CROPGRO is generic, in that it has one common FORTRAN code and will simulate three different legumes (soybean, peanut, and common bean). Early prototype versions of CROPGRO for simulating all three

M.J. Kropff et al. *(eds.): Applications of Systems Approaches at the Field Level, 113–133.*

legumes with added N balance were described by Hoogenboom et al. (1991, 1992, 1993). There is a read-in species file (*.SPE) for each legume that describes species characteristics such as tissue composition and partitioning traits, and includes sensitivity of processes to temperature, light, plant water deficit and plant N deficiency. There is also a read-in file describing typical cultivars within each legume (*.CUL), giving information on life cycle phases, vegetative traits, leaf traits, potential seed fill duration, seed size, and seed composition. Documentation on the species, ecotype, and cultivar files and the standard input/output files for the CROPGRO model and the other models of the Decision Support System for Agrotechnology Transfer (DSSAT V3.0) is available (Jones 1993; Jones et al. 1994; Hoogenboom et al. 1994a). A description of the processes in CROPGRO, modifications from prior model versions, and a model sensitivity analysis to temperature, planting date and row spacing for these three legumes is provided by Boote et al. (1996).

Certain component submodels were improved or added to CROPGRO so the model would be applicable over a wider range of conditions (e.g. row geometry effects, soil nitrogen effects, improved vegetative and reproductive development) than the original SOYGRO V5.42 (Jones et al. 1989), PNUTGRO V1.02 (Boote et al. 1989), and BEANGRO V1.00 and V1.01 (Hoogenboom et al. 1990, 1994b) models. See Boote et al. (1996) for more details on processes modified in CROPGRO. Improved response to row spacing, plant density, CO_2, and climatic conditions was facilitated by addition of two options for canopy photosynthesis: (i) the original daily canopy assimilation approach, modified from SOYGRO V5.42, and (ii) canopy photosynthesis computed at hourly time steps using leaf-level photosynthesis parameters and hedgerow light interception calculations (Boote and Pickering 1994). The leaf-level photosynthesis submodel allows more mechanistic response to climatic factors and CO_2, as well as facilitating model analysis relative to plant physiological factors. Response to soil N was included by addition of a complete soil-plant N balance, with N uptake and N_2-fixation, as well as N deficiency effects on photosynthesis, vegetative, and seed growth processes. The soil N balance and root N uptake processes use the same code as the CERES models and are described by Godwin and Jones (1991). CROPGRO has improved phenology prediction based on newly estimated relationships between development and temperature and daylength. It has a more flexible approach that allows crop development during various growth phases to be differentially sensitive to temperature, photoperiod, water deficit and N stresses.

The objective in this paper is to evaluate the CROPGRO model Version 3.1 (released in 1996) for its ability to simulate soybean growth, above-ground biomass, seed yield, flowering dates, and season lengths for specific experiments covering a wide range of conditions.

Material and methods

Soybean data (weather, soil characteristics, management conditions, growth, and yield) were assembled from various locations in the USA. Some of these data sets were used in developing the CROPGRO-Soybean model (and SOYGRO V5.42

Table 1. Experimental conditions, measured yield and measured biomass at maturity for seven treatments with Bragg soybean grown at Gainesville, FL, on Arredondo fine sand. Simulated yield and biomass are shown in parentheses. Row spacing was 91 cm in 1978 and 1979, 76 cm in 1984, and 30 cm in 1976. Plant population was 29.9, 47.0, 31.1 and 12.9 plants m^{-2} in 1978, 1979, 1984 and 1976 respectively. Nitrogen fertilizer at 40 kg N ha^{-1} was applied before planting in 1984.

Year	Treatment	Planting date (m-d)	Seasonal rainfall (mm)	Total irrigation (mm)	Mean temp. (°C)	Seed yield (kg ha^{-1})	Biomass at maturity (kg ha^{-1})
1976	Irrigated	5-5	776	75	26.12	3439 (3537)	6848 (7418)
1978	Irrigated	6-15	534	196	27.12	3041 (2826)	6068 (6012)
1978	Rainfed	6-15	534	0	27.12	1178 (1089)	3491 (3200)
1979	Irrigated	6-19	695	113	26.95	2891 (3178)	5781 (6400)
1979	Rainfed	6-19	695	0	26.95	2883 (3127)	5534 (6111)
1984	Irrigated	6-12	469	382	26.34	3732 (3420)	6689 (6552)
1984	Rep. stress	6-12	469	259	26.34	2114 (2281)	5867 (4889)

before it) and were not considered independent. Additional independent data were obtained to evaluate how well the CROPGRO-Soybean version simulates these experiments. Experimental data sets were grouped to illustrate model response to different factors, such as irrigation, planting date and the same cultivar grown across different sites and seasons. The range of treatment and environmental conditions (temperature, rainfall, cultivar, etc.) are described in Tables 1, 2 and 3.

The experiments described above were simulated with the CROPGRO-Soybean model, and simulated yields, above-ground biomass, leaf area index, flowering dates, beginning pod dates and maturity dates were compared to measured values. The soybean phenological development coefficients for different soybean maturity groups are those obtained by Grimm et al. (1993, 1994). For all simulations, the CROPGRO model was run with water balance, soil N uptake, N_2-fixation, plant N balance and leaf-level hedgerow photosynthesis options turned on. The Priestley-Taylor method was used to compute the evapotranspiration. The relative soil fertility parameter was set to 0.92 for all Florida experiments, 0.95 for North Carolina, and 1.00 for Iowa and Ohio. The 'soil fertility' parameter is a multiplicative modifier of canopy assimilation, that accounts for non-nitrogen soil fertility factors that the model is not sensitive to. The standard of 1.0 is set by the highly fertile soils in the midwestern USA.

The primary model aspects tested in these simulations include: carbon balance (ability to predict dry matter increase over time and final yields), soybean crop life cycle prediction, and to a limited extent, water balance, as affected by temperature, photoperiod, solar irradiance, rainfall, irrigation and cultural management inputs. Water balance was tested indirectly via water deficit effects in many studies, and soil

Table 2. Experimental conditions, measured and predicted phenology, yield and biomass for eight treatments with Williams soybean cultivar grown in Ohio, Iowa and Florida. Simulated values are shown in parentheses. Row spacing was 19 cm in Ohio, 70 cm in Iowa, and 35 cm in Florida. Plant population was 31.1, 41.2, 27.2, 19.2 and 62.0 plants m^{-1} for Ohio-88, Ohio-90, Iowa-88, Iowa-90, and the four Florida treatments respectively. Stages R3 and R7 are beginning pod and beginning maturity respectively (Fehr and Caviness 1977).

Year	Treatment	Planting date (m-d)	Seasonal rainfall (mm)	Total irrig. (mm)	Mean temp. (°C)	R3 date (days)	R7 date (days)	Seed yield ($kg\,ha^{-1}$)	Biomass at maturity ($kg\,ha^{-1}$)
1988	Ohio-88	5-01	370	557	20.61	91	145	3976	8090
						(89)	(140)	(3626)	(7673)
1990	Ohio-90	4-30	581	0	19.24	96	150	3149	7113
						(90)	(143)	(3222)	(7088)
1988	Iowa-88	5-11	381	0	23.26	71	129	2368	6617
						(75)	(126)	(2632)	(5721)
1990	Iowa-90	5-08	776	0	21.50	–	139	3168	6674
						(81)	(133)	(3576)	(7400)
1979	Fla-Rainfed*	3-15	413	0	22.86	48	86	1243	2856
						(48)	(96)	(1462)	(2630)
1979	Fla-Vstress*	3-15	372	115	22.94	48	92	1539	2831
						(48)	(97)	(1965)	(2998)
1979	Fla-Rstress*	3-15	393	131	22.98	48	86	2085	4068
						(48)	(98)	(2117)	(3575)
1979	Fla-Irrigated*	3-15	420	144	23.01	48	93	2474	3893
						(48)	(99)	(2906)	(4399)

*Independent treatments not previously evaluated with the model.

Table 3. Experimental conditions and measured yield and biomass for LeFlore (MG 6) and Coker 6847 (MG 7) soybean cultivars grown in 1988 and 1989 in on-farm tests on Troup fine sand soil at Greenwood, FL. Simulated yield and biomass are shown in parentheses. Row spacing was 91 cm. Plant population was 31.8 plants m^{-2} in 1988, and 29.9 and 25.1 plants m^{-2} in irrigated and rainfed plots respectively in 1989.

Year	Treatment	Planting (date) (m-d)	Seasonal rainfall (mm)	Total irrig. (mm)	Mean temp. (°C)	Seed yield ($kg\,ha^{-1}$)	Biomass at maturity ($kg\,ha^{-1}$)
1988	LeFlore-Irr*	6-03	372	188	26.60	2424	5329
						(2331)	(4635)
1989	CO-6847-Irr*	5-09	996	48	26.09	1835	5480
						(1888)	(5620)
1989	CO-6847-Rf*	5-09	996	0	26.09	1312	4728
						(1522)	(4573)

*Independent treatments not previously evaluated with the model.

water content predictions were directly tested in one case. Lastly, predictions of N_2-fixation and total N accumulation were evaluated for a 1984 experiment.

Results and discussion

Bragg soybean response to water availability and seasons in Florida

In our research at Gainesville, Florida, we conducted a number of experiments on Bragg soybean in several growing seasons that often differed widely in rainfall and drought conditions (Table 1). Although these data sets were used in developing both

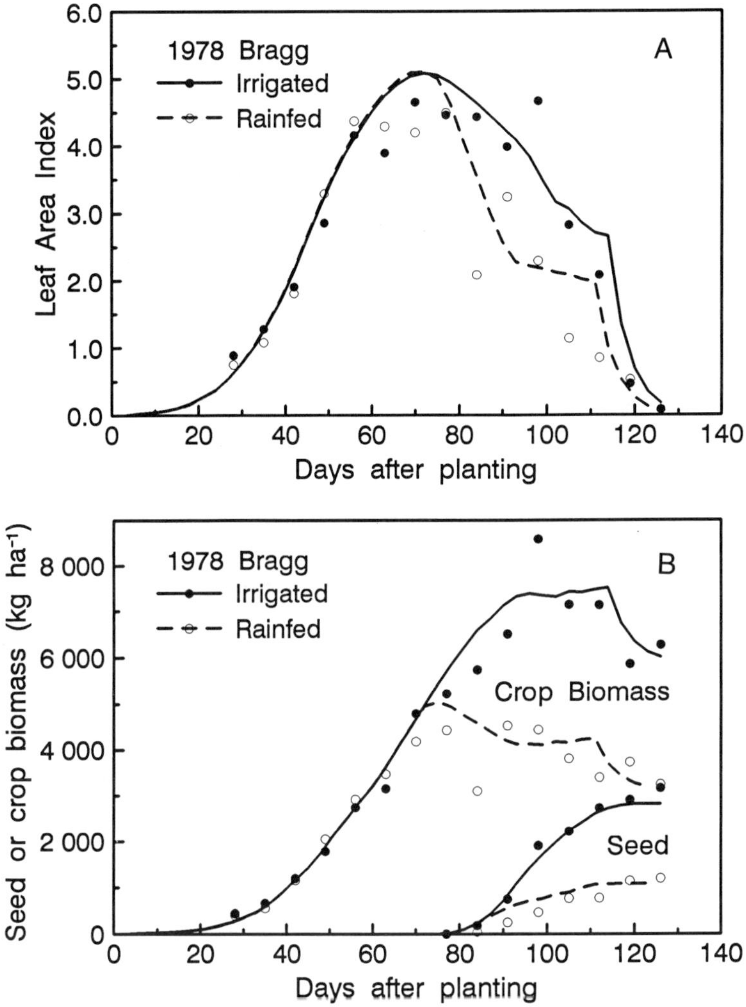

Figure 1. (A) Simulated (lines) and observed (symbols) leaf area index, and (B) simulated and observed crop biomass and seed mass, of Bragg soybean grown under irrigated versus rainfed conditions at Gainesville, FL, USA, 1978.

the SOYGRO V5.42 and CROPGRO-Soybean models, they are important to illustrate CROPGRO's in-season response to water deficit, as well as seed yield response to water supply. The 1978 season had an extended drought during podfill, with 2 cm of rainfall occurring during a 75-day period (1.6 cm occurring during the 58-day period from beginning pod (R3 on day 66) to 95% mature pods (R8)). On this Arredondo fine sand soil (a loamy siliceous hyperthermic Grossarenic Paleudult), a soybean crop frequently begins to wilt within 7 days without rain. Under extreme drought stress in 1978, the rainfed crop began to lose leaf area (Figure 1A), and ceased net dry matter

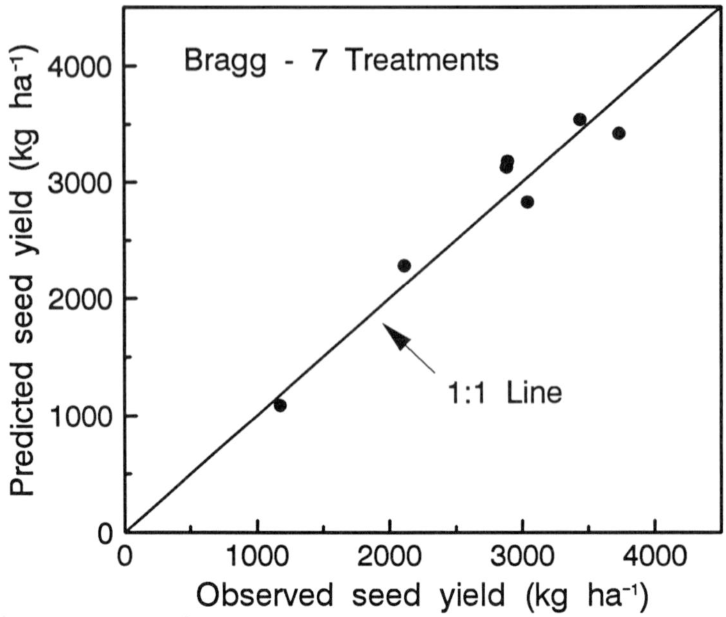

Figure 2. Comparison of simulated versus observed seed yield for seven Bragg soybean treatments grown in 1976, 1978, 1979 and 1984 under irrigated and rainfed conditions.

accumulation (Figure 1B). In the model, water-stress can cause leaf area loss up to 6% per day, as the ratio of actual plant transpiration to potential climatic transpiration declines from 1.0 to 0. The maximum cumulative leaf area loss to water deficit is capped at a stage-dependent value (generally 60%). Despite severe continuous water deficit, the rainfed soybean crop successfully set seeds and produced a yield of 1178 kg ha^{-1}, as it slowly depleted deep soil water to 2 m depth. With irrigation, a seed yield of 3041 kg ha^{-1} was produced. The CROPGRO model was able to successfully mimic the water deficit effects on leaf area loss, decreased dry matter accumulation, and decreased seed yield. The 1978 experiment was an important data set used in model calibration for the Gainesville site.

Additional growth analyses and final seed yield data were available for the Bragg cultivar grown in the 1976, 1979, and 1984 seasons (Table 1). The 1979 season had uniform rainfall and there were only minor differences between rainfed and irrigated yields. The 1984 season suffered a sustained period of water deficit during which 1.3 cm of rainfall was received over a 28-day period beginning shortly after the beginning pod (R3) stage. The rainfed crop in 1984 yielded 2114 kg ha^{-1} compared to 3732 kg ha^{-1} for the irrigated crop. The 1976 crop was planted much earlier than the other three experiments, and there was no rainfed treatment. The 1976 crop was grown in nearly equidistant spacing a 12.9 plants m^{-2} and produced a good yield.

An overall comparison of simulated versus observed seed yield is possible using data from the four years (1976, 1978, 1979 and 1984) with rainfed and irrigated treatments in three years. Figure 2 illustrates that CROPGRO-Soybean yield predictions over the seven treatment conditions were close to observed values and to the 1:1 line. Yield differences here are mostly attributed to water availability.

Williams soybean – predictions in Florida, Iowa and Ohio

Table 2 lists model simulations of experiments conducted on Williams soybean grown in 1979 at Gainesville, Florida (K.J. Boote, 1979, unpubl. data), in 1988 and 1990 at Ames, Iowa (R.M. Shibles, 1988, 1990, unpubl. data, Iowa State University), and in 1988 and 1990 at Wooster, Ohio (R.L. Cooper, 1988, 1990, unpubl. data, USDA-ARS, Wooster, OH). The Iowa and Ohio studies had been used to evaluate both model versions, but the Florida experiment was never before evaluated. It is important to note that the cultivar traits for Williams remain the same across years and locations, thus the growth and yield responses are attributed to differences in soil water availability, air temperature, daylength, solar radiation, planting date, row spacing, and plant population. There are no soil fertility differences between Iowa and Ohio, but the Florida site has a relative fertility effect of 0.92. The Iowa site was not irrigated. The 1988 season in Iowa was warm and the crop suffered severe water deficit, whereas the 1990 season was cool and wet. In Ohio, the 1988 crop was irrigated. In Ohio as well, the 1988 season was warm and 1990 was cool and wet. The 1990 season was about 2°C cooler than 1988, in both Iowa and Ohio (Table 2), with the greatest difference occurring during the period from flowering to beginning seed.

The CROPGRO model appeared to capture the warm versus cool season phenomena quite well at both locations. The predicted and observed LAI developed more slowly in the cool 1990 season in Ohio than in the 1988 season (Figure 3A), and accumulation of dry matter in crop and seed was also delayed in the cooler 1990 season (Figure 3B). Based on the Ohio irrigated conditions, it appears that 1988 was optimum for high yield, but the 1990 season was too cool for optimum yields. In Iowa, the 1988 season was also warmer than 1990, and the simulated and observed LAI and dry matter accumulation were initially delayed by cool weather in 1990 (Figure 4A and B). In 1988, however, the rainfall deficit eventually caused drought stress on the crop, resulting in decreased LAI (senescence of leaf area) and slower biomass accumulation after 80 days. Simulated and observed seed yields were lower in Iowa as a result of drought in 1988. Low solar radiation contributed to the slower growth in both locations in 1990 (21.90 and 17.93 $MJ\,m^{-2}\,d^{-1}$ in 1988 and 1990 in Iowa, and 19.64 and 17.84 $MJ\,m^{-2}\,d^{-1}$ in Ohio).

The 1979 Gainesville, Florida, experiment with the Williams cultivar represents a highly different environment in which the crop was planted on 15 March in 35-cm row spacing at 62 plants m^{-2}, and grown under short daylengths. Because the soil was sandy and the rainfall was sparse, we obtained a good yield response to water supply. The CROPGRO phenology model predicted much earlier flowering and a shorter life cycle in Florida than in Ohio or Iowa (Table 2). The short daylength (12.45 h compared to 14.73 and 14.92 h in Ohio and Iowa) caused early flowering (36 days), early pod formation (48 days), and resulted in only 9–10 nodes produced.

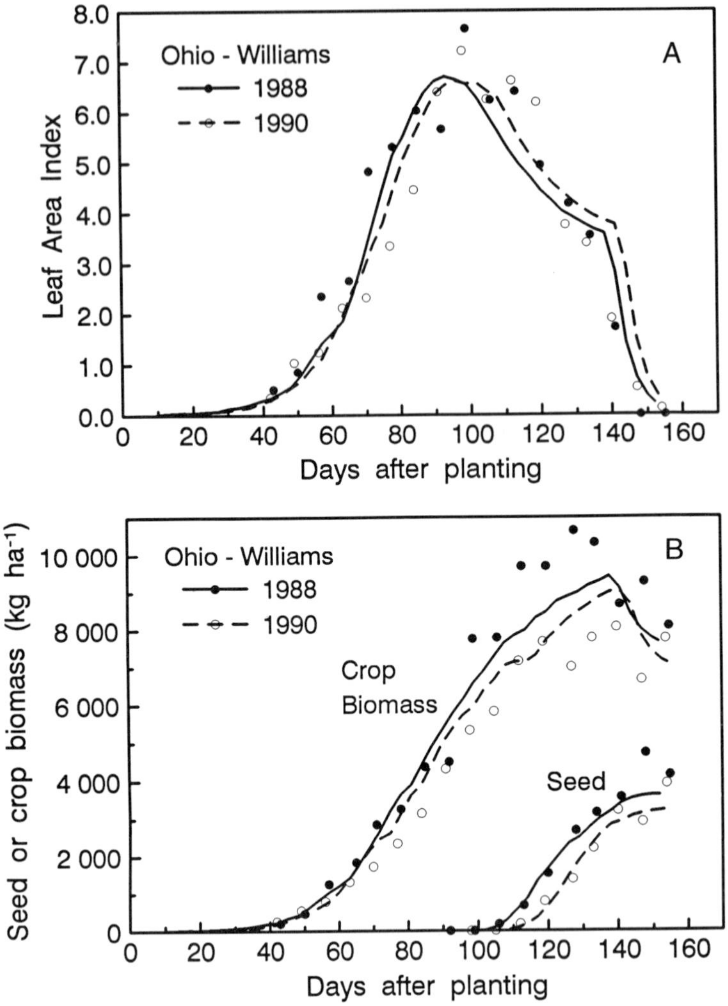

Figure 3. (A) Simulated (lines) and observed (symbols) leaf area index, and (B) simulated and observed crop biomass and seed mass, of Williams soybean cultivar grown in 1988 and 1990 at Wooster, OH, USA (data courtesy of R.L. Cooper and American Soybean Association).

Predictions of flowering and beginning pod stage (R3) were close to observed (Table 2, Figure 5A). The 1979 experiments had not been evaluated with either CROPGRO or SOYGRO and the coefficients for Williams had been derived from other data sets. Maturity predictions were also good (Figure 5B), but the model did not progress rapidly enough from beginning pod to physiological maturity at the Florida location. The somewhat longer life cycle predictions caused the model to somewhat over-predict yields in Gainesville, but rainfall and irrigation effects on yield were well predicted (Table 2). When considered as a whole, the ability to simulate the Williams

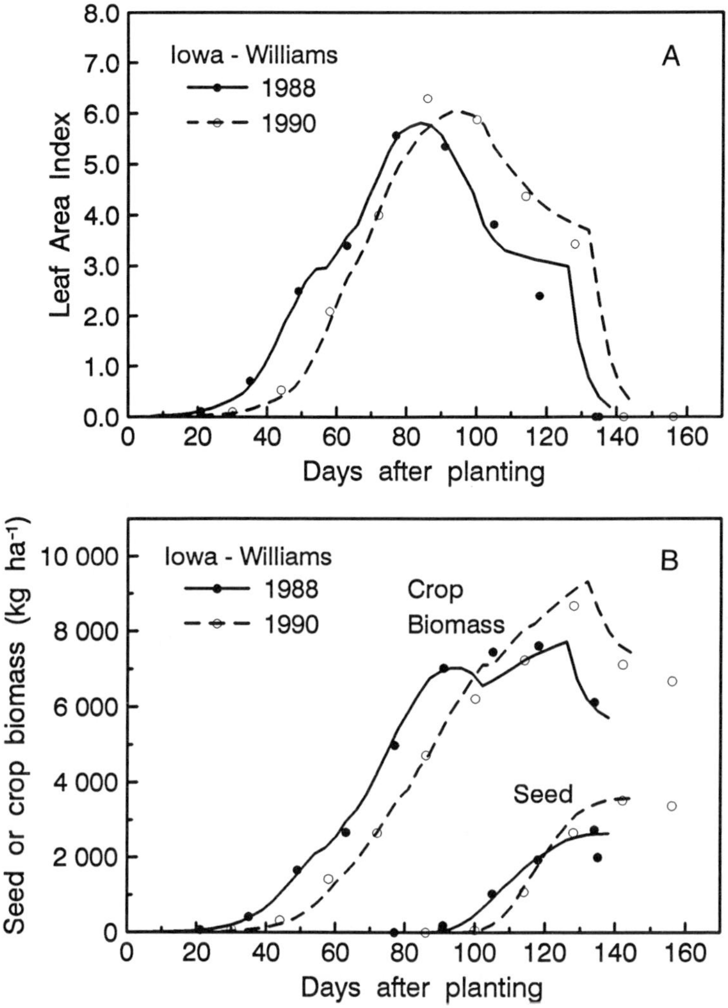

Figure 4. (A) Simulated (lines) and observed (symbols) leaf area index, and (B) simulated and observed crop biomass and seed mass, of Williams soybean cultivar grown in 1988 and 1990 under rainfed conditions at Ames, Iowa (data courtesy of R.M. Shibles and American Soybean Association).

life cycle and yield in Florida, Iowa and Ohio illustrates good transportability of the soybean crop model and its phenological and growth coefficients across climatic regions.

These eight available treatments on Williams soybean (2 years in Iowa, 2 years in Ohio and 4 irrigation treatments in Florida) were plotted together to compare simulated seed yield versus observed yield (Figure 6A). The simulated results are slightly above the 1:1 line. A plot of simulated biomass versus observed biomass shows predictions to be close to the 1:1 line (Figure 6B). Differences in Figure 6 are attributed

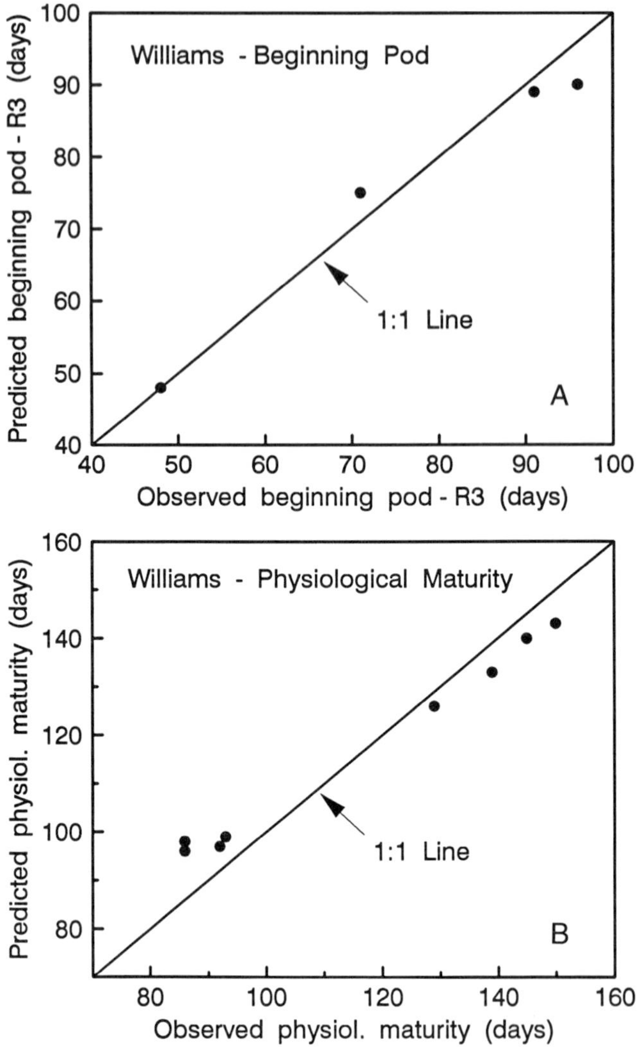

Figure 5. Comparison of simulated versus observed: (A) days to beginning pod (R3, Fehr and Caviness 1977), and (B) days to physiological maturity (R7), for Williams soybean cultivar grown under eight different treatment conditions in Florida, Iowa and Ohio. Phenology coefficients are from Grimm et al. (1993, 1994) and these phenology data are independent data not used for model calibration.

to combinations of water availability, temperature and life cycle (the latter caused by shorter daylength in Florida). Simulated versus observed values for seed yield and biomass fell along the 1:1 line regardless of whether the cause for lower yield was shorter daylength in Florida (irrigated treatment), or water supply occurring in Iowa or Florida. This was a good test because the treatment spanned nearly two-thirds of the range from zero to maximum yield and biomass.

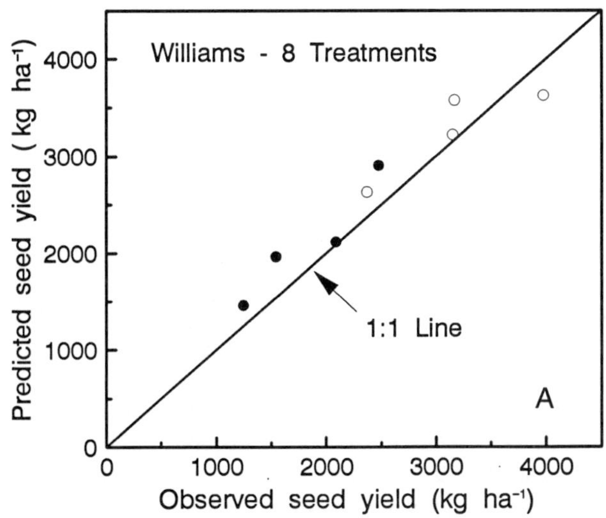

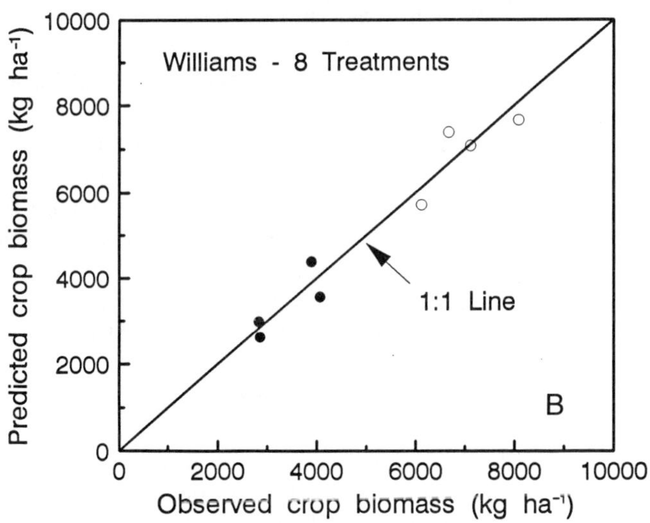

Figure 6. Comparison of simulated versus observed: (A) seed yield, and (B) total crop biomass at maturity, for Williams soybean grown under eight different treatment conditions in Florida, Iowa and Ohio. Closed circles indicate the four independent treatments.

Model 'validation' tests

Additional independent data were obtained to evaluate how well the new CROPGRO-Soybean version simulates growth and yield in new locations and environments. Five experiments (Williams cultivar in Florida in 1979, two on-farm trials, and two planting date trials in North Carolina) represent independent data never used in model calibration and illustrate the ability of CROPGRO to predict growth and yield in new locations and environments.

Williams cultivar in Florida. The 1979 Williams soybean study in Gainesville was never before used for soybean model calibration, and, thus, constitutes an independent model test. Except for delayed maturity (by 5 to 12 days), the CROPGRO-Soybean model satisfactorily predicted phenology of Williams grown at a location in which the cultivar is not normally grown. CROPGRO also satisfactorily predicted seed yield response to irrigation treatments on a sandy soil.

On-farm tests with LeFlore and Coker 6847 cultivars in Greenwood, Florida. In 1988 and 1989, we conducted on-farm tests with a soybean grower in the panhandle of Florida. We set up a LI-COR LI-1200 minimum weather station and collected solar radiation, rainfall, maximum temperature, and minimum temperature. The soil type was taken from county soil survey maps. Percent sand, silt, clay, and organic matter and other information was taken from the soil survey, and used with the DSSAT V3.0 program (Tsuji et al. 1994) to estimate soil water supplying characteristics. We also collected soybean growth and soil water data at 3 to 4 week intervals. CROPGRO had never been run with these data sets, so this comparison also constitutes an independent model test. We did not have cultivar characteristics for LeFlore or Coker 6847, but their maturity falls into MG (Maturity Group) 6 and 7 respectively. Therefore, we used our generic cultivar traits for MG 6 and 7 available with the model (values had been solved by Grimm et al. (1993, 1994) for other cultivars in MG 6 and 7). We used the same soil fertility parameter (0.92) as for Gainesville's Arredondo fine sand, because the Troup soil is also a fine sand. Rooting traits were collected by sampling root length density versus depth on soybean grown on the Troup soil.

In the 1988 trial at Greenwood, Florida, the model somewhat underpredicted biomass accumulation, but accurately predicted seed dry matter accumulation (Figure 7). The final simulated seed yield was 2331 kg ha^{-1} compared to observed yield of 2424 kg ha^{-1} (Table 3). The generic MG 6 coefficients allowed CROPGRO to correctly predict the observed dates of R1 (beginning bloom) and R3 (beginning pod) stages to within 1 day, and to predict R7 (beginning maturity) only 4 days early (stages according to Fehr and Caviness 1977). Model predictions of soil water status were quite satisfactory when compared to measured soil water content (Figures 8A–C), and both confirm that the crop suffered soil water deficit during the middle of the season. The 1988 season in Greenwood, Florida, was dry (372 mm of rainfall) and the grower apparently did not fully satisfy the crop's water needs despite irrigating 8 times with a total of 188 mm.

In the 1989 test, CROPGRO satisfactorily predicted the time course of crop biomass and seed (Figure 7B), and slightly overpredicted final seed yield (Table 3). The abundant rainfall was unevenly distributed and the crop suffered significant water deficit during seedfill, even on the irrigated part of the field.

Planting date-cultivar studies in North Carolina. Data were collected on yield and phenology of three soybean cultivars, Forrest (MG 5), Centennial (MG 6), and Ransom (MG 7), planted on eight sowing dates during 1987 at Rocky Mount, North Carolina and on six sowing dates during 1988 at Clayton, North Carolina (Tables 4

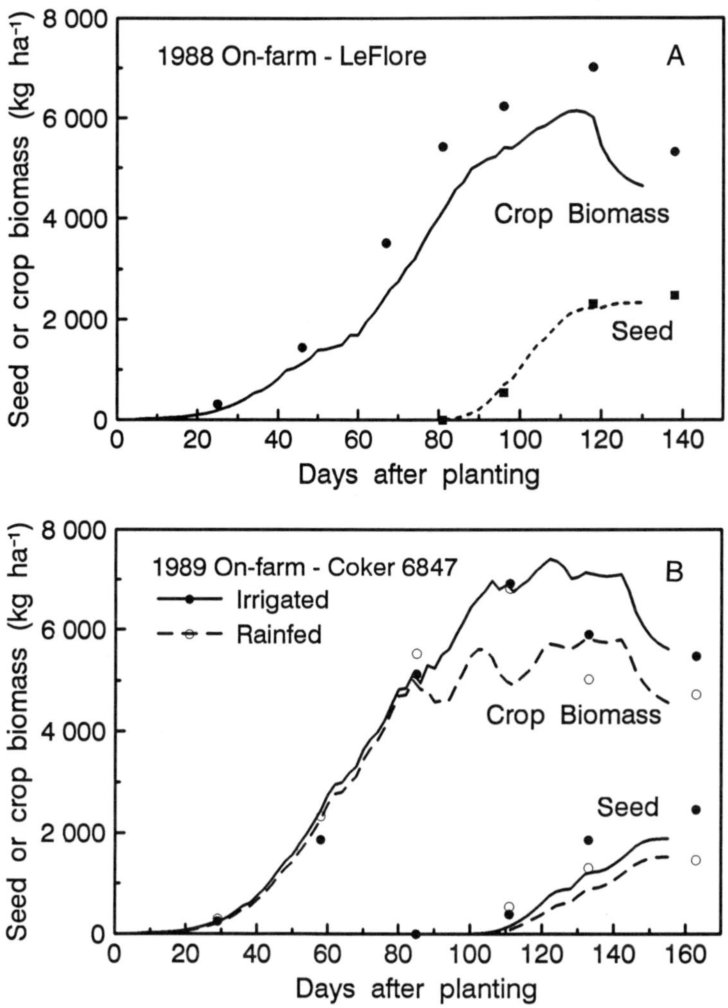

Figure 7. Simulated and observed crop biomass and seed mass of (A) LeFlore soybean cultivar grown in 1988 under irrigated conditions, and (B) Coker 6847 cultivar grown in 1989 under irrigated and rainfed conditions, in on-farm trials at Greenwood, FL.

and 5) (G.G. Wilkerson, unpubl. data, 1987, 1988). These experiments represent independent data not used in development of the CROPGRO model, and were very helpful to evaluate CROPGRO's ability to accurately respond in phenological timing and yield response to planting date.

The CROPGRO model satisfactorily predicted yield response to planting date for the 1987 study at Rocky Mount, but underestimated the high yields (nearly 4000 kg ha^{-1}) obtained in early to mid-season dates for the 1988 season at Clayton, North Carolina (Figure 9). In both seasons, the model was satisfactory in its relative yield responses to planting date, and showed that early planting was best, with rapid

Table 4. Observed and predicted reproductive stages, observed and predicted seed yield for three soybean cultivars sown at eight planting dates at Rocky Mount, NC, 1987 (unpubl. data, G.G. Wilkerson, North Carolina State University, Raleigh, 1987). Stages R1, R5 and R7 are beginning bloom, beginning seed and beginning maturity respectively (Fehr and Caviness 1977).

Planting date (day number)	R1 stage Obs. (DAS)*	R1 stage Sim. (DAS)	R5 stage Obs. (DAS)	R5 stage Sim. (DAS)	R7 stage Obs. (DAS)	R7 stage Sim. (DAS)	Seed yield Obs. ($kg\,ha^{-1}$)	Seed yield Sim. ($kg\,ha^{-1}$)
Forrest (MG5)								
128	51	56	94	92	135	133	2229	2158
138	51	56	87	88	132	127	2912	2046
159	47	51	78	76	123	115	2932	2269
173	44	46	68	70	113	108	2943	2410
182	44	43	65	65	107	103	2632	2073
196	39	38	61	58	100	97	1947	1507
208	39	34	60	52	99	93	1637	945
226	34	32	56	55	84	98	459	166
Centennial (MG6)								
128	65	63	119	103	162	145	2740	2287
138	63	62	109	99	150	140	2842	2365
159	54	56	88	85	129	125	2982	2320
173	54	50	80	76	116	116	2533	2343
182	48	46	72	71	111	111	2449	2199
196	44	40	65	62	101	103	1905	1721
208	39	36	60	56	95	98	1377	1084
226	35	31	56	58	131	–	463	171
Ransom (MG7)								
128	70	70	119	109	166	150	2233	2479
138	67	69	109	103	157	143	2526	2358
159	59	60	88	88	136	127	2760	2402
173	53	53	81	78	123	117	2400	2426
182	49	48	72	72	114	110	1747	2243
196	44	42	65	63	105	101	1426	1734
208	39	37	60	56	99	95	1218	1100
226	35	31	57	53	84	94	244	198

*DAS: days after sowing.

drop-off in yield for planting dates after early June (after day 160), approaching nearly zero yield for sowing in early August (after day 220).

Dates of beginning bloom and physiological maturity were satisfactorily predicted by the phenology submodel of CROPGRO (Table 5, Figures 10A and B) over a wide range in sowing dates for four cultivars, Essex, Forrest, Centennial and Ransom in 1988. Results for 1987 (not shown) were also well predicted. For three of the cultivars, Figure 10 represents an independent test, because the cultivar coefficients were derived from other trials by Grimm et al. (1993, 1994) using a wide range of experiments (locations and planting dates) to solve for the coefficients. Forrest phenology data did constitute a small part of the data Grimm used. Based on Figure 10 (also true for 1987), the phenology module appears to have a slight bias because it predicted maturity too early for the long life cycle cases (earliest planting dates). We possibly need to consider additional delaying effects of both long days and warm temperatures on the soybean life cycle, particularly after flowering and before seed growth.

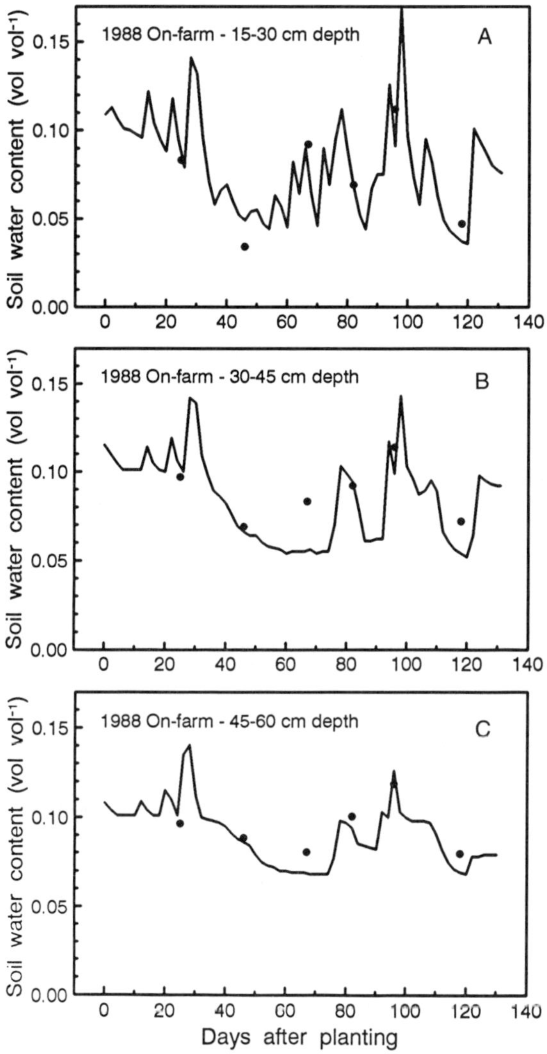

Figure 8. Simulated and observed volumetric soil water content at (A) 15–30 cm, (B) 30–45 cm, and (C) 45–60 cm depths of Troup fine sand soil upon which LeFlore soybean cultivar was grown under irrigated conditions in farmer trial at Greenwood, FL, 1988.

A comparison of CROPGRO-Soybean relative to the SOYGRO V5.42 model

Our primary reason for modifying and developing the CROPGRO model was to add new features that SOYGRO V5.42 did not address, such as soil N uptake, N_2-fixation, leaf-level photosynthesis for climate change studies, and pest-coupling aspects. Failure of SOYGRO V5.42 to adequately predict growth and yield played only a minor role in encouraging these modifications. Pest-coupling aspects and climate

Table 5. Observed and predicted reproductive stages, observed and predicted seed yield for three soybean cultivars sown at six planting dates at Clayton, NC, 1988 (unpubl. data, G.G. Wilkerson, North Carolina State University, Raleigh, 1988). Stages R1, R5 and R7 are beginning bloom, beginning seed and beginning maturity, respectively (Fehr and Caviness 1977).

Planting	R1 stage		R5 stage		R7 stage		Seed yield	
date	Obs.	Sim.	Obs.	Sim.	Obs.	Sim.	Obs.	Sim.
(day number)	(DAS)*	(DAS)	(DAS)	(DAS)	(DAS)	(DAS)	($kg\,ha^{-1}$)	($kg\,ha^{-1}$)
Forrest (MG5)								
109	75	72	114	108	161	152	4431	3198
130	62	62	101	94	144	135	4082	2946
152	53	54	83	81	123	120	3913	3065
176	43	45	66	67	108	105	3389	2230
193	37	38	59	58	98	97	2229	1625
214	33	32	54	50	94	90	889	754
Centennial (MG6)								
109	87	79	135	120	180	163	4074	3323
130	72	69	116	104	159	146	3821	3158
152	62	59	95	90	137	129	3580	2974
176	48	48	74	73	114	113	2924	2208
193	40	41	64	63	101	103	1996	1802
214	33	33	54	53	94	95	1067	896
Ransom (MG7)								
109	84	86	131	126	180	167	4269	3168
130	72	75	112	109	159	148	3060	3006
152	62	64	92	93	137	131	3185	2877
176	48	51	73	76	113	113	2902	2234
193	40	42	61	64	102	102	2208	1890
214	33	34	54	53	94	91	1100	923

*DAS: days after sowing.

change evaluations of CROPGRO have been previously described; however, the N_2-fixation and N accumulation features of CROPGRO have not been adequately tested and presented. An example is given in Figure 11 to illustrate that the CROPGRO model satisfactorily simulates the N_2-fixation observed by De Vries (1986) in a 1984 study on irrigated Bragg soybean. The model was not calibrated to give this result, but simulated mineral N uptake on this fine sand soil was small (39 $kg\,ha^{-1}$ for 1984), thus resulting in good nodulation and most of the plant N needs were apparently supplied by N_2-fixation. These model features are needed to evaluate uptake of 'unused' mineral N remaining after non-legume crops and subsequently to predict any added N contribution of legumes to subsequent non-legumes in rotations.

Statistical evaluation of prediction ability is shown in Table 6 for CROPGRO versus SOYGRO. When tested with 'calibration' data such as Bragg treatments at the site of initial model development (Gainesville), both models performed well. However, the CROPGRO model appeared to be more broadly applicable to other regions, while SOYGRO appeared to be a more 'site-specific' model because it performed less well for diverse sites. For example, CROPGRO generally had lower RMSE, higher index of agreement, lower percent systemic error, and better slope and intercept values for comparisons involving Williams across Iowa, Ohio and Florida, or the four cultivars at North Carolina. This was particularly apparent for the

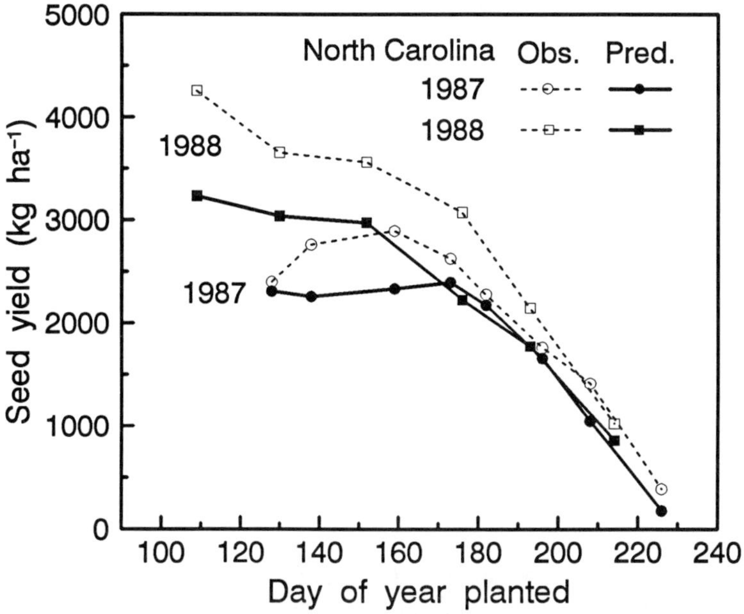

Figure 9. Predicted and observed seed yield of three soybean cultivars in response to sowing date, grown at Rocky Mount, NC, in 1987, and at Clayton, NC, 1988 (unpubl. data, G.G. Wilkerson, North Carolina State University, Raleigh, 1987, 1988).

phenology predictions where the RMSE values for CROPGRO were usually less than half that of the SOYGRO phenology predictions. Although SOYGRO satisfactorily predicted seed yield of Williams, it failed to adequately predict yield response to planting date in North Carolina. These problems were attributed to SOYGRO's problems in predicting phenology, such as incorrectly simulating slow delayed maturity (and too high yields) for later planting dates. CROPGRO satisfactorily accounted for planting date effects on yield and phenology and consistently had better slope and intercept values, lower RMSE values, and higher r^2 values. CROPGRO performed better than the SOYGRO model, particularly when compared with independent data such as the North Carolina planting date studies.

Conclusions

In this paper, we have evaluated the new CROPGRO-Soybean model for its ability to simulate soybean growth, above-ground biomass, seed yield, flowering dates and season lengths using data collected from 60 different soybean crops grown over a wide range of conditions. The model responds satisfactorily to varying water inputs, and accurately simulates growth and yield response of both determinate Bragg and indeterminate Williams across multiple years and climatic environments. Particularly for the Williams cultivar, we were able to demonstrate substantial transportability of crop and cultivar coefficients across climatic regions (Florida, Iowa and Ohio) and

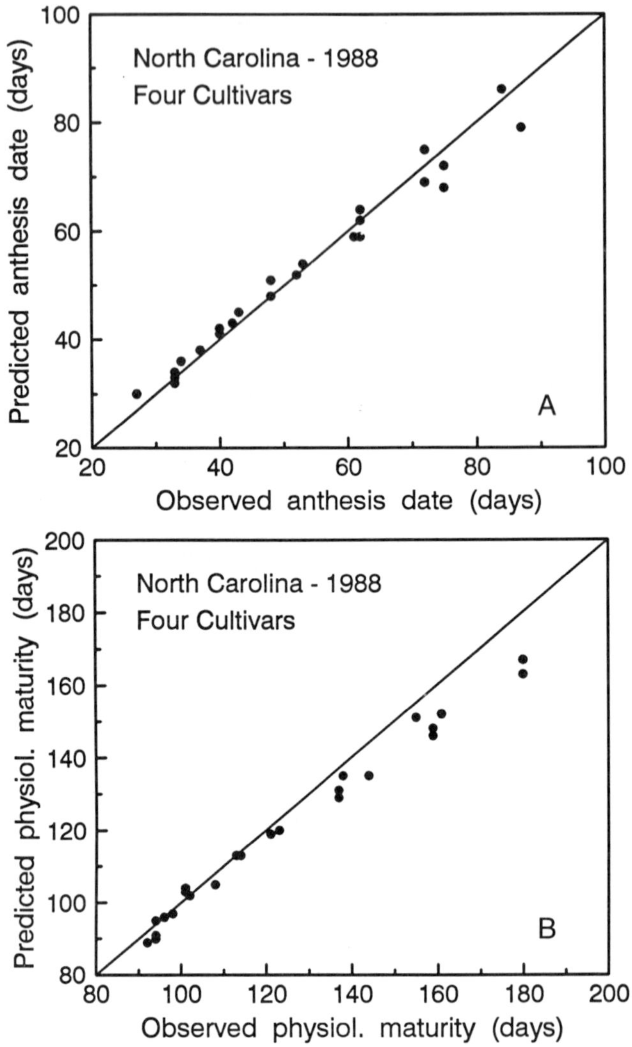

Figure 10. (A) Predicted versus observed anthesis (R1, Fehr and Caviness 1977) and (B) predicted versus observed physiological maturity dates (R7) for four soybean cultivars in response to sowing date, at Clayton, NC, 1988 (unpubl. data, G.G. Wilkerson, North Carolina State University, Raleigh, 1988).

across warm, droughty years to cool, wet seasons. CROPGRO-Soybean appeared to predict crop performance satisfactorily in three independent tests (planting date trial in North Carolina, Williams cultivar under irrigation variation in Florida, and two years of on-farm trials in Greenwood, Florida). Planting date studies in North Carolina showed the model's ability to predict typical yield response to sowing date. We conclude that CROPGRO-Soybean gives good predictions of crop behaviour under a wide range of environmental conditions.

Table 6. Evaluation of CROPGRO-Soybean and the SOYGRO V5.42 model with comparisons of simulated versus observed seed yield, total biomass at harvest, and phenology stage predictions for experimental data sets. Statistics include intercept (*a*), slope (*b*), and r^2 of the regression line; root mean square error of the predicted minus observed (RMSE); percent of error that is systematic (%S); and index of agreement (*d*) (Willmott 1982)

Trait/model	Regression coefficients					
tested	*a*	*b*	r^2	RMSE	%S	*d*
Bragg experiments (n=7)						
Seed yield (kg ha^{-1})						
CROPGRO	147.1	0.956	0.927	217.42	4.0	0.981
SOYGRO V5.4	369.0	0.867	0.894	259.73	16.5	0.970
Biomass at harvest (kg ha^{-1})						
CROPGRO	−673.2	1.125	0.825	548.25	6.0	0.941
SOYGRO V5.4	−242.2	1.028	0.920	323.11	6.8	0.976
Williams experiments (n=8)						
R3, Beginning pod (days)						
CROPGRO	4.4	0.923	0.985	2.83	33.8	0.998
SOYGRO V5.4	22.7	0.686	0.992	6.58	96.6	0.961
R7, Beginning maturity (days)						
CROPGRO	28.5	0.738	0.993	7.28	94.9	0.975
SOYGRO V5.4	43.1	0.591	0.962	11.29	92.4	0.928
Seed yield (kg ha^{-1})						
CROPGRO	597.9	0.836	0.921	312.47	56.0	0.961
SOYGRO V5.4	−129.7	0.949	0.940	319.70	65.9	0.964
Biomass at harvest (kg ha^{-1})						
CROPGRO	72.2	0.982	0.952	423.74	0.9	0.988
SOYGRO V5.4	−423.0	1.001	0.951	603.18	47.9	0.976
North Carolina experiments						
R1, Beginning bloom (days)						
1987 (n=30)						
CROPGRO	−6.3	1.113	0.952	2.74	25.1	0.982
SOYGRO V5.4	25.6	0.433	0.633	6.49	76.9	0.792
1988 (n=24)						
CROPGRO	−4.8	0.907	0.978	2.86	31.9	0.992
SOYGRO V5.4	22.3	0.516	0.925	9.43	92.8	0.876
R7, Beginning maturity (days)						
1987 (n=29)						
CROPGRO	23.2	0.752	0.940	7.39	70.1	0.962
SOYGRO V5.4	82.5	0.293	0.198	18.87	61.9	0.668
1988 (n=24)						
CROPGRO	13.3	0.843	0.989	6.79	85.9	0.983
SOYGRO V5.4	59.1	0.493	0.638	17.85	65.5	0.842
Seed yield (kg ha^{-1})						
1987 (n=24)						
CROPGRO	50.1	0.844	0.843	423.21	50.5	0.927
SOYGRO V5.4	411.2	1.069	0.684	813.95	46.7	0.825
1987 (n=18)						
CROPGRO	191.7	0.731	0.924	710.84	89.3	0.879
SOYGRO V5.4	3256.1	−0.428	0.446	1919.76	92.5	0.206

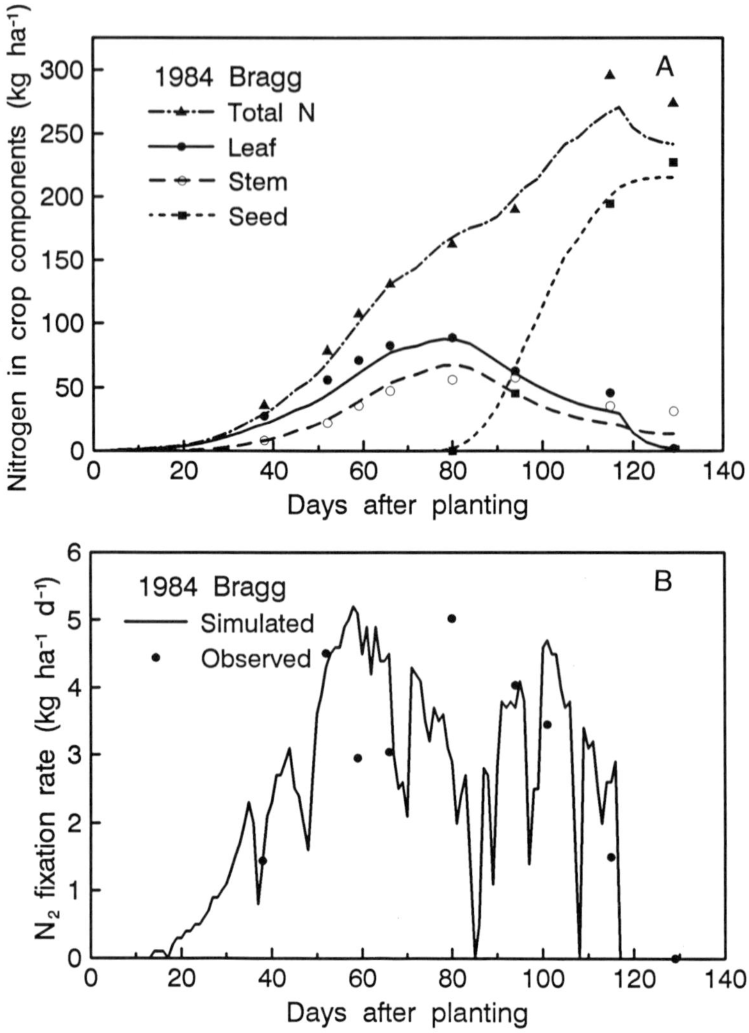

Figure 11. Comparison of simulated (lines) and observed (symbols): (A) nitrogen accumulation in plant components, and (B) N_2-fixation rate in irrigated Bragg soybean grown at Gainesville, FL, 1984. Observed data from De Vries (1986).

Acknowledgements

The American Soybean Association is acknowledged, for their overall general support of soybean modelling activities, and specifically, for their support for collecting data for the Gainesville 1978 and 1979 experiments and the Ohio and Iowa experiments in 1988 and 1990.

References

Boote K J, Pickering N B (1994) Modeling photosynthesis of row crop canopies. Hort. Science 29: 1423–1434.

Boote K J, Jones J W, Hoogenboom G, Wilkerson G G, Jagtap S S (1989) PNUTGRO V1.02, Peanut Crop Growth Simulation Model, User's Guide. Florida Agric. Exp. Sta., Journal No. 8420. University of Florida, Gainesville, FL, USA. 76 p.

Boote K J, Jones J W, Hoogenboom G, Pickering N B (1996) CROPGRO model for grain legumes. In Tsuji G Y, Hoogenboom G, Thornton P K (Eds.) International Benchmark Sites Network for Agrotechnology Transfer: A systems approach to research and decision making. Kluwer Academic Publishers, Dordrecht, The Netherlands (in press).

De Vries J D (1986) Water relations and nitrogen fixation of three grain legumes during water stress. MSc thesis, University of Florida, Gainesville, FL, USA.

Fehr W R, Caviness C E (1977) Stages of soybean development. Iowa State Agric. Exp. Station, Special Report No. 80, Ames, Iowa, USA. 12 p.

Godwin D C, Jones C A (1991) Nitrogen dynamics in soil-plant systems. Pages 287–321 in Hanks R J, Ritchie J T (Eds.) Modeling soil and plant systems. American Society of Agronomy (ASA), Monograph 31, Madison, WI, USA.

Grimm S S, Jones J W, Boote K J, Hesketh J D (1993) Parameter estimation for predicting flowering date of soybean cultivars. Crop Sci. 33:137–144.

Grimm S S, Jones J W, Boote K J, Herzog D C (1994) Modeling the occurrence of reproductive stages after flowering for four soybean cultivars. Agron. J. 86:31–38.

Hoogenboom G, White J W, Jones J W, Boote K J (1990) BEANGRO V1.0: Dry bean crop growth simulation model. User's Guide. Florida Agric. Exp. St., Journal No. N-00379. University of Florida, Gainesville, FL, USA. 120 p.

Hoogenboom G, Jones J W, Boote K J (1991) Predicting growth and development of grain legumes with a generic model. Paper No. 91-4501. Am. Soc. Agr. Engr., St. Joseph, MI 49085, USA.

Hoogenboom G, Jones J W, Boote K J (1992) Modeling growth, development, and yield of grain legumes using SOYGRO, PNUTGRO, and BEANGRO: A review. Trans. ASAE 35:2043–2056.

Hoogenboom G, Jones J W, Boote K J, Bowen W T, Pickering N B, Batchelor W D (1993) Advancement in modeling grain legume crops. Paper No. 93-4511. Am. Soc. Agr. Engr., St. Joseph, MI 49085, USA.

Hoogenboom G, Jones J W, Wilkens P W, Batchelor W D, Bowen W T, Hunt L A, Pickering N B, Singh U, Godwin D C, Baer B, Boote K J, Ritchie J T, White J W (1994a) Crop models. Pages 95-244 in Tsuji G Y, Uehara G, Balas S (Eds.) DSSAT Version 3, Vol. 2. University of Hawaii, Honolulu, HI, USA.

Hoogenboom G, White J W, Jones J W, Boote K J (1994b) BEANGRO, a process-oriented dry bean model with a versatile user interface. Agron. J. 86:182–190.

Jones J W (1993) Decision support systems for agricultural development. Pages 459–471 in Penning de Vries F W T, Teng P S, Metselaar K (Eds.) Systems approaches for agricultural development. Kluwer Academic Publishers, Dordrecht, The Netherlands.

Jones J W, Boote K J, Hoogenboom G, Jagtap S S, Wilkerson G G (1989) SOYGRO V5.42, soybean crop growth simulation model. User's guide. Florida Agric. Exp. Sta., Journal No. 8304, University of Florida, Gainesville, FL, USA.

Jones J W, Hunt L A, Hoogenboom G, Godwin D C, Singh U, Tsuji G Y, Pickering N B, Thornton P K, Bowen W T, Boote K J, Ritchie J T (1994) Input and output files. Pages 1–94 in Tsuji G Y, Uehara G, Balas S (Eds.) Decision Support System for Agrotechnology Transfer (DSSAT) Version 3, Vol. 2. University of Hawaii, Honolulu, HI, USA.

Tsuji G Y, Uehara G, Balas S (Eds.) (1994) Decision Support System for Agrotechnology Transfer (DSSAT) Version 3, Volume 1. University of Hawaii, Honolulu, HI, USA. 164 p.

Wilkerson G G, Jones J W, Boote K J, Ingram K T, Mishoe J W (1983) Modeling soybean growth for crop management. Trans. ASAE 26:63–73.

Willmott C J (1982) Some comments on the evaluation of model performance. Bull. Amer. Meteorological Soc. 63:1309–1313.

Adaptation of the CROPGRO model to simulate the growth of field-grown tomato

J.M.S. SCHOLBERG[1], K.J. BOOTE[2], J.W. JONES[3] and B.L. McNEAL[1]

[1]*Soil and Water Science Department, 106 Newell Hall, P.O. Box 110510, University of Florida, Gainesville, FL 32611-0510, USA*

[2]*Agronomy Department, 304 Newell Hall, P.O. Box 110500, University of Florida, Gainesville, FL 32611-0510, USA*

[3]*Agricultural and Biological Engineering Department, 110 Frazier Rogers Hall, P.O. Box 110570, University of Florida, Gainesville, FL 32611-0510, USA*

Key words: crop modelling, CROPGRO, tomato, model calibration

Abstract

Modelling the growth of field-grown tomato (*Lycopersicon esculentum* Mill.) should assist growers and extension workers throughout the world to outline optimal crop management strategies for specific locations and production systems. In previous use of a greenhouse tomato model (TOMGRO), effects of nutrient and water stress on the growth of field-grown tomato were not accounted for, and it was decided to modify a more generic growth model (CROPGRO) that does address these issues. It proved feasible to capture growth and production features typical of field-grown tomato by modifying existing parameter files previously used for peanut (*Arachis hypogaea* L.), without changing the FORTRAN code or model structure. Parameter estimation and model calibration involved use of data sets for three seasons of field-grown tomato at the Gulf Coast Research and Education Centre (GCREC) in Bradenton, Florida, USA. Photosynthetic parameters were calculated by comparison to the TOMGRO model. Some of the more important parameter modifications, and their implications with respect to modelling results, are presented. Results from these initial modifications using the CROPGRO generic crop model show both the versatility and the robustness of the model. The general procedure presented here may also be employed as a 'blueprint' for future CROPGRO adaptation to other vegetable crops.

Introduction

For the past few decades, researchers have developed dynamic crop-growth simulation models. These models are generally process-oriented and based on the flow of carbon, water, and nutrients within the soil-plant-atmosphere continuum (Hoogenboom et al. 1992). This allows the simulation of plant growth under well-defined environmental conditions (Jones and Ritchie 1992). Over time some of these models have evolved to fairly robust versions that can provide inputs to management decisions for commercial crops like peanut (Boote et al. 1989) and tomato (Jones et al. 1989).

With the establishment of the International Benchmark Sites Network for Agrotechnology Transfer (IBSNAT), research efforts of a number of modelling groups were linked, resulting in the development of the Decision Support System for Agrotechnology Transfer (DSSAT) modelling package (Hoogenboom et al. 1992). A generic crop model (CROPGRO) included in this DSSAT package allowed

M.J. Kropff et al. *(eds.): Applications of Systems Approaches at the Field Level, 135–151.*

the simulation of plant growth over a wide range of field crops and production systems.

A number of models have been developed for tomato in order to predict harvesting date (Wolf et al. 1986), dry matter production (Bertin and Heuvelink 1993), or the effect of CO_2 enrichment on plant production (Jones et al. 1989). However, the use of each model has usually been confined to a small group of researchers, and model application has been limited to either greenhouse and processing tomato.

The TOMGRO model (Jones et al. 1991), a growth model developed for greenhouse tomato, was designed to respond to dynamically changing temperature, solar radiation and CO_2 concentrations, assuming single-stemmed plants and no water or nutrient stress. The large number of photosynthetic and plant growth measurements used during its calibration resulted in a well-calibrated model for greenhouse tomato over a wide range of environmental conditions. However, McNeal et al. (1995) found that TOMGRO did not adequately describe the growth of field-grown tomato. Typically, node formation of field-grown tomato ceases after about twenty nodes whereas, for greenhouse tomato, node formation continues throughout the entire growth period (Scholberg 1994). The profuse branching typically observed for field crops also results in a more rapid build-up of leaf canopy mass, while the more deterministic growth pattern also causes an earlier and more rapid decline of the leaf canopy. Similar plant growth patterns have been observed for other field crops including peanut (Boote et al. 1986). It was thus hypothesized that it may be more appropriate to adapt an existing growth model for field crops to describe the growth of tomato, especially if that model could also account for the effects of both water and nutrient stress on plant growth. Since canopy development for field-grown tomato resembles that of peanut, modifying an existing crop model for peanut was considered to be a good starting point for the development of a tomato growth model. The CROPGRO model was chosen because of its generic nature and its wide use and acceptance in the modelling and agricultural research community. User-friendly interfaces and graphics output, the use of standardized input files, the quality of its documentation, and the linking of the growth routine to both water and nutrient balances also appeared to be useful features of the CROPGRO model (Hoogenboom et al. 1992). Such features should facilitate the adaptation and/or testing of the CROPGRO model for field-grown tomato by scientists in other regions of the world. It was hypothesized that, *if* adaptation of the model could be realized for tomato *without* changing the overall structure of the CROPGRO model, a similar approach might be used for other vegetable crops as well.

Objectives of research presented here were to (i) provide information on the general growth characteristics of field-grown tomato, (ii) outline some of the more important modifications to crop- and cultivar-specific parameters in order to model tomato plant growth, (iii) show the importance of these parameters on simulation results through the use of sensitivity analysis, (iv) present results of initial model calibration for a number of data sets, and (v) present some general guidelines for future modifications of the CROPGRO crop model.

Material and methods

Data were collected from experiments conducted at the Gulf Coast Research and Education Center (GCREC) in Bradenton, Florida during the spring seasons of 1991, 1992, and 1994 on an Eaugallie fine sand (sandy, siliceous, hyperthermic, Alfic Haplaquods). Tomato (*Lycopersicon esculentum* Mill.) transplants (cv Sunny) were planted on raised (fumigated) beds covered with black polyethylene mulch on 18 March, 3 March, and 1 March in the respective years. Plant spacing was 1.5 m×0.6 m and plants were subirrigated with drip-lines buried at a depth of 40 cm (termed locally the fully enclosed seepage irrigation system). Plants were typically well-fertilized and no nutrient stress was observed, except for apparent N-stress near the end of the 1992 growing season at Bradenton. The incidence of pests and diseases was controlled by frequent applications of agrochemicals.

Plant growth measurements for representative plants (node number; plant height; dry weights of roots, stems, leaves and fruits; and leaf area) were collected biweekly, with a compiled data set being entered into time-course data files (TMT files). All information pertinent to crop, fertilizer and irrigation management was entered into experimental-detail files (TMX files) using the DSSAT V3 standardized format (Jones et al. 1994). Similarly, specific soil profile descriptions for the experimental locations were added to the SOIL.SOL file of the DSSAT shell. Daily values of minimum and maximum temperature, solar radiation, and rainfall for both locations were placed in the weather data files (WTH files).

A systematic approach was used during the calibration process of the CROPGRO model for tomato. During initial modifications of crop-specific parameters, the phenological development coefficients were calibrated in order to match node development rate, date of flowering, fruit development and fruit characteristics. An overview of the definition of these and other coefficients, with their respective values and units, is presented in Table 1. In order to calibrate the model for the cost of tissue synthesis, plant composition values were entered using tissue analysis for tomato and/or values reported in the literature. Protein levels of tomato plant parts were derived from actual nitrogen concentrations of plant parts and are presented in Table 2. Subsequent steps included defining look-up function for tomato seedling potential leaf area during the sink-limited development phase; characterizing specific leaf properties such as leaf size and specific leaf area (*SLA*); defining partitioning of assimilates to plant organs as a function of node number; and estimating values of photosynthetic parameters.

An expression that describes the instantaneous production of photosynthates (Pg) as a function of photosynthetic photon flux density ($PPFD$) and leaf area was developed by Acock et al. (1978). A modified version was developed by Jones et al. (1989) to simulate canopy Pg for greenhouse tomato in the TOMGRO model:

$$Pg = \frac{D \times LF_{max}}{K} \ln \frac{(1-m)LF_{max} + Q_e \times K \times PPFD}{(1-m)LF_{max} + Q_e \times K \times PPFD \times \exp(-K \times LAI)} \quad PGRED(T) \tag{1}$$

Table 1. Definitions of modified photosynthetic, phenological, vegetative, and reproductive coefficients with units and numeric values used in the TMGRO940 species- and cultivar-specific input files.

Parameter	Definition	Units	Value
Photosynthetic coefficients			
PAR	Photosynthetically Active Radiation	$mol\ m^{-2}\ d^{-1}$	–
PHT_{max}	Daily photosynthetic production capacity for fully developed canopy intercepting all of incident *PAR* at optimal temp., leaf N concentration, and other conditions	$g\ (CH_2O)\ m^{-2}\ d^{-1}$	75.54
K_p	Empirical coefficient that defines curvature of exponential light response function	–	0.0386
K_{can}	Light extinction coefficient	–	0.65
Phenological coefficients			
TRIFL	Rate of node appearance	$nodes\ d^{-1}$	0.45
VSSINK	Maximum node number to which plant remains sink-limited	–	5.5
Vegetative development coefficients			
SIZLF	Maximum size of fully expanded leaf	cm^2	300
SLAVAR	Specific leaf area of cultivar under standard optimum growing conditions	$cm^2\ g^{-1}$	340
LFMAX	Light-saturated leaf photosynthesis rate at 30°C, 350 vpm CO_2, and high light	$mg\ CO_2\ m^{-2}\ s^{-1}$	1.18
FRLFF	Fraction of daily increase in vegetative weight that goes to leaves during 'final' growth stage	–	0.23
FRSTMF	Fraction of daily increase in vegetative weight that goes to stems during 'final' growth stage	–	0.45
Reproductive coefficients			
THRSH	The maximum ratio of seed/(seed + shell)	%	4
WTPSD	Maximum weight per seed	g	0.0035
SDPDV	Potential seed number per fruit	–	250
LNGSH	Time required for shell growth	photothermal days	30
SD-PM	Time between first seed and physiological maturity	photothermal days	50
XFRT	Maximum fraction of daily growth partitioned to seed and shell	–	0.75

Table 2. Maximum and minimum protein concentrations of root, stem, leaf blade, and fruit tissue for tomato during growth and after N-mining (g protein g^{-1} tissue). Values were obtained from tissue analysis for the 1991 experiment at Bradenton. Protein values were calculated by multiplying total-N concentrations by 6.25.

	Plant tissue protein concentration		
	During growth		After N-mining
	Max.	Min.	
Roots	0.144	0.119	0.060
Stem	0.256	0.156	0.112
Leaf blade	0.370	0.263	0.154
Fruit	0.245	0.227	0.147

in which:

D = coefficient to convert Pg from $\mu M\ (CO_2)\ m^{-2}\ s^{-1}$ to $g\ (CH_2O)\ m^{-2}\ h^{-1}$;
K = light extinction coefficient (–);
m = leaf light transmission coefficient (–);

LF_{max}	= maximum leaf photosynthetic rate ($\mu M\, m^{-2}\, s^{-1}$);
Q_e	= leaf quantum efficiency ($\mu M\, CO_2$ fixed μM photon^{-1});
PPFD	= photosynthetic photon flux density ($\mu M\, m^{-2}\, s^{-1}$);
PGRED(T)	= function to modify *Pg* under different daytime temperatures (°C); and
LAI	= canopy leaf area index ($m^2\, m^{-2}$).

This model was used to simulate hourly values of *Pg* for days with different amounts of solar radiation. Hourly values of *Pg* were integrated to give total daily *Pg*. Since this model was well-calibrated for a wide range of environmental conditions, it was used to simulate *Pg* for a range of total daily photosynthetically active radiation (*PAR*) values from 0 to 54 mol $m^{-2}\, d^{-1}$ for a tomato crop with an *LAI* of 5. These simulated daily *Pg* and solar radiation values were used to parameterize the simple equation in CROPGRO that relates daily canopy *Pg* to daily solar radiation:

$$Pg_{max} = PHT_{max}(1 - \exp(-K_p \times PAR)) \quad (2)$$

where Pg_{max} is the daily photosynthetic production capacity for a fully developed canopy intercepting all of the incident photosynthetically active radiation (*PAR*) at optimal temperature, N tissue content, and other conditions. Non-linear regression analysis (SAS Institute, Cary, NC) was used to deduce the values for PHT_{max} (maximum photosynthetic production capacity under non-limiting conditions) and K_p (an empirical coefficient that defines the curvature of the exponential light response function). The fitted values of K_p and PHT_{max} were used in the TMGRO940.SPE file during the initial calibration of *SLAVAR*, *XFRT* and K_{can}. Subsequently, the value of K_{can} was adjusted to fit simulated values to those observed in the field (Table 1).

To show the effect of parameter modifications on modelling results, sensitivity analyses were conducted for K_p, PHT_{max}, K_{can}, *SLAVAR* and *XFRT*, since each of these parameters had been shown to greatly affect modelling results during calibration. During these sensitivity analyses, the 1994 Bradenton data set was used and all parameters, except the one being tested, were held constant. During this analysis, results were shown for first increasing only PHT_{max} (from 61 to 75) and subsequently increasing both PHT_{max} and K_p (from 0.025 to 0.040). For the other coefficients increments of ±5 – 10% of the estimated value of the parameter were used, resulting in steps of 0.10, 25 and 0.10 for K_{can}, *SLAVAR* and *XFRT* respectively. Using the calibrated model, further simulations were made for the 1991 and 1992 growing season at Bradenton to evaluate the overall fit of the calibrated model to the experimental data. This is *not* a validation of the model, however, since these data sets were also used during the calibration of the model.

Results and discussion

Model adaptation

Photosynthesis. The fitting of the exponential photosynthetic function used by CROPGRO to the light-responsive curve generated by TOMGRO (Figure 1, solid line) resulted in estimates for PHT_{max} and K_p of 75.54 and 0.0386 respectively. Using

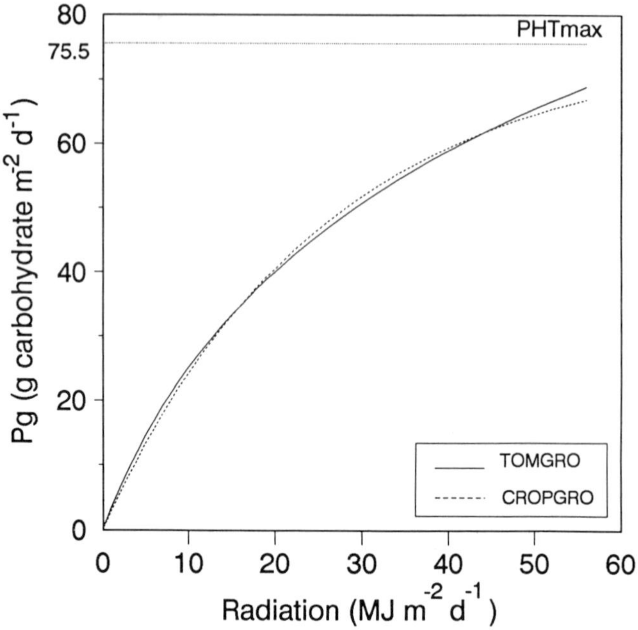

Figure 1. Simulated daily *Pg* values (g CH_2O $m^{-2}d^{-1}$) as a function of radiation for a tomato crop (*LAI*=5) using TOMGRO (Jones et al. 1989) and the matching prediction of *Pg* values by CROPGRO assuming values for the fitted photosynthetic parameters PHT_{max} and K_p of 75.54 and 0.0386 respectively.

these values in equation (2) resulted in the CROPGRO light-response curve for tomato (Figure 1, dotted line). The value of K_{can}, the light extinction coefficient, was reduced from 0.75 to 0.65, this new value being obtained by fitting simulated values of crop parameters to those measured in the field. The reduction in the value of K_{can} is related to the relatively large discrepancy between plant spacing in the row and spacing between the plant rows that is commonly observed in commercial tomato production fields. The leaf N concentrations below which a reduction in *Pg* occurs (*FNPGN*) were reduced from 4.5% to 4.0% N. As a result, the rate of photosynthesis became less readily reduced by low concentrations of N in the leaf tissue for tomato than for the leguminous peanut plant. A value of 1.18 mg CO_2 $m^{-2}s^{-1}$ for the light-saturated leaf photosynthesis rate (*LFMAX*) for tomato was obtained from the literature (Acock et al. 1978). This value is used for the leaf-level canopy photosynthesis option available in the CROPGRO model. The values reported here for PHT_{max}, K_p, and K_{can} are being used in the current version of the growth model in the TMGRO940.SPE file of CROPGRO.

Phenology. During initial parameter modification, an existing peanut ecotype file (PNGRO940.ECO) and cultivar file (PNGRO940.CUL) were adapted for tomato. However, the basic concepts of vegetative and reproductive development stages as defined for peanut by Boote et al. (1986) were maintained. Since fresh-market tomato is commonly transplanted, modification was required in parameterization of the initial crop development stages. The dry weight of the plant at transplanting was added as an input parameter in the *TMX* file, and is used by the model to initialize leaf area and plant weight at the time of planting. The time required from seeding till the first true leaf stage was subsequently set to zero. The rate of node appearance (*TRIFL*) for tomato is also slightly faster than that of peanut, with respective *TRIFL* values of 0.45 and 0.39 nodes d^{-1}. Rate of early leaf area development for tomato was assumed to be sink-limited, though the node number to which plants remain sink-limited (*VSSINK*) was reduced from 7.5 to 5.5. The *XVGROW* look-up function, which relates node number to maximum possible leaf area during the sink-limited phase, was recalibrated for tomato.

Vegetative growth and partitioning. Leaf characteristics of tomato also differ appreciably from those of peanut. To accommodate these differences, a large number of modifications were made in the crop- and cultivar-specific parameters. The value of the leaf size parameter (*SIZLF*) was increased from 20 to 300 cm^2, to account for the much larger relative leaf size of tomato. The value of the specific leaf area (*SLAVAR*) was also increased from 280 to 340 $cm^2\,g^{-1}$, to accommodate the relatively higher area-weight ratio of tomato leaves.

Partitioning of assimilates among vegetative plant parts differs between crops and also depends on the growth shortage of the crop itself (Boote et al. 1986). The CROPGRO model uses partitioning coefficients to calculate the allocation of photosynthates to the new growth of leaves, stems (including leaf petioles) and roots. Since these values change over time, the partitioning to leaves (*FRLF*) and stems (*FRSTM*) is calculated for a number of successive growth intervals and these values are then linked to 'mid-point' node numbers and listed in a look-up table in the species (SPE) file.

Measured dry weights of leaves and stems for the growth studies conducted were used to calculate the values of *FRLF* and *FRSTM* and predicted root weights. The rationale for using predicted root values (based on partitioning values used in TOMGRO) was the relatively large discrepancy between the fraction of photosynthates that are partitioned to the roots and the actual root biomass that was recovered by excavation of root systems formed under production systems with shallow watertable (McNeal et al. 1995). The root partitioning coefficients used in TOMGRO were based on data sets with more complete recovery of photosynthates partitioned to the roots, and appeared to give more realistic values for root dry weights. The measured accumulation of dry matter in the roots was of the order of 3–5% of the total biomass, which appears to be low compared to values reported for production systems with a deep watertable by Jackson and Bloom (1990). Low values may be attributed to shallow and/or fluctuating watertable, which may enhance root respiration, root sloughing and root die-back (Smucker 1984). Using actual root data would, therefore, result in an *under*prediction of dry matter partitioning to the roots and an *over*prediction of dry matter partitioning to stems and leaves.

To calculate *FRLF* for a specific node number, the increment in leaf weight (d*BL*)

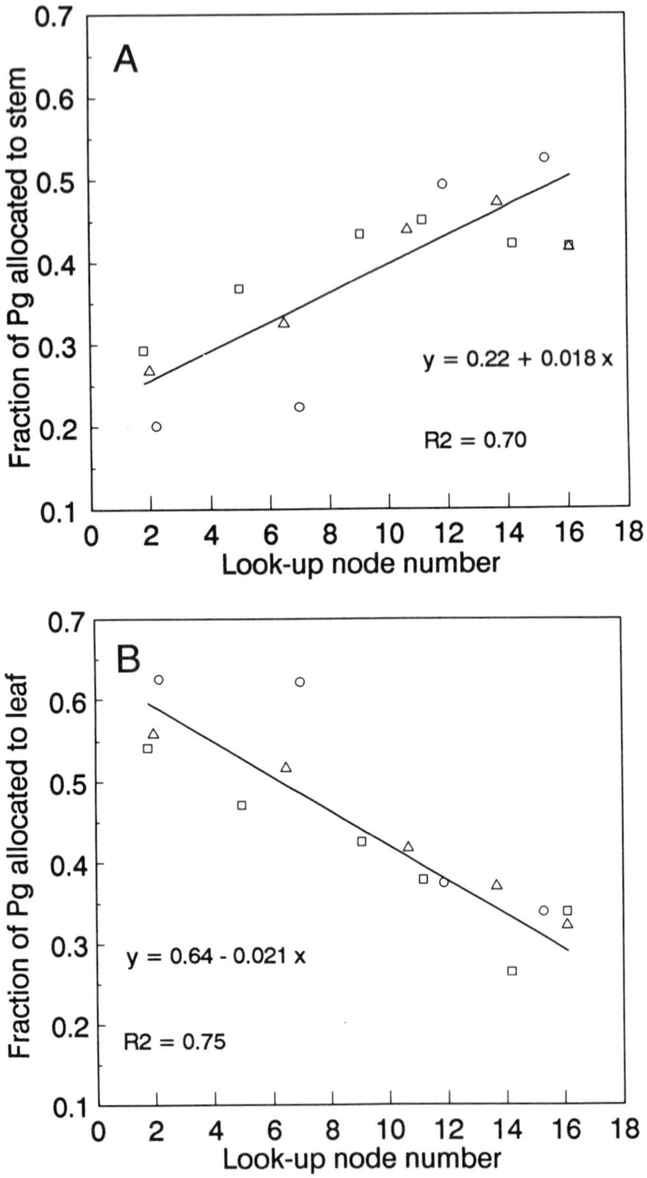

Figure 2. Partitioning of photosynthates to stems (A) and leaves (B). The calculated data points for the growing seasons 1991 (circles), 1992 (squares), and 1994 (triangles) were fitted by using linear regression (linear model, solid line), expressing dry weight partitioning to stems (y) or leaves (y) as a function of the node number (x).

Table 3. Maximum potential leaf area and partitioning of photosynthates to roots, stems, and leaves of tomato as a function of vegetative stage (node number).

Look-up node no.	Max. potential leaf area (cm^2)	Partitioning of *Pg*		
		Roots	Stems (fraction)	Leaves
0.0	15.4	0.15	0.25	0.60
1.7	28.1	0.15	0.25	0.60
4.3	83.4	0.16	0.30	0.54
5.9	210.0	0.16	0.33	0.51
8.0	–	0.18	0.36	0.46
10.0	–	0.18	0.40	0.42
12.0	–	0.19	0.43	0.38
15.0	–	0.20	0.49	0.31

between two sampling dates was divided by the increment in total vegetative biomass (d*VB*). Similarly, *FRSTM* was calculated by dividing d*STM* by d*VB*. Using linear regression, a functional relation was established between 'mid-point' node number and partitioning coefficients. Results for both *FRLF* and *FRSTM* are shown in Figures 2A–B. An outline of the partitioning coefficients of photosynthates to roots, stems, and leaf blades, along with the parameterization of the *XVGROW* look-up function (discussed in *Phenology*), are presented in Table 3.

Towards the end of the growing season, weight increments between sampling dates became relatively small compared to the overall variation between plants. Observed differences between subsequent sampling dates may, thus, be caused by inherent variation within the plant population and also by senescence and retranslocation of assimilates to the fruits. As a result, partition coefficients can no longer simply be deduced from observed biomass values of different organs, since other processes such as senescence and reallocation of photosynthates affect dry matter accumulation. Increases in vegetative biomass over time eventually approach zero and may even become negative in some circumstances.

From Table 3, it may be concluded that tomato plants initially invest most of their assimilates into a rapid build-up of photosynthetic production capacity (leaves). Over time, however, plants increase the partitioning of photosynthates into structural stability (stems) required to support the subsequent fruit load.

Reproductive growth. It was conceptualized that a tomato fruit could be perceived as a seeded pod with a very low seed to total fruit dry weight ratio. Based on experimental data, the value of *THRSH* (which expresses this ratio as a percentage) was reduced from 80% (for peanut) to 4%. The weight per seed (*WTPSD*) value used for tomato was 0.0035 g (compared to 0.57 g for peanut), while the potential seed number per fruit (*SDPDV*) was increased from 1.7 to 250.

Based on values reported in the literature, the time required for tomato fruit growth to be completed (*LNGSH*) was estimated to be 30 days (Wolf et al. 1986). This is appreciably longer than the 20 days used for peanut. Based on experimental obser-

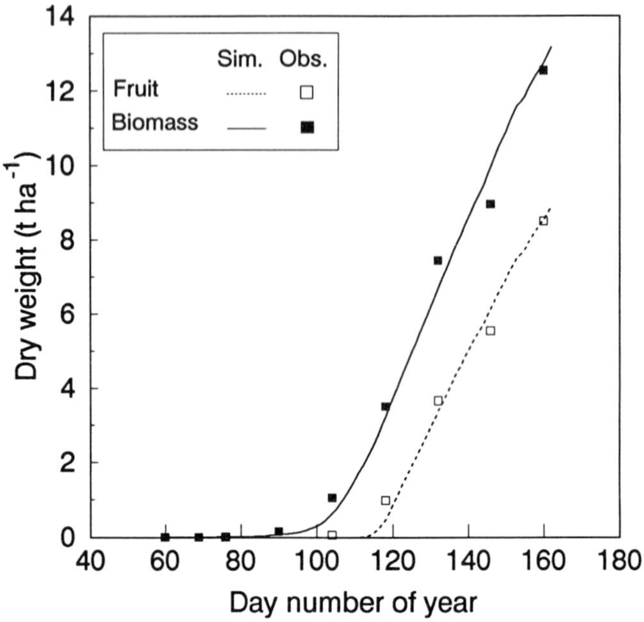

Figure 3. Simulated and observed fruit dry weight and total above-ground dry weight (biomass) for Sunny tomato at Bradenton for the spring 1994 growing season.

vations, the time between first seed formation and physiological maturity (*SD-PM*) was reduced from 83.7 to 50.0 days. After fruit initiation, fruits increasingly competed with vegetative parts for assimilates. Unlike more determinant processing-tomato cultivars or crops like soybean, fresh-market tomato shows a more gradual shift towards reproductive growth.

The average value of the partitioning coefficient to the fruits during the time of maximum total sink strength of the fruits (node number >16) was of the order of 0.75. This value subsequently was used as an estimate for the *XFRT* parameter, which is the maximum limit for partitioning of assimilates to fruit. Later in the growing season, values close to or even in excess of 1.0 may occur. This is related to the 'shut-down' of both leaf and stem growth, and the retranslocation of photosynthates from these organs to the fruits (Scholberg 1994).

Model comparison with 1994 field data

Calibration was continued until the predicted time-series values of leaf nodes, *LAI* and *SLA*, and the dry weights of stems, leaves, fruits and above-ground biomass, fitted actual field data. After this calibration of the model, predicted fruit dry weight and total above-ground biomass closely matched actual field measurements (Figure 3). Similar results were observed for most of the other variables (Scholberg 1994).

Sensitivity analysis

Increasing PHT_{max} from 61 to 75 (estimated value for tomato) resulted in a sharp increase in leaf area index and fruit dry weight (Figures 4A–B, dotted and solid lines). Increasing K_p from 0.025 (value used for peanut) to 0.0386 (estimated value for tomato) resulted in still further increases in simulated values (Figures 4A–B, solid and bold lines). The value of PHT_{max} is indicative of the capacity of the photosynthetic apparatus at infinitely high radiation levels, whereas K_p reflects the efficiency of the system at lower radiation levels.

The value of the canopy light extinction coefficient (K_{can}) is typically 0.75 for equidistant plant spacings but, for crops with greater row spacings, the value may be more realistically reduced to 0.65. Decreasing values of K_{can} resulted in reduced values of *LAI*, and also in reduced leaf dry weight (Figure 5) and fruit dry weight (Scholberg 1994). The CROPGRO model appears to be very responsive to changes in K_p, PHT_{max} and K_{can}. This is probably related to the impact of these parameters on the total available assimilate pool, which appears to be one of the main driving forces for the model.

Increasing the value of *SLAVAR* (model input value for *SLA* under standard conditions) resulted in increased simulated values for *SLA* (Figure 6). For the 1994 data set, a value of 340–350 seemed to give the best results. The model appeared to be moderately sensitive to changes in this parameter.

The effects of changing the fraction of photosynthates partitioned to the fruit (*XFRT*) on dry weights of stems and fruits are shown in Figures 7A–B. With increased values of *XFRT*, canopy expansion during the onset of fruit development was shut down completely (data not shown), whereas only a reduction in stem growth occurred (Figure 7A). Fruit dry weight increased as *XFRT* increased, with a value of roughly 0.70 giving the best fit to the data (Figure 7B). It may be concluded that the accumulation of dry matter in stems and leaves after initial fruit set is greatly affected by the value of *XFRT*. Low values result in prolonging the vegetative growth phase, whereas high values result in over-prediction of fruit yields (Figure 7B).

Model simulation versus other data sets

Simulated fruit weights were generally over-predicted for the 1991 and 1992 field data for Bradenton (Figures 8A–B). However, the onset of predicted fruit production for the 1992 season was too late. This 'timing' effect may be related to earliness induced by nitrogen stress and possibly also to the lower temperatures in 1992. This may require additional calibration of both nutrient and temperature effects on the onset and development of fruits. In 1992 simulated total above-ground biomass fit the data fairly well, except that observed biomass started to 'trail off' near the end of the season for most experiments (Figures 8A–B). This phenomenon may be related to nutrient stress (Bradenton 1992), and/or uncertainties in phenology characterization. Research efforts to validate the model for a number of independent data sets and for a range of fertility levels in both Florida and other regions of the world, are under way.

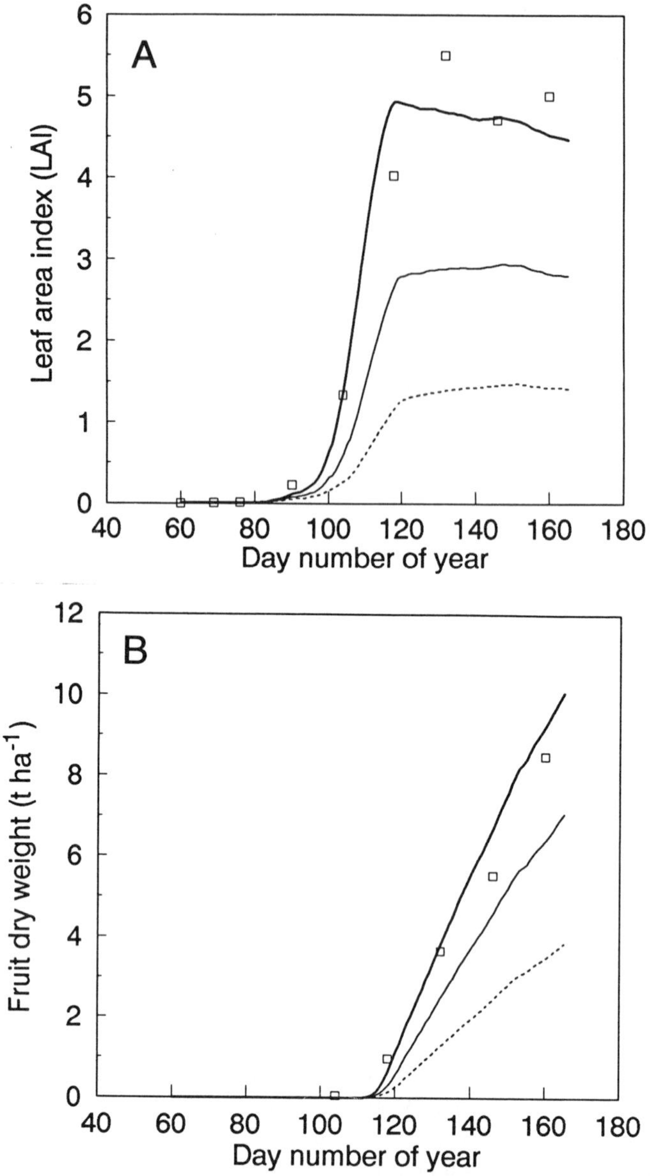

Figure 4. Simulated effects of changes in K_p (curvature coefficient) and PHT_{max} (maximum daily photosynthetic production capacity under optimal conditions) on (A) predicted *LAI* values and (B) fruit dry weight in comparison to 1994 field data at Bradenton (symbols). Upper line: K_p=0.040, PHT_{max}=75; middle line K_p=0.025, PHT_{max}=75; dashed line K_p=0.025, PHT_{max}=61.

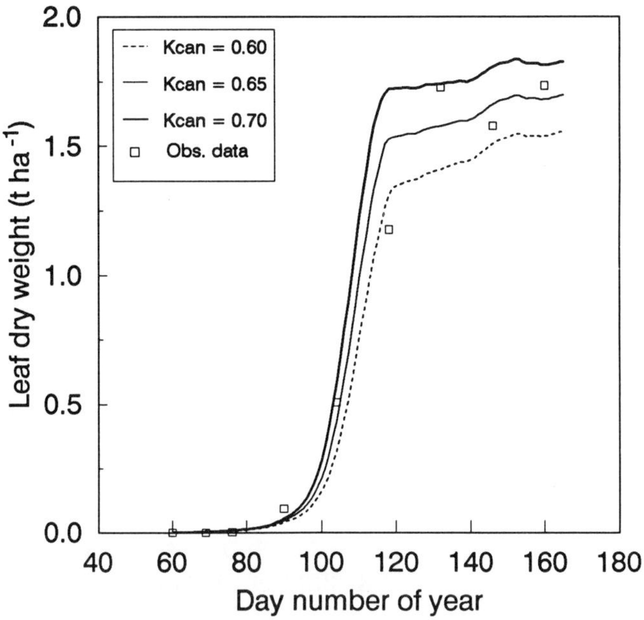

Figure 5. Simulated effects of changes in K_{can} (light extinction coefficient) on predicted leaf dry weight in comparison to 1994 field data at Bradenton.

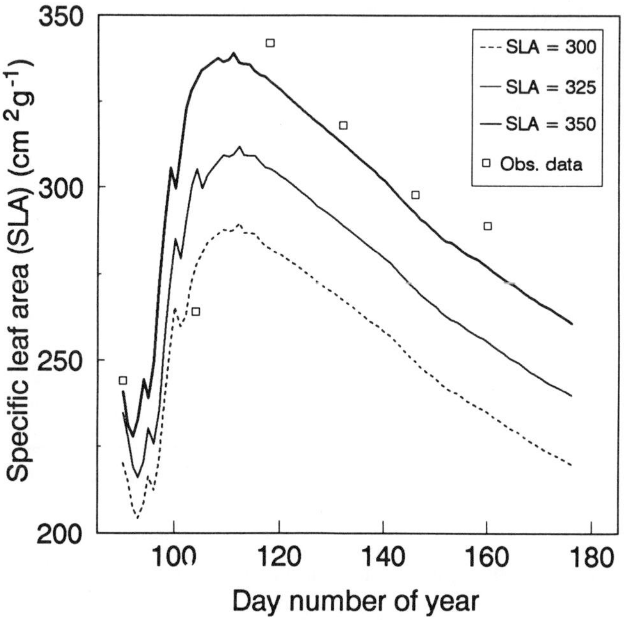

Figure 6. Simulated effects of changes in *SLAVAR* (the *SLA* under standard condition) on predicted *SLA* values in comparison to 1994 field data at Bradenton.

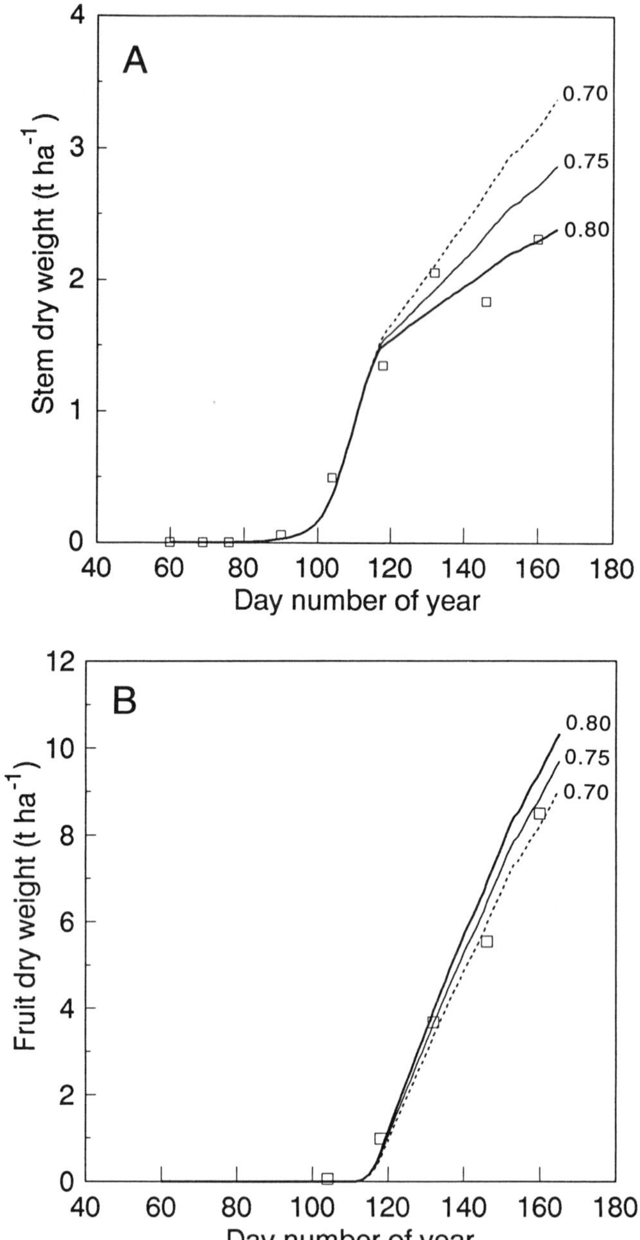

Figure 7. Simulated effects of changes in *XFRT* (fruit partitioning coefficient) on predicted stem dry weight (A) and fruit dry weight (B) in comparison to 1994 field data at Bradenton (symbols).

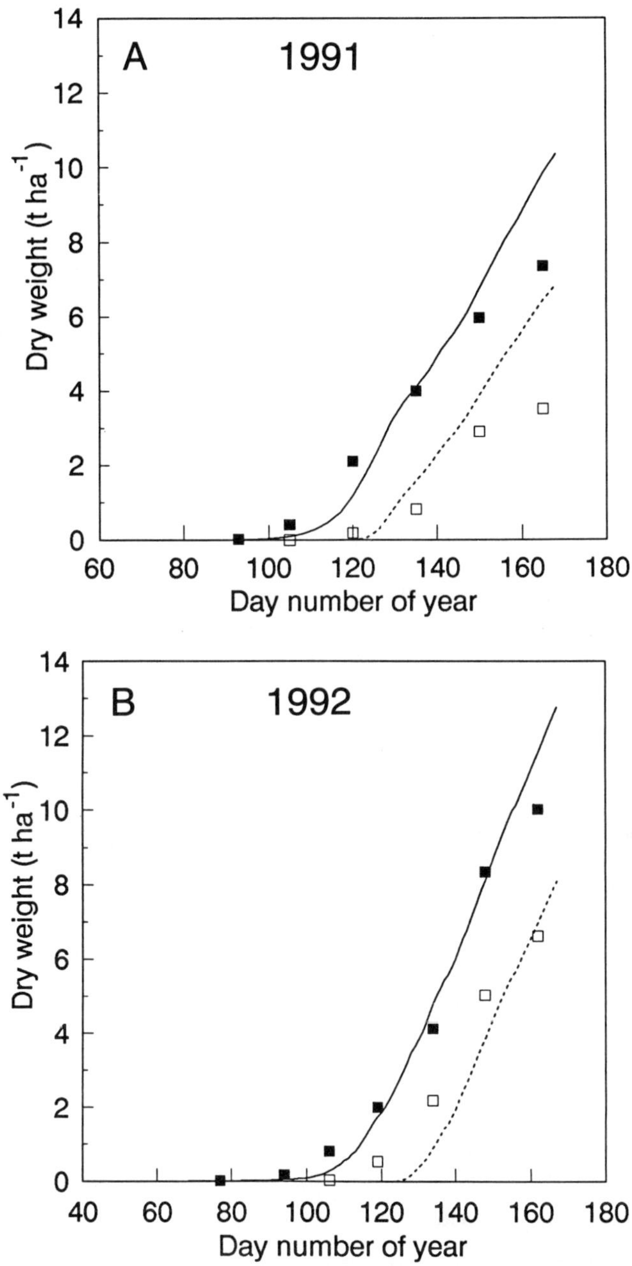

Figure 8. Simulated and observed fruit dry weight (open symbol, dashed line) and total above-ground dry weight (closed symbol, solid line) for Sunny tomato at Bradenton for the spring 1991 (A) and 1992 (B) growing season.

Conclusions

Results from initial modifications of the generic CROPGRO model to describe field-tomato growth show both the adaptability and the robustness of the model. The procedure presented here may also be employed as a 'blueprint' for future CROPGRO adaptations to other tomato varieties, and/or other vegetable crops. From the results presented above, it may be concluded that the calibrated model adequately describes plant growth of field-grown tomato for conditions where plant growth is not limited by nutrient or water-stress. Currently, research efforts to calibrate the model at different levels of water and nutrient stress are ongoing. This involves compilation of data sets for a range of irrigation and fertilizer treatments; further characterization of root development for tomato; and possibly the development of a (pseudo) two-dimensionsal water and nutrient transport routine to facilitate modelling the flow of water and nutrients for subirrigated plastic-mulched crops.

A model component is needed which will predict total fresh fruit yield and will separate total fruit yield into marketable and non-marketable fruit yield, based on fruit size at the mature-green fruit stage. If the model is to be used in both tropical and subtropical regions, it may also be desirable to include a routine that will predict fruit set as a function of day- and/or night-time temperature. Since heat-tolerance is a genetic trait (Scott et al. 1989), it would be desirable to include parameterization of this function in the cultivar file.

References

Acock B, Charles-Edwards D A, Fitter D J, Hands D W, Ludwig L J, Wilson J W, Withers A C (1978) The contribution of leaves from different levels within a tomato crop to canopy net photosynthesis: An experimental examination of two canopy models. J. Exp. Bot. 29:815–827.

Bertin N, Heuvelink E (1993) Dry-matter production in a tomato crop: Comparison of two simulation models. J. Hort. Sci. 68:995–1011.

Boote K J, Jones J W, Mishoe J W, Wilkerson G G (1986) Modeling growth and yield of groundnut. Pages 243–254 in Agrometerology of groundnut. ICRISAT, Patancheru, India.

Boote K J, Bennett J M, Jones J W, Jowers H E (1989) On-farm testing of peanut and soybean models in Florida. Paper No. 89-4040, Am. Soc. Agric. Engr. (ASEA), St. Joseph, MI, USA. 54 p.

Hoogenboom G, Jones J W, Boote K J (1992) Modeling growth, development, and yield of grain legumes using SOYGRO, PNUTGRO, and BEANGRO: A review. Trans. ASAE 35:2043–2056.

Jackson L E, Bloom A J (1990) Root distribution in relation to soil nitrogen availability in field-grown tomatoes. Plant and Soil 128:115–126.

Jones J W, Ritchie J T (1992) Crop growth models. Pages 63–89 in Hoffman G J, Howell T A, Howell K H (Eds.) Management of farm irrigation systems. Am Soc. Agr. Engr. (ASEA), St. Joseph, MI, USA.

Jones J W, Dayan E, Jones P, Seginer I, Allen L H, Zipori I (1989) On-line computer control system for greenhouses under high radiation and temperature zones. Final research report, BARD Project US-871-884, University of Florida, Gainesville, FL, USA. 203 p.

Jones J W, Dayan E, Allen L H, Van Keulen H, Challa H (1991) A dynamic tomato growth and yield model (TOMGRO). Trans. ASAE 35:663–672.

Jones J W, Hunt L A, Hoogenboom G, Godwin D C, Singh U, Tsuji G Y, Pickering N B, Thornton P K, Bowen W T, Boote K J, Ritchie J T (1994) Input and output files. Pages 1–94 in Tsuji G Y, Uehara G, Balas S (Eds.) Decision support system for agrotechnology transfer (DSSAT) Version 3, Volume 2. University of Hawaii, Honolulu, HI, USA.

McNeal B L, Scholberg J M S, Jones J W, Stanley C D, Csizinszky A A, Obreza T A (1995) Application of a greenhouse tomato-growth model (TOMGRO) to field-grown tomato. Soil Crop Sci. Soc. Fla. Proc. 54:87–93.

Scholberg J M S (1994) Initial modification of CROPGRO to develop a tomato growth routine. Compiled results of growth studies and parameter estimations. Progress report, Department of Soil and Water Science, University of Florida, Gainesville, FL, USA. 76 p.

Scott J W, Olson S M, Bryan H H, Howe T K, Stofella P J, Bartz J A (1989) Solarset: A heat tolerant fresh market tomato hybrid. Florida Agric. Exp. Station Circular S-359, Gainesville, FL, USA.

Smucker A J M (1984) Carbon utilization and losses by plant root systems. Pages 27–47 in Barber S A, Bouldwin D R (Eds.) Roots, nutrient, and water influx, and plant growth. ASA Special Publication No. 49, SSSA, CSSA, and ASA, Madison, WI, USA.

Wolf S, Rudich J, Marani A, Rekah Y (1986) Predicting harvesting date of processing tomatoes by a simulation model. J. Am. Soc. Hort. Sci. 111:11–16.

A modified version of CERES to predict the impact of soil water excess on maize crop growth and development

J.I. LIZASO[1] and J.T. RITCHIE[2]

[1]*Universidad Central de Venezuela, Facultad de Agronomía, Departamento de Agronomía, Apartado 4579, Maracay 2101, Venezuela*

[2]*Michigan State University, Crop and Soil Sciences Department, East Lansing, MI 48824, USA*

Key words: simulation model, waterlogging, maize

Abstract

Simulation models may be a powerful tool for agricultural development in developing countries, when they are able to reasonably forecast crop performance under a wide set of environmental constraints. CERES is a dynamic daily-step simulation model that predicts growth and development of five user-selected cereals: wheat, maize, sorghum, pearl millet and barley. The model predicts the effects of limited soil water and nitrogen on crop growth. The maize model in CERES was modified to include the effects of limited soil aeration on crop growth and development. The modified model includes the newest routines available to simulate soil temperature and the water and nitrogen balances. The model can run on three different modes: (i) without watertable simulation, (ii) simulating a fluctuating watertable and (iii) simulating a controlled watertable. A new input file is required only if a controlled watertable option is selected. The new input file provides input on the depth of the controlled watertable for each day of simulation. The modified model produced logical and reasonable results when flooding was simulated. Under well-aerated conditions, growth and yield was somewhat underestimated, although estimates were always closer to field measured values than estimates of the original CERES-Maize model. Predictions of crop phenology required adjustment. The new model provides an alternative for predicting maize crop performance when the crop is affected by waterlogging.

Introduction

Mechanistic models have been proposed to describe oxygen diffusion through individual roots (e.g. Armstrong and Beckett 1987). These models, however, do not attempt to predict whole-root system behaviour under conditions of limited soil aeration. Several models have been developed to simulate whole-root systems (see Luxmoore and Stolzy 1987 for a review). None of these models considers the limitations imposed by deficient soil aeration on root growth. Jones et al. (1991) proposed a model to predict root growth that includes soil constraints affecting root growth. The model uses the fraction of water filled pore space as an aeration index to affect root attributes such as rooting depth, branching and senescence.

The generic CERES model simulates growth and development of wheat, maize, sorghum, pearl millet and barley under adequate and limited soil water. In this paper, we modify the maize model in CERES to extend the predictions into environments with excessive soil water.

M.J. Kropff et al. *(eds.): Applications of Systems Approaches at the Field Level, 153–167.*

Material and methods

Field data to support our model modifications were obtained from a field experiment (Lizaso 1993), where shoots and roots were simultaneously monitored. Field information to adjust some model parameters, however, was not available and we used values yielding reasonable predictions. The experiment was planted in a field facility where the watertable can be managed independently on each of five 2.1 m×1.5 m plots (Lizaso and Ritchie 1995). Pairs of horizontal mini-rhizotrons located at 15, 40, 60 and 90 cm depth provided an opportunity for non-destructive root growth monitoring. The five treatments consisted of single flooding periods of four or eight days when maize was at the six or the 12 leaf tip stage, plus the non-flooded control. Flooding was imposed by raising the watertable between 2–5 cm above the soil surface.

The generic version of CERES was used to simulate the effects of flooding on maize growth, development and yields. The maize model in generic CERES is CERES-Maize, described by Jones and Kiniry (1986). The predictions of the modified generic CERES were compared with the predictions of the original Version 2.1 of CERES-Maize.

Modified generic CERES

An improved version of generic CERES, with a new water balance routine (J.T. Ritchie, Michigan State University, 1992, pers. commun.), was modified to include a different logic when simulating with daily watertable depths as an input for the model. New routines to simulate up-flow and down-flow of water and nitrogen were obtained from D.C. Godwin and J.T. Ritchie (1992, pers. commun.) and linked into this modified version of CERES. These routines consider the effects of the watertable and a restricting soil layer in the drainage. Simulation of nitrogen movement in the soil considers two fractions with different mobilities. A new soil temperature model (Dadoun 1993) was also linked into the modified version of CERES to provide better estimation of the effect of soil water content on soil temperature. These changes will affect predictions of flooded and non-flooded plants.

Soil aeration index

The routine that calculates root growth (ROOTGROW) was modified to include the calculation of an aeration factor on each soil layer. The aeration factor is passed to a new routine (OXSTRESS) that computed a soil aeration index. Aeration indexes are calculated in terms of soil porosity since oxygen diffusion depends on the air-filled pore space. Total porosity ($TPORE_i$) and water-filled porosity ($WFPS_i$) are calculated for each layer i as:

$$TPORE_i = 1 - (BD_i/2.65) \tag{1}$$

$$WFPS_i = SW_i/TPORE_i \tag{2}$$

where BD_i is the bulk density of layer i, assuming a soil particle density of 2.65 g cm^{-3} and SW_i is the volumetric water content of the layer in $cm^3 cm^{-3}$. The water-filled pore space at saturation is assumed to have a maximum value ($XWFPS$) of 0.93 to account for trapped air in the soil profile. Whenever the water-filled pore space ($WFPS_i$) in a soil layer is larger than a critical value ($CWFPS$), a layer aeration factor (LAF_i) was calculated as:

$$LAF_i = (1 - WFPS_i)/(1 - CWFPS) \quad (3)$$

where $CWFPS$ is equal to 0.45 and LAF_i cannot be larger than 1. A soil layer with low air-filled porosity may limit oxygen diffusion into deeper layers. In this model, a layer with $WFPS$ of 0.9 or larger limits the aeration of lower layers. In the OXSTRESS routine, root length density (RLV_i) of each soil layer is used to weight the layer aeration factor in the calculation of a zero (total stress) to one (no stress) whole rhizosphere aeration factor ($WRAF$):

$$WRAF = \left(\sum_{i=1}^{LRD} LAF_i \times RLV_i\right) / \left(\sum_{i=1}^{LRD} RLV_i\right) \quad (4)$$

where LRD is the soil layer where the maximum rooting depth occurs. Two dimensionless aeration stress indexes, ASD and $ASD2$, were calculated using $WRAF$ to simulate the cumulative aeration status experienced by the plant. ASD will affect root growth and plant development, and $ASD2$ root depth and biomass production and partitioning. Both indexes increased following the same calculation as the soil aeration decreased but they differed during the recovery period. The model compares yesterday's and today's values of $WRAF$ to decide if the soil aeration improved. When the aeration in any day is more limiting for root growth than the previous day, the aeration indexes (ASD and $ASD2$) are increased according to:

$$ASD_{DAY} = ASD_{DAY-1} + (1 - WRAF) \quad (5)$$
$$ASD2_{DAY} = ASD2_{DAY-1} + (1 - WRAF)$$

During the recovery period, ASD will decrease as:

$$ASD_{DAY} = ASD_{DAY-1} - WRAF \times (WRAF - WRAFY) \times 0.855 \times ISTAGE^2 \quad (6)$$

This formulation implies that the decrease in the stress index ASD will be modulated by the current status of soil aeration ($WRAF$), by the daily increase in soil aeration ($WRAF - WRAFY$), and by the phenological stage ($ISTAGE$), assuming faster recovering later in the season. $ISTAGE$ is an integer variable (1 to 9) identifying the phenological stage of the crop and $XSTAGE$ is a non-integer (0 to 6) cumulative growth stage indicator. $ASD2$ recovers early in the vegetative growth stage ($XSTAGE < 2.5$) in two steps. When $ASD2$ is large (> 4.0):

$$ASD2_{DAY} = ASD2_{DAY-1} - WRAF \times ASD2_{DAY-1} \times 0.8 \quad (7)$$

When $ASD2$ is small (< 4.0):

$$ASD2_{DAY} = ASD2_{DAY-1} - WRAF \times ASD2_{DAY-1} \times 0.1 \quad (8)$$

During the remainder of the season (*XSTAGE*>2.5) *ASD*2 will recover as:

$$ASD2_{DAY} = ASD2_{DAY-1} - WRAF \times ASD2_{DAY-1} \times 0.5 \tag{9}$$

Both aeration indexes are maintained in the range between zero and 10, assuming no additional damage beyond that limit. A cumulative aeration stress index (*CASD*), similar to the soil water and nitrogen stress indexes in CERES-Maize, is calculated when *WRAF* changes:

$$CASD_{DAY} = CASD_{DAY-1} + (1 - WRAF) \tag{10}$$

CASD is divided by the duration of the phenological stage, at the end of each stage, to provide an average aeration stress index per phenological stage to be printed.

Predicting plant growth

CERES-Maize calculates daily dry matter production by multiplying the potential dry matter production (*PCARB*) by the most limiting of three zero-to-unity stress factors due to temperature, limited soil water, and nitrogen. A new soil aeration deficit factor (*SADEF*) was computed to affect the actual dry matter production under conditions of excessive soil water. The calculation of *SADEF* assumes that a maize crop may endure up to two days of complete soil submergence without a noticeable disruption in the ability of the crop to accumulate biomass. A period between two and four days is critical in affecting dry matter production. Beyond four days of complete soil submergence, no additional damage is predicted. A slow recovery after drainage is assumed to occur only during the vegetative growth stage. When the maximum registered value of *ASD*2 (*XASD*2) is between two and four, *SADEF* is calculated as follows:

$$SADEF = 1.45 - 0.226 \times AAFF \tag{11}$$

where *AAFF* is made equal to *XASD*2 and is updated every time *XASD*2 increases. In successive days, *AAFF* is reduced by 0.095 units each day during the vegetative growth stage, while *XASD*2 maintains the maximum *ASD*2. When *XASD*2 is larger than four, *SADEF* is calculated as:

$$SADEF = \exp(-AAF \times 0.15) \tag{12}$$

where *AAF* has an initial value of four and every day is decreased by 0.095 units during the vegetative growth stage. Actual dry matter production (*CARBO*) is then calculated as:

$$CARBO = PCARB \times \mathrm{AMIN1}(PRFT,\ SWDF1,\ NDEF1,\ SADEF) \tag{13}$$

where *PCARB* is potential dry matter production and AMIN1 is a FORTRAN function that selects the most limiting stress factor due to temperature (*PRFT*), soil water deficit (*SWDF*1), nitrogen (*NDEF*1) and soil aeration (*SADEF*).

CARBO is partitioned among the different organs growing in any phenological stage. The CERES-Maize model allocates the biomass in above-ground growing organs according to several algorithms and the remaining biomass is assigned to root growth. The model checks, however, if a minimum proportion of *CARBO* is going to roots. The transport of photosynthates to the roots during the vegetative growth stage is impaired under waterlogging. For that reason, above-ground biomass may increase, while leaf area immediately decreases during inundation, as compared with non-flooded plants. CERES-Maize calculates leaf area as a function of leaf biomass. The program was modified to calculate a labile fraction of the biomass (*LABIO*) being allocated into the roots (*g*) which will be maintained temporarily above-ground until soil aeration improves. This labile biomass was prevented from increasing the calculated plant leaf area (*PLA*). *LABIO* was calculated during the vegetative growth stage as a function of the aeration stress index *ASD*2:

$$SADF1 = \exp(-ASD2 \times 0.6) \qquad (14)$$
$$LABIO = GRORT \times (1 - SADF1)$$

where *GRORT* is the biomass (*g*) being allocated into the roots. *LABIO* was subtracted from *GRORT* and added to leaf biomass (*GROLF*) early in the season ($ISTAGE = 1$ or $ISTAGE = 2$). Late in the vegetative growth stage ($ISTAGE = 3$), *LABIO* will be added to stem biomass (*GROSTM*) instead of leaf biomass. *LABIO*, allocated in leaves and stems, is accumulated in temporary variables (*TLABIOL* and *TLABIOS*) which are used to keep track and to return the labile biomass to the roots after drainage. When soil aeration improves, the labile biomass temporarily stored in leaves returns to roots at a fraction of the biomass allocated that day into leaves according to the following equations:

$$LFWT_{DAY} = LFWT_{DAY-1} - GROLF \times PROP \qquad (15)$$
$$GRORT_{DAY} = GRORT_{DAY-1} + GROLF \times PROP$$
$$TLABIOL_{DAY} = TLABIOL_{DAY-1} - GROLF \times PROP$$

The proportion (*PROP*) was made equal to 0.85 early in the season ($XSTAGE < 1.75$). For the remainder of the season, the proportion is equal to one. Similar equations are used to return labile biomass from stems to the roots:

$$STMWT_{DAY} = STMWT_{DAY-1} - GROSTM \times PROP \qquad (16)$$
$$GRORT_{DAY} = GRORT_{DAY-1} + GROSTM \times PROP$$
$$TLABIOS_{DAY} = TLABIOS_{DAY-1} - GROSTM \times PROP$$

Predicting root growth

The routine that calculates root growth was modified to predict the effects of excessive soil water. Two root attributes received attention: root depth and number of roots per layer. In CERES-Maize, root depth increases through the whole season and the maximum daily increase in root depth is calculated as 0.22 cm per unit of thermal time accumulated in that day. This growth may be slowed down by limited soil water or stopped when the maximum depth in the soil profile is reached. In the modified routine, the growth in root depth occurs only up to the silking stage unless the rooting depth by that time is less than 50 cm. If the rooting depth by the beginning of the linear phase of grain filling is still less than 30 cm, additional growth in root depth is allowed. In any case, root depth growth was stopped when the rooting front arrived at a saturated soil layer or at the watertable. If the rooting front stayed in a saturated layer for more than five days, root death was assumed and the new root depth was set equal to the middle of the nearest non-saturated upper layer. The modified formulation, however, maintained a minimum rooting depth to simulate the presence of aerenchymatous roots (Drew et al. 1985; Thomson et al. 1990). A minimum distance of 20 cm from the rooting front to the soil surface or to the ponding water surface was maintained. In the new formulation, increases in root depth were slowed down by limited soil aeration according to:

$$SADF4 = LAF_i \times 3 \tag{17}$$

$$RTDEP_{DAY} = RTDEP_{DAY-1} + DTT \times 0.22 \times \mathrm{AMIN1}(SWDF1 \times 2, SWDF, SADF4)$$

$$SADF4 = \exp^{(-ASD2 \times 0.3)}$$

$$RTDEP_{DAY} = RTDEP_{DAY-1} + DTT \times 0.22 \times \mathrm{AMIN1}(SWDF1 \times 2, SWDF, SADF4)$$

where *RTDEP* is in cm, *DTT* is the thermal time (°C day) accumulated in the day, *SWDF*1 and *SWDF* are dimensionless water deficit coefficients for the whole rhizosphere or for the layer where root depth is calculated, and *SADF*4 is a dimensionless aeration deficit factor. LAF_i represents the aeration status in the soil layer where the rooting front is growing, and *ASD*2 estimates the current status of the plant in terms of root aeration. The first equation to calculate *SADF*4 is used if the plants have not experienced any significant period of soil submergence ($XASD2 < 2$). Otherwise, the second equation is selected.

To calculate the effect of inadequate soil aeration on the number of roots in each soil layer, a new variable (*RLVLOSS*) was introduced. *RLVLOSS* ($\mathrm{cm\,cm^{-3}}$) will reduce a fraction of the root length density of each soil layer (RLV_i in $\mathrm{cm\,cm^{-3}}$) according to the aeration status of that layer (LAF_i). During the vegetative growth stage, two equations were used to calculate *RLVLOSS*: the first was used before tassel initiation ($ISTAGE = 2$) and the second after ($ISTAGE = 3$):

$$RLVLOSS = (1 - LAF_i) \times 0.075 \times ASD \times RLV_i \tag{18}$$

$$RLVLOSS = (1 - LAF_i) \times 0.025 \times ASD \times RLV_i$$

These equations imply that during the vegetative growth stage the fraction of roots lost in each soil layer due to insufficient soil aeration depends upon the aeration in the soil layer (LAF_i) and the current rhizosphere aeration (*ASD*). During the reproductive stages, the calculation of *RLVLOSS* differs according to the existence or non-existence of a previous submergence period (maximum *ASD*2 recorded <2):

$$RLVLOSS_i = (1 - LAF_i) \times 0.01 \times TRLVY \times RLV_i \tag{19}$$

$$RLVLOSS_i = (1 - LAF_i) \times 0.02 \times TRLVY \times RLV_i$$

Root loss, in this case, is a function of the aeration of the soil layer (LAF_i) and of the accumulated root length density of the previous day (*TRLVY*). Root length density (RLV_i) is updated every day considering the new component *RLVLOSS* and increasing root senescence from 0.5% during the vegetative growth stage to 5% during the reproductive stage.

Predicting plant development

Plant development is estimated in CERES-Maize in two routines: PHENOL and PHASEI. PHENOL determines the beginning of a new growth stage and PHASEI initializes and calculates variables used in the following growth stage. Several changes were introduced to simulate the delay of phenological events as a consequence of soil submergence. Thermal time, calculated as the accumulation of degrees centigrade (°C) above a base temperature, was calculated using the mean soil temperature, instead of the mean air temperature, from sowing to tassel initiation. During this period, the base temperature was set 0.5°C above the values used in CERES-Maize. After tassel initiation, the shoot meristem is assumed to be above the soil surface (Ritchie and Hanway 1984). Thus, air temperature was used to calculate the thermal time. The base temperature was set to 8.0°C as in CERES-Maize.

The actual thermal time that the modified model accumulates may be reduced early in the vegetative growth stage ($XSTAGE<2$) as a consequence of inadequate soil aeration ($ASD>1$):

$$SADF5 = \exp(-ASD \times 0.4) \qquad (20)$$
$$TT = DTT \times SADF5$$

where *DTT* is the thermal time (°C day) accumulated in the day, *ASD* is the aeration stress index calculated earlier and *TT* is the actual thermal time used to estimate plant development.

CERES-Maize estimates that a new leaf emerges every time a certain amount of thermal time, a phyllochron, is accumulated. A new dimensionless variable (*LFDELAY*) was introduced to account for the delay in leaf emergence associated with limited soil aeration. *LFDELAY* is calculated as a function of the soil aeration early ($XSTAGE<2$) or late in the season as:

$$SADF6 = \exp(-ASD \times 0.01) \qquad (21)$$
$$SADF6 = \exp(-ASD \times 0.15)$$
$$LFDELAY = SADF6$$

LFDELAY is used later in the calculation of the phyllochron:

$$CUMPH_{DAY} = CUMPH_{DAY-1} + (TI \times LFDELAY) \qquad (22)$$
$$CUMDELAY_{DAY} = CUMDELAY_{DAY-1} + DTT \times (1 - LFDELAY)$$

where *CUMPH* is the accumulated number of phyllochrons (unitless), which corresponds with the number of emerged leaves, and *TI* is the fraction of the phyllochron accumulated in the day. *CUMDELAY* is used during the reproductive growth stage to modify the calculation of the normal plant senescence (*SLAN*). In CERES-Maize, *SLAN* during the reproductive stage is calculated as a function of the accumulated thermal time during the phenological stage. The modified version calculates *SLAN* as a function of the accumulated thermal time plus *CUMDELAY*.

Additional inputs

The new additions allowing the simulation of a watertable by considering three possible scenarios: without a watertable, with a fluctuating watertable updated daily and with a controlled watertable which depth is read from an input file. In the third option, used here for our simulations, the model calculates the depth of the watertable according to the inputs and outputs of the system, as in the second option, and then computes subirrigation or drainage as required to move up or down the simulated depth of watertable to match the read depth.

Each experiment and treatment simulated using the controlled watertable scenario must have an additional input file with two variables per line: day of the year (*RDOY*) and soil layer number where the watertable is (*RWTLAYR*). Both variables are integers with four and three fields (I4 and I3 format code in FORTRAN). The file name has six characters and the extension MZ0. The six characters are: WT for watertable, two digits for the experiment number and two digits for treatment number. Thus, a file named 'WT0201.MZ0' corresponds to the watertable input file for experiment two, treatment one.

Results and discussion

Field data obtained in 1991 (Lizaso 1993) were used to compare the predictions of CERES-Maize (Version 2.1) and the modified version. Figure 1 shows the leaf area index predicted for both models and the measured values in the control plot. CERES-Maize underestimated leaf growth because it predicted a major nitrogen deficiency after tassel initiation (*ISTAGE*=3). The nitrogen stress factor that affected the simulation of leaf growth (*CNSD*2) during this period was 0.47 as compared with 0.30 in the modified version of generic CERES. A value of 0.47 means that the plant had on the average 47% of the required nitrogen for normal growth during this period. The modified version underestimated leaf expansion only during the last week of leaf growth.

When an early flooding was simulated (Figure 2), CERES-Maize predicted larger leaf area when the crop was flooded for eight days rather than when it was flooded for four days. No effect was predicted on the time of occurrence of maximum leaf surface. The modified version predicted a delay of one week in the occurrence of maximum leaf area between the four and eight days flooding that was similar to the field measured plants, but the peaks of both curves fell below the measured leaf area. Figure 3 shows the leaf area predicted when a flooding was simulated late in the vegetative growth stage. CERES-Maize predicted the same leaf area when four or eight days of waterlogging were simulated. The modified version, on the other

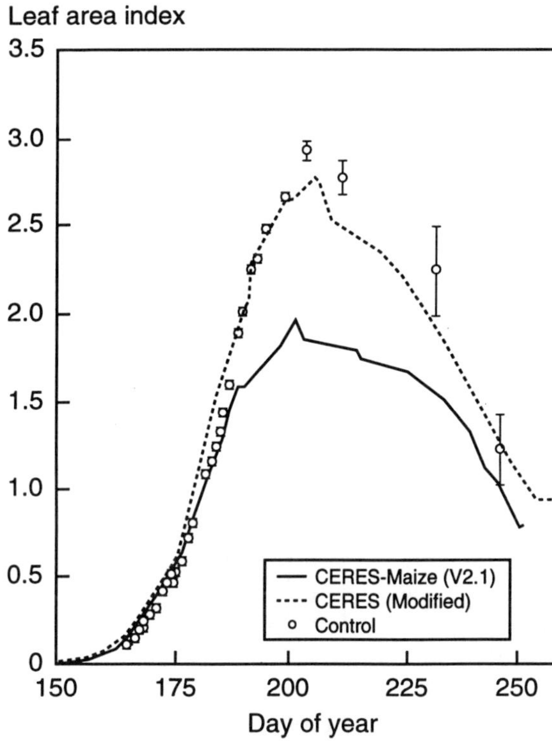

Figure 1. Leaf area index (m^2 m^{-2}) predicted by CERES-Maize and the modified version, and measured values for maize (*Source:* Lizaso 1993). Vertical lines are standard errors of measured values.

hand, yielded excellent predictions with a slight overprediction of the maximum leaf area when eight days of flooding was simulated.

One of the most important reasons to develop simulation models is to forecast crop yields accurately under a wide range of field conditions. Table 1 shows the predicted yields by CERES-Maize and the modified version of generic CERES, together with the measured values. CERES-Maize predicted yields between 38% and 80% of the field-measured yields. This poor performance was associated with the predicted nitrogen deficiency as discussed earlier. The modified generic CERES accurately predicted grain yields of the flooded plants. Yields of non-flooded plants were underestimated by 35% due to underestimates of leaf area and plant biomass (data not shown).

The prediction of root growth in CERES-Maize is insensitive to excessive soil water and also seems to overpredict root growth under well-aerated conditions. Figure 4 shows the pattern of root growth as predicted at 15 cm depth by CERES-Maize and by the modified generic CERES as compared with the field-measured root number. At this depth, CERES-Maize predicted that the maximum root length density of 5 cm cm^{-3} was reached in 25 days and the same value was maintained through the

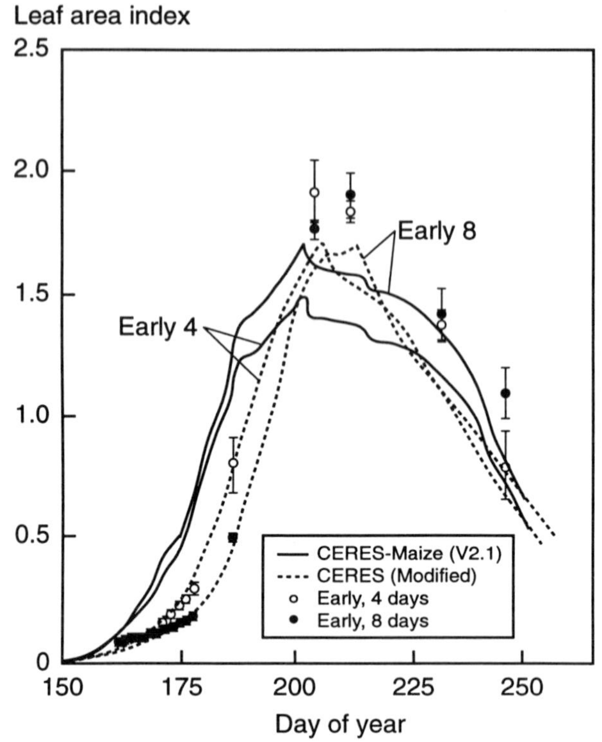

Figure 2. Leaf area index predicted by CERES-Maize and the modified version, and measured values in maize plants that were flooded four or eight days at the 6 leaf tip stage. Vertical lines are standard errors of measured values.

season. The measured values, on the contrary, showed a slower rate of root growth, with a mid-season plateau around 4–5 cm cm^{-3} and decreasing root numbers during the grainfilling. The modified generic CERES followed the described trend, although it seemed to overpredict root growth early in the season.

When an early season flooding was simulated (Figure 5), CERES-Maize predicted the same root growth as without soil submergence. The modified model predicted a delay in root growth, associated with an early soil submergence, before a fast proliferation of roots. This predicted delay, especially in the longer inundation, was not observed in the roots measured in the field. When the flooding period was simulated late during the vegetative growth stage (Figure 6), field data showed a fast but incomplete recovery in root number after drainage in the short flooding (four days) and no recovery in the long flooding (eight days). The modified model accurately predicted root growth in the short flooding, but predicted a recovery after the long flooding that did not occur.

The day when floral stigmata first appear (silking date) has received attention to evaluate crop phenology in simulation models. Table 2 shows the predicted and observed silking dates in the experiment here evaluated. CERES-Maize predicted

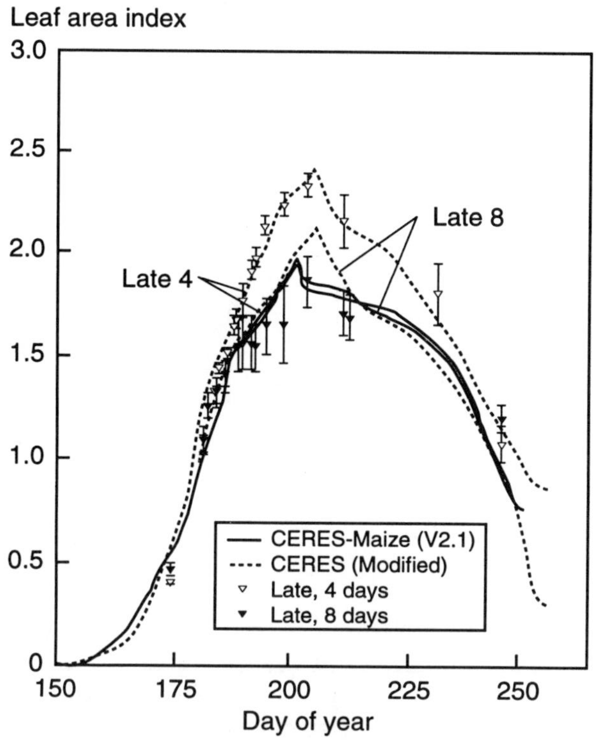

Figure 3. Leaf area index predicted by CERES-Maize and the modified version, and measured values in maize plants that were flooded four or eight days at the 12 leaf tip stage. Vertical lines are standard errors of measured values.

the same silking date for all the treatments. The predicted appearance of silks was three days earlier than the observed date in the field for non-flooded plants. The modified version of generic CERES was only sensitive to the larger delay observed in the early flooding when plants were in saturated soil for eight days.

The modified version of generic CERES produced reasonable estimations of above-ground plant growth in flooded conditions. Under well-aerated conditions, however, predictions seemed to be consistently less than field measured values. Combined underestimates of leaf area and biomass accumulation (data not shown) resulted in inaccurate crop yield forecasts.

The measured root numbers in mini-rhizotrons were converted to root length densities using the relation described by Melhuish and Lang (1968). The model proposed by Upchurch and Ritchie (1983) provided values of root length density that were too low. Bland and Dugas (1988) reported better correlations between root length densities calculated using the relation of Melhuish and Lang (1968) and roots washed from soil cores working with cotton. In sorghum, however, they could not find a simple relation (Bland and Dugas 1988). The modified version of generic CERES seemed to provide reasonable estimates of root growth, although the model

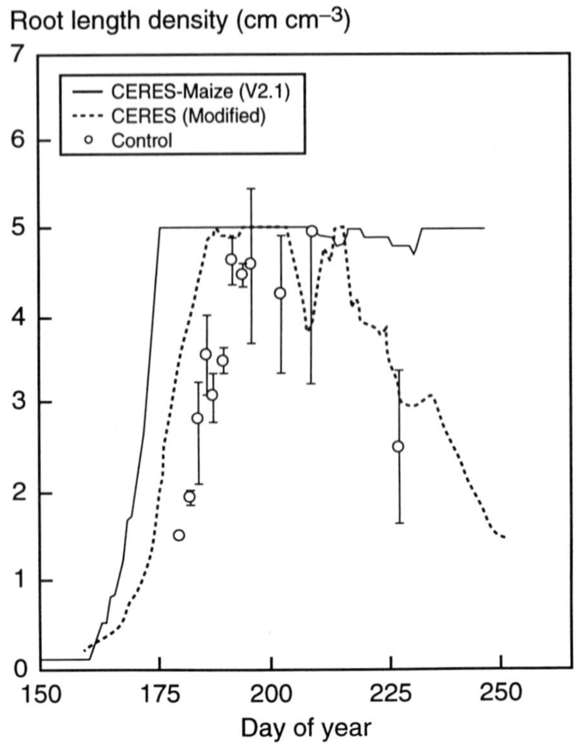

Figure 4. Root length density at 15 cm depth predicted by CERES-Maize and the modified version, and measured values. Vertical lines are standard errors of measured values.

still forecasts faster root growth than observed in well-aerated conditions. More accurate predictions of crop development will require further model adjustments.

Conclusions

Significant periods of soil submergence, during the vegetative growth stage of a maize crop, introduce changes in the soil-plant atmosphere relationships that extend well beyond the period of limited root aeration, and affect plant growth and development. Timing of occurrence of soil submergence had a significant effect on plant responses. To the best of the authors' knowledge, this is the first attempt to address comprehensively the complex array of factors and interactions involved in forecasting the whole-season performance of a maize crop affected by waterlogging. Therefore, this effort requires additional refinements.

The availability of field data with seasonal information of crop growth and development associated with periods of limited soil aeration is a major limitation for future model improvements. Two areas of information are critically required: phenological changes and root growth. Converting root counts obtained with mini-rhizotrons into root densities requires additional review before a standard method can be proposed.

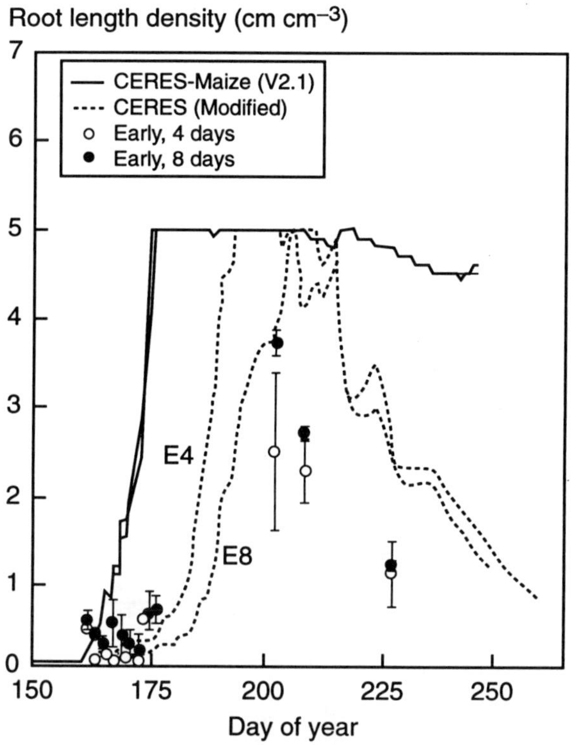

Figure 5. Root length density at 15 cm depth predicted by CERES-Maize and the modified version, and measured values. Maize was flooded four or eight days at the 6 leaf tip stage. Vertical lines are standard errors of measured values.

Table 1 Yields (kg ha^{-1}) predicted by CERES-Maize and the modified version, and measured data. Maize was flooded for zero, four or eight days at the 6 (early) or 12 (late) leaf tip stage. Values in parentheses are standard errors.

Treatment	Measured	Predicted	
		CERES-Maize (V. 2.1)	Generic CERES (modified)
Control	8605 (687.8)	3310	5594
Early, 4 days	4700 (710.6)	1760	4254
Early, 8 days	4387 (340.1)	2347	4225
Late, 4 days	4417 (477.0)	3415	4396
Late, 8 days	4421 (509.9)	3500	4589

Table 2 Silking date (day of year) predicted by CERES-Maize and the modified version, and measured data. Maize was flooded for zero, four or eight days at the 6 (early) or 12 (late) leaf tip stage.

Treatment	Measured (Day of year)	Predicted	
		CERES-Maize (V. 2.1)	Generic CERES (modified)
Control	205	202	206
Early, 4 days	210	202	206
Early, 8 days	213	202	214
Late, 4 days	206	202	206
Late, 8 days	208	202	206

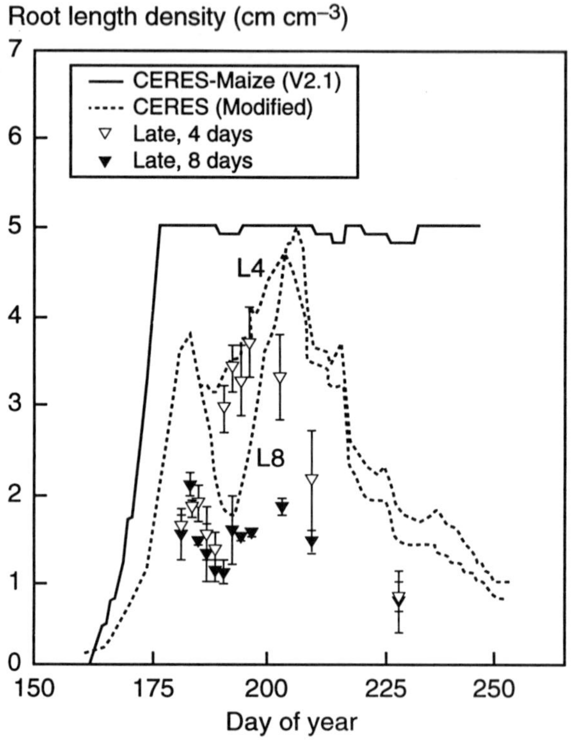

Figure 6. Root length density at 15 cm depth predicted by CERES-Maize and the modified version, and measured values. Maize was flooded four or eight days at the 12 leaf tip stage. Vertical lines are standard errors of measured values.

References

Armstrong W, Beckett P M (1987) Internal aeration and the development of stelar anoxia in submerged roots: A multishelled mathematical model combining axial diffusion of oxygen in the cortex with radial losses to the stele, the wall layers and the rhizosphere. New Phytol. 105:221–245.

Bland W L, Dugas W A (1988) Root density from minirhizotron observations. Agron. J. 80:271–275.

Dadoun F A (1993) Modeling tillage effects on soil physical properties and maize (*Zea mays* L.): Development and growth. PhD dissertation, Michigan State University, East Lansing, MI, USA. 194 p.

Drew M C, Saglio P H, Pradet A (1985) Larger adenylate energy charge and ATP/ADP ratios in aerenchymatous roots of *Zea mays* in anaerobic media as a consequence of improved internal oxygen transport. Planta 165:51–58.

Jones C A, Bland W L, Ritchie J T, Williams J R (1991) Simulation of root growth. Pages 91–123 in Hanks R J, Ritchie J T (Eds.) Modeling soil and plant systems. American Society of Agronomy (ASA), Monograph 31, Madison, WI, USA.

Jones C A, Kiniry J R (1986) CERES-Maize: A simulation model of maize growth and development. Texas A & M University Press, College Station, Tex., USA. 194 p.

Lizaso J I (1993) Flooding and field grown maize: Above- and below-ground responses and a simulation model. PhD dissertation, Michigan State University, East Lansing, MI, USA. 217 p.

Lizaso J I, Ritchie J T (1995) A field facility for monitoring root systems as affected by water table management. Pages 445–451 in Belcher H W, D'Itri F M (Eds.) Subirrigation and controlled drainage. Lewis Publishers, Boca Raton, FL, USA.

Luxmoore R L, Stolzy L H (1987) Modeling belowground processes of roots, the rhizosphere and soil communities. Pages 129–153 in Wisiol K, Hesketh J D (Eds.) Plant growth modeling for resource management. II. Quantifying plant processes. CRC Press, Boca Raton, FL, USA.

Melhuish F M, Lang A R (1968) Quantitative studies of roots in soil. I. Length and diameters of cotton roots in a clay loam soil by analysis of surface-ground blocks of resin-impregnated soil. Soil Sci. 106(1): 16–22.

Ritchie S W, Hanway J J (1984) How a corn plant develops. Iowa State University, Coop. Ext. Serv. Spec. Rep. 48, Ames, Iowa, USA.

Thomson C J, Armstrong W, Waters I, Greenway H (1990) Aerenchyma formation and associated oxygen movement in seminal and nodal roots of wheat. Plant Cell Environ. 13(4):395–403.

Upchurch D R, Ritchie J T (1983) Root observations using a video recording system in mini-rhizotrons. Agron. J. 75:1009–1015.

Mitigating climate change effects on rice yield

S. MOHANDASS and T.B. RANGANATHAN
Tamil Nadu Rice Research Institute, Tamil Nadu Agricultural University, Aduthurai 612 101, India

Key words: climate change, simulation, ORYZA1 model, General Circulation Models, rice yield reduction, mitigatory measures, India

Abstract

The ecophysiological crop model ORYZA1 was used to simulate the effect of climate change on rice production in India under the three General Circulation Models (GCMs), the General Fluid Dynamics Laboratory (GFDL), the Goddart Institute for Space Studies (GISS) and the United Kingdom Meteorological Office (UKMO). There was a reduction in grain yield in all the three GCMs when temperature effect was considered, especially under the doubled CO_2 level, the decrease being 9.4%, 12.5% and 12.0% for GFDL, GISS and UKMO scenarios respectively. Hence, the ORYZA1 model was used to identify important mitigatory options for reducing the negative effects of temperature in the GISS (comparatively more disadvantageous) model for the future climates. Options such as the use of temperature-tolerant varieties and adjusting the planting dates are known to mitigate the temperature-induced possible rice yield reduction in the changing climates of the GCM.

Introduction

Crop growth and yield under optimal conditions are largely determined by weather during the growing season. Climate change via increasing atmospheric CO_2 concentration can affect global agricultural production through changes in photosynthesis (Lemon 1983; Cure and Acock 1986) and transpiration rates (Goudriaan and Van Laar 1978; Sionit et al. 1980). Climate change is likely to have greater impact on rice production depending upon the magnitude of variation in CO_2 and temperature. Increased temperature significantly reduced the spikelet fertility and also modified the crop duration (Kropff et al. 1994; Matthews et al. 1995; Mohandass et al. 1995). An adverse climate change scenario and its effect on rice production has also been studied (Matthews et al. 1995). In this paper, we explored the possibilities of employing different mitigating options to alleviate the climate change effects on rice production under different climate change scenarios defined by different General Circulation Models (GCMs).

Material and methods

The rice crop growth model ORYZA1 (Kropff et al. 1994) was used in the present study to evaluate the effect of climate change on rice production in the major rice growing regions in India. The model simulates the potential yield of rice, mainly driven by solar radiation and temperature and varietal characteristics. It is assumed to be free of insect pest and disease effects and to have optimal nutrient supply without

M.J. Kropff et al. *(eds.): Applications of Systems Approaches at the Field Level, 169–176.*

Table 1. Mean simulated potential rice yields (kg ha^{-1}) of variety IR36 and their coefficients of variation (CV%) for current climatic conditions in selected locations in India.

Location	Main season		Second season	
	Yield	CV%	Yield	CV%
Aduthurai	8307	4.4	9817	5.3
Bijapur	8121	6.1	5342	37.1
Coimbatore	9023	8.1	10277	2.9
Cuttack	7129	12.6	8958	13.2
Hyderabad	9007	–	–	–
Kapurthala	10503	–	–	–
Pattambi	7736	0.0	8541	9.5
Madurai	8823	6.1	9513	10.8
Patancheru	8408	5.3	8365	24.4

water stress. For the present study, a calibrated ORYZA1 model was used with the crop data sets of rice cultivar IR36, obtained from multilocation experiments conducted at Tamil Nadu Rice Research Institute, Aduthurai, Tamil Nadu, India, during the dry season (June–September 1993), in which the observed and simulated yields showed good agreement (Mohandass et al. 1995).

Simulation analysis was done using historical weather data (Mohandass et al. 1995) for Aduthurai, Bijapur, Coimbatore, Hyderabad, Madurai, Pattambi and Patancheru (in southern India), Cuttack (in eastern India) and Kapurthala (in northern India). Dates of sowing and transplantation for each site were based on the local practice.

General circulation models (GCMs)

Three GCM scenarios used in this study were those of the General Fluid Dynamics Laboratory (GFDL, Wetherald and Manabe 1988), the Goddard Institute for Space Studies (GISS, Hansen et al. 1988), and the United Kingdom Meteorological Office (UKMO, Wilson and Mitchell 1987) models.

The GCMs describing processes in the earth's atmosphere, have predicted that a doubling of the current CO_2 level will produce an accompanying warming of the earth's surface by 1.9–5.2°C (Cubash and Cess 1990). For crop growth, it is not the global mean that is important, but more the changes in climate during the growing season in the region where the crop is grown. Probable changes in precipitation and solar radiation under the climate change scenarios were not taken into consideration in this analysis.

Results

Current climate

The average potential rice yields for the nine sites under current climatic conditions along with their coefficients of variation (CV%) are shown in Table 1. Simulated yields ranged from 5342 kg ha^{-1} (second season in Bijapur) to 10503 kg ha^{-1} (at Kapurthala).

Table 2. Simulated effects of climate change on selected crop growth parameters of rice variety IR36 for two CO_2 levels and temperature increase according to different GCM models. Values (%) represent average of nine selected locations in India.

CO_2 level	Plant parameters	Temperature scenario			
		Present	GFDL	GISS	UKMO
330 ppm	Grain yield (kg ha^{-1})	8328	7588	7405	7493
	Temperature effect	100%	91.1%	88.9%	90.0%
	Maturity duration (d)	127	128	133	138
	Temperature effect	100%	100.8%	104.7%	108.7%
	Total above-ground biomass (ton ha^{-1})	13.55	12.80	13.01	13.46
	Temperature effect	100%	94.5%	96.0%	99.3%
	Harvest index	0.615	0.598	0.574	0.561
	Temperature effect	100%	97.2%	93.2%	91.2%
660 ppm	Grain yield (kg ha^{-1})	12444	11275	10891	10953
	Temperature effect	100%	90.6%	87.5%	88.0%
	Maturity duration (d)	127	128	133	138
	Temperature effect	100%	100.8%	104.7%	108.7%
	Total above-ground biomass (ton ha^{-1})	20.03	18.85	19.00	19.56
	Temperature effect	100%	94.1	94.9%	97.6%
	Harvest index	0.622	0.604	0.578	0.565
	Temperature effect	100%	97.1%	92.9%	90.8%

The higher yield in Kapurthala might be due to the low temperature and higher solar radiation prevalent during the growing season in that area. The mean potential production of rice across all sites and years fell within the predicted level of potential rice grain yield (around 10 ton ha^{-1}) in previous simulations for India (Penning de Vries 1993).

Simulation of climate change impact for GCM scenarios

The UKMO scenario predicted higher temperature increase (+4.19°C) than GFDL (+1.86°C) or GISS (+3.75°C) during the respective growing seasons across locations (data not shown). The results of simulation for three GCM scenarios are presented in Table 2. Simulated grain yield of rice for different GCM scenarios ranged from 88.9% (in the GISS) to 91.1% (in the GFDL) of crop yield under current CO_2 level (330 ppm), while it was from 87.5% (in the GISS) to 90.6% (in the GFDL) under the doubled CO_2 (660 ppm) scenario. The UKMO temperature gave the highest reduction in Harvest Index (HI) of 8.8% and 9.2% respectively for the current and doubled CO_2 conditions. The total above-ground biomass was very much affected under the GFDL scenarios in both levels of CO_2. The UKMO scenarios predicted longer duration of maturity in both current and double CO_2 levels (Table 2).

All the GCM scenarios (under 660 ppm CO_2 level) predicted decreased yields (Table 3) for almost all locations. Mean reduction across different GCM scenarios ranged from −8.77% in Patancheru to −14.90% in Madurai. The marginal increase (+8.3%) was noticed at Bijapur with GFDL scenario, while higher decrease (−21.9%) was simulated at the same Bijapur site for the UKMO scenario.

Table 3. Changes in potential rice grain yields (ton ha^{-1}) for three GCM scenarios in nine selected sites in India.

Site	CO_2* 660 ppm	GFDL Yield	%	GISS Yield	%	UKMO Yield	%	Mean
Aduthurai	12.9	12.0	−6.3	11.3	−12.5	11.4	−11.6	−10.13
Bijapur	9.6	10.4	+8.3	7.9	−17.7	7.5	−21.9	−10.43
Coimbatore	12.8	11.7	−9.6	11.2	−12.5	11.0	−14.8	−9.42
Cuttack	11.3	10.1	−10.7	9.9	−12.4	9.1	−19.5	−14.25
Hyderabad	13.9	13.1	−5.8	12.2	−12.2	12.0	−14.7	−10.90
Kapurthala	16.2	13.9	−14.2	14.7	−9.3	15.4	−5.0	−9.50
Pattambi	10.2	9.2	−9.8	8.9	−12.7	9.0	−12.7	−11.73
Madurai	13.2	10.4	−21.2	11.5	−12.9	11.8	−10.6	−14.90
Patancheru	11.8	10.6	−10.2	10.4	−11.9	11.3	−4.2	−8.77
Mean	12.4	11.3	−9.4	10.9	−12.5	11.0	−12.0	−11.30

*In this column, the temperature effect is not simulated and the yields are for the 660 ppm CO_2 alone.

Table 4. Effect of temperature tolerance of new ideotype on rice yields under the GISS climate change scenario in different locations in India.

Location	Day of year	Current (kg ha^{-1})	% Change due to temperature tolerance*		
			+0°C	+2°C	+4°C
A: Main planting season					
Aduthurai	161	8101	+39.9	+49.5	+49.0
Bijapur	182	7177	+9.5	+25.1	+31.9
Coimbatore	161	8915	+25.8	+27.1	+28.9
Cuttack	213	7005	+41.1	+46.2	+41.1
Hyderabad	152	9410	+29.7	+29.8	+29.8
Kapurthala	191	10323	+42.5	+42.5	+42.5
Pattambi	201	7148	+24.4	+24.4	+24.4
Madurai	161	8414	+37.3	+53.2	+53.2
Patancheru	152	8452	+22.5	+22.5	+22.5
Mean		8327	+30.3	+35.6	+35.9
B: Second planting season					
Aduthurai	244	7127	+25.0	+25.0	+25.0
Bijapur	121	8351	+21.1	+26.7	+26.7
Coimbatore	244	9003	+13.7	+13.7	+25.9
Cuttack	001	8038	−49.8	+30.8	+18.5
Pattambi	121	8138	+21.3	+21.3	+21.3
Madurai	244	7994	+35.1	+35.1	+35.1
Patancheru	105	8158	+26.2	+26.2	+26.2
Mean		8116	+13.2	+16.7	+25.5
Overall mean		8235	+22.8	+27.3	+31.4

*Sensitivity of spikelet sterility is shifted by the respective temperature increase taking the ORZYA1 as base.

Thus, the mean grain yield was very affected under the GISS scenario for both current and double CO_2 levels. Hence, the assessment of the options to mitigate the negative effects of climate change was discussed with special reference only to the GISS scenario.

Mitigating measures

Varietal tolerance to temperature

It is likely that due to the predicted climate change, there is a possibility for reduction in grain yield due to the increased temperature depending on the nature of the GCM used. Therefore, it is felt more pertinent to develop some mitigatory measures in order to overcome or reduce the potential negative effect of climate change on rice production in India.

The present simulation analysis is mainly based on the assumption that cultivar characteristics are almost the same in the future as at present. In reality, it is true that the plant breeding programme will develop newer varieties more closely adapted to the predicted changing climatic conditions, thus alleviating the climate change impact (Matthews et al. 1995). One possibility that might be promising as of now is the use of varieties that are more tolerant to high temperature. Considerable variation among varieties to high temperature tolerance is already available (Satake and Yoshida 1978).

In the current ORYZA1 model, spikelet fertility is sensitive to high temperature until 36.6°C is reached.

As an example of the possible effect of an increase in tolerance to high temperature, Matthews et al. (1995) assumed that the sensitivity of spikelet sterility is shifted by 2°C, taking the ORYZA1 response for mitigating the negative effects of climate change. In the present study, simulation analysis has been done by shifting the sensitivity of spikelet sterility by different temperature increments (0, +2 or +4°C) considering the predicted temperature increase (of +3.75°C) in the GISS scenario. The simulation runs are made for the adaptation to these temperature tolerance (TT) increments for both main and second planting seasons in all nine locations (Table 4).

The results indicated that though there was a general increase in grain yield due to temperature tolerance (TT) from 0°C to 4°C, the effect was more phenomenal for the main planting season only, the increase being from 30.3% to 35.9% from 0°C to 4°C respectively in the main season. For locations, Madurai seemed to respond more favourably (from 37.3 to 53.2%) for the temperature tolerance up to 4°C. However, in sites like Hyderabad, Pattambi and Patancheru, the adaptation did not have any effect. Nevertheless, a temperature tolerance of 2°C is found sufficient to mitigate the climate change effect in places such as Aduthurai, Cuttack and Madurai.

Planting date adjustments

Suitable adjustments in planting dates are necessary for efficient utilization of natural resources under the climate change scenarios. Towards this, an attempt has been made to simulate the temperature tolerance effects under varied planting dates for the GISS scenario, the most disadvantageous GCM model, starting from day number 1 to 365 in at least eight locations, and the results are depicted in Figure 1. The simulation indicated that for Aduthurai the planting date may be advanced from June to the first week of April, so that the high temperature tolerance could take care of the climate

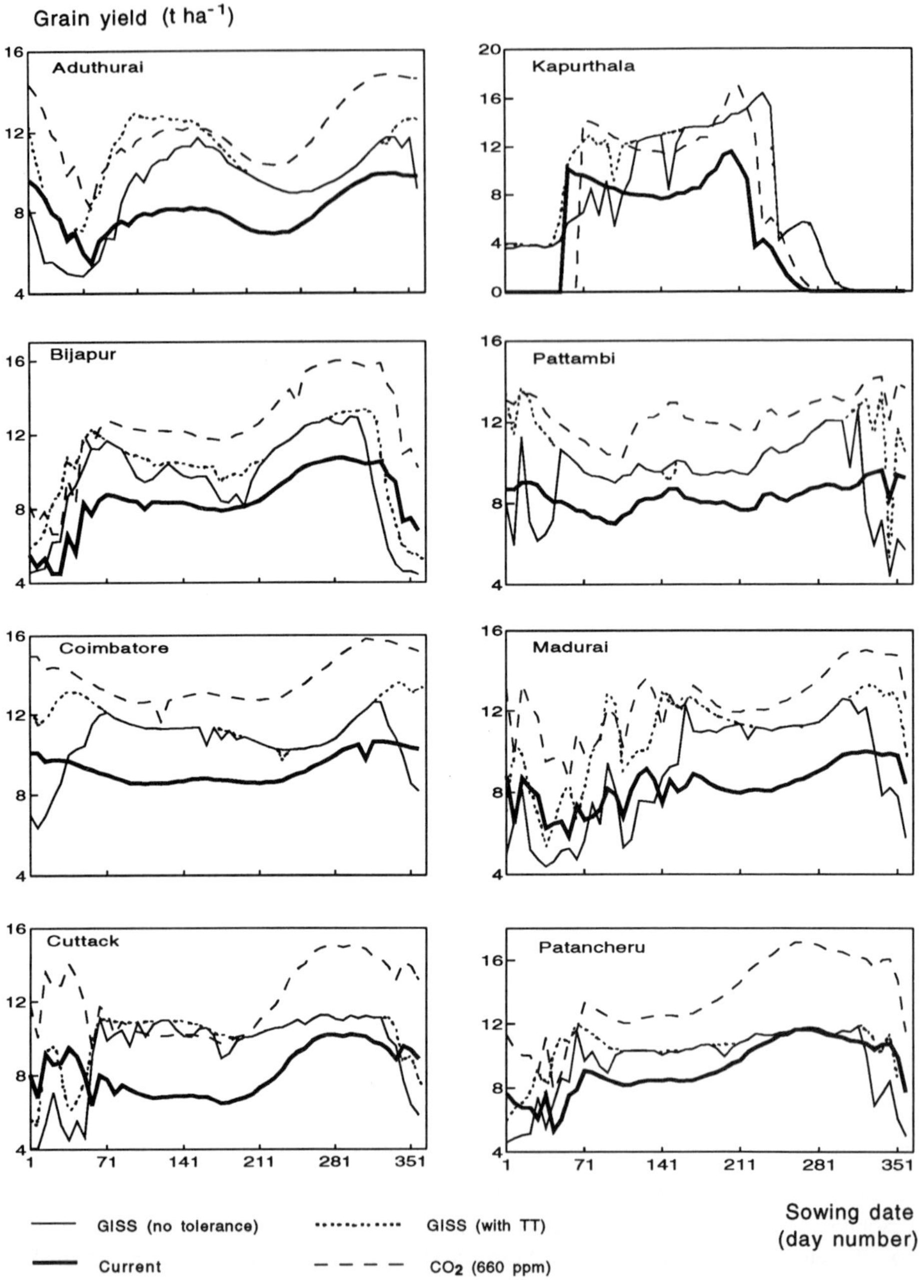

Figure 1. Effect of mitigating options (TT = temperature tolerance) on climate change effect on grain yield (using GISS scenario) in selected sites in India: Aduthurai, Bijapur, Coimbatore, Cuttack, Kapurthala, Pattambi, Madurai and Patancheru.

change impact. This is true also for Coimbatore, with the shift in planting date to the first week of January.

Nevertheless, the planting date has to be delayed in other places (like Kapurthala) for mitigating the temperature effect of climate change. If this is to be adopted, it might not be possible to plant one more crop in the second season, thus leading to overall reduction in the food production in these regions. Therefore, advancing or delaying the planting dates must be decided based on the rainfall pattern and its distribution in that particular region.

Conclusions

The model gives reasonable yield predictions in almost all locations chosen for the study. The predicted effects of climate change on grain yield by three different GCMs were of the order of −9.4%, −12.5% and −12.0% for GFDL, GISS and UKMO scenarios respectively from no temperature change under double CO_2 concentration.

Some mitigating measures are known to alleviate the high temperature induced climate change effect. Varietal tolerance to high temperature is able to mitigate the detrimental effect of temperature increase in currently high temperature regions. Also, changes in the management strategies such as advancing or delaying the planting dates is found favourable for increased yields.

Acknowledgements

The research described in this article has been funded by the IRRI/SARP Project on Simulation of Impact of Climate Change (SIC2) on Rice Production.

References

Cubash U, Cess R D (1990) Processes and modelling. Pages 69–91 in Houghton J T, Jenkins G J, Ephraums J J (Eds.) Climate change. Cambridge University Press, Cambridge, UK.

Cure J D, Acock B (1986) Crop responses to carbon dioxide doubling: A literature survey. Agric. For. Meteorol. 38:127–145.

Goudriaan J, Van Laar H H (1978) Relations between leaf resistance, CO_2 concentration and CO_2 assimilation in maize, beans, lalang grass and sunflower. Photosynthetica 12:241–249.

Hansen J, Fung I, Lacis A, Rind D, Lebedeff R, Russell R (1988) Global climate changes as forecast by Goddard Institute for Space Studies three dimensional model. J. Geophys. Res. 93:9341–9364.

Kropff M J, Van Laar H H, Matthews R B (Eds.) (1994) ORYZA1: An ecophysiological model for irrigated rice production. SARP Research Proceedings, International Rice Research Institute, P.O. Box 933, Manila, Philippines. 110 p.

Lemon E R (1983) CO_2 and plants: The response of plants to rising levels of atmospheric carbon dioxide. Westview Press, Boulder, CO., USA.

Matthews R B, Kropff M J, Bachelet D, Van Laar H H (Eds.) (1995) Modelling the impact of climate change on rice production in Asia. CAB International, Wallingford, UK. 289 p.

Mohandass S, Ranganathan T B, Kareem A A (1995) Simulating the effect of varietal adaptation to climate change. Pages 87–89 in Aggarwal P K, Matthews R B, Kropff M J, Van Laar H H (Eds.) Application of systems approaches in plant breeding. SARP Research Proceedings, International Rice Research Institute, P.O. Box 933, Manila, Philippines.

Penning de Vries F W T (1993) Rice production and climate change. Pages 174–189 in Penning de Vries F W T, Teng P S, Metselaar K (Eds.) Systems approaches for agricultural development. Kluwer Academic Publishers, Dordrecht, The Netherlands.

Satake T, Yoshida S (1978) High temperature-induced sterility in indica rices at flowering. Jap. J. Crop Sci. 47:6–17.

Sionit N, Hellmers H, Strain B R (1980) Growth and yield of wheat under CO_2-enrichment and water stress. Crop Sci. 20:456–458.

Wetherald R T, Manabe S (1988) Cloud feedback processes in a general circulation. J. Atmos. Sci. 45: 1397–1415.

Wilson C A, Mitchell J F B (1987) A doubled CO_2 climate sensitivity experiment with a GCM including a simple ocean. J. Geophys. Res. 92:13315–13343.

Competition for light in windbreak-millet systems in the Sahel

M. MAYUS[1], H. VAN KEULEN[2] and L. STROOSNIJDER[1]

[1]*Department of Irrigation and Soil and Water Conservation, Wageningen Agricultural University, Nieuwe Kanaal 11, 6709 PA Wageningen, The Netherlands*

[2]*DLO-Research Institute for Agrobiological and Soil Fertility Research, and Department of Animal Husbandry, Section Animal Production Systems, Wageningen Agricultural University, Marijkeweg 40, 6709 PG Wageningen, The Netherlands*

Key words: millet, windbreaks, competition, model

Abstract

Windbreaks, although beneficial for wind erosion control, compete with crops for light, nutrients and soil water, which may affect crop yields at the tree-crop interface. A modelling approach was chosen to optimize the design of a windbreak-millet system at the ICRISAT Sahelian Centre, Niger. Soil water content, soil nutrients, radiation and millet growth were studied. An existing plant growth model was tested for millet (*Pennisetum glaucum* L.) during the 1993 growing season. Thereafter, the tested model was extended to evaluate the influence of shading on the production of the adjacent millet crop. The required minimum plant and soil data sets were either field determined or obtained from literature. A comparison of simulated and field data for the growing season of 1993 suggests that 50% of the reduction in dry matter production adjacent to the shelter *Bauhinia rufescens* is due to shading and 50% is due to below-ground competition.

Introduction

Millet (*Pennisetum glaucum* L., R. Br.) is the most important source of food in the Sahel. In this region, production is limited by low soil fertility, low and variable rainfall and wind erosion. Windbreaks (WB), although beneficial for wind erosion control, may compete with crops for space, light, water and nutrients. On the other hand, also positive effects such as water conservation, and by-products providing useful material or cash to farmers, are often mentioned.

WB-crop systems have been extensively studied all over the world, including the Sahel (Brenner 1991; Long 1989; Onyewotu et al. 1994). Whereas some physical processes are reasonably well understood, insight in effects on soil nutrients, soil water, radiation and crop growth is limited. The reason is that a windbreak-crop system is very complex with many interacting factors. A model could help analysing such systems with the aim to design optimal WB-crop systems in dependence of site and objectives of the system.

Several models on wind erosion do exist, i.e. EPIC (Erosion Productivity Impact Calculator; Williams 1990) and WEPS (Wind Erosion Prediction System; Hagen 1991). Both models focus more on the erosion process than on crop growth, which is described rather simplistically, lacking the required detail to describe competition effects between windbreaks and a crop. Plant growth models are available that account

M.J. Kropff et al. *(eds.): Applications of Systems Approaches at the Field Level, 177–187.*

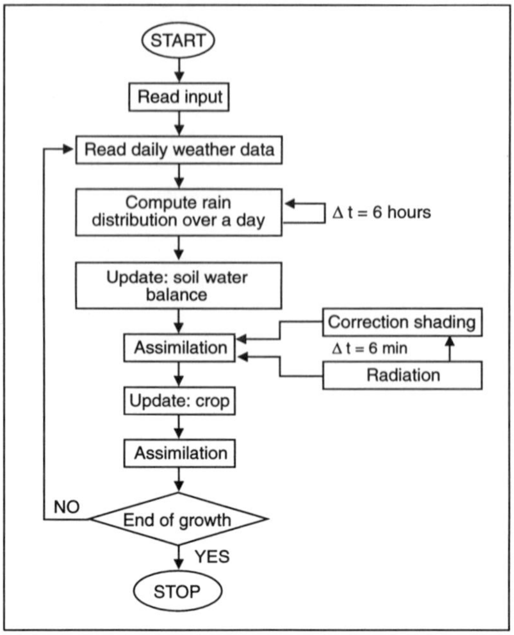

Figure 1. Task flow diagram for the windbreak-millet shade model.

for the presence of single trees, RECAFS (Conijn 1995), or weeds, INTERCOM (Kropff and Van Laar 1993), but currently no models exist that simulate competition between windbreak and millet.

Objectives of this study were: (i) to validate an existing millet crop growth model and (ii) to integrate major effects of WBs on the crop as established in a prior experimental study into this tested model: competition for light and soil water, and reduced speed of 'normal' winds. The model WB-mil-shade, described here, is the first step and accounts for the shading effect of a *Bauhinia* windbreak of given height and porosity on a millet crop in the Sahel.

The model WB-mil-shade

This model is based on the crop growth model CP-BKF3 (Cultures Pluviales-Burkina Faso), which accounts for the soil water and nitrogen balance (Van Kraalingen and Van Keulen 1988; Verberne et al. 1995). The modified model WB-mil-shade simulates the soil water balance, radiation without or with shading and crop growth for each row parallel to the windbreak separately (Figure 1). The nitrogen part has been reduced, in a way that nitrogen is not limiting the production. The option 'without shading' is obtained by simply setting WB height to zero in the model. The growth of WB trees is not simulated.

Shading is introduced by reducing the amount of radiation available for the adjacent

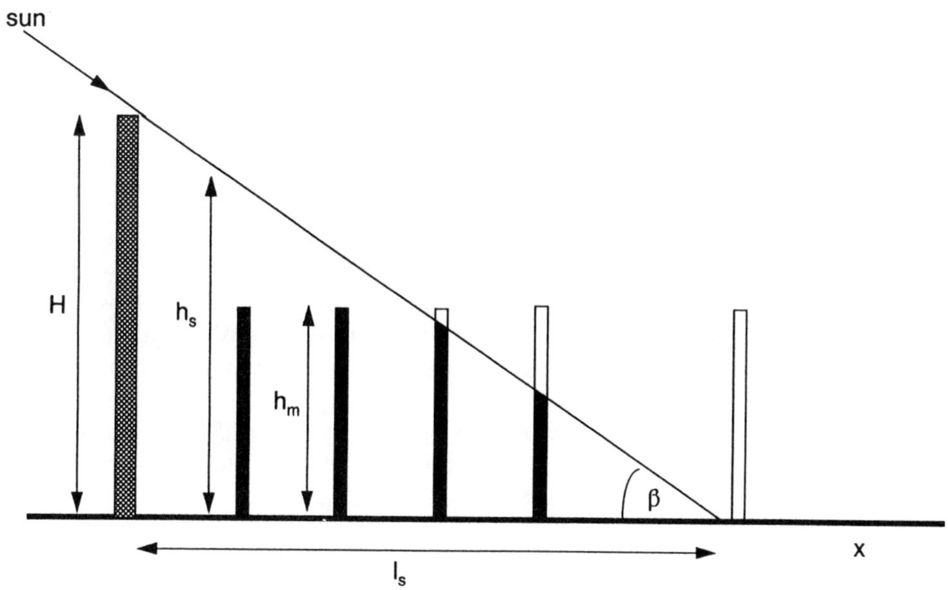

Figure 2. Shaded area behind a porous barrier.

millet crop and, consequently, its assimilation. In the model, the WB is regarded as a two-dimensional barrier of a certain height (H, from ground level) and porosity (Figure 2), equal to the measured values for *Bauhinia* WBs in the experiment. The length of the WB is considered infinite, i.e. WB edge effects are neglected. Shading of millet from above is not considered; it played a minor role in the experimental design (only the first row was partly covered by tree branches at the end of the growing season). Reflection by the tree canopy is neglected and absorption is implicitly accounted for by the porosity factor. Interception of radiation by the WB is the amount of light shielded, minus that passing through the barrier (porosity). Interception differs for the direct and diffuse radiation fluxes.

Direct light (S_{dir}) comes from the direction of the sun, and can throw shade, whereas that is not possible with diffuse light (S_{dif}), coming from all directions. The level of intercepted S_{dir}, as well as the position and extension of the shade, depend on solar elevation (β), the angle of the sun beam with the WB line (γ) and WB height (H) in m.

Form and position of the shadow change continuously with the position of the sun. Total length of the shadow $l_s(t)$ and its height $h_s(x, t)$, both in m, for each distance x are calculated as function of time t (Figure 2):

$$l_s(t) = \sin\gamma(t)\frac{1}{\tan\beta(t)}H(t) \tag{1}$$

$$h_s(x, t) = (l_s(t) - x)\frac{H(t)}{l_s(t)} \tag{2}$$

The shaded fraction of a plant pocket (f_s) depends on millet height (h_m) and its distance from the windbreak (x) (Figure 2). Tree and plant height, as well as porosity, are changing in the course of the cropping season. Their values are incorporated as linear functions of time. For each plant row (indicated by i), f_s is computed as:

$$f_s(x_i, t) = \begin{cases} h_m(x_i, t) > h_s(x_i, t): & \dfrac{h_s(x_i, t)}{h_m(x_i, t)} \\ h_m(x_i, t) \leq h_s(x_i, t): & 1 \end{cases} \quad (3)$$

The WB intercepts also part from the diffuse light. By means of fitting, it was found that a coefficient of 0.5 multiplied by the porosity factor (ε) of the WB for S_{dif} reaching the adjacent crop row, resulted in good agreement between measured and simulated global radiation.

The shaded fraction of the crop receives only the fraction of S_{dif} that penetrates through the WB ($f_{dir} \times \varepsilon$); however, the unshaded fraction of the crop ($1 - f_s$) receives the total amount of S_{dir}. Within the shade, S_{dir} is assumed to be constant in space. Now the fraction of $S_{dir}(t)$ and $S_{dif}(t)$ adjacent to the WB for shaded and unshaded millet can be given for each row at distance x_i as a function of time t:

$$\begin{aligned} f_{dir}(x_i, t) &= f_s(x_i, t)\varepsilon S_{dir}(t) + (1 - f_s(x_i, t))S_{dir}(t) \\ f_{dif}(x_i, t) &= f_s(x_i, t)0.5\varepsilon S_{dif}(t) + (1 - f_s(x_i, t))S_{dif}(t) \end{aligned} \quad (4)$$

These fractions of $S_{dir}(t)$ and $S_{dif}(t)$ are used for the calculation of the available amount of direct respectively diffuse light above the canopy as function of the distance from the WB.

Since radiation conditions change continuously, a time step of 6 minutes is applied for radiation, and for the light-dependent assimilation. However, the conversion of CO_2 into dry matter and its partitioning among organs is calculated in daily intervals.

Prior to calculation of crop growth, based on light interception, the soil water balance is updated using a parametric model (Stroosnijder 1982). Downward water flow after infiltration is computed according to the tipping bucket principle with the total soil depth subdivided in 12 layers. The first 10 soil layers correspond with the zone from the soil surface till the maximum rooting depth, while the last very thick ones were added to cover the depth to the groundwater table. Redistribution of water between layers in the profile takes place each time step. The rate of flow of redistribution is set to the maximum (saturated) conductivity. For a sandy soil, this is about 14 cm h^{-1} (Bley 1990). Calibration test runs showed satisfactory results for the soil water balance with a time step of 6 h, which was therefore used. Potential evapo-transpiration is calculated using Penman (1948). The total daily rate of potential evapo-transpiration then is partitioned on the basis of the plant leaf area index into potential soil evaporation and potential plant transpiration.

Millet crop, windbreak, soil and weather data were derived from WB-millet system experiments, conducted during the 1992 and 1993 rainy seasons at the Sahelian Centre of ICRISAT in Niger. Results of 1993 (with more detailed field data) were used for calibration, and those of 1992 for testing the model. The model CP-BKF3 has been specifically designed and calibrated for maize, sorghum and millet growth in the Sahel (Burkina Faso). For the specific environment (southwest Niger) and millet cultivar, the following parameters have been calibrated again for the development of WB-mil-shade; plant parameters: initial biomass, development rate, specific leaf area, leaf age, growth rate of grains, days from sowing till emergence, and soil parameters: soil moisture retention curve, N and organic carbon content per soil layer, number of soil layers.

Material and methods of field experiments

During the 1992 and 1993 rainy seasons, field experiments were conducted at the Sahelian Centre of ICRISAT in Niger (13° 14′N; 2° 16′E). The windbreaks, oriented north-south, consisted of a double row of alternately planted trees, with 1.5 m between rows and 3 m between trees in the row. The windbreaks were pruned to a height and width of about 2 and 3.5 m respectively before the onset of the rainy season. Pearl millet (*Pennisetum glaucum* L., R. Br. var. CIVT) was grown at a density of 30000 plants (10000 pockets) per hectare, starting at a distance of 1 m from the WB lines. Phosphorus was broadcast applied before sowing (45 kg P_2O_5 ha^{-1}) as single super-phosphate. Nitrogen was applied at seeding in each hill as calcium-ammonium nitrate (45 kg N ha^{-1}). The experiment was a split plot design in randomized blocks with three replicates, each consisting of 9 plots with various shelter species. Studies presented here refer either to means of plots with windbreaks of *Bauhinia rufescens Lam.*, to means of 10 pockets of the plot with a WB of *Ziziphus mauritiana*, or to means of all plots. The plots were 28.5 m×50 m. All measurements were performed along transects perpendicular to the WB.

Rainfall and other weather data were collected at the central weather station of ICRISAT located 1 km westward of the windbreak site. Global radiation was measured continuously in the lee of the *Bauhinia* WB with 4 tube solarimeters (Delta T, UK) at 20 cm above the soil, oriented perpendicular to the plant rows. The middle of the tubes was placed at a distance of 1, 2, 3 and 10 m from the WB.

Soil water content was measured with a neutron probe (Troxler, model 3332) at weekly intervals at 0.30 m increments (between 0.10 and 1.90 m soil depth) and distances of 1, 2, 3, 5, 7 and 10 m from the base line of trees of *Bauhinia*. Data were analysed as repeated measurements over lateral distances from WBs (three replicates).

Porosity and height of WB trees were observed at the beginning and end of each growing season in all three blocks. Estimates of porosity, which is a ratio of perforated area to total area, were derived from black and white photography. Estimated porosity was then checked by simulating total global radiation on day numbers 233, 201 and 171 at 1, 2, 3 and 10 m from the windbreak and comparing results with measured global radiation.

Millet parameters for calibration of the model, i.e. plant weight, leaf area index (LAI) and development stage (DVS) were measured on 10 pockets every 10 days. These pockets were located at a distance of 10 m and more from a WB of *Ziziphus mauritiana* trees (*Ziziphus* plot), where the WB influence on the crop is assumed negligible.

Millet parameters for validating the model were determined in plots of *Bauhinia* and other WB fields; yield and dry matter production were determined at physiological maturity in all plots. Millet panicles and straw were harvested from 20 pockets in each millet row. Panicles were oven-dried at 60°C and threshed manually. Stover was left in the field and weighed after sun-drying for three weeks. All data reported are based on the effectively harvested area, not including the area occupied by windbreaks. Mean values were calculated separately for each distance. Height of millet plants was measured at weekly intervals on *Bauhinia* plots only.

Results

Millet-Sahel: not shaded

Dry matter accumulation during the 1993 growing season is shown in Figure 3A. Simulated (after calibration) shoot dry matter corresponds well with the field measurements with the exception of early growth and final harvest. However, the seasonal simulated maximum value (2224 kg ha^{-1}) was similar to that measured on *Ziziphus* plot (2229 kg ha^{-1}). In 1992 (tested) simulated and measured final shoot dry matter (of all plots) were also similar with 2637 and 2620 kg ha^{-1} respectively.

Comparison of simulated (after calibration) and measured leaf area index (*LAI*) for the growing season 1993 shows that simulated maximum *LAI* is within the range of the observations (Figure 3B), but in the early and late growing season the model overestimates *LAI*, as well as the weight of the living leaves.

Simulated seasonal water content of the soil follows the course and level of the data measured at a distance of 10 m from the WB in 1993 (calibration) and 1992 (validating) (Figure 4).

WB-Millet Sahel: shaded

The north-south-oriented WB of *Bauhinia* trees was 2–3 m high with a porosity of 0.9–0.2 at the beginning and end of the growing season respectively. The radiation module gives good estimates of light reduction as function of the distance from the WB.

Simulated millet shoot dry matter to the west of such a shelter shows an asymptotic course with distance from the *Bauhinia* belt (Figure 5). Directly adjacent to the shelter (1 m) production is about 400 kg ha^{-1} lower than at 15 m, where shading is virtually absent. The absolute values for shoot dry matter in the experimental *Bauhinia* plot (1993) are about 50% lower than the simulated values (Figure 5). Biomass production varied considerably along the transect perpendicular to the *Bauhinia,* nevertheless the tendency of an increase in dry matter with distance is also visible in the measured data.

The relative yield reduction at a distance of 5 m behind the *Bauhinia* WB compared to the unshaded zone (15 m behind the WB) is similar for measured (7%) and

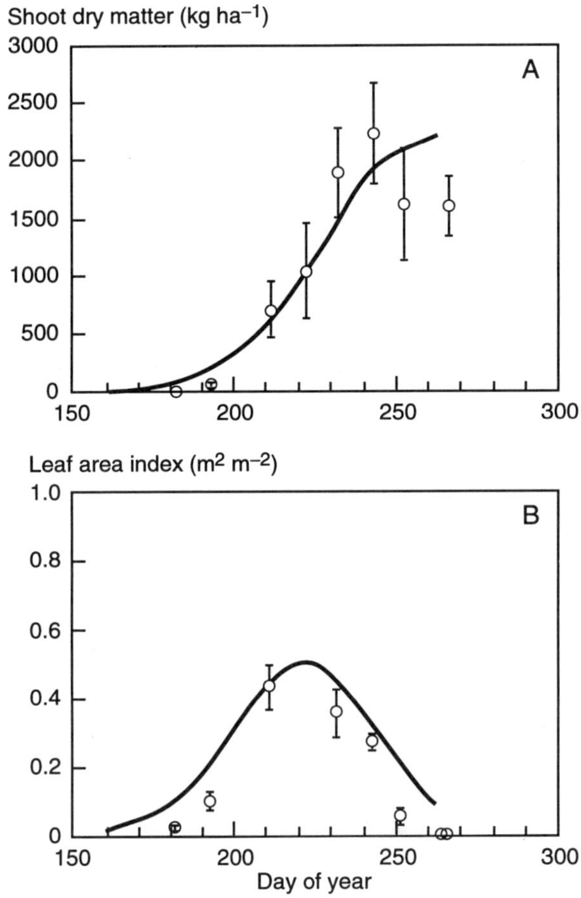

Figure 3. Simulated (line) and measured (symbols) (A) shoot dry matter and (B) leaf area index during the growing season 1993. The measured values were used for calibration. Bars show standard error of the mean of 10 samples.

simulated data of 1993 (6%). However, at 1 m, the simulated reduction (20%) was only half that found in the field (40%). The other half must be due to below-ground competition (water and/or nutrient) in the field, which was not accounted for in the model WB-mil-shade.

Discussion

In 1993, in the *Bauhinia* plots observed yield was much lower than simulated yield (Figure 5), mainly due to damage by the millet head caterpillar in these plots, while the crop growth module was calibrated with data derived from the undamaged *Ziziphus* plots. For means of *Ziziphus* plots (1993) and means of all plots (1992), the correspondence with simulation is satisfactory considering the generally high spatial

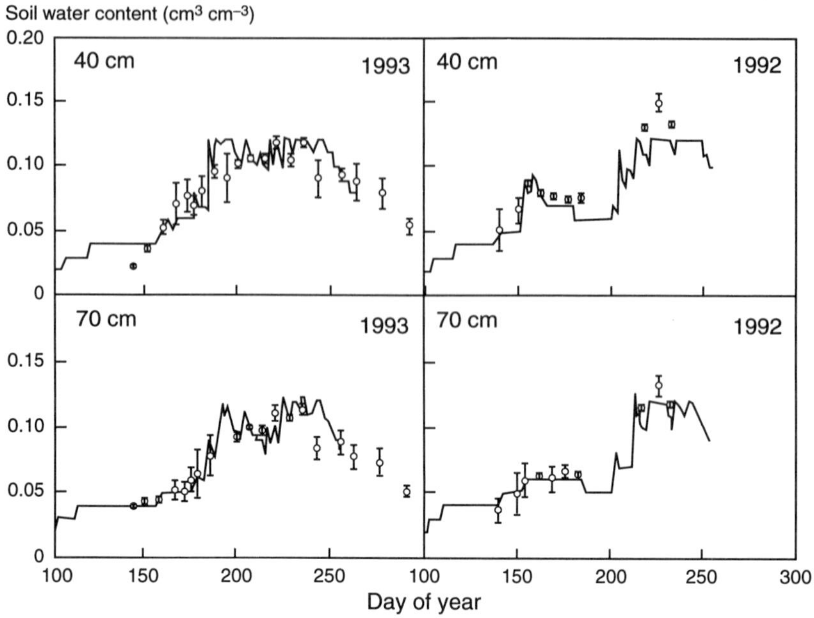

Figure 4. Simulated (line) and measured (symbols) seasonal soil water content at 40 and 70 cm soil depth in 1993 (calibration) and 1992 (testing). Bars show standard error of the mean of three samples.

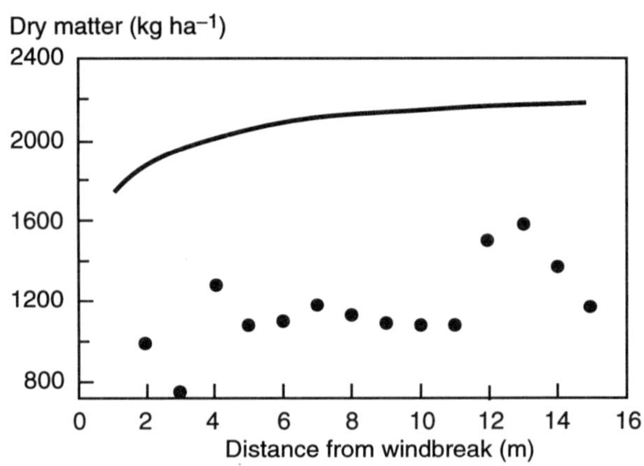

Figure 5. Simulated (line) and measured (symbols) shoot dry matter as function of distance from the windbreak in 1993.

variability of millet production on poor sandy soils in the Sahel. However, in the beginning and end of the growing season the simulation of *LAI* and biomass should still be improved by using a more appropriate concept to describe leaf area dynamics which, in turn, determines crop growth. With respect to the present model two weak points may be mentioned.

(1) In WB-mil-shade, leaf area was calculated from leaf dry matter using Specific Leaf Area (*SLA*, m^2 leaf kg^{-1} leaf). Hence, simulated results are extremely sensitive to the following parameters: *SLA*, physiological parameters related to CO_2 assimilation, and partitioning coefficients for the leaves. In the field, however, temperature is the overriding factor determining leaf area development until mutual shading occurs (*LAI*=1) (Horie et al. 1979). A better approach, suggested by Kropff and Van Laar (1993), is to estimate *LAI* for the early growth period (until *LAI*=1) as follows:

$$LAI_{ts} = N\,LAI_0 \exp(Rl_{ts}) \quad (5)$$

where LAI_{ts}=leaf area index (m^2 leaf m^{-2} surface) at a specific temperature sum (*ts*, °Cd) after emergence; *N*=number of plants per m^2; LAI_0=initial area per plant at seedling emergence (m^2 $plant^{-1}$) and Rl_{ts}=relative leaf area growth rate at a specific temperature sum ($°C^{-1}\,d^{-1}$).

(2) After day 250, the simulated values of *LAI* were higher than the measured ones, partly due to the underestimation of the dying rate of leaves after flowering (based on data of Verberne et al. 1995). Another reason is the underestimation of living green area in the field: small green parts of mainly brown leaves could not be measured for technical reasons. For the same reason, those data are not sufficiently accurate to allow formulation of a new concept for the dying rate of leaves.

The time step for calculations in process-based simulation models should cover the period of the relevant physical process, and in more simplified models the time step should be smaller than the time constant of the system. In crop growth models, the relative growth rate characterizes the time constant of the conversion of assimilates into dry matter of the various plant organs. A daily time step for integration in crop growth models appears sufficiently small (De Wit 1982). However, for accurately describing the functional relation between CO_2 assimilation and diurnal course of temperature and irradiation, a time step of one hour is needed. For situations of shading by a windbreak, light interception and CO_2 assimilation are changing in intervals of minutes, while the conversion of assimilates in dry matter can be described satisfactorily with a time step of one day.

A soil water submodel, linked to a crop growth module must provide an accurate estimate of the distribution of soil moisture in the rooting zone over the period of growth calculations. Redistribution of soil water between layers appeared to be an important process in this connection and in WB-mil-shade it was improved by reducing the time step from 24 to 6 h.

Rosenberg (1974) stated that the effect on north-south-oriented WBs on reduction in radiation and associated plant growth is negligible on a daily basis, since only small areas are shaded for only short periods and an area shaded in the morning by a windbreak to the east will receive additional reflected energy off that windbreak in the later afternoon. The latter was not observed in our WB experiment. Furthermore, the results of the WB-mil-shade model show a clear reduction in dry matter production within 3 m from the WB, which may be negligible for wide cropped fields between WBs, but substantial for narrow cropped strips. Brenner et al. (1993) found in a study on north-south-oriented WBs of neem (*Azadirachta indica a. Juss*) that the reduction in millet stover yield corresponded to the reduction in photosynthetic quantum flux density.

A future extension of WB-mil-shade will account for the competition for soil water between WB trees and the millet crop.

Conclusions

The model WB-mil-shade adequately simulates millet growth and soil water balance for conditions in the Sahel and is suitable as a basis for a WB-millet Sahel system model.

The simulation of dry matter production behind a WB (Figure 5) suggests that close to the *Bauhinia* WB the reduction in millet production could for 50% be attributed to competition for light. The remaining part might be due to below-ground competition.

Acknowledgements

The research project was financed by the 'German Research Foundation' (DFG). The C.T. De Wit Graduate School Production Ecology financed participation in SAAD2.

References

Bley J (1990) Experimentelle und modellanalytische Untersuchungen zum Wasser-und Nährstoffhaushalt von Perlhirse (*Pennisetum americanum* L.) im Südwest-Niger. PhD dissertation, Universität Hohenheim, Stuttgart, Germany. 132 p.

Brenner A J (1991) Tree-crop interactions within a Sahelian windbreak system. PhD dissertation, University of Edinburgh, Edinburgh, UK. 234 p.

Brenner A J, Van Den Beldt R J, Jarvis P G (1993) Tree-crop interface competition in a semi-arid Sahelian windbreak. Pages 15–23 in Proceedings of the 4th International Symposium on Windbreaks and Agroforestry, 26–30 July 1993, Hedeselskabet, Denmark.

Conijn J G (1995) RECAFS: A model for resource competition and cycling in agroforestry systems. Model description and user manual. Rapports Production Soudano-Sahélienne (PSS) No. 12. AB-DLO, P.O. Box 14, 6700 AA Wageningen, The Netherlands. 101 p.

De Wit C T (1982) Coordination of models. Pages 26–32 in Penning de Vries F W T, Van Laar H H (Eds.) Simulation of plant growth and crop production. Simulation Monographs, Pudoc, Wageningen, The Netherlands.

Hagen L J (1991) A wind erosion prediction system to meet user needs. J. of Soil and Water Conservation 46:106–111.

Horie T, De Wit C T, Goudriaan J, Bensink J (1979) A formal template for the development of cucumber in its vegetative stage. Proceedings of the Koninklijke Nederlandse Akademie van Wetenschappen (Amsterdam, The Netherlands) Series C, 82(4):433–479.

Kropff M J, Van Laar H H (Eds.) (1993) Modelling crop-weed interactions. CAB International, Wallingford, UK, and the International Rice Research Institute, Philippines. 274 p.

Long S P (1989) Influence of neem windbreaks on yield, microclimate, and water use of millet and sorghum in Niger, West Africa. Masters thesis, Texas A & M University, Tex., USA. 160 p.

Onyewotu L O Z, Ogigirigi M A, Stigter C J (1994) A study of competitive effects between a *Eucalyptus camaldulensis* shelterbelt and an adjacent millet (*Pennisetum typhoides*) crop. Agric. Ecosystems and Environment 51:281–286.

Penman H L (1948) Natural evaporation from open water, bare soil and grass. Proceedings of the Royal Society of London, Series A 193:120–146.

Rosenberg N J (1974) Windbreaks and shelter effect. Pages 238–264 in Rosenberg N J (Ed.) Microclimate: The biological environment. John Wiley and Sons, New York, USA.

Stroosnijder L (1982) Simulation of the soil water balance. Pages 175–193 in Penning de Vries F W T, Van Laar H H (Eds.) Simulation of plant growth and crop production. Simulation Monographs, Pudoc, Wageningen, The Netherlands.

Van Kraalingen D W G, Van Keulen H (1988) Model development and application for the 'Project pilote en agrométéorologie'. Report prepared for submission to World Meteorological Organization. Agricultural University Wageningen and DLD-Research Institute for Agrobiological and Soil Fertility (AB-DLO), Wageningen, The Netherlands.

Verberne E, Dijksterhuis G, Jongschaap R, Bazi H, Sanou A, Bonzi M (1995) Simulation des cultures pluviales au Burkina Faso (CP-BKF3): Sorgho, mil et mais. DLD-Research Institute for Agrobiological and Soil Fertility (AB-DLO), Wageningen, The Netherlands, Bureau National des Sols INERA and Institut d'Etudes et de Recherches Agricoles, Nota 18, Burkina Faso. 53 p.

Williams J R (1990) The erosion-productivity impact calculator (EPIC) model: A case history. Phil. Trans. R. Soc. Lond. B 329:421–428.

Crop models and precision agriculture

M.Y.L. BOONE[1], M. KIKUSAWA[2] and J.M. McKINION[3]

[1]*Department of Plant and Soil Sciences, Mississippi State University, P.O. Box 9555, Mississippi State, MS, USA*

[2]*Centre for Information Science, Fukui Prefectural University, 4-1-1 Kenjojima Matsuoka-cho, Fukui 910-11, Japan*

[3]*USDA-ARS Crop Simulation Research Unit, P.O. Box 5367, Mississippi State, MS, USA*

Key words: crop model, GOSSYM-COMAX, cotton simulation model

Abstract

Crop simulation models, geographic information systems (GIS), global positioning systems (GPS), intelligent implements and site-specific management (SSM) farming techniques are the technologies behind precision agriculture. Mechanistic and process-oriented crop models using real-time daily climatic data obtained from a local, on-farm weather station accessed by the user's computer via a modem can provide relevant information for the day-to-day or weekly management of the crop. Since most of these models cannot predict all the disturbances occurring in the field, periodic plant mapping data and visual field observations are used to adjust simulation results. Results obtained from models can be used to manage row spacing; population density; nitrogen fertilizer, irrigation water and the application of plant growth regulators; and harvest timing, to name a few. Many of these crop attributes are important considerations directly impacted by precision agriculture methodologies. The case for cotton production using GOSSYM-COMAX will be presented and demonstrated.

Introduction

US agriculture is one of the most productive in the world, where 'farms have become more capital intensive, mechanized, specialized and intensified' (Stauber 1994). However, 'this high level of productivity has hidden costs – loss of rural communities, degradation of the resource base, increased reliance on purchased inputs and increased regulations' (Bezdicek and DePhelphs 1994). Modern western agricultural production is characterized by its 'heavy use of chemical nitrogen, pesticides and monoculture, with the resulting soil erosion, loss of organic matter and water pollution' (Stauber 1994). Urban migration of rural people has become common because of a lack of rural economic opportunity. And as Stauber (1994) has reported, 'biotechnology and infotechnology may cause even greater changes than those produced by the combined mechanical and chemical innovations of the twentieth century'. With reduced rural population and increased concerns about food safety and the environment, a new paradigm for agriculture is emerging.

Precision, prescription, site-specific crop management – all refer to a management system of production agriculture using diverse technologies to increase field productivity and protect the environment. These technologies improve input efficiency, maximize farm profitability and provide computerized field histories. For example,

M.J. Kropff et al. *(eds.): Applications of Systems Approaches at the Field Level, 189–199.*

GPS can guide: (i) in grid sampling to produce an accurate soil data base, weed and insect pest population data; and (ii) use of variable rate farm machineries to increase input application efficiency as a function of soil properties, fertility or pest population density.

However, specific data on crop growth and development status is an essential requirement of precision agriculture technologies. Field monitoring, in discrete time steps, can provide some of this information, but at considerable cost in time and labour. Crop models can continuously provide this information with substantial savings in time and labour. Crop models running in real time can furnish data on irrigation and nitrogen stress status; and monitor the growth, development and yield progression of the crop. Therefore, results obtained from simulation models, coupled with scouting reports, can provide the data essential for achieving greater accuracy in the field and increased efficiency in the use of inputs. The focus of this paper is on the use of simulation in precision agriculture applications, using a cotton crop model as an example.

Crop model GOSSYM-COMAX

Farmers need information that minimizes the risks of making wrong decisions. They need to know the occurrences, among others, of water and nutrient stresses, as well as the presence of pests and the extent of any resulting infestations. Substantial amounts of time and labour are invested in gathering scouting reports to help farmers and farm managers decide on the timing and amount of fertilizer, irrigation and other chemical applications. However, most of their recommendations are prescriptive in nature. Their decisions are based on an average value for the whole farm – failing to consider the inherent heterogeneity in the field whether it be the soil type, or the insect density or the phenological/physiological status of the crop.

Cotton is an intensively managed crop requiring several in-season decisions to attain maximum yield at minimum production costs. The number of in-season decisions a typical US, mid-south, cotton farmer will make are: 1 to 3 for fertilizer application, 4 to 7 for irrigation, 8 to 12 for insect control, 2 to 4 for plant growth regulators, 2 to 3 for weed control and 2 to 3 for harvest aids (H.F. Hodges 1996, pers. commun.). Many of these decisions are based on years of experience, and the best information the farmer can find on the subject. However, even with the best available information, there are subjective aspects impacting the decision which cannot be readily determined. For example, the status of the crop is sometimes difficult to discern. How fast is it growing? Where are the roots relative to the nutrients? How much water is available in the soil and where is the water relative to the roots? None of these answers can be answered with confidence even by the best producers. Fortunately, answers to these kinds of questions and numerous others can be evaluated with a mechanistic, physiologically based crop model.

Mechanistic and process-oriented crop models like GOSSYM-COMAX are being used by cotton farmers across the US cotton belt to make intelligent management decisions. GOSSYM-COMAX, a cotton simulation model and cotton management expert system, is a result of continuing research efforts by a multidisciplinary team from Mississippi State University, Clemson University and the USDA-ARS Crop Simulation Research Unit. How GOSSYM works can best be described using the model's initial developers' (Baker et al. 1983) own statements:

> GOSSYM simulates the plant's response to environmental factors as follows: photosynthesis depends on light intensity and canopy light interception, and is reduced by water stress. Respiration depends on temperature and plant biomass. Growth is a function of temperature, tissue turgor and metabolite supply. Thus, the plant water status is a determinant of both supply and demand for metabolites. Water stress reduces photosynthesis, transpiration and nitrogen uptake. It also reduces growth and the demand for nutrients. The supply/demand ratios for carbohydrates and nitrogen are used as indices of stress-induced time delays for morphogenetic events. We assume that the metabolic supply/demand status of the plant determines hormone balances which alter morphogenetic rates. This status also determines or shifts the balance in hormone systems which results in the abscission of fruits. Therefore, while the morphogenetic rate is driven by temperature, it is affected indirectly by those factors determining the supply and demand for carbohydrates and nitrogen. Severe moisture stress and a heavy boll load may combine to stop new node formation, while a mild moisture stress which reduces growth more than it reduces carbohydrate supply may have no effect on new node formation or may cause a relative increase in the morphogenetic rate.

GOSSYM is also a materials-balance model that keeps track of the carbon, nitrogen and water in the soil-plant system.

Because GOSSYM is a dynamic process-level model, it requires constants and rate coefficients that are obtained under closely controlled and monitored environmental conditions, called the Soil Plant Atmosphere Research (SPAR) units (Phene et al. 1978; McKinion 1986). Growth rates obtained from these SPAR experiments define the potential growth rates of the crop responsive to changes in temperature and CO_2 levels (Reddy et al. 1996). Actual growth rates are then estimated from the potential growth after accounting for the effects of nitrogen, water and carbohydrate-stresses.

GOSSYM simulates the basic biological and physical processes involved in the growth, development and yield of cotton over a wide range of soils and climates. It uses the farmer's cultural practices (i.e. planting density, row spacing, variety, preplant fertilizer applications, date of crop emergence, in-season fertilizer application, irrigation and plant growth regulator applications), soil physical characteristics (i.e. bulk density, hydraulic properties and initial fertility and moisture status) and daily weather data (i.e. maximum and minimum temperature, total solar radiation, rainfall and wind run) obtained from a weather station accessed by the user's computer via a modem. It provides daily status reports on plant height; number of fruiting nodes and vegetative nodes; number of squares, green bolls, open bolls and abscised fruits; and carbohydrate, nitrogen and water stresses. A summary table at the end of a full-

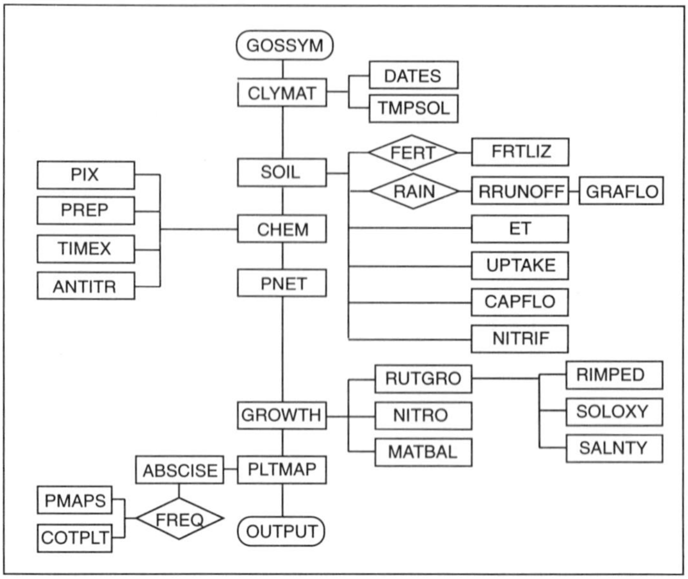

Figure 1. Flow chart of GOSSYM cotton crop simulation model showing the organization of the model and program flow.

season run presents the date of occurrence, plant height, LAI, yield and the number of nodes, squares, green bolls and open bolls at each designated developmental event.

GOSSYM can be subdivided into two major parts: the above-ground (primarily plant physiological) and the below-ground (soil) processes. Figure 1 presents the structural framework of GOSSYM. The major subroutines are:

CLYMAT	Reads all the weather information and calls the subroutines that keep track of the time (DATES) and that calculate the soil temperature (TMPSOL).
SOIL	The controlling program for all the soil subroutines.
CHEM	The controlling program for all the plant growth regulator subroutines.
PNET	Calculates the photosynthesis rate/dry matter production.
GROWTH	Calculates potential and actual daily growth rates of each of the organs on the plant including roots (RUTGRO).
PLANTMAP	Simulates plant morphogenesis and abscission of leaves and fruits (ABCISE) (Baker et al. 1983; Baker and Landivar 1991).

The SOIL subroutine is basically derived from the early work of J.R. Lambert (Clemson University) and D.N. Baker (USDA-ARS Crop Simulation Unit), known as RHIZOS (Boone et al. 1995). It uses a two-dimensional geometry to simulate root growth, water and nitrogen uptake and redistribution, and microbiological processes involving nitrogen. It provides three parameters important to the management of the above-ground crops, namely: effective soil water potential, used in calculating plant water potential; estimates of metabolite sink strength in roots; and the rate of nitrogen uptake. SOIL consists of the following subprograms:

FRTLIZ	Initializes nitrogen and organic matter content of the profile at planting and distributes applied fertilizers in the profile.
RRUNOFF	Estimates the amount of run-off from a rain or irrigation for the day.
GRAFLO	Moves the water into the profile after a rain or an irrigation event by gravitational flow and moves nitrogen in solution by mass flow.
ET	Estimates evapotranspiration.
UPTAKE	Calculates the amount of soil water taken up from the soil region where the roots are present.
CAPFLO	Rewets the soil by capillary flow in response to soil moisture gradients and moves nitrogen by mass flow.
NITRIF	Calculates the mineralization of organic matter and urea and the nitrification of ammonia.

In RUTGRO, the potential and actual growth of roots and average soil water potential is calculated, and calls RIMPED, which estimates the effects of bulk density on the capability of the root to elongate; and TMPSOL calculates the soil temperature by layer.

The model is continuously being improved and updated as new information becomes available. For example, as new cultivars are introduced and widely used by the farmers, GOSSYM has to be recalibrated. GOSSYM's capability to simulate various cultivars resides in the VARIETY file which contains variety-dependent, temperature-dependent and other parameters that reflect knowledge gaps or research needs. These calibration parameters are used as limits of genetic potentials and water stress values, modifiers and arguments in the equations describing stem elongation and fruit loss equations, to name a few (Boone et al. 1993).

Calibration of a model like GOSSYM is time-consuming and expensive considering the detailed information necessary for the process. Data used for the calibration are: plant height, number of nodes, number of squares, bolls and open bolls, weights of leaves, stems, bolls and squares, LAI and yield. In addition, complete records of the soil and weather data; timing of various developmental stages, cultural inputs and practices (name of cultivar, row spacing, planting pattern and density, application schedules, rates and methods of irrigation and chemicals, etc.) are needed. Sequeira et al. (1994) have demonstrated that employing genetic algorithm can speed up or automate the calibration of GOSSYM.

Once calibrated, GOSSYM is then routinely validated using data from several years (at least 8 years), location (10 US states) and soil types (at least 47), cultivars and cultural management practices (Boone et al. 1994). These data sets represent diverse conditions of cotton production from all over the US cotton belt. To avoid confusing validation with calibration, it is worth noting that validation is the 'subsequent testing of a calibrated model to additional field data preferably under different scenarios to further examine model validity' (Thomann and Mueller 1987). GOSSYM has been extensively validated in Mississippi and Arizona, and Israel (Fye et al. 1984; Reddy

Table 1. Linear regression equations of measured (Y) versus predicted (X) values for stem height (Z), number of nodes ($NODZ$), stem weight ($STEMWT$), number of squares ($SQRZ$), number of bolls ($BOLZ$), and leaf weight ($LFWT$) per cotton plant.

Variable	n	Equation	r^2	F-test
Z	8	$Y= 5.885+0.920\times X$	0.958	Not significant
$NODZ$	8	$Y=-3.739+1.176\times X$	0.917	Not significant
$STEMWT$	8	$Y= 3.820+0.739\times X$	0.892	Highly significant
$SQRZ$	6	$Y= 6.872+0.736\times X$	0.956	Highly significant
$BOLZ$	8	$Y= 2.107+0.847\times X$	0.842	Not significant
$LFWT$	8	$Y= 2.966+0.881\times X$	0.920	Not significant

Note: Actual yield was 1520 kg ha^{-1} compared to simulated yield of 1385 kg ha^{-1}.

Table 2. Validation of GOSSYM in terms of yield prediction expressed in terms of average absolute ratio of actual over simulated yield.

Cultivar class	n	Yield ratio*
GC 510	8	0.86
DES 119	16	0.83
EARLY	15	0.82
FULL	8	0.80
MID	36	0.81
ST1	3	0.92
ST2	3	0.87
OVERALL	89	0.83

*Ratio of maximum (actual or predicted) over minimum (actual or predicted) yield.

et al. 1985; Marani and Baker 1981). Jennings (1989) reported that the model accurately simulated the dry matter partitioning and the N concentration of different plant parts when the crops were grown in different N fertility conditions. The validation is an integral and vital part of model development.

Figures 2A–C and Table 1 present a comparison of the performance of GOSSYM to actual field data. Linear regressions of the actual values with respect to the simulated ones are used to evaluate the goodness of fit. Yield data comparison for several variety classes across the US is shown in Table 2.

COMAX (CrOp MAnagement eXpert) is a rule-based expert system consisting of a knowledge base, an inference engine, a file management system and a graphical user interface (Lemmon 1986; McKinion et al. 1989; AGBIT 1995). COMAX applies a set of rules to formulate different management strategies; executes the GOSSYM program and evaluates the impacts of certain management decisions. Results from a series of optimization runs are specific recommendations on water usage, nitrogen usage and timing of crop termination chemicals. COMAX, in a way, has captured the expertise of not only the modelling team's ability to run and interpret GOSSYM results, but also the local 'expertise' supplied by the users.

A mechanistic crop model realistically predicting crop response to its environment allows a producer to make simulations of that crop grown with various combinations of soil, weather and cultural practices. GOSSYM estimates cotton growth, development and yield, and updates its physiological status daily. Consequently, it

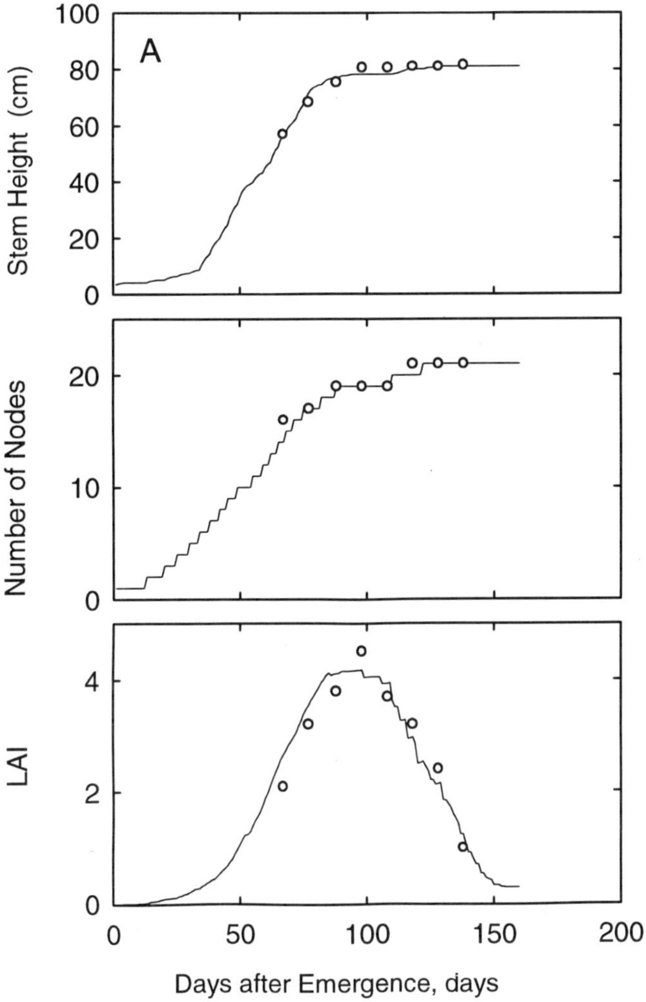

Figure 2. Time course of simulated (solid line) and measured (symbols) values for (A) stem height, number of nodes and LAI on a per cotton plant basis; (B) number of squares, bolls and open bolls per plant; and (C) weight of leaves, stems and open bolls per plant.

provides the producer the location and distribution of roots, water and soluble nutrients in the soil at any time which can be displayed graphically by a computer. Availability of this information makes it possible for a producer to avoid excessive applications of nutrients, thus minimizing pollution risks.

At the very start of the model development in the early 1970s, the developers have envisioned the model to serve as an educational, research and management tool. The model has been immensely successful in helping users better understand the physiology of the cotton crop. It has enabled producers make 'what-if' simulation

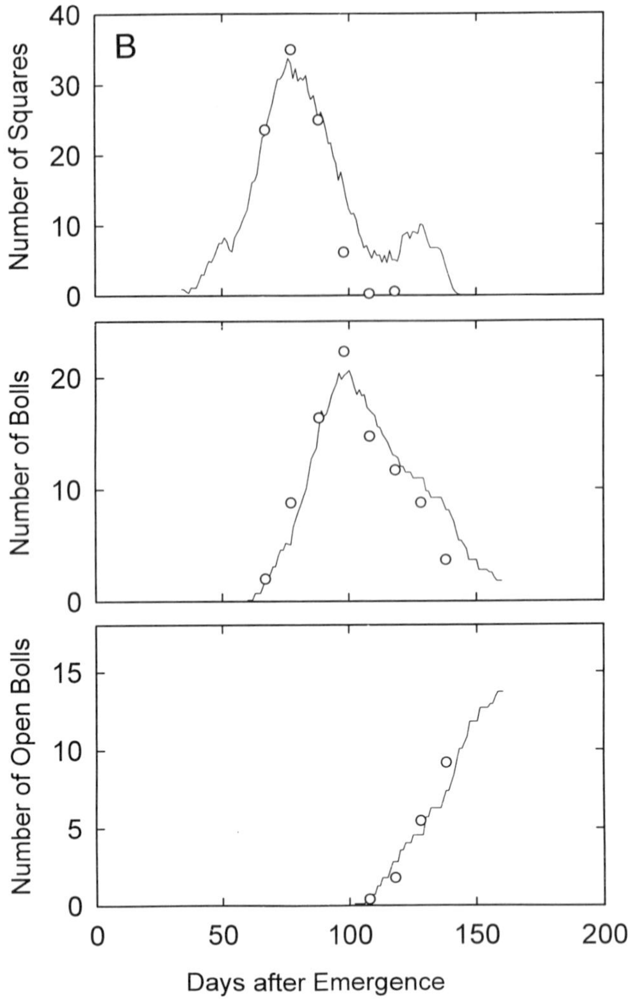

Figure 2. continued.

runs before and/or during the cropping season to guide their decisions on the application of fertilizer, insect control practices, crop growth regulator applications, irrigation timing, and timing of harvest-aid chemicals with greater precision. It has also been used to evaluate, retrospectively, the impact of environmental conditions and cultural practices on cotton yield trends (Reddy et al. 1987, 1989, 1990; Whisler et al. 1982).

GOSSYM is currently used by more than 300 farmers and consultants on over 400 000 ha of commercial cotton in all 14 cotton growing states in the US. The value of GOSSYM-COMAX in the US has been estimated by two independent surveys conducted by the Agricultural Economics Department of Texas A & M University to

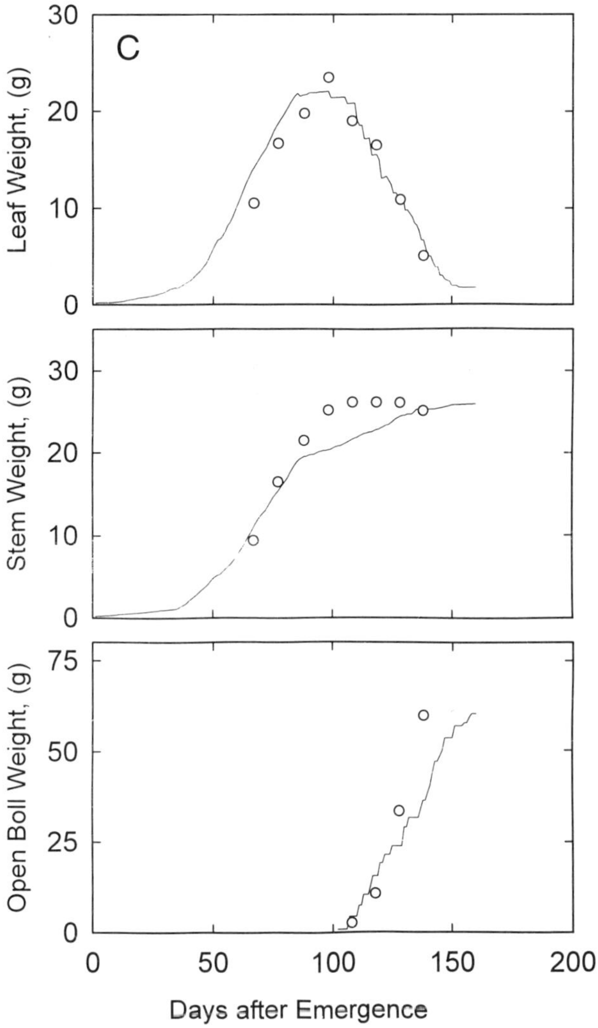

Figure 2. continued.

be in the range of $100 per hectare per year (McKinion et al. 1989). The value assigned to different types of decisions has depended on user experience with the model and the willingness to test its capabilities and applicability in various areas.

The technical specifications of the GOSSYM-COMAX system are: 66 MHz or higher, P5 or 80486 processor, 8 megabyte RAM, 170 megabyte hard disk, SVGA graphics board and SVGA colour monitor, high-density floppy disk drives (1.2 or 1.44 megabyte), serial and parallel ports for communications and printer, 9600 baud Hayes-compatible modem, 80 column Epson or IBM compatible printer, MS-DOS

Version 5.0 or later, Windows 3.1 or higher, and mouse, Microsoft compatible (AGBIT 1995).

Despite the widespread use of GOSSYM-COMAX, it is still far from complete or perfect. Factors unaccounted for by the model can influence the growth of the crop. These include herbicide injury, nematodes, plant pathogens, insect pests, mineral nutrients other than nitrogen, and weather events (e.g. high winds which cause lodging and hail). Consequently, whenever any of these factors impact crop growth, the model should not be expected to simulate the crop correctly. Therefore, periodic plant mapping data and visual field observations are necessary to adjust the simulation results and interpret the appropriate management needs.

A primary goal of GOSSYM is to effectively manage production inputs – irrigation, nitrogen fertilizers and plant growth regulators. In conjunction with COMAX, it can provide recommendations on optimal applications of irrigation and fertilizers, timing of plant growth regulator applications and crop termination for maximal economic growth production and minimum nitrogen residuals in the soil profile at the end of the growing season. The ability of GOSSYM-COMAX to anticipate the impact of current proposed management tactics or future weather influences is essential in precision agriculture.

Conclusions

GOSSYM-COMAX simulates cotton growth and provides relevant information for the day-to-day or weekly management of the crop. Its use in combination with other technologies such as geographic information systems (GIS), variable rate machineries and global positioning systems (GPS) could further optimize site-specific farming practices to address some of the social, ethical and environmental issues that face the declining number of farmers. The technical details of integrating the model with other technologies is the focus of current research efforts.

References

AGBIT Inc. (1995) GOSSYM-COMAX user's manual. Starkville, Miss., USA. 336 p.

Baker D N, Landivar J A (1991) The simulation of plant development in GOSSYM. Pages 153–170 in Hodges T (Ed.) Predicting crop phenology. CRC Press, Boston, MA, USA.

Baker D N, Lambert J R, McKinion J M (1983) GOSSYM: A simulator of cotton crop growth and yield. S.C. Agric. Exp. Stn. Bull. 1089, South Carolina, USA. 135 p.

Bezdicek D F, DePhelphs C (1994) Innovative approaches for integrated research and educational programs. Am. J. of Alternative Agric. 9:3–8.

Boone M Y L, Porter D O, McKinion J M (1993) Calibration of GOSSYM: Theory and practice. Computers and Electronics in Agric. 9(3):193–204.

Boone M Y L, Porter D O, McKinion J M, Turner S (1994) Calibration and validation of the cotton simulation model GOSSYM. Pages 689–696 in Salokhe V M, Singh G (Eds.) Proceedings of the International Agricultural Engineering Conference, 6–9 December 1994, Asian Institute of Technology, Bangkok, Thailand.

Boone M Y L, Porter D O, McKinion J M (1995) RHIZOS 1992: A simulator of row crop rhizosphere. National Technical Information Service, US Department of Commerce, Springfield, VA, USA. 175 p.

Fye R E, Reddy V R, Baker D N (1984) The validation of GOSSYM: Part I, Arizona conditions. Agric. Syst. 14:1–21.

Jennings T A (1989) Managing cotton fertilization using GOSSYM. MSc thesis, Mississippi State University, Miss., USA.

Lemmon H (1986) COMAX: An expert system for cotton crop management. Science 233:29–33.

McKinion J M (1986) SPARNET: A data acquisition/analysis computer network. Computers and Electronics in Agric. 1:163–172.

McKinion J M, Baker D N, Whisler F D, Lambert J R (1989) Application of the GOSSYM/COMAX system to cotton crop management. Agric. Syst. 31:55–65.

Marani A, Baker D N (1981) Development of a predictive dynamic simulation model of growth and yield in Acala cotton. Science Report to US–Israel Binational Foundation, 30 April 1981, P.O. Box 7677, Jerusalem, Israel. 177 p.

Phene C J, Baker D N, Lambert J R, Parsons J E, McKinion J M (1978) SPAR: A soil-plant-atmosphere-research system. Trans. ASAE 21:924–930.

Reddy V R, Baker D N, Jenkins J N (1985) The validation of GOSSYM: Part II, Mississippi conditions. Agric. Syst. 17:133–154.

Reddy V R, Baker D N, Whisler F D, Fye R E (1987) Application of GOSSYM to yield decline in cotton. I. Systems analysis of effects of herbicides on growth, development and yield. Agron. J. 79:42–47.

Reddy V R, Baker D N, McKinion J M (1989) Analysis of atmospheric carbon dioxide and ozone on cotton yield trends. J. Environ. Qual. 18:427–432.

Reddy V R, Baker D N, Whisler F D, McKinion J M (1990) Analysis of the effects of herbicides on cotton yield trends. Agric. Syst. 33:347–359.

Reddy V R, Hodges H F, McKinion J M (1996) Crop modeling and applications: A cotton example. In Advances in Agronomy. ASA, Madison, WI, USA (in press).

Sequeira R A, Olson R L, Willers J L, McKinion J M (1994) Automating the parameterization of mathematical models using genetic algorithms. Computers and Electronics in Agric. 11:265–290.

Stauber K N (1994) The future of agriculture. Am. J. Alternative Agric. 9:9–15.

Thomann R V, Mueller J A (1987) Principles of water quality modeling and control. Harper and Row, New York, USA.

Whisler F D, Lambert J R, Landivar J A (1982) Predicting tillage effects on cotton growth and yield. Pages 179–198 in Predicting tillage effects on soil physical properties and processes. ASA, SSSA, Madison, WI, USA.

A conceptual model for sodium uptake and distribution in irrigated rice

F. ASCH[1,2], M. DINGKUHN[3], M.C.S. WOPEREIS[1], K. DÖRFFLING[2] and K. MIÉZAN[1]

[1]*West Africa Rice Development Association (WARDA), B.P. 96, St. Louis, Senegal*

[2]*Institut für Allgemeine Botanik und Botanischer Garten, Universität Hamburg, Ohnhorststraße 18, 22609 Hamburg, Germany*

[3]*West Africa Rice Development Association (WARDA), B.P. 2551, Bouaké 01, Ivory Coast*

Key words: salinity, potassium, root filter, stem retention, stomatal reactions, transpiration, varietal differences

Abstract

Sodium uptake and distribution in rice was investigated in various field and screenhouse trials at WARDA's research station in Ndiaye, Senegal. A conceptual model for sodium uptake and distribution was developed on the basis of the following results: (i) sodium uptake to the plant was transpiration driven; (ii) varieties differed in the way they regulated their stomata in relation to relative humidity and salt stress; (iii) sodium uptake was modulated at the root level by a 'root filter'; and (iv) sodium was taken out of the transpiration stream and retained in the sheaths with the daily retention capacity as a varietal constant. The model describes the passive uptake of sodium to the plant and its distribution as a function of several varietal constants and transpiration. The root filter function for sodium, the maximum stem sodium retention and the sodium toxicity threshold in the leaves vary among varieties. Varietal differences in stomatal reactions to relative humidity and salt stress were implemented in the ORYZA_W evapotranspiration routine to simulate these reactions. The interactions between sodium and potassium are discussed in relation to available information from literature and preliminary results obtained from screenhouse studies. A concept of potassium distribution in the plant and its interaction with sodium uptake and distribution was included in the model structure.

Introduction

Irrigated rice production in river deltas and former floodplains in semi-arid and arid climates is prone to salinity. Even with proper irrigation and drainage techniques, topsoil salinity at the beginning of the planting season can hardly be avoided. Naturally saline watertables which are close to the surface (1–1.5 m) can rise throughout the season into the rootzone or cause upward salt transport through capillary rise in the off-season, resulting in relatively high soil solution electrical conductivity (Dingkuhn et al. 1993).

Poorly constructed irrigation systems without drainage facilities may lead to salinization of the soil and farmers may abandon such fields within a few seasons. In the process of recovering these soils, rice can play a major role (Neue et al. 1990). Rice tolerates standing water, allowing salt to flush out of the topsoil through irrigation and drainage, while providing instant revenues for the farmer to offset the cost of reclamation. Thus, farmers need varieties that are productive under moderate salinity.

M.J. Kropff et al. *(eds.): Applications of Systems Approaches at the Field Level, 201–217.*

Salt stress affects the entire plant at all development stages and rice sensitivity to salt varies between growth stages (Fageria 1985). Salt resistance is composed of avoidance and tolerance components. In non-halophytes, avoidance is defined as the ability to restrict the entry of ions into the shoot (Yeo and Flowers 1982), resulting in lower sodium concentration in the leaves. However, tolerant rice varieties can differ much in their leaf sodium concentration (Greenway and Munns 1980); therefore, a general association between high leaf sodium concentration and salt sensitivity cannot be established. Salt resistance in rice is probably based on a combination of avoidance and tolerance traits. Sodium uptake to the rice plant is thought to be passive and related to the transpirational volume flow (Yeo et al. 1987). Thus, mechanisms of sodium exclusion may fall into one or more of three categories: (i) physical or chemical blockage of sodium at the root surface; (ii) excretion from the roots back to the outside medium; and (iii) low sodium uptake through low transpiration (combined with a high water use efficiency for high-yielding tolerant varieties). No evidence for root excretion of sodium in rice has been reported. Therefore, salt uptake to the plant and distribution in the plant seems a priori to be regulated by root properties and transpiration (Yeo et al. 1984, 1985, 1987). Transpiration rates depend on canopy properties and stomatal conductance. Stomatal conductance depends on the leaf's physiological status, e.g. leaf nitrogen content (Dingkuhn et al. 1992) and water status (Dingkuhn et al. 1989). Transpiration is driven by temperature, intercepted solar radiation, wind speed and relative air humidity.

In growth chamber studies, shoot water potentials of rice varieties decreased when salinity increased (Asch et al. 1995), but field studies showed no significant difference in water potentials between fresh water and salt water treatments (Dingkuhn et al. 1993). Stomatal conductance was more influenced by low relative air humidity than by soil electrical conductivity in the chamber study (Asch et al. 1995), and no significant difference was found between varieties and treatments in the field study.

Sub-Saharan irrigated rice production uses two cropping seasons. The main wet or rainy season (end-July to mid-November) has high air humidity (60–100%), the bulk yearly rainfall and air temperatures between 22°C and 40°C. The dry season (mid-February to July) is characterized by low air humidity (10–65%), almost no precipitation and air temperatures between 10°C (cold dry season) and 45°C (hot dry season) (Dingkuhn et al. 1994).

To date, most rice research in the tropics has been conducted in Asia in humid environments with low evapo-transpiration. Humidity mitigates the effects of drought, salt and temperature stresses (Capell and Dörffling 1989; Lauter and Munns 1987; Mizrahi et al. 1971). Humidity effects on salt tolerance were also described (e.g. Neue et al. 1990; Yeo et al. 1985), but variety reactions to air humidity have not been studied as a component trait for salt tolerance.

A continuous salinity trial at WARDA's research station in Ndiaye, Senegal, showed that the material from INGER or IRRI salt tolerance nurseries could, in many cases, not maintain its salt tolerance under Sahelian dry season conditions (Asch et al. 1995).

Due to the extremely low relative air humidity in the dry season, water demand and water turnover of the rice crop can be very high (Raes et al. 1995). This results in high potential sodium uptake in saline environments. Tests with the modern semi-dwarf salt-tolerant variety IR4630-22-2 on saline clay soil at IRRI, Philippines (IRRI 1979), resulted in an accumulation of 6.5 mg Na per g dry matter. Under similar saline conditions at WARDA's research farm (Senegal), the same variety accumulated 24 mg Na per g dry matter (Asch, unpublished data).

Starting in 1992, mechanisms of salt tolerance in irrigated rice varieties were studied at WARDA's research station at Ndiaye, Senegal. We chose a modelling approach because salinity effects on rice as described in literature and from our own experience form a complex fabric of physiological responses to climate and salt stress. Although systems analyses cannot replace varietal development through conventional breeding, underlying principles are easier to understand through modelling. The characterization of component traits for salt tolerance and, therefore, the identification of potential donors are accelerated through modelling and may shorten the time it needs to release a salt-tolerant variety (8–10 years). After preliminary studies on sodium distribution in the rice plant, we developed a simple transpiration-driven sodium uptake model in basic (Dingkuhn and Asch 1994). Early model runs showed that a simple 'open tube' approach to transpiration-driven sodium uptake was not adequate as simulated leaf sodium concentrations exceeded by far those observed in field trials. This exercise raised the following questions: (i) is sodium uptake to the plant exclusively transpiration driven, and is the transpiration stream the exclusive vehicle for sodium distribution in the rice plant? (ii) is stomatal regulation of transpiration a major component in salt resistance, and do varieties show different stomatal behaviour under variable evaporative demand and salt stress? (iii) is sodium uptake modulated by a 'root filter', and is this filter genotype and environment dependent? and (iv) is sodium removed from the transpiration stream as it passes through the leaf sheaths, and would such retention and storage depend on genotype and environment?

Field and screenhouse experiments on sodium uptake and distribution and climate influences on varietal performance under salt stress were conducted to address these issues. This paper presents a conceptual model for sodium uptake and distribution in rice based on the results.

Material and methods

Experimental site, plant material and initial growing conditions

All experiments were conducted at WARDA's research farm in Ndiaye, Senegal (16° 14′N; 16° 14′W) in 1994 and 1995. Three varieties were used for all experiments:

- I Kong Pao (IKP), local short-duration salt tolerance check, origin Taiwan;
- IR4630-22-2, (IR4630) medium-duration, salt-tolerant, selected by WARDA specifically for the Senegal River delta, origin Philippines; and
- IR31785-58-1-2-3-3, (IR31785) short-duration, well adapted to Sahelian climatic conditions, WARDA's salt sensitivity check, origin Philippines.

Seed was obtained from WARDA, Senegal. Seed was soaked and pregerminated for 48 h and sown in an improved dapoc system in mid-February for the dry season field trials (28 days' seedbed duration), mid-July for the wet season field trials (21 days) and mid-May for the screenhouse trials (21 days).

Field experiments

Two treatments were replicated thrice in a randomized strip plot design. The treatments were irrigation with fresh water (EC 0.5–0.9 mS cm^{-1}), and irrigation with saline water (EC 3.5 mS cm^{-1}). Per variety, treatment and replication 100 hills were transplanted at a 0.2 m×0.2 m spacing. To achieve salinization, evaporative salt residues scraped from nearby soil were added to a clay water reservoir. Floodwater EC was kept constant throughout the season. The soil was a heavy, slightly acid Vertisol clay (percolation < 1 mm d^{-1}). Initial soil samples were taken for each treatment. Topsoil (0–0.25 m) salinity depended on the treatment (1:5 extraction: 0.26 and 0.67 mS cm^{-1} respectively).

In the hot dry season 1994, samplings for leaf positions, dead tissue, stems and roots were taken in 7-day intervals to determine dry matter and morphology. Leaf area (destructive with Li3000) and specific leaf weight for one hill were taken in 10-day intervals. Leaf Area Index (LAI) curves were calculated from leaf dry matter development and specific leaf area measurements. Calculated LAI curves and daily weather data (temperature, solar radiation, wind speed and relative humidity) were used as input for the evapo-transpiration routine of the ORYZA_W model (Wopereis et al. 1996) to calculate daily potential canopy transpiration.

In the wet season 1994, the same set-up was used, but only one sampling of the late vegetative stage close to panicle initiation is reported here. Daily meteo data were obtained from a micromet station, installed next to the experiment.

Screenhouse experiments

Trial 1. In a completely randomized design individual 21 day old plants were transferred to styrofoam floats in water-cooled buckets containing 4 l of nutrient solution (Yoshida et al. 1976). Plants were grown in solution for four weeks before they were subjected to the different treatments. Sodium was given as NaCl (p.a.). Five plants per variety and treatment were subjected to two levels of salinity: 0 mM NaCl (EC 0.8 mS cm^{-1}) and 60 mM NaCl (6.5 mS cm^{-1}).

After 24 h, root systems were cut off from the plant just below the stembase. The root was dried superficially on blotting paper. The entire root was inserted into a Scholander-type pressure bomb. Superficial water was removed by applying to 0.2 MPa for 30 s. Xylem sap was sampled for 2 min at 1 MPa. Root length and volume were also determined.

Trial 2. In a randomized complete block design, three 21 day old plants each were transferred to styrofoam floats in buckets containing 8 L of nutrient solution. Buckets were cooled in a concrete waterbath using water circulated from an adjacent underground reservoir. Plants were grown in solution for 7 days before 18 plants per variety, treatment and replication (three) were continuously subjected to the salt treatments as described under trial 1. Daily transpiration rate was measured by difference weighing and leaf area (destructive with Li3000; Li-Cor, Lincoln, Nebraska) was determined in 10-day intervals on a one-hill basis. Temperature and relative humidity were recorded with a thermohygrograph (Cole Parmer, Illinois, 7-day revolution). In both trials nutrient solution was changed weekly.

Measurements and analyses

Oven-dried ground plant tissue samples were analysed for K^+ and Na^+ content using acid extraction (Yoshida et al. 1976) and flame photometry (Digital Flame Analyzer, model 2655-00, Cole Parmer, Illinois). Xylem sap and nutrient solution samples were diluted with Cole Parmer standard diluent and were directly flame photometrically analysed for K^+ and Na^+.

Results and discussion of observations

Transpiration and sodium uptake

In a whole-season kinetic study, the salt uptake and distribution, dry matter, growth, leaf area and root dimension evolution over the growth period were followed with weekly samplings for three varieties. In Figure 1, the salinity influence on green leaf dry matter development is illustrated. Salinity-induced dry matter reduction was least in IR4630, intermediate in IKP and highest in IR31785. The decline in leaf dry matter due to monocarpic senescence towards the end of the season was not simultaneous between treatments because time to flowering changed under salinity.

We simulated the potential canopy transpiration for each variety, using the evapotranspiration routine of ORYZA_W, and on its basis we calculated the cumulative canopy transpiration per ground area over the season. The observed cumulative sodium uptake was then plotted against transpiration (Figure 2A). Sodium accumulation was proportional to cumulative transpiration; the slopes of the relationship, however, depended on the variety and treatment. Susceptible check IR31785 had the steepest slope, or highest rate of sodium uptake per unit of water transpired.

Plant sodium uptake per unit ground area, was similar for the three varieties, if irrigated with salt water (Figure 2B). This was the result of two effects that neutralized each other in terms of total Na uptake: the reduction in growth (Figure 1) and the increased sodium concentration in the shoot (Figure 2B). Tissue concentrations increased throughout the season in IR31785 under salinity, but they changed little in the tolerant check IR4630. IKP was intermediate.

In a crop canopy, changes in LAI during the season, as dependent on developmental stage and treatment, can be expected to affect leaf area based transpiration rates through mutual shading of leaves. In order to estimate time courses of transpiration

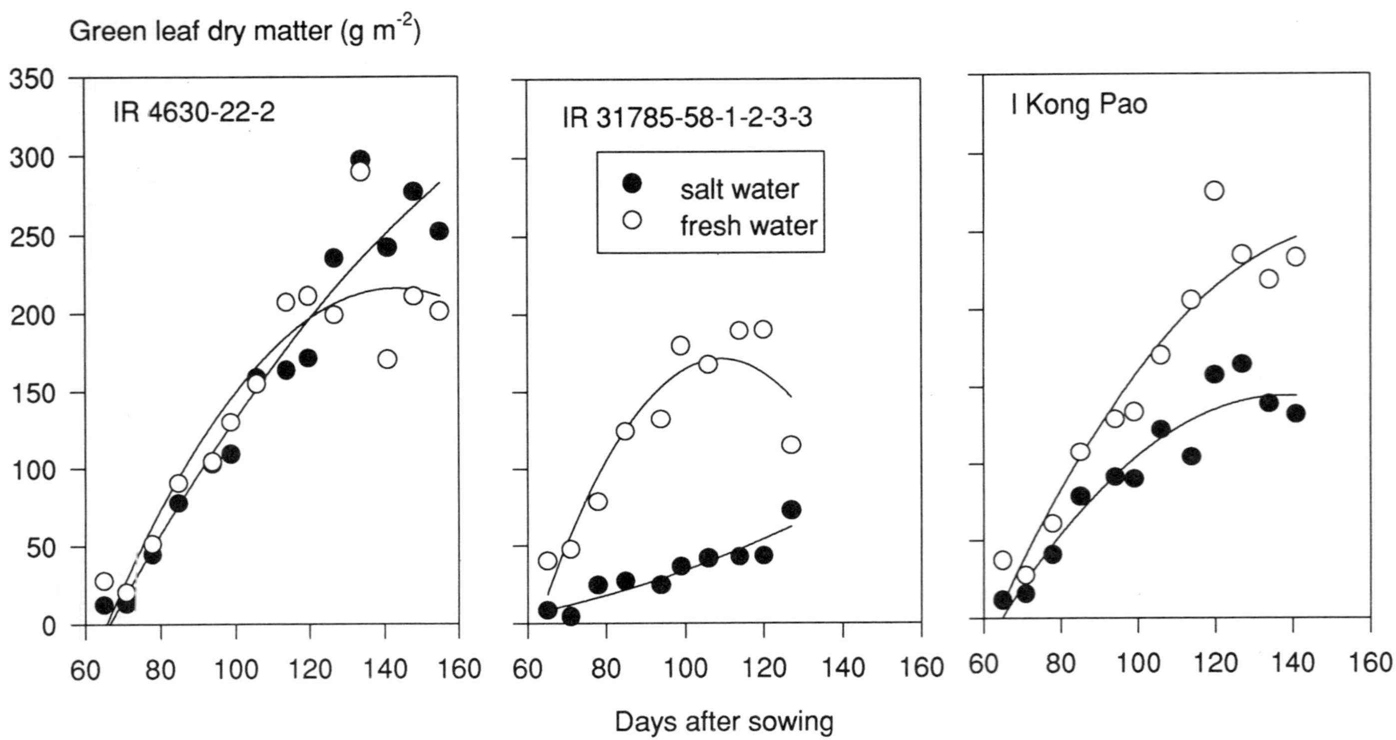

Figure 1. Green leaf dry matter development of three varieties from a field trial over the hot dry season 1994. Fresh water treatment was between 0.5 and 0.9 mS cm^{-1}; salt water treatment was 3.5 mS cm^{-1}. EC was kept constant over the season.

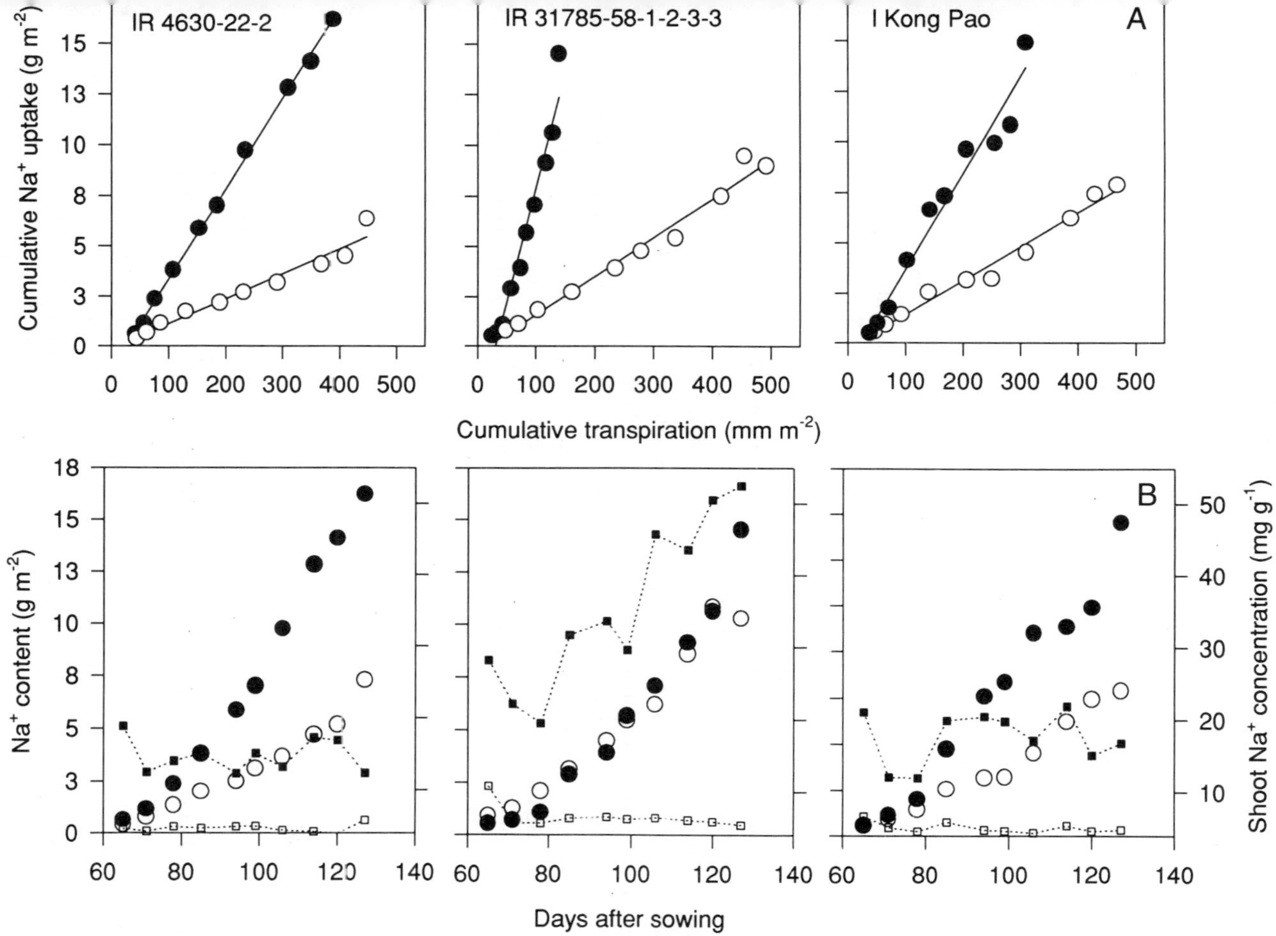

Figure 2. (A) Simulated cumulative potential transpiration per ground area versus cumulative sodium accumulation of the varieties from a field trial in the hot dry season 1994. To simulate transpiration, the evapotranspiration routine from ORYZA_W was used. (B) Cumulative sodium uptake and shoot sodium concentration over time from the same trial. Treatments are the same as in Figure 1. Open symbols = fresh water treatment, filled symbols = salt water treatment, circles = sodium content and squares = sodium concentration.

and Na uptake per unit leaf area, we divided the ground-area based data by the LAI (Figure 3). Sodium uptake rates (Figure 3B) were calculated from non-linear regression analyses of the time courses in Figure 2B. The simulated leaf transpiration rates decreased throughout the season (Figure 3A), which was mainly due to the increasing density of the leaf canopy. Salt-treated plants, which developed a lower leaf area than the controls, maintained higher leaf transpiration rates.

Mean daily sodium uptake rates per unit of leaf area, as calculated from weekly observations of plant sodium content and LAI, were dramatically higher in the susceptible IR31785 than in the two other varieties (Figure 3B).

Humidity effects on transpiration

The analyses presented till here, as far as they concerned transpiration, were based on simulation and assumed neither salt treatment nor air humidity specific effects on stomata. In order to determine air humidity (minimum for 08.00 to 20.00 h) × salinity effects on transpiration, the daily transpiration rates and leaf area were measured over an entire season for the three varieties grown in culture solution (screenhouse trial 2). Transpiration rates were expressed on a leaf area basis (Figure 4).

Transpiration rates of IR4630 were generally low ($1.5-4\,kg\,m^{-2}\,d^{-1}$), and were not influenced by air humidity or salt treatment. Since a decrease in air humidity is associated with an increase in evaporative demand, the results indicate that IR4630 stomata must have responded to air humidity under both saline and control conditions. In IKP, transpiration rates markedly increased with decreasing air humidity under non-saline conditions, but did not respond under salinity. For IR31785, transpiration rates increased with decreasing air humidity under both saline and control treatments, although transpiration rates were generally lower under salinity. It appears, therefore, that the tolerant check IR4630 controlled its transpiration under both saline and control treatments, IKP controlled its transpiration specifically under salinity, and the susceptible check IR31785 controlled its transpiration least.

Sodium uptake by the roots

Sodium enters the shoot with the transpiration stream via the xylem. In order to estimate how much sodium will get into the shoot at a given outside salinity, root xylem sodium concentrations were compared with culture solution sodium concentration (Figure 5). In the fresh water controls, no differences were observed between root xylem and culture solution sodium concentrations in any of the varieties. With 60 mM NaCl applied, varieties differed in their root xylem sodium content. Only 33% of the external sodium concentration were found in the root xylem of IR4630, 60% in IR31785 and 80% in IKP. These results indicate that there is no direct link between transpiration rate and sodium exclusion at the root level as tolerant check IKP showed the highest sodium concentration in the xylem sap.

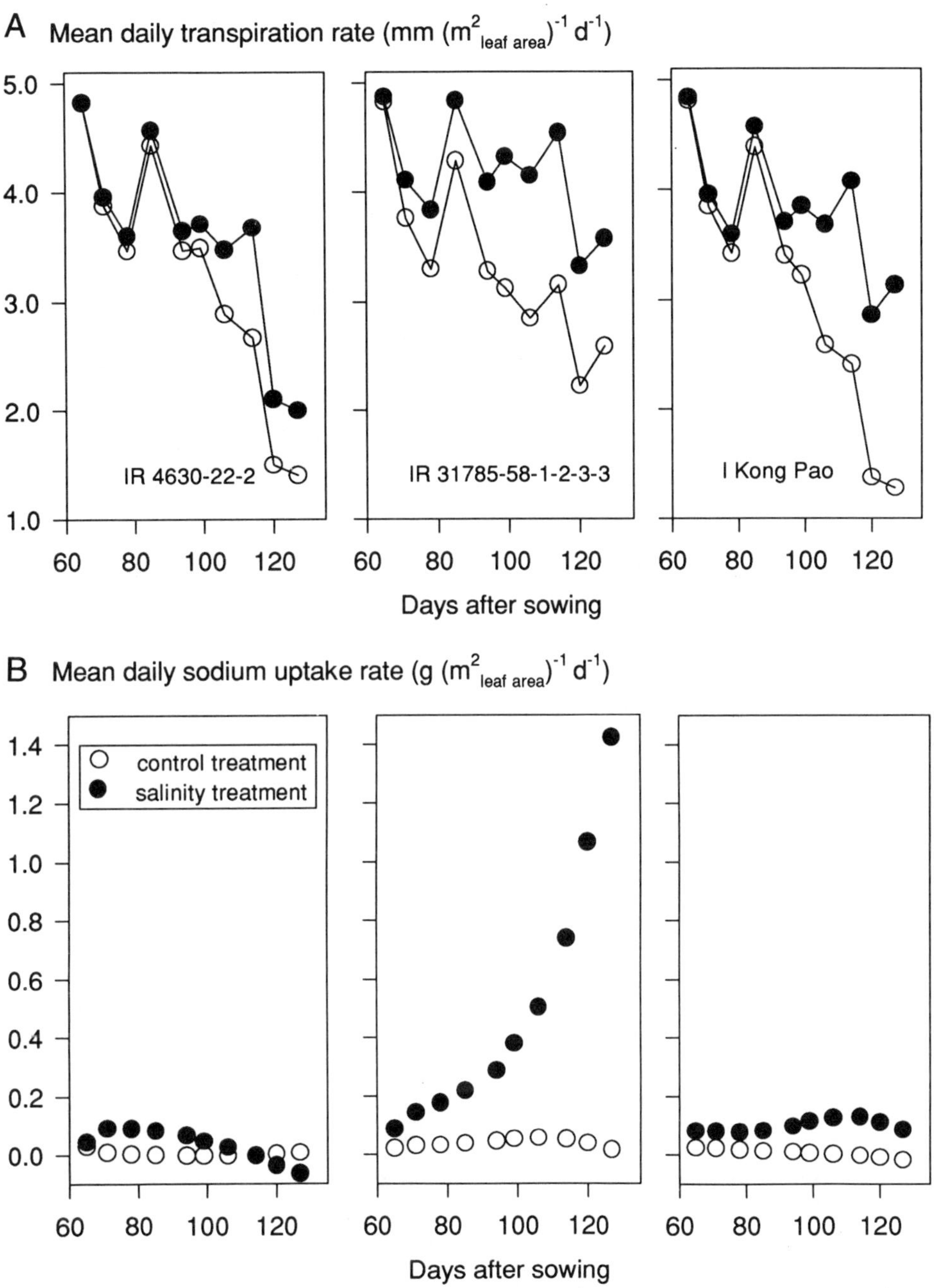

Figure 3. Mean daily transpiration rates (A) and mean daily sodium uptake rates (B) per leaf area for three varieties subjected to the same treatments as in Figure 1. Sodium uptake rates were calculated from non-linear regressions of the data in Figure 2B.

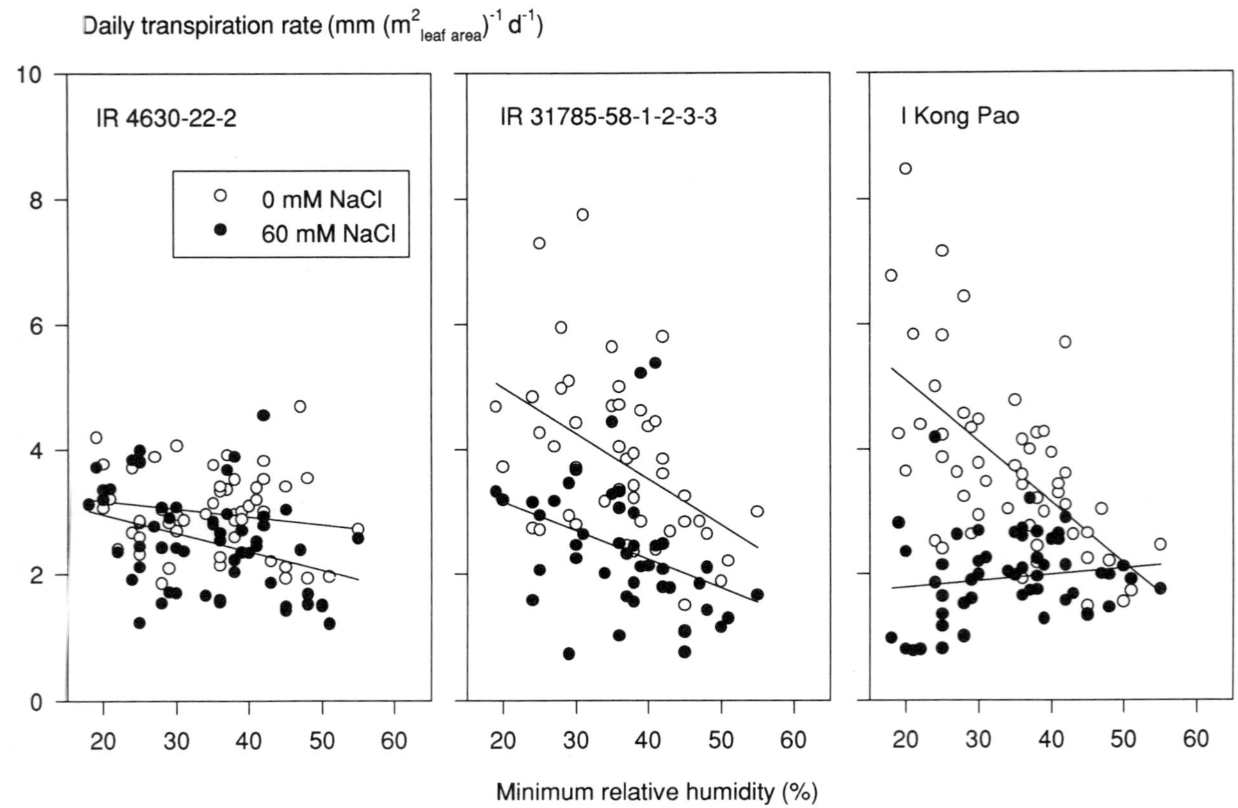

Figure 4. Daily transpiration rates versus relative air humidity of three varieties from a screenhouse trial in the WS 1994. Daily minimum relative humidity between 8 and 20 h. The upper regression line represents the fresh water treatment, the lower the salt treatment (60 mM NaCl in the culture solution). Regressions: IR4630 fresh water: $y=0.013x-3.4$, $r=0.17$; IR4630 salt water: $y=0.03x-3.6$, $r=0.35$*; IR31785 fresh water: $y=0.073x-6.5$, $r=0.45$**; IR31785 salt water: $y=0.046x-4.1$, $r=0.37$*; IKP fresh water: $y=0.1x-7.1$, $r=0.63$**; IKP salt water: $y=0.01x+1.6$, $r=0.14$. * $=p<0.05$; ** $=p<0.01$.

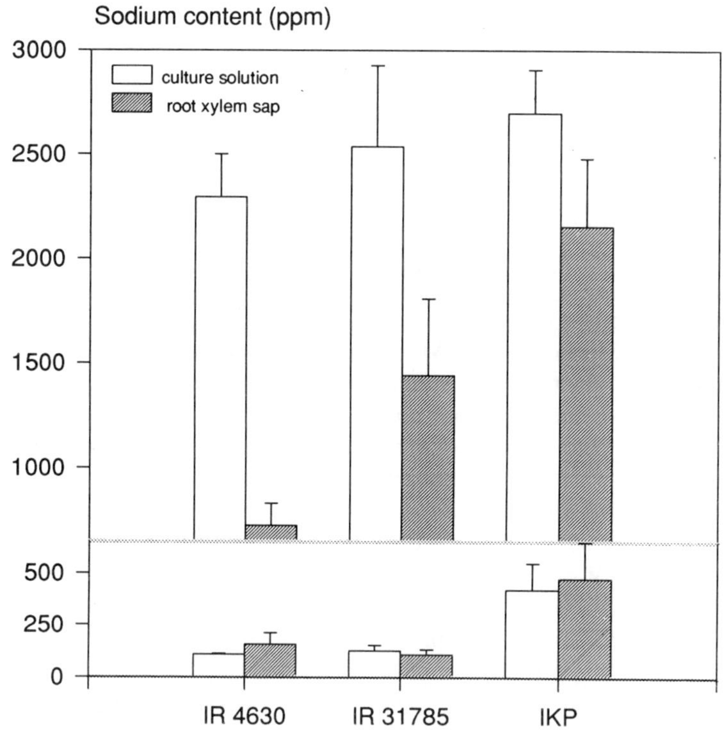

Figure 5. Root xylem sodium concentrations of three varieties in a screenhouse trial in the wet season 1995 after 24 h of exposure to 60 mM NaCl. Lower part: control treatment; upper part: salinity treatment. Error bars = SE over 5 replications.

Sodium and potassium distribution in the plant

The sodium and potassium distribution within the plant at about panicle initiation stage is shown in Figure 6 for three varieties, grown under salinity in the field during wet season 1994 (WS) and the hot dry season 1995 (HDS). Generally, sodium uptake was lower and potassium uptake was higher in the WS than in the HDS. The highest sodium and potassium concentrations were found in the stems, which consisted mainly of leaf sheaths at this development stage. Root sodium and potassium concentrations did not differ much between seasons and varieties, the sodium concentration being generally much higher than the potassium concentration.

Varietal differences in cation concentration were most pronounced in the leaf blades, the sodium concentration increasing by leaf position (descending), and potassium decreasing. IR4630 had the lowest sodium and the highest potassium concentrations, and the reverse was true for IR31785. The strongest seasonal effect on leaf sodium concentration was observed in IKP, which had low concentrations in the wet and high concentrations in the dry season. IKP also showed the most consistent cation gradient across leaf positions. The susceptible check IR31785 stood out by showing very low potassium concentrations in all plant tissues during both seasons, as compared to the other varieties.

As a general trend, leaves showed similar combined concentrations of sodium and potassium, particularly in IR4630 and IKP. In IR31785, the combined cation concentrations were higher than in the other two varieties. Stems, however, showed highly variable and comparatively high combined concentrations. This observation might indicate that stems (sheaths) store these cations.

Development of the conceptual model

Leaf transpiration as the driving force for sodium uptake to the shoot has been subject to several studies. Cheeseman and Wickens (1986) found a direct relationship between salt uptake to the plant and transpiration in *Spergularia marina*, but no relation between leaf sodium accumulation and transpiration. Yeo and Flowers (1984) found a clear dependency of sodium uptake and transpiration in rice. Bohra and Dörffling (1991) found a reduction in Na uptake to the shoot of salt-sensitive rice plants after treating the leaves with abscisic acid, a plant hormone known to reduce transpiration (Zhang and Davies 1991). The present results are in line with these reports (Figure 2).

Our simulation of canopy transpiration depended on plant properties (LAI) on one side and climatic factors on the other side. Air humidity influences stomatal conductance (Asch et al. 1995; Kappen and Haeger 1991). Humid conditions reduce sodium uptake to the shoot and increase dry matter production in chickpea (Lauter and Munns 1987) and faba bean (Salim 1987). Direct measurements of transpirational water consumption in this study (Figure 4) indicated varietal differences in the ability to regulate transpiration under different air humidity and salinity levels, which are probably due to stomatal behaviour. The results presented in Figure 4 reflected well the salt susceptibility of the three test varieties observed in the field: tolerant IR4630 strongly regulated transpiration, susceptible IR31785 did not. IKP, which is also tolerant but does not exclude sodium particularly well (Figure 5), regulated its transpiration strongly only when salinity and low air humidity occurred together.

The simulations of canopy transpiration did not take into account stomatal regulation. The simulations were probably approximately accurate for IR31785 (salinity and controls) and IKP (controls) (Figures 2A and 3A). In the other varieties and treatments, transpiration was probably overestimated because partial stomatal closure controlled transpiration. In order to make the model sensitive to these phenomena, it is necessary that the transpiration routine of ORYZA_W calculates actual (leaf conductance dependent), instead of potential transpiration (Figure 7). The simulation of stomatal conductance, the main component of leaf conductance, requires besides meteorological data information on leaf sodium and leaf nitrogen content. For the model, information is required on interactive effects of the climatic and physiological factors on stomatal conductance, modulated by varietal constants used in genotypic parameterization.

Our observation that the sodium concentration in the xylem sap can be significantly lower, but not higher, than that of the ambient solution (Figure 4) indicates that a ‘root filter’ for sodium exists. This filter can be passive (exclusion) or active

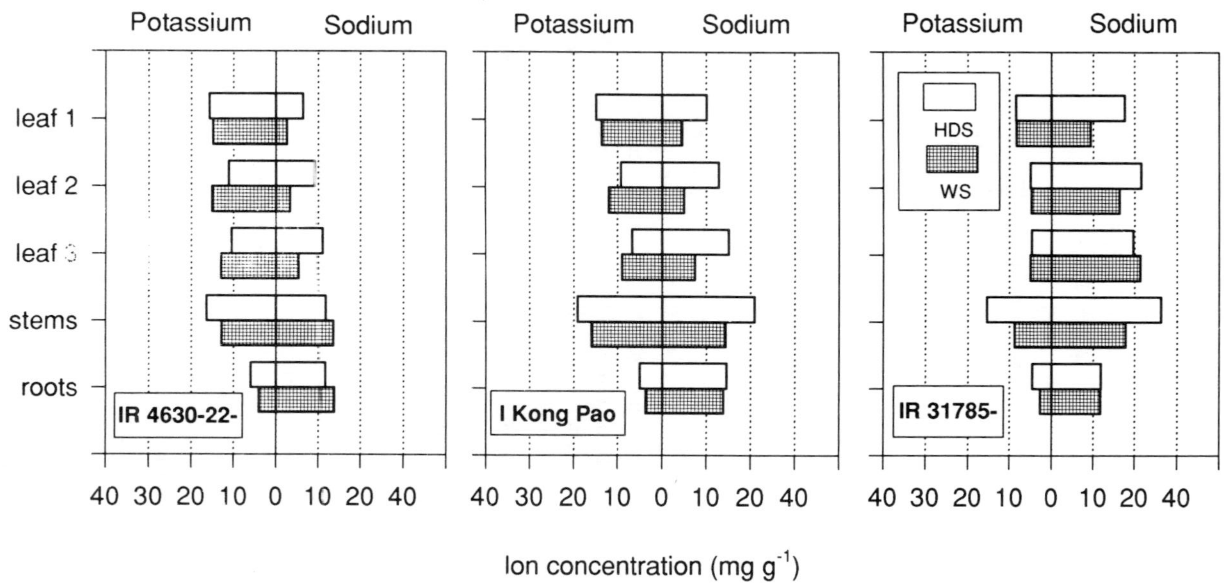

Figure 6. Sodium and potassium distribution in three varieties under salinity (3.5 mS cm^{-1} flood water EC) from field trials in wet season (WS) 1994 and hot dry season (HDS) 1994. Both samplings were conducted close to panicle initiation in the vegetative stage.

(e.g. excretion of absorbed Na at the root surface), and evidently represents an important avoidance component of rice salt resistance. The filter appears to be most efficient in IR4630 and least efficient in IKP. We included this result in our conceptual model by introducing an additional varietal constant (Figure 7, uptake value for sodium).

Yeo et al. (1985) reported that individual leaf transpiration rates are not related to individual leaf sodium uptake. Older leaves transpired less but accumulated more sodium than younger leaves. Thus, the net accumulation in the leaf blades was independent of their cumulative transpiration. This phenomenon might be explained by a partial interception of sodium as it passes through the leaf sheaths. The concentration in the stems (sheaths) was mostly higher than in the leaf blades (Figure 6), although blades are known to transpire much more than sheaths. The leaf sheaths may act as a storage tissue in which sodium is removed from the transpiration stream (see also Greenway and Munns 1980). The difference in sodium concentration between the youngest leaf and the stems was higher in the wet season than in the hot dry season. Due to climate, transpiration was lower in the WS and salt uptake slower. It seems the retention in stem tissue was more effective, when less salt was taken up and the water flow rate in the plant was lower. Therefore, sodium uptake to individual leaf sheaths would be higher: (i) when transpirational volume flow in the xylem vessels was low, allowing for more time to outbalance diffusion gradients in the surrounding tissues. Sodium replaces potassium in the vacuoles (Greenway and Munns 1980); (ii) when the sodium concentration in the sheath tissue was low, permitting a higher sodium absorption rate; and (iii) when the sodium retention capacity of the sheath is not saturated, where the sodium retention capacity is defined as the difference between xylem sap sodium concentration and the sodium concentration of the tissue directly interfacing with the xylem vessel (Jacoby 1964).

We included this theory into the concept of the model (Figure 7). Sodium is taken up via the transpiration stream and moves towards the leaf, passing through the leaf sheath. The maximum daily absorption of sodium by the sheath is defined by a varietal constant, reflecting the potential rate of sodium immobilization in the leaf sheath tissue. A fraction of the sodium in the transpiration stream passes through to the leaves. If the leaf sodium concentration exceeds a varietal limit, leaf senescence accelerates, which decreases the leaf N content and, therefore, stomatal conductance. At 0.5% leaf N, the leaf is considered dead and the transpiration, therefore, zero. This concept assumes that sodium in the leaf does not directly restrict stomatal conductance, but instead through stress-induced senescence and N (and probably, K) mobilization. The varietal upper limit for leaf sodium uptake corresponds to varietal differences in leaf tissue sodium tolerance (Yeo and Flowers 1982).

The role of potassium

The interactions of potassium and sodium are widely discussed in literature. Potassium is selectively and actively taken up by the root and genotypes might differ in their root selectivity for K^+ in relation to their salt tolerance (Yeo and Flowers 1984).

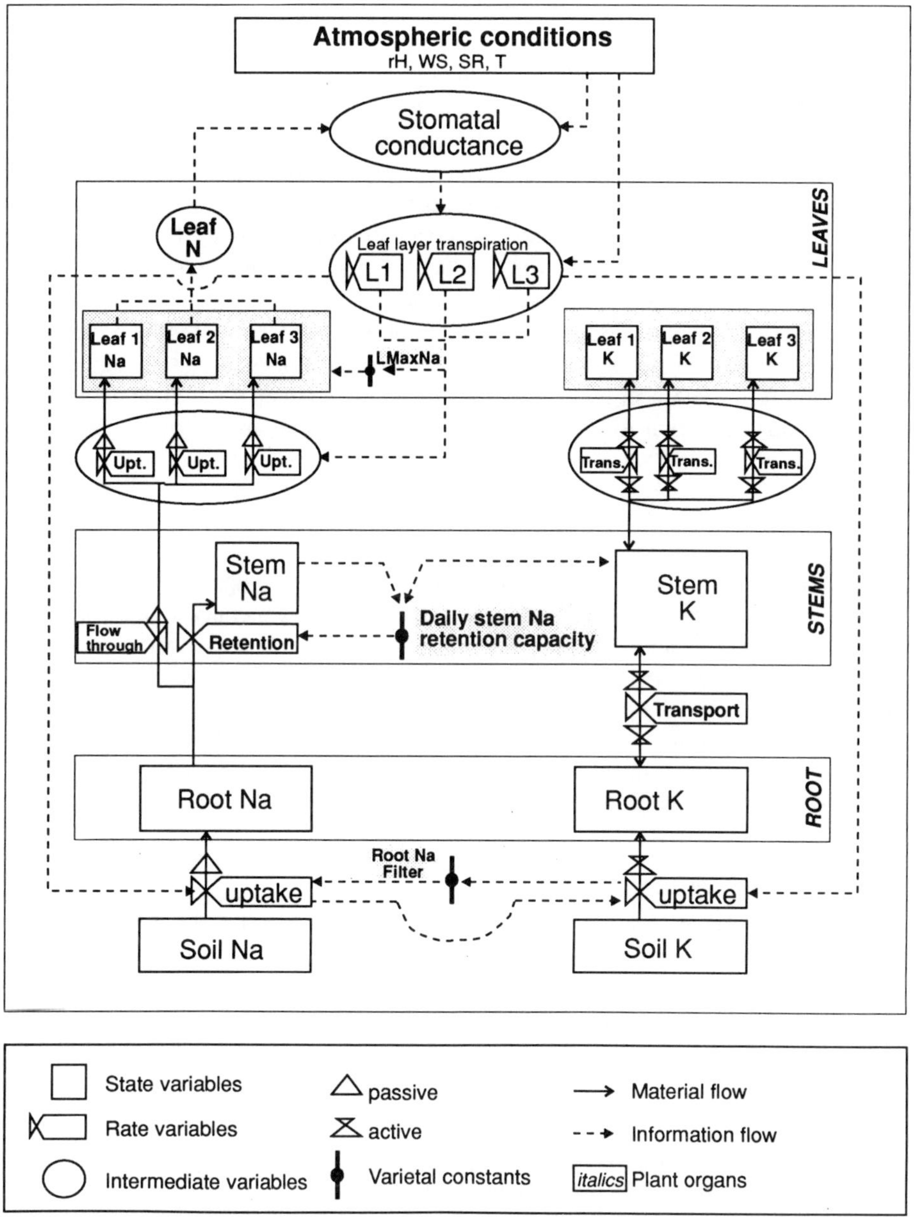

Figure 7. A relational diagram of sodium and potassium uptake and distribution.

Potassium can move freely between xylem and phloem vessel and, thus, be transported to any site in the plant (Wolf and Jeschke 1987). Potassium can be exchanged for sodium in the vacuoles, thus immobilizing sodium (Greenway and Munns 1980). Our data suggest that rice leaves have an upper limit for the sum of Na^+ and K^+ (Figure 6). The extent to which this combined upper limit restricts sodium accumulation in the leaf would depend on the rate at which potassium is mobilized from ageing leaves. By contrast, a high potassium concentration in the stems (sheaths) seems to be coupled with a high sodium concentration (Figure 6). Preliminary data from ongoing trials indicate a strong reduction in sodium uptake to the root if the root xylem potassium concentration is high (Asch, unpubl. data).

Our conceptual model assumes an active uptake of potassium by the root, distribution in the plant via the xylem and relocation via the phloem. Potassium is actively mobilized from senescent leaves and fed into an incremental pool. Each plant organ has a demand function for K^+, which is modified by the amount of sodium it has accumulated.

Conclusions and perspectives

We presented a conceptual model for sodium and potassium uptake and distribution in irrigated rice. The empirical findings reported here are generally in line with other reports, but provide additional detail on the physiological linkages between transpiration and the uptake and distribution of sodium in the plant. More research is needed to validate our model assumptions that concern potassium and its interactions with sodium uptake, particularly with respect to sodium exclusion at the root-soil interface and the storage of both cations in stem/sheath tissues. Follow-up studies are in progress.

Acknowledgements

The work presented here forms part of a project on salinity tolerance of rice in the Sahel, which is funded by the Gesellschaft für Technische Zusammenarbeit (GTZ), Bonn, Germany. This article is based in part on a doctoral study by F. Asch in the Faculty of Biology, University of Hamburg. Thanks are due to Christine Wittstock and Stephan Häfele for critical comments and stimulating discussions.

References

Asch F, Dörffling K, Dingkuhn M (1995) Response of rice varieties to soil salinity and air humidity: A possible involvement of root-borne ABA. Plant and Soil 177:11–19.

Bohra J S, Dörffling K (1991) Effect of abscisic acid application on salt tolerance of rice varieties. IRRI Research News 16(5):15–16.

Capell B, Dörffling K (1989) Low temperature-induced changes of abscisic acid contents in barley and cucumber leaves in relation to their water status. J. Plant Physiol. 135:571–575.

Cheeseman J M, Wickens L K (1986) Control of Na^+ and K^+ transport in *Spergularia marina*. I. Transpiration effects. Physiol. Plant. 67:1–6.

Dingkuhn M, Asch F (1994) Testing hypotheses on rice salt tolerance by simulation. Pages 85–86 in Annual Report 1993. West Africa Rice Development Association (WARDA), BP 2551, Bouaké 01, Ivory Coast.

Dingkuhn M, Cruz R T, O'Toole J C, Dörffling K (1989) Net photosynthesis, water use efficiency, leaf water potential and leaf rolling as affected by water deficit in tropical upland rice. Aust. J. Agric. Res. 40:1171–1181.

Dingkuhn M, De Datta S K, Pamplona R, Javellana C, Schnier H F (1992) Effect of late-season N fertilization on photosynthesis and yield of transplanted and direct-seeded tropical flooded rice. II. A canopy stratification study. Field Crops Res. 28:235–249.

Dingkuhn M, Asch F, Miézan K (1993) Salt tolerance of rice varieties under irrigated conditions in the Sahel. Pages 3–11 in Annual Report 1992. West Africa Rice Development Association (WARDA), BP 2551, Bouaké 01, Ivory Coast.

Dingkuhn M, Sow A, Samb A, Diack S, Asch F (1994) Climatic determinants of irrigated rice performance in the Sahel. I. Photothermal and micro-climatic responses of flowering. Agric. Syst. 48:385–410.

Fageria N K (1985) Salt tolerance of rice cultivars. Plant and Soil 88:237–243.

Greenway H, Munns R (1980) Mechanisms of salt tolerance in nonhaplotypes. Ann. Rev. Plant Physiol. 31:149–190.

IRRI – International Rice Research Institute (1979) Annual Report for 1978. International Rice Research Institute, P.O. Box 933, Manila, Philippines.

Jacoby B (1964) Function of bean roots and stems in sodium retention. Plant Physiol. 39:445–449.

Kappen L, Haeger S (1991) Stomatal responses of *Tradescantia albiflora* to changing air humidity in light and darkness. J. Exp. Bot. 42:979–986.

Lauter D J, Munns D N (1987) Salt sensitivity of chickpea during vegetative growth and at different humidities. Aust. J. Plant Physiol. 14:171–180.

Mizrahi Y, Blumenfeld A, Bittner S, Richmond A E (1971) Abscisic acid and cytokinin contents of leaves in relation to salinity and relative humidity. Plant Physiol. 48:752–755.

Neue H U, El-Naggar M I A, Rashid M (1990) Responses and tolerance mechanisms of rice to saline soil conditions. Transaktions IV:50–55.

Raes D, Sy B, Feyen J (1995) Water use in rice schemes in the Senegal river delta and valley. Irrig. Drain. Syst. 9:117–128.

Salim M (1987) Effects of salinity and relative humidity on growth and ionic composition of faba bean. Faba-Bean-Information-Service, ICARDA 1987, Aleppo, Syria, FABIS-Newsletter 18:11–14.

Wolf O, Jeschke W D (1987) Modeling of sodium and potassium flow via phloem and xylem in the shoot of salt-stressed barley. J. Plant. Physiol. 128:371–386.

Wopereis M C S, Bouman B A M, Tuong T P, Ten Berge H F M, Kropff M J (1996) ORYZA_W: Rice growth model for irrigated and rainfed environments. SARP Research Proceedings, IRRI, P.O. Box 933, Manila, Philippines. 159 p.

Yeo A R, Flowers T J (1982) Accumulation and localisation of sodium ions within the shoots of rice (*Oryza sativa* L.) varieties differing in salinity resistance. Physiol. Plant. 56:343–348.

Yeo A R, Flowers T J (1984) Mechanisms of salinity resistance in rice and their role as physiological criteria in plant breeding. Pages 151–170 in Staples R E (Ed.) Salinity tolerance in plants: Strategies for crop improvement. John Wiley and Sons, New York, USA.

Yeo A R, Yeo M E, Flowers T J (1984) Nonosmotic effects of polyethylene glycols upon sodium transport and sodium-potassium selectivity by rice roots. Plant Physiol. 75:298–303.

Yeo A R, Yeo M E, Caporn S J M, Lachno D R, Flowers T J (1985) The use of ^{14}C-ethane diol as a quantitative tracer for the transpirational volume flow of water and an investigation of the effects of salinity upon transpiration, net sodium accumulation and endogenous ABA in individual leaves of *Oryza sativa* L. J. Exp. Bot. 36:1099–1109.

Yeo A R, Yeo M E, Flowers T J (1987) The contribution of an apoplastic pathway to sodium uptake by rice roots in saline conditions. J. Exp. Bot. 38:1141–1153.

Yoshida S, Forno D A, Cock J H, Gomez K A (1976) Laboratory manual for physiological studies of rice. International Rice Research Institute (IRRI), P.O. Box 933, Manila, Philippines.

Zhang J, Davies W J (1991) Antitranspirant activity in xylem sap of maize plants. J. Exp. Bot. 42:317–321.

Using decision support systems to optimize barley management on spatial variable soil

H.W.G. BOOLTINK and J. VERHAGEN

Department of Soil Science and Geology, Wageningen Agricultural University, P.O. Box 37, 6700 AA Wageningen, The Netherlands

Key words: precision agriculture, simulation, CERES, geostatistics, GIS, economic analysis

Abstract

Increasing awareness of possible pollution, due to agricultural practices, has resulted in 'precision agriculture'. Precision agriculture allows the farmer to use soil spatial variability to site-specifically adjust farm management strategies, e.g. spatial variable fertilization or irrigation applications. The aim of precision agriculture is to optimize crop yields and to make more efficient use of agrochemicals by reducing leaching. The CERES-Barley model from the decision support system DSSATv3 (Decision Support System for Agrotechnology Transfer) was used to simulate crop and soil responses on the experimental farm, the 'Professor van Bemmelenhoeve' in The Netherlands. DSSAT was subsequently used to simulate and analyse different farm management scenarios. Barley yields, economic benefits and possible leaching of fertilizers (as a result of the different management scenarios) were quantified. The use of this type of decision support system, for analysing agroecological systems, allows farmers, extension services and scientists to define agricultural practices for different crops which balance production and ecological interests in various climatological zones.

Introduction

Soil spatial variability is often seen as a nuisance which distracts values from the average. However, this variability can also be the start of a new type of farm management: Precision Agriculture (Bouma 1995; Bouma et al. 1996). Precision agriculture or site-specific management refers to an agricultural management system that promotes variable management practices within a field according to site or soil conditions (Robert et al. 1994). Verhagen et al. (1995) showed that spatial variability of soil conditions resulted in potato yields varying from 30 to 45 ton ha^{-1} in The Netherlands on a 6 ha field. Sadler et al. (1994) found differences in maize yields in the USA ranging from 214 to 4849 kg ha^{-1} (at 15.5% water content) on an 8 ha field. Soil spatial variability may also lead to variable leaching patterns as was illustrated by Dagan and Bresler (1983). Finke and Stein (1994) quantified these variable leaching patterns, using Disjunctive Co-kriging, and expressed leaching risks in terms of probabilities of exceeding critical levels instead of absolute values.

To comprehend all aspects of precision farming at the field scale and to understand the output of simulation models, a systems approach is necessary. Decision Support Systems (DSS) are interactive computer-based systems that help decision makers utilize data and models to solve unstructured problems (Sprague and Carlson 1982). One such system is DSSAT Version 3, Decision Support System for Agrotechnology

M.J. Kropff et al. *(eds.): Applications of Systems Approaches at the Field Level, 219–233.*

Transfer (Jones 1993; Uehara and Tsuji 1993). DSSAT3 was designed to allow users to (i) input, organize and store data on crops, soils and weather, (ii) retrieve, analyse and display data, (iii) validate and calibrate crop growth models and (iv) evaluate different management practices (Jones 1993). A linkage between DSSAT3 and ArcView (ESRI Ltd) was established by Engel and Jones (1995), allowing the creation of thematic maps based on polygon features. Linking DSSAT3 crop models with GIS to support decision making was described by Singh et al. (1993). Both of these spatial links are especially suitable for farm or regional level simulation studies. At the field scale, small-scale variability can hardly be described by these systems. An approach where individual point information can be included in the spatial analyses is, therefore, required. Geostatistics provide the possibility of creating maps based on simulation results. In this study, geostatistical tools were included in the DSSAT3 shell and made compatible with DSSAT3 data and file formats. This extended DSSAT version was applied on experimental data on site-specific farming with the objective of exploring the feasibility of simulation models in optimizing site-specific farm management strategies, including minimization of fertilizer losses. Main goals of the scenario exploration study were: (i) to define optimum time windows and amounts for nitrogen application, (ii) to optimize crop yields and (iii) to minimize nitrogen losses to the groundwater.

Material and methods

Soil

At the Van Bemmelenhoeve experimental farm in the Wieringermeer, The Netherlands, a field of approximately 6 ha was selected. A soil survey was conducted and the soil was classified as a Typic Udifluvent (Soil Survey Staff 1975). Geo-referenced data of the 65 soil profile descriptions within the experimental field were stored in a data base and generalized using functional horizons. Functional horizons consist of combinations of genetic horizons which show identical behaviour (Wösten et al. 1986; Finke 1993). Four functional layers could be distinguished within the experimental field. Basic characteristics of the four functional layers are presented in Table 1. Use of functional horizons is attractive because the vertical sequence of each point observation can be represented by only a few functional horizons, rather than a relatively high number of pedological horizons (Verhagen et al. 1995). Basic soil data can be used for simulation modelling for each point, which will be discussed later.

Farm

One of the objectives of this study is to compare current, conventional farm management with site-specific farm management through simulation studies. Environmental and economical results were also used to judge farm management strategies. Use of realistic farm management data and practices is, therefore, essential.

Table 1. Description of the functional layers and corresponding textural and hydraulic characteristics. ρ refers to the dry bulk density, sd to standard deviation, K_{sat} to the saturated hydraulic conductivity and θ_{sat} to the saturated water content.

Layer	ρ (kg m^{-3})	sd	Organic matter (%)	Clay (%)	K_{sat} (m d^{-1})	θ_{sat} (m^3 m^{-3})	sd
F1	1480	39	0–2	0–4	1.83	0.40	0.02
F2	1210	97	0–2	4–11	1.28	0.53	0.03
F3	1150	231	0–2	11–23	0.07	0.57	0.05
F4	1300	45	0–2	4–23	3.35	0.52	0.02

Table 2. Basic economic data for barley, price level 1995 (*Source:* F.G. Wijnands and A. Krikke, PAGV 1995, pers. commun.; 1 Dfl = 0.7 US$).

	Units	Distribution	Average	sd
Grain yield	Dfl 1000 kg^{-1}	Normal	350.00	30.00
Harvest by-product	Dfl 1000 kg^{-1}	Normal	85.00	8.00
N-fertilizer cost	Dfl kg^{-1}	Fixed	1.04	–
N-costs per application	Dfl appl^{-1}	Fixed	24.00	–
Irrigation costs	Dfl mm^{-1}	Fixed	4.00	–
Irrigation costs per application	Dfl appl^{-1}	Fixed	25.00	–
Seed costs	Dfl kg^{-1}	Fixed	1.22	–

Fertilization and management. Fertilizer amounts in The Netherlands are based on the Dutch fertilizer recommendation, which uses the total mineral nitrogen level (N_{min}) in early spring, present in the rootable zone before fertilizing, to determine the application amount. The parameters for this fertilizer recommendation were derived from national trials. At the experimental field, the field average N_{min} level, measured in early spring, was 35 kg N ha^{-1}, which results in a fertilizer recommendation of 80 kg ha^{-1} to be applied in early spring, and 60 kg ha^{-1} as a second application during the growing season. According to the actual farm management practices in 1994, the first application of 80 kg ha^{-1} was applied on 18 February (day number 49) and the second application of 60 kg ha^{-1} on 18 May (day number 138). Sowing of barley in 1994 was on 27 March and harvest on 12 August (day number 86 and 224 respectively). No irrigation was applied.

Economic data. A full economical analyses of the two farming systems is beyond the scope of this study. We, therefore, made a comparison based on the simulated differences between the conventional and site-specific farming systems. Basic economic data for this comparison were supplied by the Research Station for Arable Farming and Field Production of Vegetables (PAGV) in The Netherlands and are summarized in Table 2.

Simulation model

Within DSSAT3, barley growth and production was simulated with CERES-Barley. A detailed description of the model can be found in Ritchie (1986), Jones and Ritchie

(1990) and Godwin and Jones (1991). Briefly, the model allows the quantitative determination of barley growth and yield.

The simulation of the soil water balance in CERES includes processes such as infiltration and surface run-off, evaporation and transpiration, and drainage at the bottom of the soil profile. Unsaturated flow of water from one soil layer to another is simulated as a capacity flow system, requiring only the soil water content at field capacity and permanent wilting point as hydraulic input parameters. The nitrogen submodel simulates organic matter turnover rates with the associated mineralization, immobilization of nitrogen, nitrification and denitrification, urea hydrolysis and ammonia volatilization for different types of nitrogen such as organic fertilizers, green manure and synthetic fertilizers. Transport of nitrogen species occurs at the same rate as flow of water. Convective and/or dispersive transport of chemicals is not included. Crop growth is simulated as a phasic development with respect to the physiological age of the plant. Potential growth is dependent on photosynthetically active radiation (PAR). Interception of PAR is influenced by Leaf Area Index (LAI), row spacing, planting density and photosynthetic efficiency of the crop. Actual biomass production is regulated by suboptimal temperatures, soil-water deficits and nitrogen deficiencies. Partitioning of the photosynthetic products is governed by crop-specific parameters and development stages of the crop. The model simulates the above mentioned processes at a daily time step.

To spatially interpret simulation results, DSSAT3 was extended with geostatistical tools which allow the calculation of semi-variances, fit various semi-variogram models and that can perform different types of kriging, such as Ordinary kriging and Disjunctive kriging for creating probability maps. The software integrated is existing software which has been used and tested widely. A brief summary of the software and the most relevant references are:

- To calculate semi-variances on simulation results: SPATANAL, developed by Staritsky et al. (1992).
- Semi-variances were modelled using WLSFIT, developed by Heuvelink (1995).
- Ordinary and Disjunctive kriging options: DISCO, developed by Finke (1993).
- GIS tools: USTED, developed by Stoorvogel (1995).

Weather data

The minimum set of weather data needed for the CERES models include, next to general parameters such as latitude, longitude and elevation of the site, daily values for precipitation, global radiation and minimum and maximum air temperature. Rainfall provides the water supply for the crop, global radiation is used for the calculation of photosynthetic products and evapotranspiration; temperature determines, among others, the developmental stage of the crop. An extended description of the weather data requirements is given by Ritchie et al. (1990). DSSAT3 is also equipped with utilities for managing and generating daily weather data (Pickering et al. 1994). The WGEN weather generator of Richardson and Wright (1984) allows stochastic time series generating of sequences of weather data when long-term weather records are not available. Stochastic parameters in WGEN were determined by feeding WGEN with a continuous time series of weather data, recorded on the Van Bemmelenhoeve and a nearby weather station, starting in 1988 and seven years onwards.

Simulation procedure

The simulation procedure in this study consists basically of four steps:

Step 1: The current conventional farm management for barley was simulated for 1994 for every one of the 65 profiles within the experimental field. Fertilizer was applied non-site-specifically according to the field average recommendation as described in the fertilization section.

Step 2: Based on the results of this step, a limited number of characteristic soil profiles were selected for further analysis and explorative simulation exercises (1 year, 1994). Most relevant and realistic management scenarios for site-specific management were selected; and

Step 3: Used in a study in which every one of each options was simulated for 10 years of generated weather data, to include temporal variability of weather over different years. Apart from environmental criteria expressed as the leaching probability of exceeding the critical $50\,g\,m^{-3}$ nitrate concentration to the groundwater, the scenarios were also judged on their economical merits.

Step 4: Finally, the optimal scenarios were site-specifically simulated for each of the 65 profiles within the experimental field with data from 1994, and results were compared to the initial simulations of conventional farming.

Results

Simulation and spatial interpolation of non-site-specific management

In this first simulation step, current non-site-specific farm management was simulated. For every one of the 65 profiles, the field average fertilizer recommendation of $80\,kg\,N\,ha^{-1}$, as a first application, and an additional $60\,kg\,N\,ha^{-1}$, in the middle of the growing season, was applied according to the current farm management practices. No spatial differentiation was made.

Simulation results for barley seed weight and the amount of N leached to the groundwater, both expressed in $kg\,ha^{-1}$, were geostatistically analysed. Semi-variances were calculated and the semi-variogram was fitted. The fitted semi-variogram was subsequently used to create maps for simulated barley seed weight (kriging) and for the probability of exceeding the critical nitrogen leaching level of $50\,g\,m^{-3}$ to the groundwater (disjunctive kriging). The nitrogen leaching level of $50\,g\,m^{-3}$ is in The Netherlands, on average, equivalent to $33\,kg\,N$ leaching loss per hectare per year.

In Figure 1, a map of simulated barley yield is presented. Yields vary from approxi-

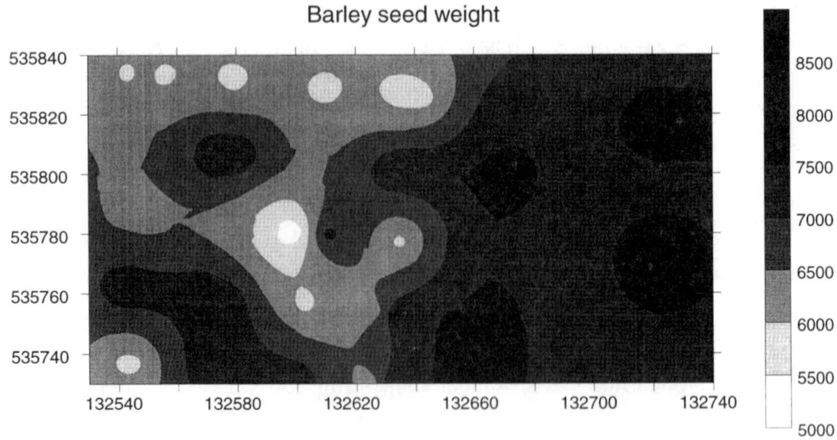

Figure 1. Kriged map showing the simulated barley seed weight in kg ha^{-1} for non-site-specific farm management. The X- and Y-coordinates refer to the Dutch national coordinate system (RDM).

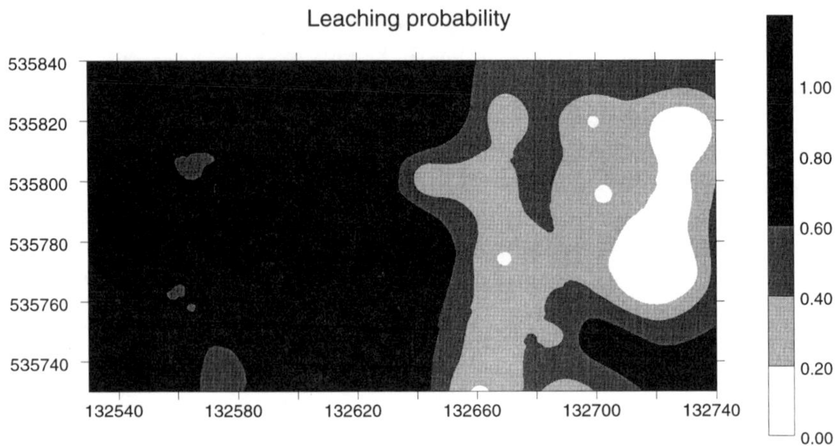

Figure 2. Leaching maps of simulated nitrogen leaching concentrations, expressed as the probability of exceeding the critical level of 50 g nitrate m^{-3} in the leachate for non-site-specific farm management. The X- and Y-coordinates refer to the Dutch national coordinate system (RDM).

mately 5500 kg ha^{-1} on the left side of the map to 8000 kg ha^{-1} on the right side of the map. The two designated areas coincide well with the textural differences in the soil data base. The left side is dominated by a more sandy texture, and the right side by a more clayey texture. Since the water-holding capacity is higher in the clayey part, N-leaching is less in winter and availability of water during the relatively dry summer is higher; resulting in higher yields.

A map of the conditional probability of exceeding the nitrogen leaching threshold value of 50 g m^{-3} is shown in Figure 2. A comparable spatial pattern compared to

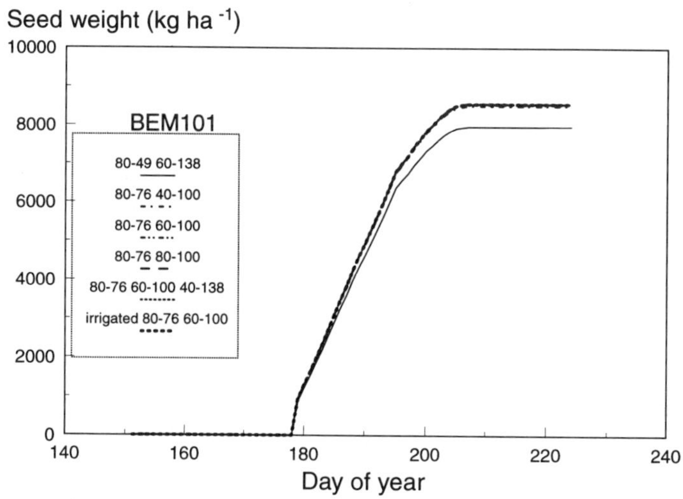

Figure 3. Simulated barley seed weights for the clayey profile (BEM101) codes refer to the day number and amount of fertilizer applied on that date, e.g. 80-49 60-138 refers to a double application of 80 and 60 kg N on day number 49 and 138 respectively. Except for the 80-49 scenario, all simulations are coinciding.

Figure 1 is visible: high leaching probabilities, ranking from 0.6 to 1.0, in the sandy part, and relatively low probabilities, from 0.0 to 0.4, in the clayey part.

The clear spatial distinction between the sandy and clayey part of the field allows the exploration of site-specific farm management practices for both soil types. Two representative profiles, BEM101 and BEM107 from the clayey and sandy area respectively, were selected for further analysis in the second step, i.e. the site-specific scenario exploration.

Site-specific scenario exploration

To define optimum time windows and optimized crop yields, the automatic fertilization option within DSSAT was used. The result from this simulation showed that application on 18 February (day number 49), 37 days prior to barley seeding, was far too early. Ten days before barley seeding on 27 March, there was a more efficient use of the fertilizer applied for both soil profiles BEM101 and BEM107, the clayey and sandy profiles respectively.

In Figures 3 and 4, the results of the most relevant explorative simulations are presented in terms of seed yield and leached N for the clayey part. Although the alternative scenarios simulate a little higher seed yield compared to the original simulation, indicated by BEM101 80-49 60-138, this is not a very significant increase. Also, for the amount of N leached to the groundwater, no significant improvement could be stimulated. The simulated values of 25 kg N ha^{-1} yr^{-1} are still far below the critical level of 33 kg N ha^{-1} yr^{-1}. For the sandy profile (BEM107), there were significant improvements. In Figure 5, seed weight increased from approximately

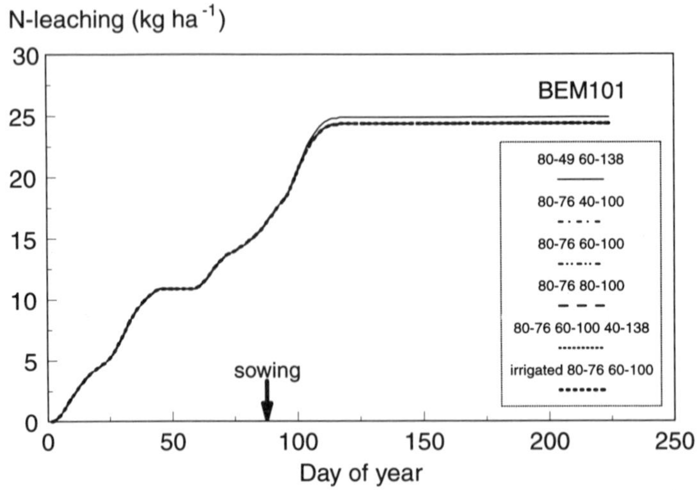

Figure 4. Simulated nitrogen leaching for the clayey profile (BEM101) codes refer to the day number and amount of fertilizer applied on that date, e.g. 80-49 60-138 refers to a double application of 80 and 60 kg N on day number 49 and 138 respectively.

5500 kg ha^{-1} to more than 7500 kg ha^{-1}. The differences among the alternative scenarios are little, indicating that optimizing the time window of application is more important than increasing the amount of fertilizer applied. With respect to nitrogen leaching, the alternative scenarios perform a little better than the original simulation (BEM107 80-49 60-138); however, the total amount of N leached is still far too high as compared to the critical level of 33 kg N ha^{-1} yr^{-1}. As can be seen in Figure 6, most leaching occurred before the start of the growing season (sowing was on 27 March, day number 86), indicating that the effect of the fertilizer applications, based on the amount of nitrogen leached, in the alternative scenarios was limited. Additional management measures are necessary to fulfil the leaching objective within this soil type. With respect to leaching, there is again little difference among the alternative scenarios, since leaching during the growing season is negligible and fertilizer was applied shortly before the sowing date for all alternative scenarios.

Stochastical testing of management scenarios

In the previous section, the scenarios were tested on data from one year, 1994. This year had an exceptionally dry summer. However, for a more reliable conclusion on the effect of alternative, site-specific scenarios, these scenarios need to be tested on sets of weather data from different years to include temporal variability of weather data. Since only a limited number of weather data sets were available, generated weather data, using the WGEN-weather generator within DSSAT, for a period of 10 years was used. Every simulation scenario for the clayey and the sandy soil profile was executed for all of the 10 years of weather data, resulting in a distributed model

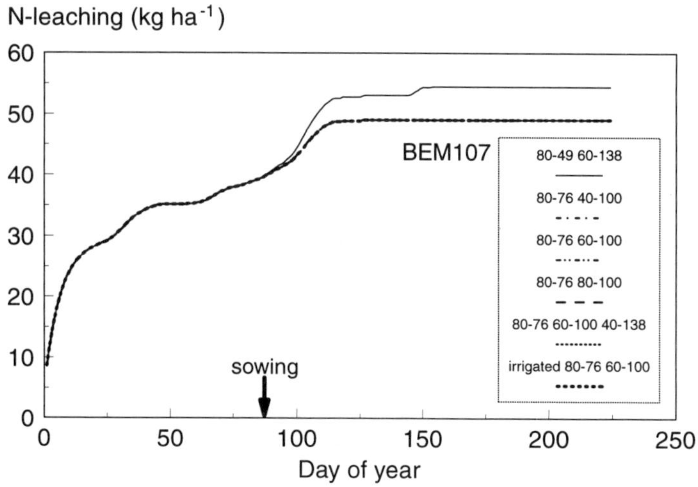

Figure 5. Simulated barley seed weights for the sandy profile (BEM107) codes refer to the day number and amount of fertilizer applied on that date.

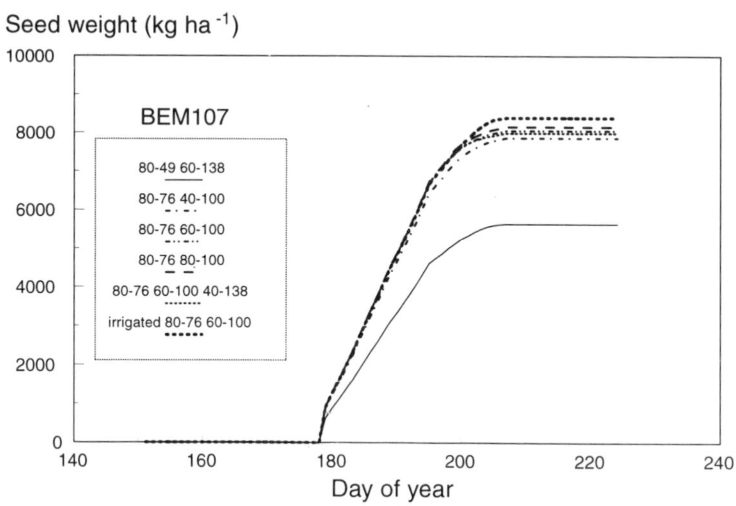

Figure 6. Simulated nitrogen leaching for the sandy profile (BEM107) codes refer to the day number and amount of fertilizer applied on that date. Except for the 80-49 scenario, all simulations are coinciding.

outcome (Tables 3 and 4). For the clayey profile simulated barley seed weights are always higher than for the sandy profile. However, differences between the base scenario (80-49 60-138) and the alternative scenarios are larger for the sandy profile, indicating a better possibility for site-specific management on the sandy profiles within the field. The sandy profile also has a higher standard deviation of the simulated seed weight, which demonstrates the higher production risks on this type of soil. Since

Table 3. Simulation results for the site-specific scenarios using 10 years of generated weather data for the clayey profile. Avg refers to the average simulated value, std dev to the standard deviation and Dfl. to Dutch guilder. Scenario codes refer to the day number and amount of fertilizer applied on that date, e.g. 80-49 60-138 refers to a double application of 80 and 60 kg N on day number 49 and 138 respectively. Irrigated refers to a scenario in which a single irrigation of 25 mm was applied.

Scenario	Seed weight ($kg\,ha^{-1}$)		N-leaching ($kg\,ha^{-1}$)		Relative economic return ($Dfl\,ha^{-1}$)
	avg	std dev	avg	std dev	
80-49 60-138	8340	979	19.3	6.5	0
80-76 40-100	8442	1040	19.0	6.2	67
80-76 60-100	8525	1035	19.0	6.2	92
80-76 80-100	8531	1035	19.0	6.2	83
80-76 60-100 40-138	8553	1021	19.0	6.2	61
Irrigated 80-76 60-100	8626	902	19.0	6.2	24

Table 4. Simulation results for the site-specific scenarios using 10 years of generated weather data for the sandy profile. Avg refers to the average simulated value, std dev to the standard deviation and Dfl to Dutch guilder. Scenario codes refer to the day number and amount of fertilizer applied on that date, e.g. 80-49 60-138 refers to a double application of 80 and 60 kg N on day number 49 and 138 respectively. Irrigated refers to a scenario in which a single irrigation of 25 mm was applied.

Scenario	Seed weight ($kg\,ha^{-1}$)		N-leaching ($kg\,ha^{-1}$)		Relative economic return ($Dfl\,ha^{-1}$)
	avg	std dev	avg	std dev	
80-49 60-138	7645	1588	43.4	6.5	0
80-76 40-100	7645	1286	41.9	5.1	58
80-76 60-100	7799	1312	41.9	5.1	92
80-76 80-100	7885	1345	41.9	5.1	124
80-76 60-100 40-138	7900	1332	41.9	5.1	48
Irrigated 80-76 60-100	8033	1283	41.9	5.1	50

most of the leaching occurs prior to the growing season, as previously mentioned, differences in the amount of nitrogen leached do not vary much between the base scenario and the alternative scenarios. Among the alternative scenarios there are no differences in N leaching. However, one can conclude that in the clayey profile the critical leaching concentration of $50\,g\,m^{-3}$ (equivalent with $33\,kg\,N\,ha^{-1}\,yr^{-1}$) will not be exceeded with a probability of more than 95%. On the sandy profile, on the contrary, the critical level is always exceeded.

The relative economic return (expressed in Dutch guilders, Dfl) is calculated with the 80-49 60-138 scenario as base level. As can be seen from Tables 3 and 4, the margins on the sandy soil are higher, ranging from 58 to 124 Dfl, compared to 67 to 92 Dfl on the clayey profile. The economically most profitable scenario for the clayey soil is scenario 80-76 60-100, and for the sandy profile scenario 80-76 80-100. The triple applications (80-76 60-100 40-138) result in a slightly higher yield for both profiles but the costs of an additional application do not meet the extra return. The last scenario (irrigated 80-76 60-100) refers to a scenario in which one single irrigation of 25 mm was applied during the dry period in early July. This scenario results in

Figure 7. Kriged map of the simulated barley seed weight in kg ha^{-1} for site-specific farm management. The X- and Y-coordinates refer to the Dutch national coordinate system (RDM).

higher yields for both profiles, but again the costs of an irrigation application do not meet the extra return of a higher yield. Irrigation does reduce the management risk as indicated by a lower standard deviation of the simulated seed weights.

On average over the sandy and clayey part of the experimental field, the simulated economical profits were about Dfl 100.-ha^{-1} yr^{-1}.

Simulation and spatial interpolation of optimized management scenarios

Based on the individual soil profile description of every one of the 65 soil profiles, the soil profiles were grouped into either a *clayey* or a *sandy* category. The economically most profitable scenarios for both profiles, as obtained in the previous step, were used in site-specific simulation for the experimental field using the actual weather data from 1994. Simulation results were geostatistically analysed and maps of yield and nitrogen leaching were produced (Figures 7 and 8 respectively). Compared to the original, non-site specific scenario (Figure 1), site-specific application results in a higher yield for the experimental field. Yields for the site-specific simulation range from 7000 to 8500 kg ha^{-1}, compared to 5500–8000 kg ha^{-1} for the non-site-specific simulation. For nitrogen leaching the site-specific simulation results in low leaching probabilities in the clayey part (lower than 0.20), but for the sandy part of the experimental field leaching probabilities are still far too high and, for large areas, close to 1.00.

Defining criteria for the effective use of catch crops

Management practices aimed at lowering the leaching probability prior to the growing season have to focus on the fixation of the available amount of mineral nitrogen. Fixation of mineral nitrogen in autumn and early winter, by catch crops, is a relatively

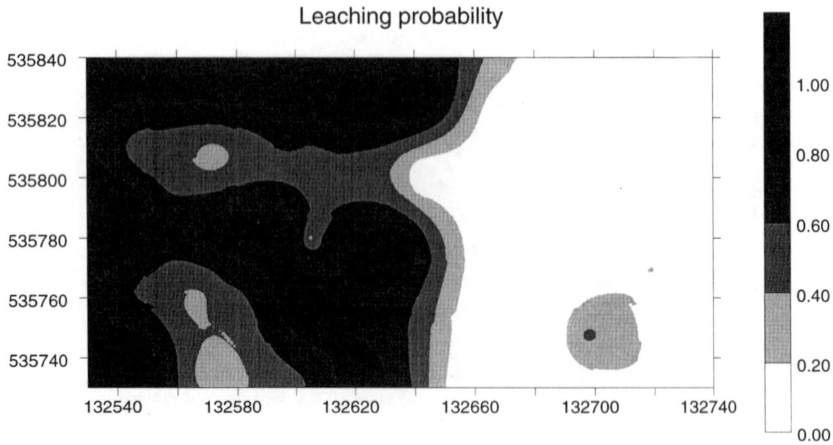

Figure 8. Leaching maps of simulated nitrogen leaching concentrations, expressed as the probability of exceeding the critical level of 50 g nitrate m^{-3} in the leachate for site-specific farm management. The X- and Y-coordinates refer to the Dutch national coordinate system (RDM).

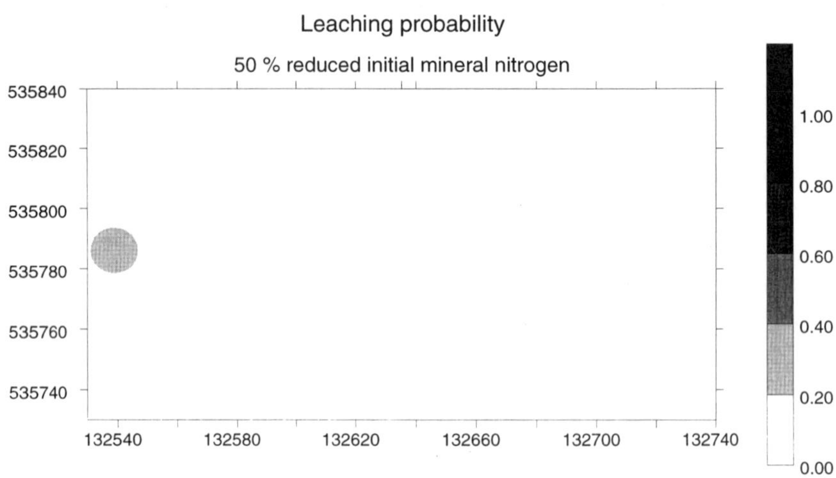

Figure 9. Leaching maps of simulated nitrogen leaching concentrations, expressed as the probability of exceeding the critical level of 50 g nitrate m^{-3} in the leachate for site-specific farm management with a 50% reduced initial amount of mineral nitrogen. The X- and Y-coordinates refer to the Dutch national coordinate system (RDM).

simple way of avoiding large amounts of free mineral nitrogen during periods with high leaching probabilities.

To explore the effect of catch crops, an exploratory simulation study was carried out. The basic question in this study was: 'what amount of mineral nitrogen should be fixed by the catch crop to reach an acceptable leaching probability (lower than 0.20)'. By trial and error a number of alternatives were simulated. In case the amount

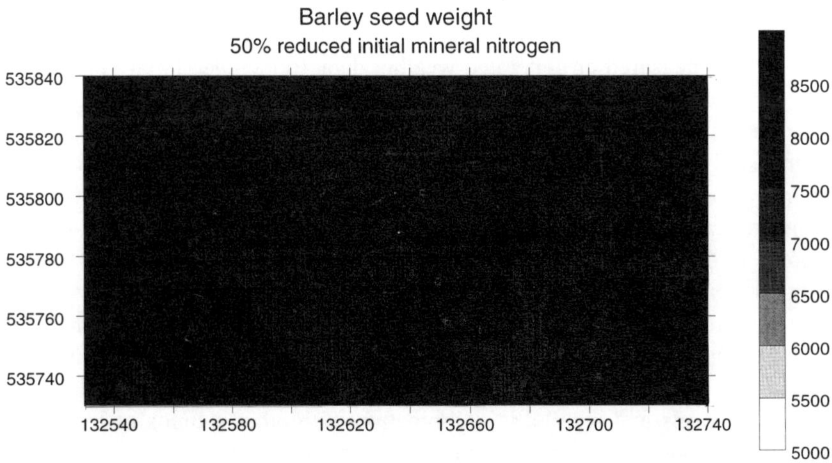

Figure 10. Kriged map of the simulated barley seed weight in kg ha^{-1} for site-specific farm management with a 50% reduced initial amount of mineral nitrogen. The X- and Y-coordinates refer to the Dutch national coordinate system (RDM).

of mineral nitrogen, prior to the leaching season, could be reduced with 50%, the leaching probability was always below 0.20 and, therefore, acceptable. In Figure 9, the leaching probability map for this scenario is presented. The crop yield in this 'catch crop' scenario is hardly affected, as can be seen from Figure 10. This substudy also demonstrates the value of simulation models in defining boundary conditions for farm management practices.

Discussion and conclusions

This study, in which simulation models were used to illustrate the effects of precision farming, shows that optimizing the use of agrochemicals, as it is one of the main objectives of site-specific management, is beneficial for both farmer and environment. By fine-tuning fertilizer amounts and the time window at which they were applied, simulated profits increased up to about Dfl 100.- on average per hectare per year for a barley crop. This might seem marginal, but knowing the small profits on cereal crops within the European Community, it is a significant amount. On economically more valuable crops, such as sugarbeet or potatoes, profits may be higher. Since it is outside the scope of this study, investment necessary for site-specific farming has not been taken into account. It is, however, expected that these investments will be significantly lower than the profits. Leaching of nitrogen could only partly be reduced by precision farming. Due to the high leaching, prior to the growing season, in the sandy parts, the critical N level was exceeded. Additional measures which reduce the amount of mineral nitrogen in the winter (leaching) season are, therefore, required.

Simulation models such as DSSAT, linked with geostatistical tools as demonstrated in this study, provide the possibility of spatially analysing simulation results and allow an easy linkage with GIS. For risk analysis, simulation results should be based on time series of measured or generated weather data, to include climatological variability. To ensure realistic simulated farm management practices, close contact and interaction with practising farmers is essential.

In this study, simulation models were also used to define boundary conditions for farm management measures. The amount of mineral nitrogen, to be fixed by a catch crop, in order not to exceed the critical nitrate concentration of 50 g m^{-3}, was simulated to be 50% of the initial amount present. Catch crops can, therefore, play an important role in preventing excessive leaching of nitrogen to the groundwater.

The next step in the research on precision agriculture or site-specific management will be the implementation of the simulated scenarios in actual farm management and the comparison, c.q. validation of simulation results with 'real-world' data. Finally, simulation models and results, in combination with other sources of information, such as yield mapping and remote sensing, will be used to define a decision support system for precision agriculture.

Acknowledgements

This study was partly financed by the European Community project EV5V-CT94-0480. The authors would like to thank H. Oosterhuis, Manager of the Van Bemmelenhoeve, for cooperation during the field experiments, and F.G. Wijnands and A. Krikke of the Research Station for Arable Farming and Field Production of Vegetables (PAGV) for providing the economic data.

References

Bouma J (1995) Methods to characterize soil research variability in space and time. Pages 201–207 in Robert P C, Rust R H G, Larson W E (Eds.) Site-specific farm management for agricultural systems. Second International Conference, Minneapolis, 27–29 March 1994, ASA, CSSA, SSSA publication, Madison WI, USA.

Bouma J, Verhagen J, Brouwer J, Powell J M (1996) Using system approaches for targeting site specific management on field level. Pages 25–36 in Kropff M J, Teng P S, Aggarwal P K, Bouma J, Bouman B A M, Jones J W, Van Laar H H (Eds.) Applications of systems approaches at the field level. Kluwer Academic Publishers, Dordrecht, The Netherlands (this volume).

Dagan G, Bresler E (1983) Unsaturated flow in spatially variable fields. 1. Derivation of models of infiltration and redistribution. Water Res. Res. 19:413–420.

Engel T, Jones J W (1995) AEGIS/WIN user's manual. University of Florida, Gainesville, FL, USA. 50 p.

Finke P A (1993) Field scale variability of soil structure and its impact on crop growth and nitrate leaching in the analysis of fertilizing scenarios. Geoderma 60:89–107.

Finke P A, Stein A (1994) Application of disjunctive co-kriging to optimize fertilizer additions on a field scale. Geoderma 62:247–263.

Godwin D C, Jones C A (1991) Nitrogen dynamics in soil-plant systems. Pages 287–321 in Hanks R J, Ritchie J T (Eds.) Modeling plant and soil systems. American Society of Agronomy (ASA), CSSA, SSSA publication, Monograph No. 31, Madison, WI, USA.

Heuvelink G B M (1995) WLSFIT manual. Landscape and Environmental group, University of Amsterdam, The Netherlands.

Jones J W (1993) Decision support systems for agricultural development. Pages 459–471 in Penning de Vries F W T, Teng P S, Metselaar K (Eds.) Systems approaches for agricultural development. Kluwer Academic Publishers, Dordrecht, The Netherlands.

Jones J W, Ritchie J T (1990) Crop growth models. Pages 63–89 in Hoffman G J, Howell T A, Solomon K H (Eds.) Management of farming irrigation systems. American Society of Agricultural Engineers Monograph, ASAE, St. Joseph, MI 49085, USA.

Pickering N B, Hansen J W, Jones J W, Wells C M, Chan V K, Godwin D C (1994) Weatherman: A utility for managing and generating daily weather data. Agron. J. 86:332–337.

Richardson C W, Wright D A (1984) WGEN: A model for generating daily weather variables. ARS-8. USDA-ARS, Washington DC, USA.

Ritchie J T (1986) A user-oriented model of the soil water balance in wheat. Pages 203–305 in Day W, Atkins R K (Eds.) Wheat growth and modelling. Plenum, New York, USA.

Ritchie J T, Godwin D C, Singh U (1990) Soil and weather inputs for the IBSNAT crop models. Pages 31–45 in Proceedings of IBSNAT Symposium: Decision Support System for Agrotechnology Transfer, Las Vegas, 16–18 October 1989. I. Symposium Proceedings, Department of Agriculture and Soil Science, University of Hawaii, Honolulu, HI, USA.

Robert P C, Rust R H, Larson W E (1994) Site-specific management for agricultural systems. Proc. of the Second International Conference ASA-SSSA-CSSA, 27–30 March 1994, Minneapolis, Minn., USA. 993 p.

Sadler E J, Bauer P J, Busscher W J (1994) Spatial corn yield during drought in the SE coastal plain. Pages 365–381 in Robert P C, Rust R H, Larson W E (Eds.) Site-specific management for agricultural systems. Proceedings of Second International Conference, ASA-SSSA-CSSA, 27–30 March 1994, Minneapolis, Minn., USA.

Singh U, Brink J E, Thornton P K, Christianson C B (1993) Linking crop models with a geographic information system to assist decision making: A prototype for the Indian semiarid tropics. International Fertilizer Development Center, Muscle Shoals, AL, USA. 39 p.

Soil Survey Staff (1975) Soil Taxonomy: A basic system of soil classification for making and interpreting soil surveys. USDA-SCS Agricultural Handbook 436, US Government Printing Office, Washington DC, USA.

Sprague R H Jr, Carlson E H (1982) Building effective decision support systems. Prentice-Hall Inc., Englewood Cliffs, NJ, USA.

Staritsky I G, Sloot P H M, Stein A (1992) Spatial variability and sampling of cyanide polluted soil on former galvanic factory premises. Water, Air and Soil Pollution 61:1–16.

Stoorvogel J J (1995) Linking GIS and models: Structures and operationalization for a Costarican case study. Neth. J. Agric. Sci. 43:19–29.

Uehara G, Tsuji Y (1993) The IBSNAT project. Pages 505–513 in Penning de Vries F W T, Teng P S, Metselaar K (Eds.) Systems approaches for agricultural development. Kluwer Academic Publishers, Dordrecht, The Netherlands.

Verhagen A, Booltink H W G, Bouma J (1995) Site-specific management: Balancing production and environmental requirements at farm level. Agric. Syst. 49:369–384.

Wösten J H M, Bannink M H, De Gruijter J J, Bouma J (1986) A procedure to identify different groups of hydraulic conductivity and moisture retention curves for soil horizons. J. Hydrology 86:133–145.

Application of SOYGRO in Argentina

S. MEIRA and E. GUEVARA

Instituto Nacional de Tecnología Agropecuaria (INTA), C.C. 31, 2700 Pergamino (BA), Argentina

Key words: soybean, SOYGRO, validation, temperature

Abstract

In Argentina, soybean (*Glycine max* (L.) Merr.) is an important crop. Eighty percent of soybean oil and 95% of soymeal are exported, which represents 50% of the country's agricultural exported product. In the last years, the sowing area with soybean crop has increased, reaching more than 6 10^6 ha in 1995. This area is planted with different maturity groups in dryland conditions with an average yield of 2.5 ton ha^{-1}. Soybean is usually planted as the first crop at the beginning of November or as second crop after wheat from the beginning of December until late December. Although the National Institute for Agricultural Research (INTA) has been developing several maturity group strips according to latitude, the farmers use different maturity groups with overlapping criteria. During the last three years, INTA has been calibrating and validating the Soybean Model included in DSSAT V2.1 and V3.0 (Decision Support System for Agricultural Transfer). Specific objectives were: (i) to calibrate the soybean SOYGRO model for cultivars used in Argentina, (ii) to predict soybean yield for Argentina during1992–93 and 1993–94 growing seasons and (iii) to determine research needed to improve the ability of the model to predict soybean yields under weather and soil conditions of Argentina. The production was calculated using satellite information in order to determine the surface cultivated with soybean, and the SOYGRO model to calculate the yield according to planting date, soil and climate condition in the main soybean region. According to the Government Department of Agriculture, soybean production was 11.05 10^6 ton, and 11.31 10^6 ton for crop seasons 1992–93 and 1993–94 respectively. By using the SOYGRO model, the production was overestimated by about 5% for both years. The results suggest that the effect of low temperature during grainfilling might cause deviations in yield which cannot be simulated by the model. The results showed that SOYGRO could be used to predict the final yield of soybean in calculating the average yield in broad conditions of soil, climate and planting date.

Introduction

In Argentina, soybean (*Glycine max* (L.) Merr.) is an important crop. Eighty percent of soybean oil and 95% of soymeal are exported, which represents 50% of the agricultural exported product. In the last years, soybean area has increased reaching more than 6 Mha. Different maturity groups are planted in dryland conditions, and the average yield is 2.5 ton ha^{-1}. Soybean is usually planted as the first crop at the beginning of November or as the second crop after wheat from the beginning of December until late December. The main constraints to crop yield are: pests, diseases, weeds, drought and, in some years, flooding. The dryland conditions, the wide range of planting dates, environmental variability in time and space, and the maturity groups cause variability in yields between years and for the same year in different areas.

Until 1990, soybean production was calculated with information provided for the extension service, agricultural cooperatives and the National Seed Board. In this way, the soybean production was known several months after harvesting. The economical

M.J. Kropff et al. *(eds.): Applications of Systems Approaches at the Field Level, 235–242.*

opportunity to know the production in time, and the opportunity to get the best price and for planning, was not possible with this procedure.

Since 1991, INTA and the National Seed Board started a project to determine the suitable areas for the main crops (wheat, corn, soybean and sunflower) and their production using the DSSAT models (Tsuji et al. 1994). Remote sensing was used to determine the cropped area. This new approach was used to determine the yield potential across the region, the crop constraints and to understand the cultivar-environment-management interactions. The long-term objective of this project was to define different management strategies for each crop-environment system.

Results reported by Egli (1993) and Gay et al. (1980) showed an association between grainfilling duration and yield. The temperature during this phenological phase can affect the rate and/or duration of grainfilling (Egli and Wardlaw 1980; Seddigh and Jolliff 1984a, b; Seddigh et al. 1988). In the case of soybean planted after wheat, the effect of maturity group and planting date needs to be studied. It is important to choose cultivars with long grainfilling duration within each maturity group that would improve the yield.

The objectives of this research were to calibrate and validate the soybean SOYGRO model (Jones et al. 1989) for use as a tool for predicting the yield of the main soybean region in Argentina, to test the hypothesis about the relationship between grainfilling duration and yield in order to determine cultivars with long grainfilling duration in different environments, and how the duration of the grainfilling is affected by low temperature.

Methods

During 1991, 1992 and 1993, 42 experiments were conducted in six experimental stations of INTA. The characteristics of weather, soil and latitude for each site are presented in Table 1.

The cultivars and maturity groups (MG) used were: Asgrow 3127 (MG 3), Asgrow 5308 (MG 5), Federada (MG 6) and RA 702 (MG 7). At least two planting dates were used in each location. The cultivars used at each site depended on latitude of each site but, in order to know the sensitivity of the SOYGRO model, the cultivars were also planted outside the optimum latitude. All experiments were irrigated, and insects and weeds were controlled. Diseases were not present. The experimental design was a randomized complete block design with three replications.

The individual plots consisted of four 6 m rows with 0.7 m row spacing. Average seeding rate was 28 plants m^{-2}. Three samples of 0.5 m^2 were taken at different phenological stages. Total above-ground dry matter, and dry matter of leaves, petioles, stems and pods were determined in each sample. The phenology was recorded according to the Fehr-Caviness scale (Fehr and Caviness 1977). At harvest, total biomass, yield and yield components were determined, expressed in dry weight. The observed values versus calculated were plotted to determine the accuracy of the model SOYGRO for the different environments at flowering (R1), physiological maturity (R7) and for yield.

Table 1. Latitude, environmental characteristics and soil properties for the experimental stations.

	Latitude (South)	Longitude (West)	Altitude (m.s.l.)	Soil Parameters		
				O.M. (%)	pH	Available water (mm × 1.60 m^{-1})
Rafaela	31° 11′	61° 33′	100	2.5	5.6	221
Manfredi	31° 49′	63° 46′	292	2.9	5.8	245
Parana	31° 50′	60° 31′	110	2.6	6.1	239
Oliveros	32° 33′	60° 51′	26	3.0	6.0	237
Pergamino	33° 56′	60° 35′	65	2.9	6.2	227
Balcarce	37° 45′	58° 18′	130	5.6	5.9	270

Mean rainfall (mm)

	Nov.	Dec.	Jan.	Feb.	Mar.	Apr.	Total
Rafaela	107.2	115.5	121.9	127.5	152.5	84.8	709.4
Manfredi	101.6	134.5	100.8	111.7	111.1	53.6	613.3
Parana	105.8	106.2	115.8	123.2	145.7	93.9	690.6
Oliveros	103.4	119.4	114.8	113.9	160.9	80.1	692.5
Pergamino	93.4	88.2	110.0	115.1	140.2	71.1	618.0
Balcarce	62.6	120.8	101.4	82.0	78.4	72.4	517.6

Mean minimum temperature (°C)

	Nov.	Dec.	Jan.	Feb.	Mar.	Apr.
Rafaela	14.4	16.8	18.3	17.7	15.7	12.5
Manfredi	14.0	16.7	17.4	16.5	14.1	11.1
Parana	15.4	17.8	19.6	18.9	16.6	13.6
Oliveros	14.4	16.9	18.5	17.7	15.5	11.9
Pergamino	13.2	15.7	17.6	16.4	13.9	10.4
Balcarce	8.9	11.4	13.6	13.2	11.6	8.1

Mean maximum temperature (°C)

	Nov.	Dec.	Jan.	Feb.	Mar.	Apr.
Rafaela	28.0	29.9	32.2	29.9	27.5	24.3
Manfredi	27.4	28.8	30.2	28.2	26.1	23.5
Parana	26.8	29.3	31.3	29.3	26.7	23.3
Oliveros	26.6	29.1	31.3	29.1	26.7	23.3
Pergamino	25.5	28.5	30.7	28.2	26.4	22.6
Balcarce	21.9	24.7	27.4	25.9	23.9	19.7

Mean radiation (MJ m^{-2})

	Nov.	Dec.	Jan.	Feb.	Mar.	Apr.
Rafaela	22.6	24.3	23.6	20.6	17.5	13.0
Manfredi	20.5	21.6	22.4	19.2	16.2	11.7
Parana	22.1	23.3	23.4	20.6	17.7	12.9
Oliveros	21.9	22.4	23.6	20.1	16.5	12.7
Pergamino	22.1	24.2	24.7	21.7	17.2	13.1
Balcarce	20.8	23.3	22.7	21.8	15.4	11.2

The genetic coefficients for each cultivar were calculated with GenCalc included in DSSAT V3.0 (Hunt and Pararajasingham 1994), using average data for all locations and each cultivar across the region. For the study of temperature effects, the yield of each experiment was plotted against grainfilling duration (days between R5 and R7; Fehr and Caviness 1977). The second step was to sort and plot the experiments in which the temperature during the grainfilling was higher than 14.5°C.

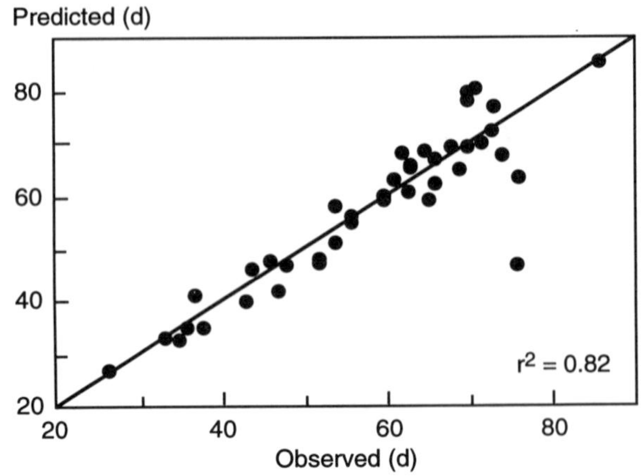

Figure 1. Relationship between observed and predicted duration (d) to flowering (R1).

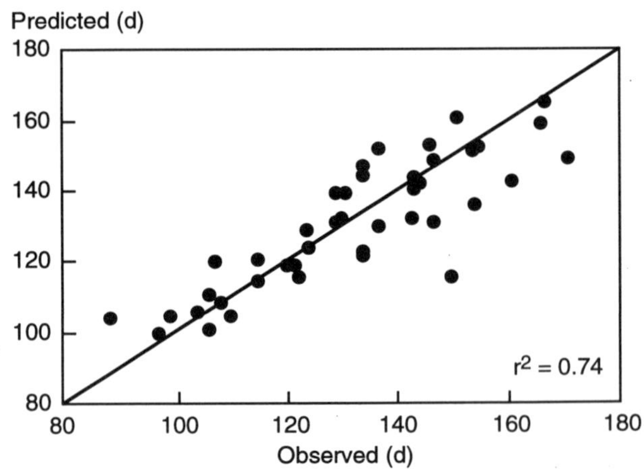

Figure 2. Relationship between observed and predicted duration (d) to physiological maturity (R7).

Results

SOYGRO model results

Figures 1 and 2 show the relationships between observed and predicted durations for flowering (R1) (r^2=0.82) and physiological maturity (R7) (r^2=0.74). The good correlation between observed and simulated values found for phenology allowed us to assume that the phenology can be predicted quite accurately, and the same genetic coefficients can be used across the region.

The need for daily weather data in different areas across the region limited the study. Weather data will be the 'bottle neck' for model use in many developing countries in the future. For our production system, water stress is the major constraint

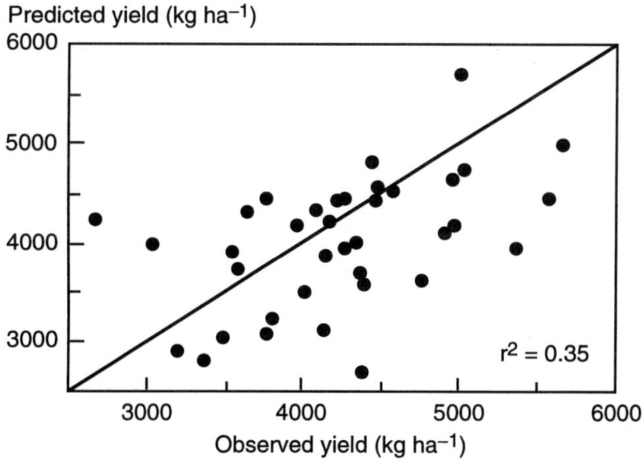

Figure 3. Relationship between observed and predicted yields (kg ha^{-1}) in all experiments.

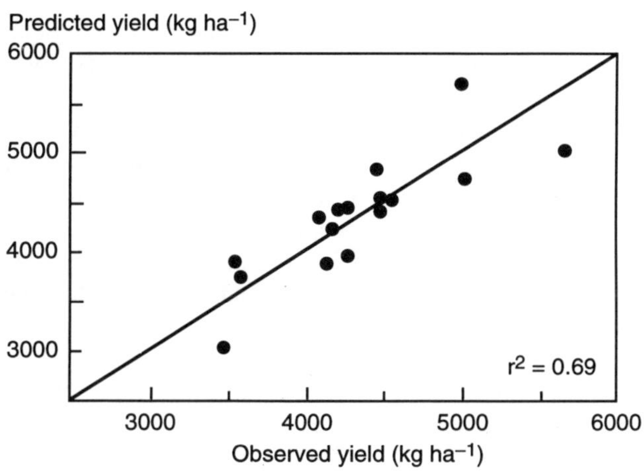

Figure 4. Relationship between observed and predicted yields (kg ha^{-1}) of cultivars recommended according to ecological areas.

for soybean, and the period of maximum demand of water should match the supply of rainfall, even though it is possible to try to improve the production system with a simple model where the critical phenological period matches with the high probability of precipitation.

Figure 3 shows the observed versus predicted values for yield with the genetic coefficients calculated with the average of each cultivar across the region. The model did not predict yield very well ($r^2 = 0.35$). This could be due to the fact that the model failed to predict yield when the cultivars are planted in areas where the latitude was not suitable for that cultivar.

The cultivars of each maturity group need to be planted in recommended areas.

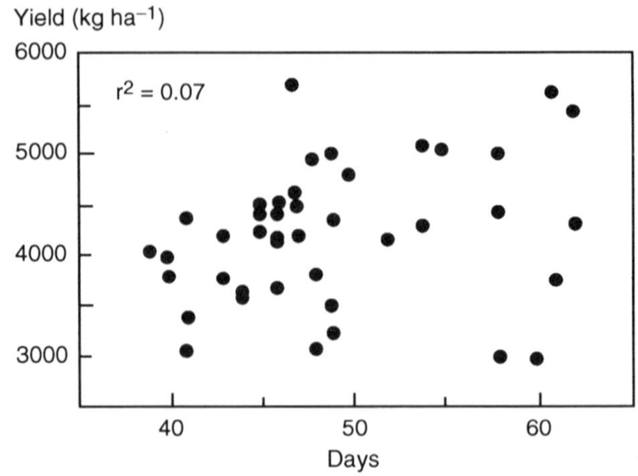

Figure 5. The relationship (observed data) between grainfilling duration (R5–R7, days) and yield (kg ha^{-1}).

Figure 4 shows the experimental observed versus predicted values for cultivars selected according to the recommendations for each ecological area, and the model performance was improved (r^2 = 0.69).

The genetic coefficients obtained in the experimental plots were used to calculate national soybean production. Common crop management practices were used for each area with corresponding soil and weather records, and the cropped area was obtained from remote sensing. Data were provided by the extension service, agricultural cooperatives and National Seed Board. The balance between overpredicted and underpredicted values across regions was good enough to predict the final national yield with an overestimation close to 5% for both years.

The yield components (grain numbers, and grain weight), and total biomass at harvesting resulted in more variability (data not shown). For on farm prediction it would be necessary to adjust the genetic coefficient of each cultivar to each production area and/or improve the sensitivity of the model to our conditions, e.g. temperature effect on seed growth period.

Temperature effect

Even when crop yield is the final result of the life cycle of the crop, and of the rates and duration of many physiological processes at all stages of development (Evans 1993), it is important to find a simple relationship between weather data and yield. Low temperature during the grainfilling period affects translocation from leaves to sinks and causes inhibition of photosynthesis and/or altering water relation due to slow stomatal response or reduced hydraulic conductivity of roots (Purcell et al. 1987; McWilliam et al. 1982). Figure 5 shows the relationship of yield versus grainfilling duration (r^2 = 0.07) for all the experiments without considering planting dates, maturity groups or low temperature during the grain growth period. The low correlation could

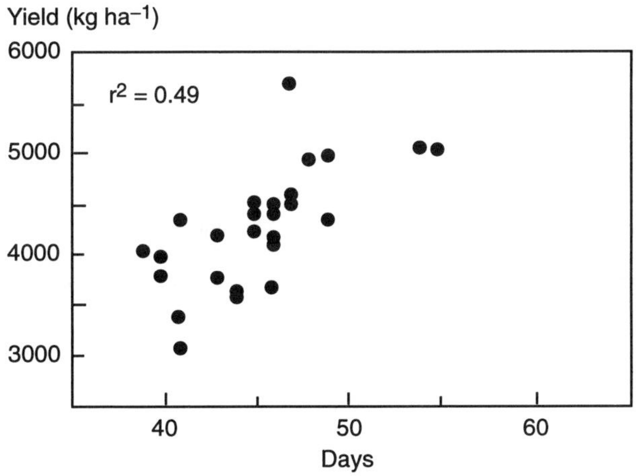

Figure 6. The relationship (observed data) between grainfilling duration (R5–R7, days) and yield (kg ha^{-1}), with temperatures higher than 14.5°C during grainfilling.

be associated with low temperature during the grainfilling period. In our case, it is common to get low temperature during the grainfilling period in soybean cultivars MG 6 and 7 planted after wheat.

According to Holmberg (1973), the critical threshold is 13–14°C for seed formation. In our south soybean region, there is a high probability of getting these temperatures or less, during the grainfilling period. Figure 6 shows how the relationship of grainfilling period improves (r^2=0.49) when only results are included from experiments where the temperatures were higher than 14.5°C during grainfilling. This indicates that in soybean planting after wheat, the low temperature during grainfilling duration might affect the yield. Seddigh and Jolliff (1984a,b), working with indeterminate soybean cultivars. suggested that low night temperatures limited seed growth rate, rather than limiting availability of assimilates for seed. Our hypothesis is that low night temperature conditions at late planting dates affect the seed growth rate and/or grainfilling period because of a modification in the translocation of assimilates to seed and inhibition of photosynthesis.

More fieldwork needs to be conducted in order to understand the effect of low temperature during the grainfilling period, and how the low temperature affects the grainfilling duration or the grainfilling rate according to the phenology of different soybean cultivars, and at different magnitudes of low temperatures. Finally, future experimental data will help to improve the temperature function during the grainfilling period in the SOYGRO model for our environmental conditions.

References

Egli D B (1993) Cultivar maturity and potential yield of soybean. Field Crops Res. 32:147–157.

Egli D B, Wardlaw I F (1980) Temperature response of seed growth characteristics of soybean. Agr. J. 72:560–564.

Evans L T (1993) Physiological aspects of crop improvement. Pages 169–268 in Evans L T (Ed.) Crop evolution, adaptation and yield. Cambridge University Press, New York, USA.

Fehr W R, Caviness C E (1977) Stages of soybean development. Iowa State Agric. Exp. Station, Special report No. 80, Ames, Iowa, USA. 12 p.

Gay S, Egli D B, Reicosky D A (1980) Physiological aspects of yield improvement in soybean. Agron. J. 72:387–391.

Holmberg S A (1973) Soybeans for cool temperate climate. Agrihort. Genetica 31:1–20.

Hunt L A, Pararajasingham S (1994) GenCalc: Genotype coefficient calculator. Pages 201–233 in Tsuji G Y, Uehara G, Balas S (Eds.) DSSAT Version 3.0 International Benchmark Sites Network for Agrotechnology Transfer (IBSNAT), University of Hawaii, Honolulu, HI, USA.

Jones J W, Boote K J, Hoogenboom G, Jagtap S S, Wilkerson G G (1989) SOYGRO V5.42, soybean crop growth simulation model. User's Guide. Journal No. 8304. Florida Agric. Exp. Stn., University of Florida, Gainesville, FL, USA.

McWilliam J R, Kramer P J, Musser R L (1982) Temperature induced water stress in chilling sensitive plants. Aust. J. Plant. Physiol. 9:343–352.

Purcell L C, Ashley D A, Boerma H R (1987) Effects of chilling on photosynthetic capacity, and leaf carbohydrate and nitrogen status of soybean. Crop Sci. 27:90–95.

Seddigh M, Jolliff G D (1984a) Effects of night temperature on dry matter partitioning and seed growth of indeterminate field-grown soybean. Crop Sci. 24:704–706.

Seddigh M, Jolliff G D (1984b) Night temperature effects on morphology, yield and yield components of indeterminate field-grown soybean. Agron. J. 76:824–828.

Seddigh M, Jolliff G D, Orf J H (1988) Field evaluation of early maturity soybean genotypes for differential adaptation to low night temperature. Crop Sci. 28:639–643.

Tsuiji G Y, Uehara G, Balas S (Eds.) (1994) DSSAT Version 3.0. International Benchmark Sites Network for Agrotechnology Transfer (IBSNAT), University of Hawaii, Honolulu, HI, USA. 491 p.

Modelling the effect of nitrogen on rice growth and development

T. HASEGAWA[1] and T. HORIE[2]

[1]*School of Agriculture, Kyushu Tokai University, Choyo, Aso-gun, Kumamoto 86914, Japan*
[2]*Faculty of Agriculture, Kyoto University, Sakyo-ku, Kyoto 606, Japan*

Key words: dry matter, nitrogen, *Oryza sativa* L., simulation model, soil N supply, spikelet number

Abstract
A dynamic crop growth model was developed to analyse irrigated paddy rice (*Oryza sativa* L.) productivity as determined by climatic factors and N availability. The model consists of submodels related to soil N processes, rice N-uptake, developmental processes, photosynthesis, dry matter production and spikelet formation. Soil N processes include N mineralization, expressed as a function of temperature and soil moisture before flooding. The balance between soil N supply and crop N demand determines rice N-uptake. The phenological developmental rate was expressed as a non-linear function of daily temperature and daylength. Submodels for canopy photosynthesis and dry matter production were based on an age-dependent relation between single-leaf photosynthesis and leaf N. Spikelet number was determined by an empirical function of dry weight and N concentration of above-ground biomass at the panicle formation stage, which was derived from a number of field experiments conducted in widely different regions in Japan. Data sets covering a range of N and climatic conditions were used for validation. The model, written in BASIC for PCs, satisfactorily simulated daily biomass growth and final spikelet number with inputs of latitude, N fertilizer input and daily climatological data, and might be useful for evaluation of N fertilizer management scenarios over sites and seasons.

Introduction

Increasing productivity of paddy rice will continue to be a major challenge because the demand of the swelling population needs to be met with a limited area of arable land. Fertilizer N has contributed immensely to the current level of productivity and will play a key role in the future rice production. The management of N, however, requires higher efficiency because of economical and environmental reasons. While N leaching from paddy fields is less than that from upland fields because most of the inorganic N exists in the form of ammonium, the fate of applied N should be quantitatively understood to minimize the negative impact on the environment. Thus, the need for simulation models that include the effects of N on the processes underlying crop growth and development, as well as the N balance in the soil, is increasing rapidly.

The present study is an attempt to model irrigated paddy rice growth and development, taking into account the effect of the crop N status in combination with the availability of N to the crop with certain simplicity. The primary goal of our model is to allow quantitative analyses and prediction of the effects of crop N status on productivity over the whole development cycle of the crop. We also tested whether the

M.J. Kropff et al. *(eds.): Applications of Systems Approaches at the Field Level, 243–257.*

simplified submodel for soil N processes can explain temporal changes in mineral soil N.

Model development

The model consists of four submodels: soil N processes, crop growth, N-uptake and allocation, and spikelet number estimation. Unless stated otherwise, parameters were mostly derived from our large-scale circulating water culture experiments conducted at Aso, Kumamoto, Japan, in 1990, 1991, 1993 and 1994, which covered a wide range of N conditions. The experimental procedures were described elsewhere (Hasegawa and Horie 1992; Hasegawa and Kuga 1996). The program of the model was written in Microsoft® Quick BASIC 4.5™ and runs on a personal computer (IBM-PC compatible or NEC). The time interval for the integration of the rate variables was 1 day. Parameter values in the model are given in Table 1.

Soil N process

The changes in soil inorganic N (*Npool*, $g\,N\,m^{-2}$) in the plough layer are described as follows:

$$dNpool/dt = dNmin/dt + Nf - dNup/dt - dNloss/dt \tag{1}$$

where *Nmin* is the net N mineralization ($g\,N\,m^{-2}$), *Nf* is the rate of fertilizer application ($g\,N\,m^{-2}$), *Nup* is the N taken up by the crop ($g\,N\,m^{-2}$) and *Nloss* is the loss of N from the inorganic N pool in the plough layer ($g\,N\,m^{-2}$). We did not take into account the vertical distribution of N because most of the roots exist in a relatively shallow plough layer. The rate of *Nmin* was calculated by Hasegawa and Horie (1994a) as:

$$dNmin/dt = Kw + Kd\,No\exp(-Kd\,t) \tag{2}$$

where *No* ($g\,N\,m^{-2}\,d^{-1}$) is the maximum (or asymptotic) mineralization as a result of soil drying expressed as a function of soil moisture weight percentage before flooding. Parameters, *Kw* ($g\,N\,m^{-2}\,d^{-1}$) and *Kd* (d^{-1}) are temperature-dependent rate constants expressed by the Arrhenius' equations:

$$Kw = Aw\exp(-Eaw/(R(Ts + 273))) \tag{3}$$

$$Kd = Ad\exp(-Ead/(R(Ts + 273))) \tag{4}$$

where *Aw* and *Ad* are constants for wet and dry soil ($g\,N\,m^{-2}$) respectively, *Ts* is soil temperature (°C) and *R* is a gas constant ($1.986\,cal\,deg^{-1}\,mol^{-1}$). *Eaw* and *Ead* represent the apparent activation energy ($cal\,mol^{-1}$) for *Kw* and *Kd* respectively. The parameters were obtained from anaerobic incubation under different temperatures with different moisture conditions before flooding (Table 1).

Loss of N occurs through several pathways such as leaching, volatilization, denitrification and run-off. In this study, however, they are lumped in one variable, proportional to the size of *Npool* as:

$$dNpool/dt = rl\,Npool \tag{5}$$

where *rl* is the relative loss rate (d^{-1}).

Crop growth

Developmental stage (DVI). A model for predicting rice development was developed by Horie and Nakagawa (1990), where the crop phenological stage was expressed by the developmental index, *DVI*, a continuous variable expressed as a non-linear function of daily photoperiod and temperature. In this study, we used their later version, in which $DVI=0$ at emergence, 1 at panicle initiation, 2 at heading and 3 at maturity (Nakagawa and Horie 1995). The effect of N on the developmental rate was not included.

Dry matter production. Canopy photosynthesis and dry matter production are based on the age-dependent relationship between single leaf photosynthetic rates and leaf N (Hasegawa and Horie 1994b). The leaf light response curve is expressed as:

$$P = \alpha Pmax\, I/(Pmax + \alpha I) \tag{6}$$

where *P* is the gross photosynthetic rate on a leaf area basis ($g\,CH_2O\,m^{-2}\,s^{-1}$), α is the initial slope of the light response curve ($g\,CH_2O\,MJ^{-1}$), *I* is light intensity per unit leaf area ($MJ\,m^{-2}$) and *Pmax* is the light saturated photosynthetic rate ($g\,CH_2O\,m^{-2}\,s^{-1}$). *Pmax* is a function of leaf N content, calculated on an area basis (*NA*, $g\,N\,m^{-2}$), and the development index (*DVI*) (Hasegawa and Horie 1992):

$$Pmax = Po(2/(1 + \exp(-A(NA - Ncr))) - 1) \tag{7}$$

$$Po = Pm \exp(-B\,DVI) \tag{8}$$

where *Pm*, *A*, *B* and *Ncr* are parameters. In this study, we assumed that the α is not influenced by *NA* or by *DVI* as an approximation.

The vertical light distribution in the canopy (*S*, $MJ\,m^{-2}$) is expressed as a function of accumulative leaf area (*F*, $m^2\,m^{-2}$) from the top of the canopy (Monsi and Saeki 1953):

$$S = So \exp(-kF) \tag{9}$$

where *So* is the light intensity above the canopy and *k* is the extinction coefficient ($m^2\,m^{-2}$). The rate of single leaf photosynthesis can be obtained by combining equations (6), (7) and (8). Daily canopy photosynthesis (*Pcan*, $g\,CH_2O\,m^{-2}\,s^{-1}$) is calculated by integrating *P* with respect to time and leaf area in the whole canopy:

$$Pcan = \int_0^{DL} \int_0^{LAI} P\, dF\, dt \tag{10}$$

where *DL* is daylength (h). The function is integrated numerically using the Gaussian three points method (Goudriaan 1986). The net growth rate, *CGR* ($g\,biomass\,m^{-2}\,d^{-1}$), after subtracting maintenance and growth respiration equals:

Table 1 List of variables and parameter values used in the model

Symbol	Description	Unit	Value (for parameters)
α	Initial slope of the light response curve	$g\,CH_2O\,MJ^{-1}$	7.84
A	Coefficient for the *Pmax-NA* relationship	$m^2\,g^{-1}\,N$	1.194
Ad	Frequency factor of Arrhenius' equation for *Nmin* from dried soil	d^{-1}	3.01×10^9
Aw	Frequency factor of Arrhenius' equation for *Nmin* from wet soil	$g\,N\,m^{-2}\,d^{-1}$	1.43×10^{17}
B	Coefficient for the *Po* dependence on *DVI*	–	0.2295
C	Conversion factor from N-uptake to leaf area	$m^2\,g^{-1}\,N$	0.98
CGR	Net crop growth rate	$g\,DM\,m^{-2}\,d^{-1}$	
D	Coefficient for the effect of *NC* on *LC*	$100\,g\,g^{-1}$	0.15
DL	DayLength	h	
DVI	Developmental Index	–	
E	Coefficient for the effect of *T* on *LC*	deg^{-1}	0.0335
Ead	Activation energy for *Kd*	$cal\,mol^{-1}$	24803
Eaw	Activation energy for *Kw*	$cal\,mol^{-1}$	14058
F	Cumulative leaf area	$m^2\,m^{-2}$	
G	Relative rate of *Ntr* for daily leaf growth	$g\,g^{-1}$	0.05
I	Light intensity per unit leaf surface	$MJ\,m^{-2}$	
k	Extinction coefficient	$m^2\,m^{-2}$	0.45
Kd	Temperature dependent rate constant for *Nmin* from dried soil	d^{-1}	
Kw	Temperature dependent rate constant for *Nmin* from wet soil	$g\,N\,m^{-2}\,d^{-1}$	
LAI	Leaf area index	$m^2\,m^{-2}$	
LC	Conversion factor from N-uptake to leaf area set by *NC* and *T*	$m^2\,g^{-1}\,N$	
LRmax	Maximal rate of *LAI* increase set by *DVI* and current *LAI*	$m^2\,m^{-2}$	
M_{25}	Maintenance respiration coefficient	$g\,g^{-1}$	
NA	Leaf N content on an area basis	$g\,N\,m^{-2}$	
Navl	Amount of inorganic N available	$g\,N\,m^{-2}$	
NC	Plant N concentration	%	
NCmax	Maximal *NC* set by *DVI*	$g\,g^{-1}$	
NCmin	Minimum *NC* set by *DVI*	$g\,g^{-1}$	
Ncr	Critical *NA* for photosynthesis	$g\,N\,m^{-2}$	0.0370
Ndem	Crop N demand	$g\,N\,m^{-2}$	
NHI	N harvest index	$g\,g^{-1}$	0.65
Nld	Amount of N that can be used for leaf growth	$g\,N\,m^{-2}$	
Nloss	Loss of N from *Npool*	$g\,N\,m^{-2}$	
Nmin	Net N mineralized	$g\,N\,m^{-2}$	
No	Maximum effect of soil drying	$g\,N\,m^{-2}$	
Npool	Soil inorganic N pool	$g\,N\,m^{-2}$	
Ntr	Amount of N that can be translocated from old tissues for new leaf growth	$g\,N\,m^{-2}$	
Nup	N taken up by the crop	$g\,N\,m^{-2}$	
Nupmax	Maximal N uptake rate set by *W*	$g\,N\,m^{-2}\,d^{-1}$	
P	Gross photosynthetic rate per unit leaf area	$g\,CH_2O\,m^{-2}\,s^{-1}$	
Pcan	Canopy photosynthetic rate	$g\,CH_2O\,m^{-2}\,d^{-1}$	
Pm	Light saturated photosynthetic rate per unit leaf area set by *DVI*	$g\,CH_2O\,m^{-2}\,s^{-1}$	1.30×10^{-3}

continued

Table 1. Continued.

Symbol	Description	Unit	Value (for parameters)
Pmax	Light saturated photosynthetic rate per unit leaf area set by *DVI* and *NA*	$g\,CH_2O\,m^{-2}\,s^{-1}$	
Po	Light saturated photosynthetic rate per unit leaf area set by *DVI*	$g\,CH_2O\,m^{-2}\,s^{-1}$	
Q_{10}	Temperature coefficient for maintenance respiration	–	1.80
R	Gas constant	$cal\,deg^{-1}\,mol^{-1}$	1.986
rl	Relative loss rate of N from *Npool*	d^{-1}	0.03
S	Light intensity in the canopy	$MJ\,m^{-2}$	
So	Light intensity above the canopy	$MJ\,m^{-2}$	
T	Air temperature	°C	
Tf	Transformation factor	$g\,DM\,g^{-1}\,CH_2O$	0.7
To	Base temperature for *LC*	°C	2.55
Ts	Soil temperature	°C	
U	*Nupmax* under a closed canopy	$g\,N\,m^{-2}\,d^{-1}$	0.65
W	Total dry weight	$g\,m^{-2}$	
Y	Spikelet number per unit land area	m^{-2}	

$$CGR = Tf\,Pcan - M_{25}Q_{10}^{(T/10-2.5)}W \tag{11}$$

where *Tf* is the transformation factor (g biomass g^{-1} CH_2O), M_{25} is the maintenance respiration coefficient at 25°C ($g\,g^{-1}$), Q_{10} is the temperature coefficient for maintenance respiration (dimensionless), *T* is air temperature (°C) and *W* is total dry weight (including roots, $g\,m^{-2}$). M_{25} is known to decrease as the crop ages and as N concentration of the crop decreases (Amthor 1989; Yamaguchi 1978). This is largely associated with the change in the weight ratio of metabolically active organs and storage organs as the plant ages (McCree 1988). Based on the change in the weight ratio, M_{25} was described by the following function of *DVI*:

$$M_{25} = 0.0088\,(1-(DVI/3.345)^{1.81}) \tag{12}$$

The above function indicated that M_{25} at maturity would be reduced by a factor of four after the panicle initiation stage.

Leaf area

The submodel for leaf area development was based on a well-documented linear relation between total N-uptake and leaf area (Miyama and Okabe 1984). In our experiments, we also observed a high linear correlation between the increase in *LAI* and N-uptake until the middle of the reproductive period but the slope was dependent on the year and the N treatment (Hasegawa, unpubl.). Based on the observation that the slope was small when temperature (*T*, °C) was low or/and plant N concentration (*NC*, %) is high, the conversion factor from N-uptake of leaf area (*LC*, $m^2\,g^{-1}$ N) was formulated as follows:

$$LC = C\exp(-DNC)(E(T-To)) \tag{13}$$

where C ($m^2 g^{-1} N$) is constant until $DVI = 1.3$, but thereafter decreases towards 0 until $DVI = 1.8$. Parameters D ($100 g g^{-1}$) and E (deg^{-1}) are coefficients for NC and temperature effects, and To is the base temperature (°C). Principally, the increase in LAI is proportional to daily N absorption, but when N supply is limited old leaves dilute their N as a result of translocation of N to developing leaves (Yoneyama and Sano 1978). Translocation can take place without dying of lower leaves as long as NC stays above the minimum value. The potential amount of N that can be transferred for leaf development (Ntr, $g N m^{-2}$) is described as:

$$Ntr = 0.55 (Nup - NCmin\, W) \tag{14}$$

where 0.55 is the fraction of leaf N over total vegetative N as described later, Nup is N taken up by the crop and $NCmin$ is the minimum N concentration of the crop on a dry weight basis ($g g^{-1}$) expressed as a function of DVI:

$$NCmin = 0.0103 \exp(-0.253\, DVI) \tag{15}$$

Note that $Ntr = 0$ when N supply is not limiting N-uptake. Nitrogen which can be utilized for leaf development (Nld, $g N m^{-2}$) can thus be defined as:

$$Nld = \mathrm{d}Nup/\mathrm{d}t + G\, Ntr \tag{16}$$

where G is a ratio of Ntr that can be utilized on a daily basis. The data by Yoneyama and Sano (1978) indicated that about 3–4% of total leaf N was translocated daily to new leaves. Here we assumed $G = 0.05$ (d^{-1}). The actual LAI increase ($m^2 m^{-2} d^{-1}$) is expressed as:

$$\mathrm{d}LAI/\mathrm{d}t = \mathrm{Min}\,(LC\, Nld,\ LRmax) \tag{17}$$

where $LRmax$ is the maximal rate of LAI increase ($m^2 m^{-2} d^{-1}$) that was set by the uppermost values of the relative leaf expansion rate at each DVI, obtained from the data under widely different N conditions.

The decrease in LAI is primarily caused by a negative carbon balance of the leaf. Low solar radiation, heavy mutual shading and low leaf N are largely responsible for the negative carbon balance (Navasero and Tanaka 1966). Assuming that the maintenance cost of the crop was shared by all the green leaves evenly throughout the canopy, the light compensation point at which P is equal to the respiration on a leaf area basis can be calculated from NA and DVI at the time (equations (7) and (8)). With equation (9), the cumulative LAI at which carbon balance is zero can be calculated. Here we hypothesized that leaf area with a negative carbon balance is susceptible to death.

In addition to a carbon balance, leaf longevity was taken into account. Since leaf duration could be shortened by the faster developmental rate, we used the DVI to express the longevity of each leaf instead of actual time. We set a DVI value of 1.5 for the leaf duration, so that LAI decrease due to senescence is assumed to take place with the rate proportional to the LAI at 1.5 in DVI before.

Nitrogen uptake and allocation

Nitrogen uptake (*Nup*, $g\,N\,m^{-2}$) is principally driven by the crop N demand (*Ndem*, $g\,N\,m^{-2}$), which is determined by the degree of N deficiency of the crop (Seligman et al. 1975):

$$Ndem = NCmax\, W - Nup \tag{18}$$

where *NCmax* is the maximum N concentration of the crop on a dry weight basis ($g\,g^{-1}$) expressed as a function of *DVI*:

$$NCmax = 0.0575 \exp(-0.429\, DVI) \tag{19}$$

Nitrogen uptake is limited by *Ndem*, N availability in the soil (*Navl*, $g\,N\,m^{-2}$), or the maximal N uptake rate (*Nupmax*, $g\,N\,m^{-2}$):

$$dNup/dt = \mathrm{Min}\,(Ndem/Tc,\ Navl/dt,\ Nupmax) \tag{20}$$

$$Navl = Npool\, f(W) \tag{21}$$

$$Nupmax = U\, f(W) \tag{22}$$

$$f(W) = 0.044 \exp(0.0465\, W), \quad 0 < f(W) \leq 1 \tag{23}$$

where *Tc* is the time constant for *Ndem* (2 d in the current model), $f(W)$ is the function representing the degree of canopy establishment and *U* is the *Nupmax* under a closed canopy. Here we set *U* at $0.65\, g\,N\,m^{-2}\,d^{-1}$ based on data from the N non-limiting condition.

Nitrogen taken up by the crop is then distributed to organs, particularly to the growing parts at each developmental stage. Nitrogen distribution in the vegetative organs, however, is quite conservative: leaf N accounts for 55% of N in vegetative organs (Hasegawa et al. 1995). This simplifies the N allocation process in our model.

Nitrogen harvest index (*NHI*), defined as the quotient of N in a harvesting organ over total N at harvest ($g\,g^{-1}$), is also known as a conservative parameter over various N treatments (Wetselaar and Farquhar 1980). Here we expressed the time course of N in panicles by the following logistic equation with *NHI* of 0.65:

$$\text{N in panicles} = NHI\, Nup/(1 + \exp(-6.89(DVI - 2.387))) \tag{24}$$

Accordingly N in the vegetative organs after panicle formation can be obtained by subtracting N in panicles ($g\,m^{-2}$) from *Nup*. Dividing leaf N by *LAI*, we get *NA* which is necessary for the calculation of the leaf photosynthetic rate. The model also calculates the above-ground N and above-ground biomass with the fixed ratio of 93% and 89% of the total respectively (Hasegawa, unpubl. data).

Spikelet number

A functional approach has been taken to express the effect of N on the number of spikelets (Hasegawa et al. 1994). A number of functions were tested to identify the appropriate relationship between N accumulation at the panicle formation stage (when panicle length reaches ca 1 mm, $DVI \approx 1.2$) and the spikelet number, using

157 data sets conducted in widely different ecological regions in Japan. The model selected was based on a linear function of *NC* and an asymptotic response to *W*:

$$Y = (1 - \exp(-0.00504\,W))(7250 + 15808\,NC) \tag{25}$$

where Y is the spikelet number (m^{-2}). Note that both W and NC here are above-ground data.

Data for model validation

Data from a field experiment conducted in 1989, on an experimental paddy field of Kyoto University, Japan (alluvial, sandy loam, $CEC = 15.6$ meq per 100 g dry soil, C = 3.09%, N = 0.224%), consisting of two planting dates, were used for model validation. Transplanting was conducted on 15 May (Crop 1) and on 5 July (Crop 2). Crop 1 had three N fertilizer treatments: no N application (0-0-0), basal only (12 g N m^{-2}, 12-0-0) and split application (6 g N m^{-2} at transplanting, 4 g N m^{-2} at the panicle formation, 6-4-0). Crop 2 had two treatments: no N application (0-0-0) and split application (6 g N m^{-2} at transplanting, 4 g N m^{-2} at the panicle formation, 2 g N m^{-2} at the full heading stage, 6-4-2). The amount of P_2O_5 and K_2O applied to all the treatments was 10 g and 12 g m^{-2}, respectively. The experimental design was a randomized complete block with two replicates. The size of each plot ranged from 21 to 27 m^2. Rice cultivar Nipponbare was used.

Plants were sampled approximately at 3-week intervals. After measuring leaf area, plants were dried at 90°C for more than 72 h and dry weight for each organ was determined. Each plant part was ground and then subjected to the Kjeldahl analysis for N determination.

Paddy soil was sampled from the plough layer by a cylindrical soil sampler (ϕ 30 mm) at the interval of ca 1 week for the first 30 days and 2 weeks thereafter. Ammonium N was extracted with 10% KCl and then measured by the indophenol colorimetric method (Nakanati 1981). Nitrate was not measured because most inorganic N in paddy conditions exists in the form of NH_4-N.

Daily climatic data were collected at the experimental site. The instantaneous light intensity above the canopy was calculated from a daily total solar radiation assuming that the diurnal change in solar radiation is expressed by a sine curve. Daily average temperatures for air and soil were used.

The relative loss rate (rl) in equation (5) could not be estimated from the water culture experiments; therefore, we used treatment 6-4-0 in Crop 1 and treatment 0-0-0 in Crop 2, randomly selected from the five treatments, for the parameterization of rl. The value of rl was estimated to be 0.03 d^{-1}.

Results and discussion

Crop growth and N uptake

Figure 1 shows the simulated and measured time course of total dry weight. The model gave a good presentation of the measurements except for the 6-4-2 treatment

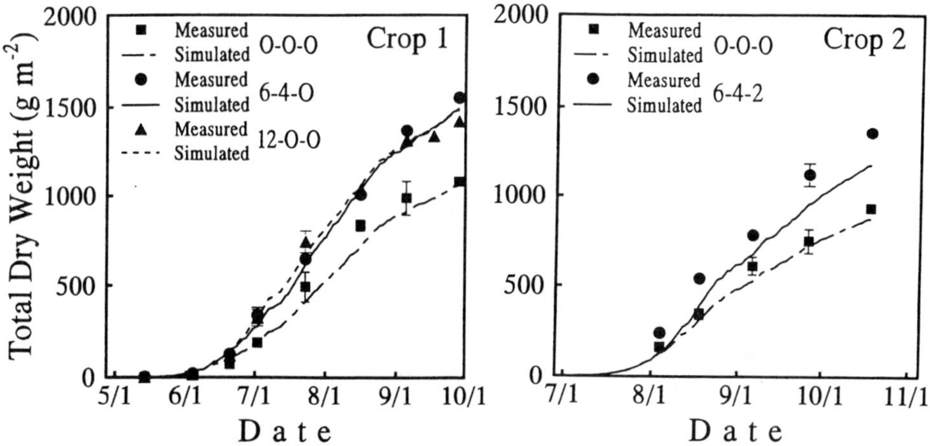

Figure 1. Time course of measured and simulated total dry weight obtained from the field experiments at Kyoto, 1989. The numerals indicate the amount of N ($g\,m^{-2}$) applied as basal, at panicle formation and at full heading respectively.

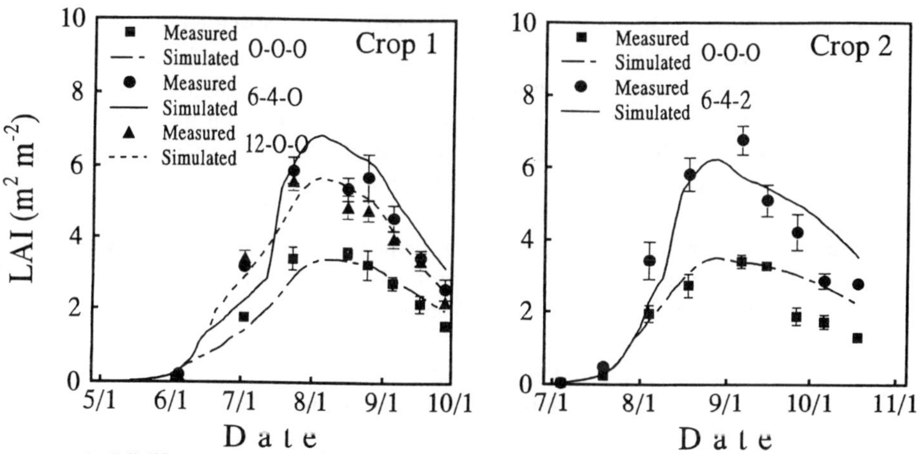

Figure 2. Time course of measured and simulated *LAI* at Kyoto, 1989. The numerals indicate the amount of N ($g\,m^{-2}$) applied as basal, at panicle formation and at full heading respectively.

in Crop 2, in which dry weight was consistently underestimated. This was largely due to the error in *LAI* estimation (Figure 2). In Crop 1, the simulated *LAI* was generally in good agreement with the measured data during the whole growing period, which resulted in successful prediction of *W*. In Crop 2, the zero-N treatment showed a good fit, but *LAI* in treatment 6-4-2 was consistently underestimated during the initial phase with the exponential increase in leaf area. Although the simulated *LAI* for treatment 6-4-2 overtook the measurements thereafter, dry weight could not catch up, indicating the importance of accurate simulation of *LAI* during the initial stage. A similar underestimation in *LAI* was observed on 3 July for treat-

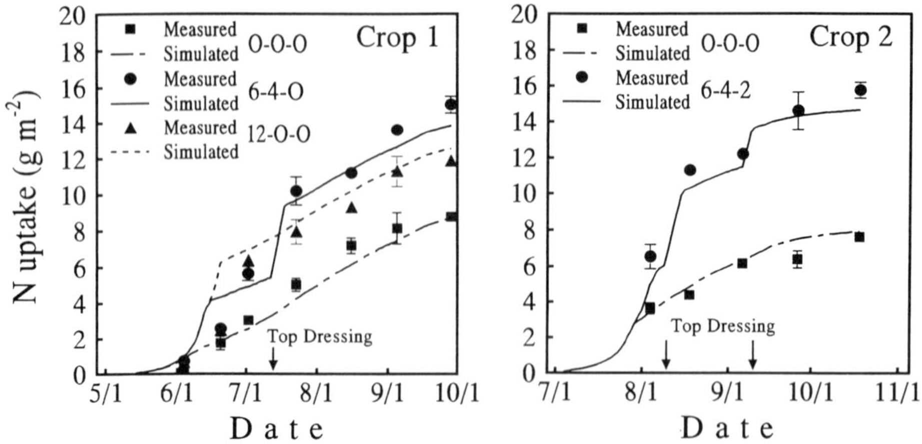

Figure 3. Time course of measured and simulated N-uptake at Kyoto, 1989. The numerals indicate the amount of N ($g\,m^{-2}$) applied as basal, at panicle formation and at full heading respectively. Arrows indicate the time of top dressing for the split application plots.

ment 6-4-0 in Crop 1, but the difference occurred after the initial exponential phase of leaf growth, resulting in a smaller accumulation of the simulation errors. *LAI* after heading was overestimated for both treatments in Crop 2, in distinction to that in the first half. In our model, *LAI* decrease is influenced by the *LAI* during 1.5 *DVI* units before, so that the early errors are carried over to the second half of the growing season. Nevertheless, the dry matter progress after heading showed a good fit.

The measured and simulated time courses of N-uptake are given in Figure 3. Nitrogen uptake increased exponentially for the first 30 days, and from then onward showed a gradual linear increase until maturity. The model was in good agreement with the measured data. However, treatments 6-4-0 and 12-0-0 in Crop 1 were slightly overestimated during the phase with the exponential increase in *Nup*, while treatment 6-4-2 in Crop 2 was slightly underestimated at the early stages. The differences in prediction errors for different planting dates might be the result of the temperature difference at the early stage: Crop 1 was exposed to relatively low temperatures, while Crop 2 grew under high temperatures. The current model does not take into account the temperature effect of N-uptake. We may also need to examine the validity of $f(W)$ that influences the early N-uptake.

Plant N concentration was obtained by dividing *Nup* by *W* (Figure 4). The model simulated the treatment differences and the ontogenetic decline in Crop 1 well, though a slight overestimation in the beginning was observed. In Crop 2, the simulated values were consistently higher than the measurements in both treatments. The deviation was particularly large in the first half.

Figure 5 illustrates the time course of *NA*, which is necessary for the calculation of *P*. In Crop 1, the overall trend was well simulated but the measured treatment difference between 12-0-0 and 6-4-0 right after the N top dressing was not apparent in the simulation. In the current model, a rapid N-uptake after N top dressing resulted

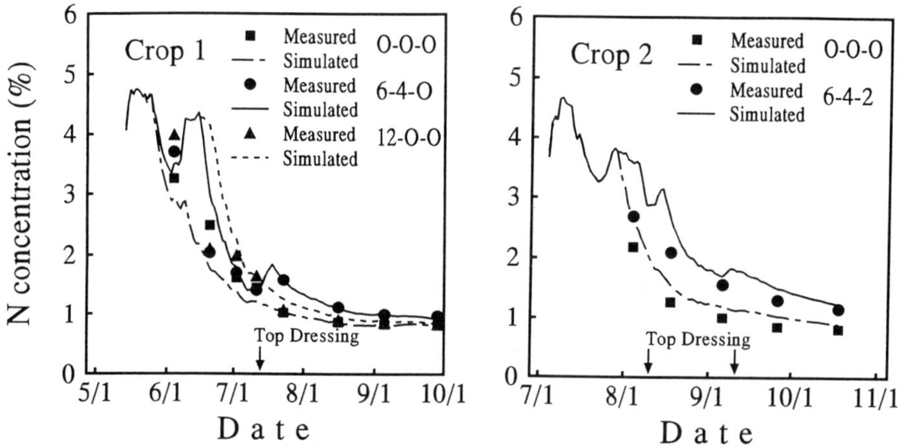

Figure 4. Time course of measured and simulated plant N concentration at Kyoto, 1989. The numerals indicate the amount of N ($g\,m^{-2}$) applied as basal, at panicle formation and at full heading respectively. Arrows indicate the time of top dressing for the split application plots.

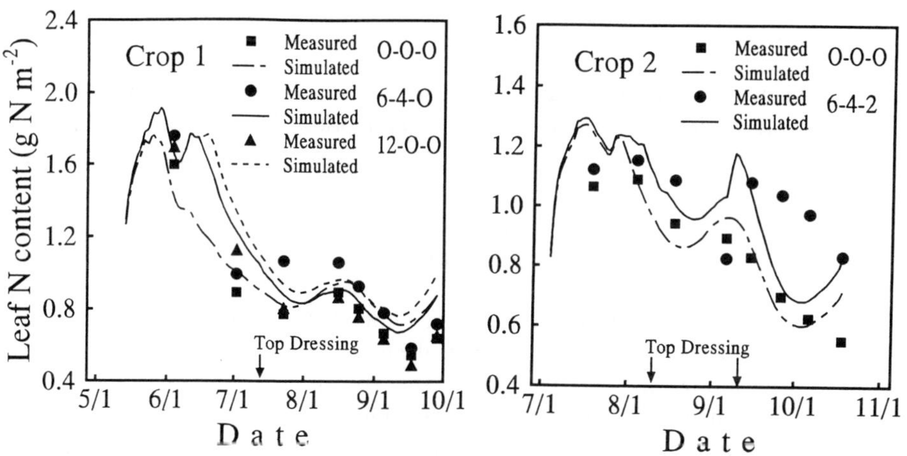

Figure 5. Time course of measured and simulated leaf N on an area basis (*NA*) at Kyoto, 1989. The numerals indicate the amount of N ($g\,m^{-2}$) applied as basal, at panicle formation and at full heading respectively. Arrows indicate the time of top dressing for the split application plots.

in an increase in *LAI* (Figure 2) but failed to simulate the increase in *NA*. In Crop 2, simulated and observed values for *NA* in treatment 6-4-2 showed a large discrepancy during the ripening stage, which is largely due to the overestimation of *LAI* and the underestimation of *Nup*.

Since *NA* is given by dividing leaf N by *LAI*, *NA* is sensitive to errors in *LAI*. When *LAI* is overestimated, *NA* tends to be underestimated if total N is well simulated. Low *NA* reduces *P*, but larger light interception resulted from the overestimates of *LAI* tends to compensate the lower *P*. The compensation is not likely to take place

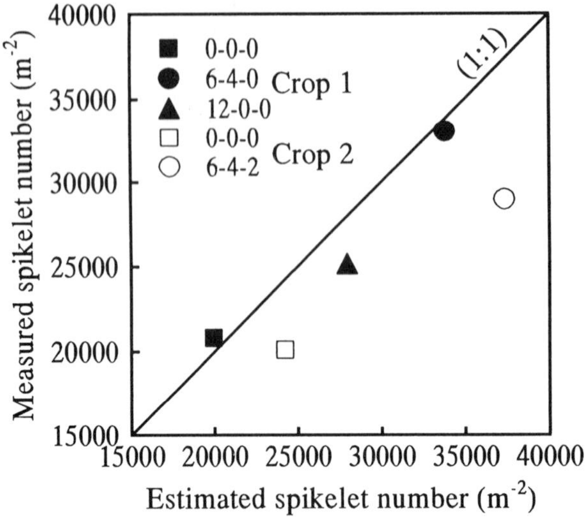

Figure 6. Relationship between the simulated and the measured spikelet number. The numerals indicate the amount of N ($g\,m^{-2}$) applied at basal, at panicle formation and at full heading respectively.

when *LAI* is underestimated, because the effect of *NA* on canopy photosynthesis or dry matter production is asymptotic (Sinclair and Horie 1989; Hasegawa and Horie 1994b). Overall, however, the current crop submodel gave a reasonable fit to the data obtained under different N regimes and climatic conditions.

Spikelet number

Figure 6 shows the relationship between the simulated and the measured spikelet number. The model showed an excellent fit to the measurements in Crop 1, that ranged from 20000 to 33000$\,m^{-2}$. On the other hand, spikelet numbers in Crop 2 were strongly overestimated. This is largely attributed to the overestimation in *NC* at the panicle formation stage ($DVI = 1.2$). While the response of spikelet number to *W* is asymptotic, that to *NC* is linear, so that the error in *NC* had a direct effect on the estimate of spikelet number. A better estimate of *NC* will improve the accuracy of spikelet number estimation.

Soil inorganic N

Figure 7 presents the simulated progress of *Nmin*. About 8 to 10$\,g\,m^{-2}$ of N was mineralized during the whole growing period. Differences between the treatments occurred basically because of soil temperature differences due to the different *LAIs*. In fact, the daily average of soil temperature in treatment 0-0-0 was higher than that in 6-4-0 (treatment with the largest *LAI*) by as much as 3°C on a clear day with a closed canopy. This confirms the importance of using soil temperature for minerali-

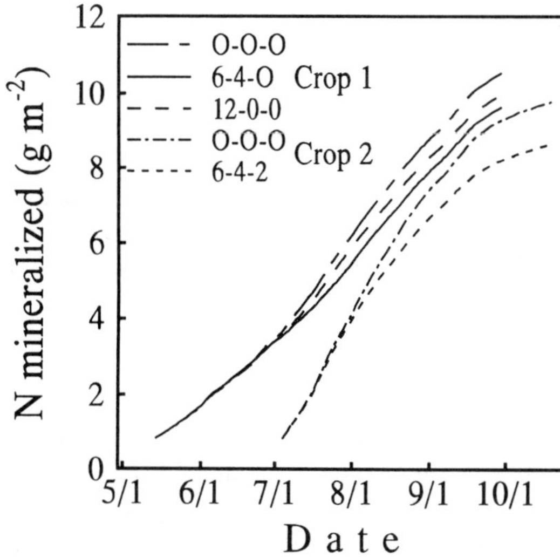

Figure 7. Time course of simulated N mineralized at Kyoto, 1989. The numerals indicate the amount of N (g m^{-2}) applied at basal, at panicle formation and at full heading respectively.

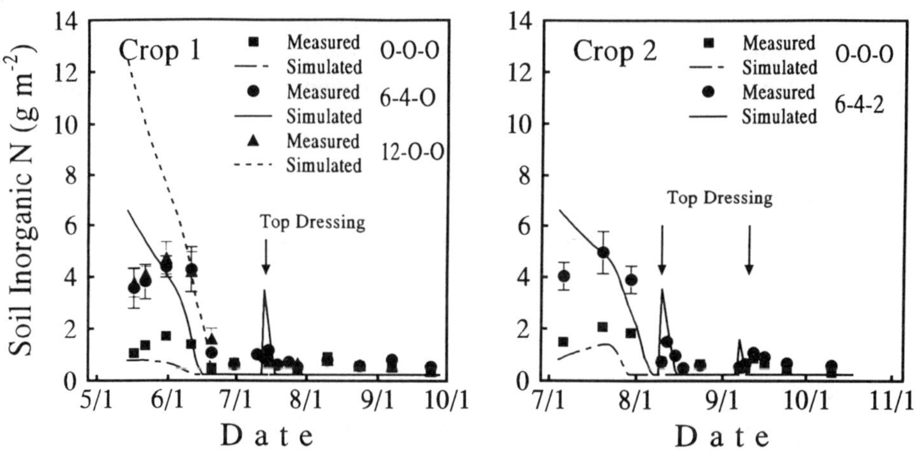

Figure 8. Time course of measured and simulated soil inorganic N at Kyoto, 1989. The numerals indicate the amount of N (g m^{-2}) applied at basal, at panicle formation and at full heading respectively. Arrows indicate the time of top dressing for the split application plots.

zation estimates. In a later version of this model, it would be necessary to incorporate a submodel for soil temperature estimation like the one developed by Takami et al. (1989), which takes into account the degree of canopy cover.

Figure 8 shows the time course of soil inorganic N. Overall, the measured trend was well simulated. The discrepancy mainly took place at the early observations and just after the N topdressing, where the measured values were far lower than the application rate, particularly for the higher dose treatment, which may also be a reason for the overestimation of *Nup* in treatments 12-0-0 and 6-4-0. Groot and De Willigen (1991) obtained a similar result in their wheat simulation, suggesting a possibility of immediate immobilization just after N application. If this is the case, our loss rate might have been exaggerated because such a change in immobilization with respect to heavy fertilization is not considered. They also suggested that recently fixed NH_4-N is not extracted with KCl solution, though it is readily taken up by the crop. This may also be the possible explanation for the current results. It is worth noting, however, that despite the errors in the N pool estimation, N-uptake was satisfactorily predicted by the model.

Conclusions

The proposed model includes important processes underlying the crop growth and N balance in the rice-soil system but each submodel was expressed in a relatively simple way. The simulated results were generally in good agreement with the data obtained from an independently conducted experiment under different N and climatic conditions. Soil processes may be too simplified to give an adequate estimate of inorganic N in the soil which is, in turn, necessary for the assessment of the environmental impact (e.g. N losses). Nevertheless, it can help in searching for an efficient use of fertilizer under various climatic N conditions with some confidence. Further validation particularly under various soil types will make the model more robust. A submodel for the grainfilling process needs to be incorporated in the whole model to evaluate the yield performance.

Acknowledgements

This study was partly financed by Grant-in-Aid for Encouragement of Young Scientists (No. 05760016) from the Ministry of Education, Science and Culture, Japan.

References

Amthor J (1989) Respiration and crop productivity. Springer-Verlag, New York, USA. 215 p.

Goudriaan J (1986) A simple and fast numerical method for the computation of daily totals of crop photosynthesis. Agric. For. Meteorol. 38:249–254.

Groot J J R, De Willigen P (1991) Simulation of the nitrogen balance in the soil and a winter wheat crop. Fert. Res. 27:261–272.

Hasegawa T, Horie T (1992) Modeling the ontogenetic change in the relationship between photosynthetic rate and leaf nitrogen in paddy rice. Jpn. J. Crop Sci. 61(2):25–26 (in Japanese).

Hasegawa T, Horie T (1994a) A simplified model for estimating nitrogen mineralization in paddy soil. Jpn. J. Crop Sci. 63:496–501.

Hasegawa T, Horie T (1994b) System analysis of nitrogen nutrition in paddy rice. (5). A model for dry matter production. Jpn. J. Crop Sci. 63(1):118–119 (in Japanese).

Hasegawa T, Kuga I (1996) Changes in the vertical distribution of leaf nitrogen concentration in the late stage of rice growth. Pages 502–503 in Ishii R, Horie T (Eds.) Crop science in Asia: Achievements and perspective. Proceedings of the 2nd Asian Crop Science Conference, 21–23 August 1995, Fukui, Japan. Crop Science Society of Japan, Asian Crop Science Association, Tokyo, Japan.

Hasegawa T, Koroda Y, Seligman N G, Horie T (1994) Response of spikelet number to plant nitrogen concentration and dry weight in paddy rice. Agron. J. 86:673–676.

Hasegawa T, Takahashi R, Tsuruta H, Kuga I (1995) System analysis of nitrogen nutrition in paddy rice. (8), A simplified approach for modeling N distribution in the crop. Jpn. J. Crop Sci. 64(2):211–212 (in Japanese).

Horie T, Nakagawa H (1990) Modelling and prediction of developmental process in rice. I. Structure and method of parameter estimation of a model for simulating developmental process toward heading. Jpn. J. Crop Sci. 59:687–695 (in Japanese with English abstract).

McCree K J (1988) Sensitivity of sorghum grain yield to ontogenetic changes in respiration coefficients. Crop Sci. 28:114–120.

Miyama M, Okabe T (1984) Optimum nitrogen content in relation to varietal characteristics in paddy rice. Jpn. J. Soil Sci. Plant Nutr. 55:1–8 (in Japanese).

Monsi M, Saeki T (1953) Über den Lichtfaktor in den Pflanzengesellschaften und seine Bedeutung für die Stoffproduktion. Jpn. J. Bot. 14:1–22 (in German).

Nakagawa H, Horie T (1995) Modelling and prediction of development process in rice. II. A model for simulating panicle development based on daily photoperiod and temperature. Jpn. J. Crop Sci. 64:33–42 (in Japanese with English abstract).

Nakanati S (1981) Ammonium nitrogen (NH_4-N). Pages 210–214 in Jpn. Soc. Anl. Chem. (Hokkaido), Mizuno Bunseki, 3rd Edition. Kgaku Dojin, Kyoto, Japan (in Japanese).

Navasero S A, Tanaka A (1966) Low-light-induced death of lower leaves of rice and its effect on grain yield. Plant Soil 24:17–31.

Seligman N G, Van Keulen H, Goudriaan J (1975) An elementary model of nitrogen uptake and redistribution by annual plant species. Oecologia (Berl.) 21:243–261.

Sinclair T R, Horie T (1989) Leaf nitrogen, photosynthesis and crop radiation use efficiency: A review. Crop Sci. 29:90–98.

Takami S, Sugaya H, Toriyama K (1989) A simple model for water and soil temperature estimate of an irrigated rice field. J. Agr. Met. 45:43–47.

Wetselaar R, Farquhar G D (1980) Nitrogen losses from tops of plants. Adv. Agron. 33:263–302.

Yamaguchi J (1978) Respiration and the growth efficiency in relation to crop productivity. J. Fac. Agric. Hokkaido Univ. 59:59–129.

Yoneyama T, Sano C (1978) Nitrogen nutrition and growth of the rice plant. II. Considerations concerning the dynamics of nitrogen in rice seedling. Soil Sci. Plant Nutr. 24:191–198.

Optimization of nitrogen fertilizer application to irrigated rice

R.N. DASH, K.S. RAO and S.K. NAYAK
Central Rice Research Institute, Cuttack 753 006, Orissa, India

Key words: ORYZA_0, irrigated rice, optimization, timing of N application, modelling

Abstract
The ORYZA_0 model was used to optimize fertilizer N application programmes for irrigated upland rice. The crop and soil input parameters for this model were derived from field experiments. Soil characteristics apparently influence these parameters. The response of the rice crop as found by numerical optimization represented the maximum attainable biomass calculated on the basis of radiation, crop parameters and soil parameters. The maximum simulated biomass accumulation was 23.6 ton ha^{-1} at 325 kg N ha^{-1} in a highly fertile soil, whereas in a soil of low fertility status the maximum attainable biomass was 15.3 ton ha^{-1} at 200 kg N ha^{-1}. The optimum dose of N application in the dry season was computed to be 120 kg N ha^{-1}. The timing of applications indicated that most fertilizer N should be applied before panicle initiation. At low N input levels (20–40 kg N ha^{-1}), results suggested to apply fertilizer N around panicle initiation. The optimization by ORYZA_0 did not suggest any basal application of N in these soils. A field experiment was conducted to test the recommended timing of fertilizer N application based on ORYZA_0. It was found that higher crop biomass, N uptake and grain yield of rice would be obtained at the optimum N application level of 120 kg N ha^{-1}, if appropriate timings suggested by the model are used. This optimum N level is higher than the level of N used in existing practices now.

Introduction

In most rice soils, application of fertilizer nitrogen (N) is necessary to exploit the genetic potential of modern rice cultivars grown with improved cultural practices. Nitrogen management should maximize N uptake at critical growth stages. At the same time, it should be ensured that N absorbed by the plant is efficiently used for the grain production. One way to increase fertilizer efficiency is to apply fertilizer when it will be used most effectively for the metabolism or the growth of the rice plant. Underfertilizing or overfertilizing can decrease grain yield. Correct doses of fertilizer N should be applied at appropriate times for specific soils and rice varieties. The source of fertilizer N and its method of application are also important. However, fertilizer urea is the main source of N to rice. It is, therefore, important to known how to optimize fertilizer N (urea) application to irrigated rice with respect to its dose and timing.

Crop growth simulation models can be used to evaluate alternative crop management strategies. They can be used to optimize systems by improved decision making. Traditional recommendations are made to farmers on the basis of factors such as e.g. location, soil type and farming systems. Management recommendations do not change from one year to the next. Field trials are conducted over a limited number of seasons in different locations, and recommendations are based on the best average perform-

M.J. Kropff et al. *(eds.): Applications of Systems Approaches at the Field Level, 259–270.*

ance. Over the last few years, there has been considerable interest to investigate production response to alternative management practices. In the present study, the SARP (Simulation and Systems Analysis for Rice Production) model ORYZA_0 (Ten Berge et al. 1994) was used to optimize N fertilizer application in irrigated rice.

Material and methods

Experiments

Field experiments were conducted during the dry seasons (January–May) of 1992, 1993 and 1995 at the Central Rice Research Institute (CRRI), Cuttack (20.5° N, 86.5° E), Orissa, India, to study the growth, yield and N uptake by irrigated rice. The location of the experiment is situated in the delta region of the Mahanadi River. The soil has a clay loam texture and is classified as Typic Haplaquept with a pH of 6.1. The average solar radiation in the dry seasons was 18 $MJ\,m^{-2}\,d^{-1}$ and 20.5 $MJ\,m^{-2}\,d^{-1}$ during pre-flowering and post-flowering periods respectively. The maximum and minimum temperatures before flowering were in the range of 20–40°C and 14–25°C respectively, and after flowering in the range of 36–40°C and 21–28°C respectively. The treatments comprised different timings and rates of N fertilizer (urea) application. The highest rate of N application was 200 $kg\,N\,ha^{-1}$. A control treatment (zero-N) was also included. The other treatments were 100 $kg\,N\,ha^{-1}$ applied at panicle initiation (PI), 13 days after transplanting (DAT) or at both 13 DAT and PI. The native soil fertility (Ramammorthy and Bajaj 1969) of the experimental site was high in 1992 based on the organic carbon status (0.8%), whereas the site in 1993 and 1995 had a low fertility status (0.45% organic C). All treatments were replicated three times in a completely randomized block design. The plot size was 30 m^2. Twenty-one day old seedlings of rice cultivar IR36 were transplanted in puddled soil in 1992, whereas seedlings were 28 days old in 1995. The seedlings were collected from a nursery which received 10 $kg\,N\,ha^{-1}$ as urea, 15 days after sowing. The crop was supplied with ample N in maximum three splits at both sites. At suitable intervals during the growing season, 10 hills were sampled and the dry weights of leaves, stems, roots, panicles and grains were determined. The N content of plant organs was determined by the micro-Kjeldahl distillation method. The treatments were designed to determine all crop and soil parameters for the ORYZA_0 rice model.

Weather inputs, such as the maximum and minimum temperature and the sunshine duration, were measured from the CRRI weather station. The daily total global solar radiation (*RAD*, $MJ\,m^{-2}\,d^{-1}$) was estimated from the recorded duration of bright sunshine using the following equation (Ångström 1924):

$$RAD = RAD_a\,(a + b \times (SSD/DAYL))$$

Where RAD_a is the extra-terrestrial radiation ($MJ\,m^{-2}\,d^{-1}$), *SSD* is the duration of bright sunshine (h), *DAYL* is the daylength (h) and *a, b* are dimensionless coefficients that depend on the prevailing climatic conditions and latitude. Indicative values for the empirical constants *a* and *b* in the Ångström formula for dry tropical zones are 0.25 and 0.45 (Frère and Popov 1979).

Table 1. Input parameters and their values for rice (cv IR36) used in ORYZA_0, for dry season conditions in Cuttack, India.

Parameter	Value 1992	Value 1993	Unit
Crop			
Fraction N in storage organs (FNSO)	0.012	0.016	$g\,g^{-1}$
Fraction total crop N present in leaves (FNCLV)	0.55	0.56	$g\,g^{-1}$
Max. amount of N in leaves (ANLVMX)	110	53	$kg\,ha^{-1}$
Max. N uptake rate (MAXUP1)	5.5	4.7	$kg\,ha^{-1}\,d^{-1}$
Max. ratio of daily N uptake to growth (NUPCO)	0.07	0.028	$kg\,kg^{-1}$
Relative N uptake rate (RUR)	0.18	0.125	d^{-1}
Max. fraction N in total biomass vs	0., 0.026	0., 0.026	DAT, $g\,g^{-1}$
days after transplanting (DAT)	22., 0.049	39., 0.025	
	45., 0.028	48., 0.019	
	62., 0.021	62., 0.017	
	93., 0.014	69., 0.016	
	108., 0.01	95., 0.013	
Harvest index (HI)	0.45	0.57	$g\,g^{-1}$
Soil			
'Native' soil N supply (SOLSUP)	0.9	0.6	$kg\,ha^{-1}\,d^{-1}$
Table of N recovery (RECT) vs	0., 0.00	0., 0.00	DAT, $g\,g^{-1}$
days after transplanting (DAT)	20., 0.55	20., 0.55	
	40., 1.00	40., 1.00	
	80., 1.00	80., 1.00	
	85., 0.00	95., 0.00	
Environmental			
Initial global radiation use coefficient (ε)	2.5	2.5	$g\,MJ^{-1}$
Initial leaf N use coefficient (p)	10	10	$g\,g^{-1}\,d^{-1}$
Site-specific match factor (FSV)	0.8	0.7	–

The model

The model ORYZA_0 (Ten Berge et al. 1994) simulates daily biomass accumulation in relation to N uptake and daily total global radiation. It is based on a limited set of relationships describing soil and crop processes. It is a simple model which includes N uptake, partitioning of N to the leaves and the utilization of leaf N in converting daily incident global radiation into dry matter. This model, when used with an optimization procedure, identifies the best fertilizer N application curve as a function of time for a total amount of N input and for specified varietal and environmental characteristics. This model generates a recommendation for the timing and split for a specified N application level that result in the maximum biomass production. Only a few empirical relations are used in this model, which makes it convenient. The model uses a fixed harvest index to convert biomass into grain production. All the input parameters for the model were derived from field experiments conducted in the dry seasons of 1992 and 1993 (Table 1) at Cuttack, India.

Optimization of N fertilizer application

Optimization techniques are used to derive parameters for the processes determining the N fertilizer response using measured inputs of crop development and weather

Table 2. Optimal timing of fertilizer application to rice (cv IR36) based on ORYZA_0, dry season, Cuttack, India.

N application ($kg\,ha^{-1}$)	Timing of N application (DAT*, 21 day old seedlings)				
	Two equal splits		Three equal splits		
	1	2	1	2	3
20	27	30	25	29	30
40	28	34	26	32	35
60	30	41	27	37	41
90	27	42	23	36	43
120	30	46	25	40	44
150	27	44	21	37	46

*Days after transplanting.

parameters. The calendar day of transplanting is the starting date in the simulation model. Date of first flowering is important because of the remobilization of N from leaves to panicles between flowering and maturity. The simulation stops at the observed date of maturity. The efficiency factor (*FSV*), which is specific to a given site and variety, is also determined by matching the simulated versus the observed time course of biomass. This match factor *FSV* reflects the overall efficiency of light and N utilization by the rice crop.

The crop and soil parameters, *FSV* values, weather data and information on planting dates were recorded in an input data file. Total N input was set to a desired level, and the optimization program was run with 100 iterations to determine the optimal timing and split at the specified N level. The optimization procedure was then repeated for different N input levels. The resulting biomass production corresponding to each N input level was used to produce a fertilizer response curve and fertilizer N application curve (Rao et al. 1995). The number of splits and their timing were assessed from the N fertilizer application curve derived by numerical optimization of ORYZA_0 and scaled relative to the total input level. This procedure resulted in a maximum biomass production level corresponding to each N input level.

Verification trial

A field experiment was conducted in the 1995 dry season at the low fertility site to verify the effect of different doses and timing of N application on growth and yield of rice. The treatments comprised different rates of N application ranging from 0–150 $kg\,N\,ha^{-1}$ and different timings and doses of N fertilizer application. The optimum timing and dose as determined by ORYZA_0 (Table 2) were compared with the current practice of fertilizer N application in three splits at basal (50% at puddling), tillering (25% at 23 DAT) and at PI (25% at 42 DAT). Seedlings of cv IR36 were transplanted in the puddled soils on 31 January 1995 (28 day old seedlings) with a spacing of 15 cm × 15 cm. All treatments were replicated thrice. Just before flowering and at maturity, 10 hills were sampled and the total dry weight of the crop biomass was determined. A 5 m^2 area was used to determine the final grain yield in each plot. The N content of the plant was determined using the Kjeldahl distillation method.

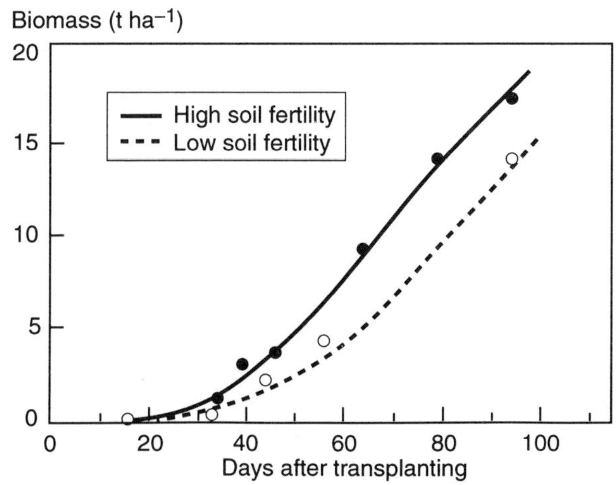

Figure 1. Total crop biomass (rice cv IR36) during the crop growth period as influenced by native soil fertility under conditions of ample N supply during the dry season.

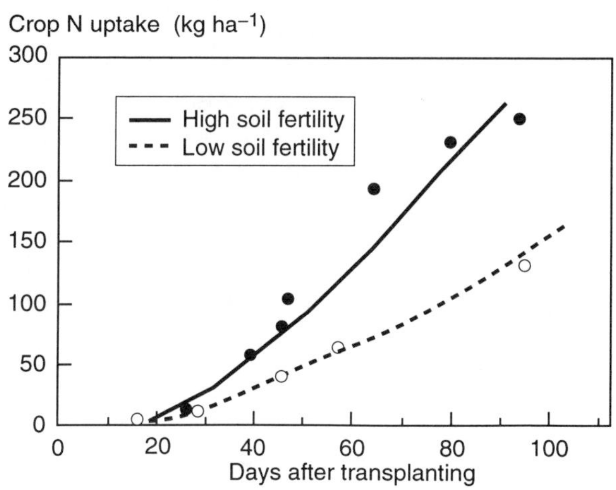

Figure 2. Total N uptake by rice (cv IR36) during the crop growth period as influenced by native soil fertility under conditions of ample N supply during the dry season.

Results and discussion

Biomass accumulation and N uptake capacity of rice

Crop biomass and N uptake of rice was influenced by the native soil fertility in spite of ample N supply from fertilizer. In low fertility soils, maximum crop biomass was 14.2 ton ha^{-1}, whereas in the high fertility soils the maximum biomass was 17.5 ton ha^{-1} at the same environmental conditions (Figure 1). However, dry matter accumulation up to 22–23 DAT was similar in both soils. The maximum N uptake capacity (Figure 2) of rice was also higher on the high fertility soils. About

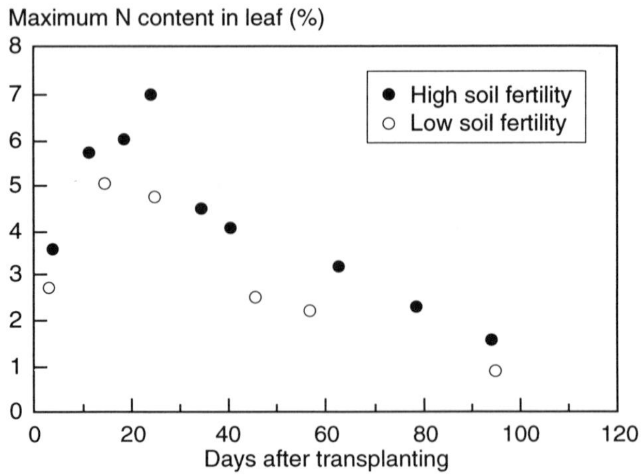

Figure 3. The maximum N concentration in rice leaves (cv IR36) during the crop growth period as influenced by native soil fertility under conditions of ample N supply during the dry season.

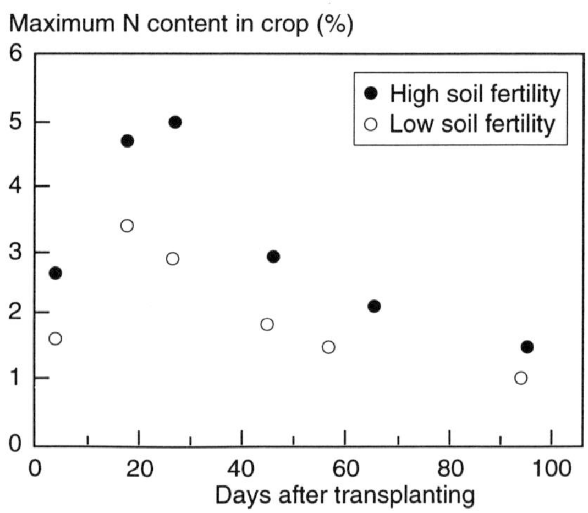

Figure 4. The maximum mean N concentration in crop biomass of rice (cv IR36) during the crop growth period as influenced by native soil fertility under conditions of ample N supply during the dry season.

250 kg N ha^{-1} was absorbed by the crop. In the low fertility soil at the same environmental conditions, only 140 kg N ha^{-1} was absorbed by rice in spite of ample fertilizer N. However, crop N uptake up to 15 DAT did not differ in these soils. The maximum amount of leaf N was 110 kg ha^{-1} in the fertile soil and 49 kg ha^{-1} in the low fertility soil. The N concentration in the leaves (Figure 3) or the mean N concentration in the biomass (Figure 4) was high throughout the growth period of the crop grown in the fertile soil. In spite of the ample N supply, the N concentration in the rice leaves or crop biomass could not be raised to this level in the low fertility soil at the same environmental conditions.

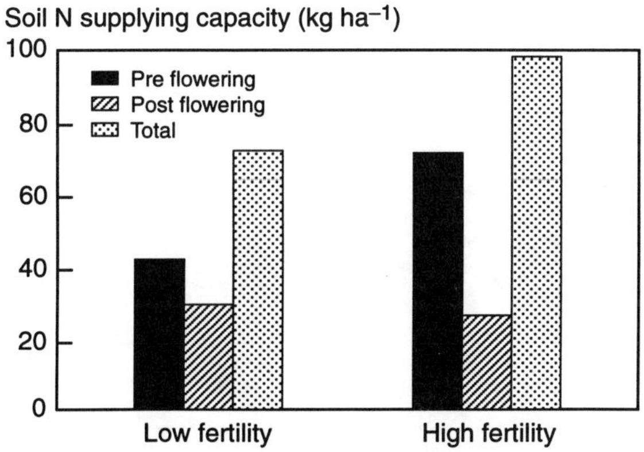

Figure 5. N supplying capacity of the soils to rice (cv IR36) as affected by native soil fertility during the dry season.

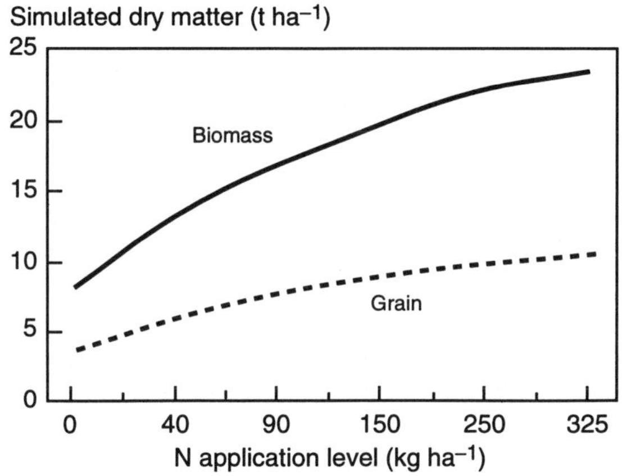

Figure 6. Simulated biomass and grain yield (cv IR36) based on optimization by ORYZA_0 in high fertility soil during the dry season.

Native N supplying capacity of both the soils (Figure 5) during the post-flowering period was not much different as is shown in the zero-N treatment. In the pre-flowering period, the fertile soil supplied more N to the crop. There was also reduced absorption of N by the crop from the fertilizer in the low fertility soil.

Response to fertilizer N application

The maximum attainable biomass accumulation and yield of rice in each total N input level in the fertile soil is presented in Figure 6. This represents the maximum attainable crop biomass and yield calculated by the optimization procedure on the basis of radiation, crop parameters and soil parameters for given N input levels. The zero-N plot yielded 8.12 ton ha^{-1} simulated biomass. The maximum simulated biomass

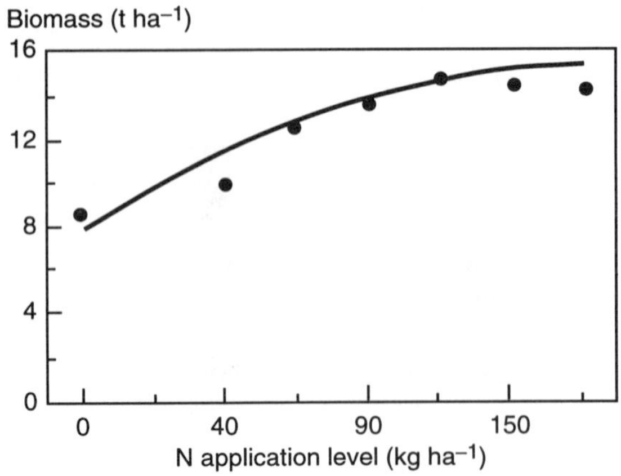

Figure 7. Comparison of simulated (line) and observed (symbols) crop biomass of rice (cv IR36) at selected N application levels in low fertility soil during the dry season.

accumulation occurred at 325 kg N ha^{-1}. The maximum simulated biomass and grain yield were 23.6 and 10.6 ton ha^{-1} respectively. Simulated biomass and yield at 120 kg N in the fertile soil were 18.6 ton ha^{-1} and 8.4 ton ha^{-1} respectively. Beyond the 120 kg N ha^{-1}, there was additional accumulation of simulated biomass which might be susceptible to pests and diseases, lodging or mutual shading. Thus, we hypothesize the optimum dose of N application for dry season could be fixed at 120 kg N ha^{-1} in the fertile soil.

In the low fertility soil a comparison was made between simulated biomass (based on parameters derived by optimization) and observed biomass at selected N input levels (Figure 7). The zero-N plot produced 7.9 ton ha^{-1} simulated biomass, whereas the observed biomass was 8.5 ton ha^{-1}. Simulated biomass was 15.3 ton ha^{-1} at 200 kg N ha^{-1}. There was steep response up to 90–120 kg N ha^{-1} where 14.0–14.5 ton ha^{-1} simulated biomass was accumulated. The observed biomass production increased up to 120 kg N ha^{-1} after which it declined. The maximum observed biomass was 14.6 ton ha^{-1} at 120 kg N ha^{-1} which closely matched with simulated biomass at the same N input level. At 40–90 kg N ha^{-1} the simulated biomass was 9.9–13.5 ton ha^{-1} as against observed biomass of 11.5–14.0 ton ha^{-1}. At 150–200 kg N ha^{-1} the simulated biomass was 14.5–15.3 ton ha^{-1} in comparison to observed biomass of 14.3 ton ha^{-1}. These results, therefore, suggest an optimal N application level of 120 kg N ha^{-1} for the low fertility soil.

Timing of fertilizer N application

The optimal timing of N application (Table 2) is based on the results of the numerical optimization of the N fertilizer application curves. Results showed that fertilizer N should not be applied after panicle initiation regardless of N input level.

At low N levels (up to 40 kg ha^{-1}), the optimal application dates of split doses were only 7–10 days apart. This implies that it would be wise to apply all fertilizer N at one time for low N levels. Except for very high N input levels, the first split dose is suggested to be applied at 3 weeks after transplanting. It might be risky to apply the first dose at such a later date because tiller production may be reduced. With two split doses, the optimal time between applications increases as the N input level increases. At 60 kg N ha^{-1} or higher, the time gap between the first two of three applications is larger than the last two applications. The optimization did not suggest a basal application of fertilizer N for Cuttack in the dry season. It suggested the application of fertilizer N at maximum tillering to panicle initiation. It also rules out the application of N at flowering.

It may be difficult for farmers to remember timings of N fertilizer application if N levels are varied. For example, timing for the third dose could be 41–46 DAT at higher levels of 60–150 kg N ha^{-1} (Table 2). Similarly, the timing of second dose in a system of three equal splits could be from 36 to 40 days. At lower levels of N application, the time gap between two or three split doses is usually a week or less. Thus, results in Table 3 simplify the recommended optimum timings of N fertilizer application for farmers. At 40 kg N ha^{-1}, all N should be applied at 32 DAT. These results suggest three equal doses at about 20, 32 and 42 DAT for N fertilizer levels of 60 kg ha^{-1} and higher if 28 day old seedlings were used for transplanting. In current farmer practice, 50% of fertilizer N input is applied as a basal dose and the remaining 50% in two equal splits at maximum tillering and panicle initiation, respectively. An application level of 80–90 kg N ha^{-1} is recommended for optimum grain production. However, our optimization results suggested different timing of fertilizer N application which was verified in the field trial.

Assessment of ORYZA_0 model based timing of fertilizer N application

Crop biomass. The total crop biomass production by rice at flowering and maturity (Table 4) was examined in the verification trial. Application of N progressively increased the crop biomass. Highest biomass was produced at 120 kg N ha^{-1} applied in three equal splits as suggested by ORYZA_0. On the basis of using existing practices, biomass response to N increased up to 90 kg N ha^{-1}, beyond which the yield declined. The zero-N treatment produced 3382 kg ha^{-1} and 8500 kg ha^{-1} crop biomass at flowering and maturity respectively. A maximum biomass of 14.6 ton ha^{-1} at maturity was obtained when N application was based on ORYZA_0. With the existing practice, the highest biomass level at maturity was 13.5 ton ha^{-1}.

At higher N application levels, the timing of N application based on ORYZA_0 proved superior to existing practice for biomass production. This was not the case at lower levels of N application (40–60 kg N ha^{-1}) where biomass was higher with existing practices.

Grain yield of rice. Application of N increased the grain yield of rice (Table 4). The zero-N treatment produced 4 ton ha^{-1} grain yield of rice, whereas the highest yield of rice (6.2 ton ha^{-1}) was obtained at 120 kg N ha^{-1} using the model derived application schedule. Grain yield responded to N up to 120 kg N ha^{-1} in the model derived practice.

Table 3. Timing of fertilizer N application in the verification trial (cv IR36, Cuttack, dry season 1995; sowing on 3 January; transplanted on 31 January; physiological maturity on 3 May).

N application ($kg\,ha^{-1}$)	Actual timing of urea application (DAT*, 28 day old seedlings)			
	Single dose	Three equal splits		
		1	2	3
40	32	–	–	–
60		23	32	42
90		16	32	42
120		23	32	42
150		16	32	42

*Days after transplanting.

Table 4. Total crop biomass at flowering and maturity and grain yield (cv IR36), as affected by dose, timing and split of N fertilizer application based on ORYZA_0 and at existing practice (Cuttack, dry season, 1995).

N application ($kg\,ha^{-1}$)	Crop biomass ($ton\,ha^{-1}$)				Grain yield ($ton\,ha^{-1}$)	
	Flowering		Maturity		Model-based practice	Existing practice
	Model-based practice	Existing practice	Model-based practice	Existing practice		
40	3.4	4.4	9.8	10.0	4.5	4.4
60	4.4	5.1	11.6	12.4	4.6	5.1
90	6.5	6.2	12.0	13.4	5.6	6.1
120	7.2	6.0	14.6	11.6	6.3	5.3
150	6.6	5.3	14.4	13.0	6.1	6.1

Table 5. Total N uptake by rice (cv IR36) at flowering and maturity as affected by dose, timing and split of N fertilizer application based on ORYZA_0 and at existing practice (Cuttack, dry season 1995).

N application ($kg\,ha^{-1}$)	N uptake ($kg\,ha^{-1}$)			
	Flowering		Maturity	
	Model-based practice	Existing practice	Model-based practice	Existing practice
40	48.4	51.4	78.4	75.9
60	62.3	65.6	85.6	102.0
90	84.0	80.4	96.6	121.7
120	95.2	80.7	128.5	112.2
150	101.1	72.5	138.1	95.7

In the existing practice, grain yield responded to N was up to $90\,kg\,N\,ha^{-1}$. At higher N application levels ($120–150\,kg\,N\,ha^{-1}$), the model recommended practice was superior and produced higher yield over the existing practice; below these levels, the existing practice was superior.

N uptake by rice. N uptake by rice continued to increase with increasing levels of N application up to $150\,kg\,N\,ha^{-1}$ when timings derived by the model were used (Table 5).

In current practice, N uptake by rice did not increase beyond the application level of 90 kg N ha^{-1}. The zero-N treatment (control) absorbed 42 and 72 kg N ha^{-1} at flowering and maturity respectively. Maximum N uptake was 138.1 kg N ha^{-1} using the model-based N application schedule compared with 121.7 kg N ha^{-1} when the current practice was followed. The schedule of N application derived by ORYZA_0 was superior to the current practice at higher N application levels (120–150 kg N ha^{-1}) but not far lower at application levels 40–60 kg N ha^{-1}.

There appear to be two stages in the growth of transplanted rice at which the efficiency of fertilizer N is high. One is soon after transplanting to encourage maximum tillering, and the second is just before or at panicle initiation to encourage maximum number of panicles and grain number per panicle. Tanaka et al. (1959) concluded that split N doses, one at transplanting and another at panicle initiation, are best to get higher grain yields of medium-duration varieties. The yield response to different timings is very complex. Crops fertilized at different timings may increase yield in different ways. Low and moderate levels of N supplied to a young crop lead to increased vegetative growth (Cook and Evans 1983), so that N concentration of tissue is diluted by the carbohydrates with a consequent decrease in assimilation rate. When low or moderate rates of N are applied at later stages of crop development, the initial effect is to increase biomass N concentration and, hence, the assimilation rate. Grain accumulates from rapid assimilation by a relatively small biomass. These processes may lead to a relatively high harvest index. Therefore, in the present experiment, the model-based timings of fertilizer N application produced lower biomass, N uptake and grain yield of rice at 40–60 kg N ha^{-1}.

At higher N levels, when 50% of the N is already applied as basal, there is no advantage over the application of three equal doses starting at a later date. It is primarily because young crops cannot contain large amounts of N in their tissues and the effective date of supply to the crop may be much later than the fertilizer application. Therefore, much of the applied N is subject to loss. Therefore, model based application of 120–150 kg N ha^{-1} at three equal splits produced higher biomass, N uptake and grain yield of rice.

References

Ångström A (1924) Solar and terrestrial radiation. Quart. J. Roy. Met. Soc. 50:121–126.

Cook M G, Evans L T (1983) Some physiological aspects of the domestication and improvement of rice (*Oryza* spp.). Field Crops Res. 6:219–238.

Frère M, Popov G F (1979) Agrometeorological crop monitoring and forecasting. Plant Production and Protection Paper No. 17. Food and Agriculture Organization, Rome, Italy. 64 p.

Ramammorthy B, Bajaj J C (1969) Available N, P, and K status of Indian soils. Fert. News 14(8):25–36.

Rao K S, Dash R N, Ten Berge H F M (1995) Numerical optimization of N fertilizer application to irrigated rice at Cuttack (India) in wet and dry seasons. Pages 9–25 in Thiyagarajan T M, Ten Berge H F M, Wopereis M C S (Eds.) Nitrogen management studies in irrigated rice. SARP Research Proceedings, International Rice Research Institute, P.O. Box 933, Manila, Philippines.

Tanaka A, Patnaik S, Abichandani C T (1959) Studies on the nutrition of rice plant. IV. Growth and nitrogen uptake of rice varieties (*O. sativa* var. indica) of different durations. Proc. Indian Acad. Sci. Soc. B. 49(4):217–226.

Ten Berge H F M, Wopereis M C S, Riethoven J J M, Thiyagarajan T M, Sivasamy R (1994) The ORYZA_0 model applied to optimize nitrogen use in rice. Pages 235–253 in Ten Berge H F M, Wopereis M C S, Shin J C (Eds.) Nitrogen economy of irrigated rice: Field and simulation studies. SARP Research Proceedings, International Rice Research Institute, P.O. Box 933, Manila, Philippines.

Simulating rice leaf area development and dry matter production in relation to plant N and weather

M. OHNISHI[1], T. HORIE[1] and Y. KORODA[2]

[1]*Faculty of Agriculture, Kyoto University, Sakyo-ku, Kyoto 606, Japan*

[2]*Hyogo Prefecture Hokubu Agricultural Institute, Wadayama, Asago-gun, Hyogo 699-52, Japan*

Key words: leaf area development, N management, modelling, rice

Abstract

A simplified dynamic model was formulated, based on the simulation model for Rice-Weather Relationships (SIMRIW), to simulate leaf area index (LAI) and dry matter production for irrigated, transplanted rice. Leaf area index was simulated, independent of its weight, in relation to mean temperature, plant nitrogen (N) content in the above-ground vegetative biomass, and developmental index. Dry matter production was simulated based on LAI, plant N and incident solar radiation. Values for the 13 parameters in the model were determined by the SIMPLEX method by fitting the simulated growth curves to those obtained from field experiments in Kyoto, consisting of 32 treatments with different N management and cropping seasons covering a wide range of plant N status and weather conditions. The model successfully explained all observed LAI and dry weight data with small errors. The values of optimized parameters and the shapes of response curves of component growth functions to environments were physiologically reasonable and similar to published values. These results suggest that this model can be combined with an N uptake model to simulate rice growth and yield in relation to soil-plant N dynamics in the future.

Introduction

Rice growth and yield varies widely among years and locations depending on environmental factors such as solar radiation, temperature and soil characteristics, and on cultivation techniques such as timing and rate of nitrogen (N) fertilizer application, sowing or transplanting time and water management. A dynamic model for simulating rice growth and yield is a very effective tool not only for understanding and predicting those environmental and cultivation effects on rice growth and yield, but also for providing timely information for the optimal management of the crop. Horie (1987) and Horie et al. (1992) proposed a simplified process-based model SIMRIW to simulate growth and yield of irrigated rice, and successfully explained variation in rice yield across locations and years in Japan and the USA, based on climate and weather. This model, however, only simulates climatic potential yield under optimal crop management. For simulating actual rice yield, effects of management factors such as N, water, pests and weeds need to be incorporated into the model. To make SIMRIW applicable for simulation of actual yields of irrigated rice, N must be incorporated into the model first, because N is the most important management factor that affects crop yield. The objective of this study was to incorporate N effects into SIMRIW and to make it applicable for N management of rice. A simplified dynamic model for simulation of leaf area development and dry matter accumulation of

M.J. Kropff et al. *(eds.): Applications of Systems Approaches at the Field Level, 271–284.*

irrigated, transplanted rice in relation to plant N and weather conditions was developed. This paper describes the structure, processes accounted for and performance of the model.

Model description

The model simulates three major processes: plant ontogenetic development, leaf area expansion and dry matter accumulation. The model follows the approach employed in SIMRIW (Horie 1987; Horie et al. 1992) in modelling these three processes. One of the most particular aspects in this approach is that leaf area development is modelled independent of leaf growth weight, by which we can avoid the modelling dry matter partitioning to leaves and specific leaf area, and keep the model simple (Horie et al. 1979). The three components of the model are described below.

Ontogenetic development

Ontogenetic development is quantified by a variable termed developmental index (*DVI*), with values being defined as 0 at seedling emergence, 1.0 at heading and 2.0 at maturity. The value of *DVI* is obtained by integrating the daily developmental rate (*DVR*), which is a function of daily mean air temperature (*T*) and daylength (*L*) (Horie and Nakagawa 1990):

$$DVI_t = \Sigma DVRi \tag{1}$$

$$DVR = \begin{cases} 1/(1+\exp(-a_1(T-Th)))/G & (DVI \leq DVI^*) \quad (2) \\ (1-\exp(a_2(L-Lc)))/(1+\exp(-a_1(T-Th)))/G & (DVI > DVI^*, L \leq Lc) \\ 0 & (DVI > DVI^*, L \leq Lc) \end{cases}$$

where G is a minimum number of days from emergence to heading of a given cultivar; Th is the temperature at which *DVR* equals half of the maximum under a given daylength; Lc is the critical daylength; DVI^* is the *DVI* value at which rice becomes sensitive to photoperiod; and a_1, a_2 are parameters which characterize the shape of the response curve. The values of these parameters for Nipponbare rice were obtained from Nakagawa and Horie (1987).

LAI growth

Leaf area index (*LAI*) is calculated from the relative expansion rate (*RER*) and relative death rate (*RDR*) of leaves:

$$dLAI/dt = (RER - RDR)LAI \tag{3}$$

RER is a function of daily mean air temperature, N content of the above-ground vegetative biomass per unit leaf area, and *DVI*. These three factors are modelled by normalized reduction functions of maximum *RER* (*RERmax*):

$$RER = RERmax\, f_L(T) f_L(N) f_L(DVI) \tag{4}$$

Temperature effects on RER. The relative growth rate increases with temperature up to an optimum and then decreases. The relationship between temperature and *RER* is given by the following equation, in which any supra-optimum temperature effects are not taken into account:

$$f_L(T)=1/(1+\exp(-b(T-Th_L))) \tag{5}$$

where Th_L is the temperature at which $f_L(T)$ becomes 0.5; and b is the parameter characterizing the shape of the response curve.

Nitrogen effects on RER. The crop growth rate is strongly dependent on plant N concentration. In the model, plant N concentration is represented by the N content of the above-ground vegetative biomass divided by leaf area. The normalized effect of N concentration on *RER* is given as:

$$f_L(N)=\begin{cases}1-\exp(-c(N-Nc)) & (N>Nc)\\ 0 & (N\leq Nc)\end{cases} \tag{6}$$

where N is the plant N concentration ($\mathrm{g\,m^{-2}}$); Nc is the critical N concentration at which $f_L(N)$ becomes 0; c is the parameter characterizing the shape of the response curve. Equation (6) implies that the effect of N concentration on leaf expansion growth does not change with crop developmental stage.

Developmental stage (DVI) effect on RER. It is generally observed that the relative growth rate of *LAI* declines during ontogenetic development. This relation is represented by:

$$f_L(DVI)=\begin{cases}1 & (DVI\leq DVIb)\\ 1-((DVI-DVIb)/(DVIc-DVIb))^{\mu} & (DVI>DVIb)\end{cases} \tag{7}$$

where *DVIb* is the *DVI* value at which *RER* begins to decline; *DVIc* is the *DVI* value at which expansion of leaves terminates and *DVI* effect becomes 0; μ is the parameter to represent the shape of the response curve. The value of *DVIc* is set to 1 (at heading), because appearance of new leaves and expansion of existing leaves are scarcely observed after this stage.

Relative death rate of leaves (RDR). The relative leaf death rate is calculated based on two processes, the relative leaf self-destruction rate (*LSDR*) and the relative leaf senescence rate (*LSSR*). Here, leaf senescence is defined as leaf death not associated with N deficiency. The highest rate is selected in the simulation model as the actual value of *RDR* at a given day:

$$RDR=\mathrm{Max}(LSDR, LSSR) \tag{8}$$

Leaf self-destruction is closely linked with the remobilization of N and other nutrients from older to younger organs and grain, and reported to occur during the grainfilling period (Sinclair and De Wit 1976) or under nutrient-deficient conditions.

In this model, *LSDR* is calculated based on plant N concentration:

$$LSDR = \begin{cases} (Nmc - N)/\tau & (Nmc > N) \\ 0 & (Nmc \leq N) \end{cases} \tag{9}$$

where *N* is the plant N concentration ($g\,m^{-2}$); *Nmc* is the critical N concentration for supporting current *LAI*; τ is the time coefficient.

Leaves may die as a result of ageing due to leaf senescence. This process is assumed to be enhanced by temperature like other metabolic processes, and modelled as:

$$LSSR = LSSR_{27} Q_{L10}{}^{(T-27)/10} \tag{10}$$

where *T* is the mean air temperature; $LSSR_{27}$ is the relative leaf senescence rate at 27°C; Q_{L10} is the temperature coefficient of the relative rate of leaf senescence, and is assumed to be 2.0.

Dry matter growth

The crop growth rate is represented by:

$$dW/dt = Pg - Rg - Rm \tag{11}$$

where *W* is the dry weight per unit ground area; *Pg* is the gross photosynthesis rate; *Rg* and *Rm* are the growth and maintenance respiration rates respectively. Since the growth respiration rate (*Rg*) is proportional to the photosynthetic rate *Pg* (McCree 1970), the term ($Pg - Rg$) in equation (11) can be converted into:

$$Pg - Rg = k_1 Pg \tag{12}$$

where k_1 is the production value (Penning de Vries et al. 1974), a conversion factor from photosynthates to biomass. In this model, the term of ($k_1 Pg$) is given by a function of intercepted radiation and radiation conversion efficiency (*RCE*) which itself is a function of the plant N concentration, and *Rm* given by a function of the daily air temperature and crop dry weight.

Radiation conversion efficiency (RCE). It is well established that the crop growth rate is proportional to intercepted or absorbed radiation when other factors are non-limiting (Shibles and Weber 1966; Monteith 1977; Horie and Sakuratani 1985). Based on this, and by employing the relation proposed by Sinclair and Horie (1989), the radiation conversion efficiency (*RCE*) of rice is represented by:

$$k_1 Pg = RCEmax\, fg(N) Si \tag{13}$$

$$fg(N) = \begin{cases} 1 - \exp(-d(N - Nrc)) & (N > Nrc) \\ 0 & (N \leq Nrc) \end{cases} \tag{14}$$

where *N* is the plant N concentration ($g\,m^{-2}$); *RCEmax* is the maximum *RCE* at saturating plant N concentration ($g\,MJ^{-1}$); *Si* is the daily intercepted radiation ($MJ\,m^{-2}$);

Table 1. Sowing and transplanting dates, and N management of 32 different experimental plots for rice cultivar Nipponbare.

Year	Sowing date	Transplanting date	Cropping season code	Nitrogen management code*							
1987	20 March	21 April	CS1		LS			VS			
	10 April	9 May	CS2		LS	LH		VS	VH	BS	OO
	11 May	4 June	CS3		LS			VS			
	22 June	9 July	CS4		LS	LH		VS	VH	BS	OO
1988	8 April	14 May	CS5	LL	LS	LH	VL	VS	VH	BS	OO
	28 May	22 June	CS6	LL	LS	LH	VL	VS	VH	BS	OO

*The N management codes are explained in Table 2.

Nrc is the critical N concentration at which *fg*(*N*) becomes 0; *d* is the parameter characterizing the shape of the response curve.

Although Sinclair and Horie (1989) used leaf N content per unit leaf area for equation (14), we used above-ground plant N content per unit leaf area to make it easy to combine this model with a plant N uptake model in the future. *Ncr* is assumed to be equal to *Nc*, the critical plant N concentration for the relative leaf expansion rate (*RER*). This means that, at $N = Nc$, both leaf area expansion and dry matter production stop simultaneously.

Radiation interception. The amount of intercepted radiation by the rice canopy (*Si*) in equation (13) is calculated as a function of daily incident radiation (*So*), *LAI* and the light extinction coefficient (k_2) by the following equation derived from by Monsi and Saeki (1953):

$$Si = So(1 - \exp(-k_2 LAI)) \tag{15}$$

Referring to Horie and Sakuratani (1985), a value of 0.45 was used for k_2.

Maintenance respiration. Maintenance respiration (*Rm*) is calculated as a function of daily mean temperature (*T*) and crop dry weight (*W*):

$$Rm = Rm_{27} Q_{m10}^{(T-27)/10} W \tag{16}$$

where Rm_{27} is the respiration rate at the reference temperature 27°C ($g\,g^{-1}\,d^{-1}$); Q_{m10} is the temperature coefficient of the respiration. Q_{m10} is assumed to be 2.0.

Material and methods

Field experiments

Experiments were conducted in 1987 and 1988 at the Kyoto University Experimental Farm, with the rice cultivar Nipponbare. The dates of sowing on the seedling beds and transplanting of each crop are shown in Table 1, and the times and rates of N application in Table 2. Transplanting density was 30 cm × 15 cm in 2 seedlings per hill. The experiments consisted of 32 treatments with different times and rates of N fertilizer

Table 2. The rate and timing of N fertilizer application in each N management plot (kg ha^{-1}). The numbers in parentheses are days before heading.

N treatment code	Basal	Tillering	Panicle formation				Ripening (+3)	Total
			(−40)	(−25)	(−15)	(−8)		
OO	0	0	0	0	0	0	0	0
BS	130	0	0	0	0	0	0	130
LL	10	10	10	15	0	15	10	70
LS	20	20	20	30	0	20	20	130
LH	20	20	40	60	0	20	30	190
VL	20	20	0	0	20	0	10	70
VS	40	40	0	0	30	0	20	130
VH	60	60	0	0	40	0	30	190

Note: The first letters of the N treatment codes are: O = no application; B = basal application only; V = high N dose during the vegetative stage; L = high N dose during the reproductive stage. The second letters of codes are: O = 0 kg N ha^{-1}; L = 70 kg N ha^{-1}; S = 130 kg N ha^{-1}; H = 190 kg N ha^{-1}.

application, and different seasons and years, in order to cover a wide range of weather conditions and N environments. Above-ground plant samples of 6 hills were taken from each plot at about 3-week intervals from transplanting to heading and at about 10-day intervals from heading to maturity. The samples were separated into green leaf blade, leaf sheath and stem, panicle and dead leaf. After measuring leaf area, each fraction was dried at 80°C for one week and dry weight was determined. The samples were ground in a vibration mill and analysed for N by the Kjeldahl method. Daily values of plant N content of above-ground vegetative organs were estimated by a linear interpolation between two adjacent measurements, and used as input values for parameters. Daily values of plant N concentration per unit leaf area were obtained by dividing the daily plant N content by simulated daily values of *LAI*. Daily solar radiation and air temperature were measured throughout the entire growth seasons at the site.

Estimation of parameters of the model

Values for the 13 parameters in the model were determined by the SIMPLEX method (Horie and Nakagawa 1990) which minimizes the sum of deviations between the measurements and model estimations for the 32 growth curves of crop dry weight and *LAI*. Measured values were used for the initial values of *LAI*, crop dry weight and plant N concentration in the model.

Goodness of model fit was evaluated by the root mean square error (*RMSE*), for *LAI* and dry weight:

$$RMSE = [1/(n-p)\Sigma(\text{observed}_i - \text{simulated}_i)^2]^{0.5} \qquad (17)$$

where n and p denote the number of measurements and estimated parameters respectively.

The coefficient of determination (r^2) and regression coefficient (α) from analysis of a linear regression through the origin ($Y = \alpha X$) between the observed data and the

Table 3. Values of fixed parameters and optimized crop parameters from simulation, and input variables.

Parameter	Value		Unit
Optimized parameters			
RERmax	0.181	Maximum relative expansion rate of *LAI*	$m^2 m^{-2} d^{-1}$
b	0.419	Temperature response curvature of *RER*	–
Th_L	19.3	Temperature at which $f_L(N)$ becomes 0.5	°C
c	0.727	N response curvature of *RER*	–
Nc	1.08	Critical plant N concentration for *RER*	$g m^{-2}$
Nrc	(1.08)	Critical plant N concentration for *RCE*	$g m^{-2}$
Db	0.185	*DVI* at which $f_L(DVI)$ started to decline	–
μ	2.81	*DVI* response curvature of *RER*	–
Nmc	1.45	Critical plant N concentration for maintaining current *LAI*	$g m^{-2}$
τ	6.74	Time constant for leaf self destruction	d
$LSSR_{27}$	6.66×10^{-3}	Relative leaf senescence rate at 27°C	$m^2 m^{-2} d^{-1}$
RCEmax	1.81	Maximum radiation conversion efficiency of crop	$g MJ^{-1}$
d	7.15	N response curvature of *RCE*	–
Rm_{27}	7.56×10^{-3}	Maintenance respiration rate at 27°C	$g g^{-1} d^{-1}$
Fixed parameters as coefficients			
DVIc	1.0	*DVI* at which $f_L(DVI)$ becomes 0	–
Q_{L10}	2.0	Temperature coefficient of *LSSR*	–
k_2	0.45	Light extinction coefficient	–
Q_{m10}	2.0	Temperature coefficient of the respiration	–
Input variables			
T		Daily mean air temperature	°C
Plant N		Daily plant N in the vegetative organ	$g m^{-2}$
DVI		Daily developmental index	–
So		Daily incident radiation	$MJ m^{-2} d^{-1}$

simulation results were also used as indicators for the adequacy of fit to measured data.

The parameter values with which the model best explained the observed data are given in Table 3, together with the fixed coefficients.

Results and discussion

The range of experimental data

The field experiments provided a wide range of growing conditions and rice yields. *LAI* at heading varied between 2.4 and 8.5 $m^2 m^{-2}$ among treatments, and dry weight at maturity varied between 7.7 and 17.1 ton ha^{-1}. Average air temperature in the vegetative phase ranged from 22.6°C to 28.2°C and during the grainfilling phase from 20.8°C to 28.0°C.

Model performance

In general, the model successfully explained variation in observed *LAI* with an *RMSE* of 0.61 $m^2 m^{-2}$. The r^2 and α values for the linear regression between observed and simulated *LAI* were 0.92 and 0.98 respectively (Figure 1). The time course of simulated *LAI* was in good agreement with the observations as well (Figure 2). Results

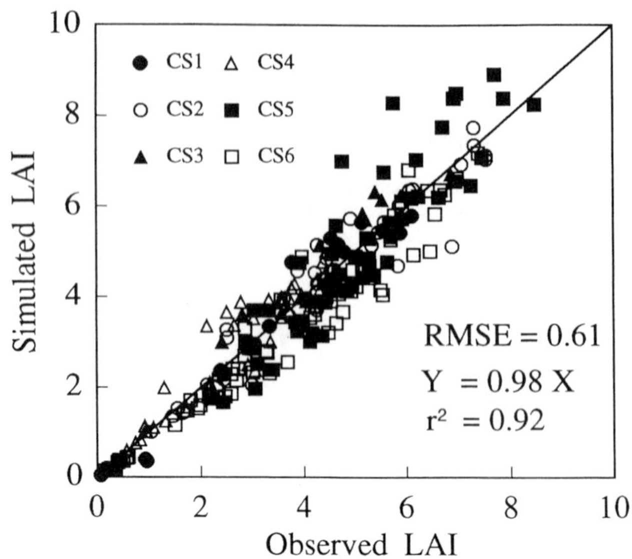

Figure 1. The relationship between observed and simulated *LAI* growth ($m^2 m^{-2}$); rice cultivar Nipponbare for 32 different field data sets obtained under widely different environmental conditions; for explanation of the codes see Table 1. RMSE denotes the root mean square error.

of the linear regression between the observed and simulated *LAI* for different treatments are shown in Table 4. All r^2 values were higher than 0.86. The standard deviations (SD) of errors were between 0.37 to 0.76 $m^2 m^{-2}$. Although the α values, for VL (α=0.83) and OO (α=0.81) N treatments, indicated an underestimation of *LAI* of more than 10% which was mainly caused by simulation errors at high *LAI* values in those treatments, SD values for those treatments were both small (0.40 for VL and 0.39 for OO). Since errors themselves are small and the errors for high *LAI* levels have a smaller influence on light interception and simulated crop dry weight than those at low *LAI*, these underestimations of *LAI* are considered to result in negligibly small effects on simulated crop growth.

The fit of the model to observed dry weight data was satisfactory (Figures 3 and 4). α values for the linear regression between the observed and the simulated dry weight for different treatments ranged from 0.96 to 1.04 and all r^2 values were higher than 0.96 (Table 4).

Parameter and response curves

Figure 5 shows the response curves of the relative leaf expansion rate (*RER*) to temperature, plant N concentration and to *DVI* determined by using optimized crop parameters as given in Table 3.

The maximum *RER* (*RERmax*) was slightly larger than the mean relative growth rate of leaf area (*LRGR*) reported for rice seedling by Murata and Togari (1975) but much smaller than their maximum *LRGR* value. This suggests that the model can

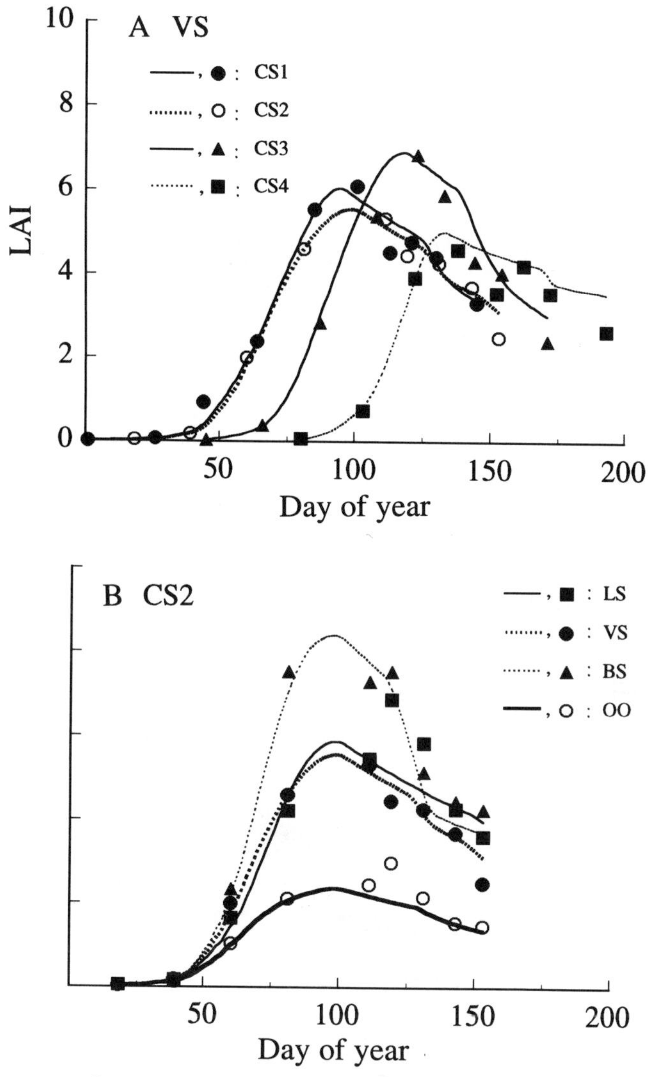

Figure 2. Simulated (lines) and observed (symbols) *LAI* growth curves; (A) for different cropping seasons (CS1, CS2, CS3 and CS4) in 1987 for the same N management (VS treatment); and (B) for different N managements (LS, VS, BS and OO) in the second cropping season in 1987 (CS2). For explanation of the codes see Tables 1 and 2.

explain *LAI* growth throughout rice growing period, except for the initial seedling stage in which a larger *LRGR* can be found. The temperature effect on *RER* attained a maximum at about 30°C and became 0 below about 7°C (Figure 5A). The optimal and the critical temperatures were similar to those reported by Yoshida (1973) for rice seedlings. The value of critical plant N concentration for leaf expansion ($Nc = 1.08$) was smaller than that for maintaining the current *LAI* ($Nmc = 1.45$), indicating that

Table 4. The goodness of model fit for each treatment. For codes see Tables 1 and 2.

Treatment	Number of data (n)	*LAI*			Dry weight		
		SD*	α**	r^2**	SD*	α**	r^2**
CS1	18	0.43	1.02	0.96	0.70	0.99	0.98
CS2	48	0.47	0.98	0.95	0.85	1.02	0.97
CS3	16	0.37	1.08	0.97	0.47	1.00	0.99
CS4	42	0.45	1.02	0.87	0.85	0.98	0.95
CS5	64	0.76	1.00	0.90	0.61	1.01	0.98
CS6	64	0.50	0.90	0.93	0.77	0.99	0.97
LL	16	0.43	0.90	0.92	0.57	1.01	0.98
LS	48	0.58	0.99	0.90	0.72	1.00	0.98
LH	31	0.58	1.09	0.94	0.66	1.04	0.98
VL	16	0.40	0.83	0.92	0.97	0.96	0.96
VS	48	0.47	0.98	0.93	0.81	0.97	0.97
VH	31	0.74	0.99	0.88	0.87	0.97	0.97
BS	31	0.49	0.95	0.95	0.72	0.99	0.98
OO	31	0.39	0.81	0.86	0.58	1.04	0.97

*SD $= [1/(n-1)\Sigma(\text{observed}-\text{simulated})^2]^{0.5}$.
**r^2 and α denote the coefficient of determination and the regression coefficient between observation and simulation, respectively.

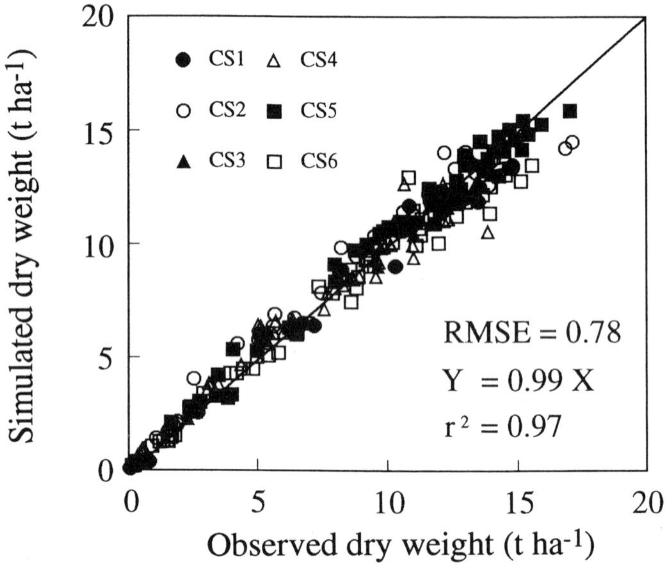

Figure 3. The relationship between observed and simulated dry weight growth (ton ha^{-1}) of rice cultivar Nipponbare for 32 different field data sets obtained under widely different environmental conditions.

leaves started self-destruction before losing their photosynthetic and expansion activities. The *DVI* effect on *RER* began to decline beyond *DVI* about 0.35 and thereafter sharply declined with progressing *DVI* (Figure 5C). To simulate leaf expansion growth for direct seeded rice, a modification of this *DVI-RER* function

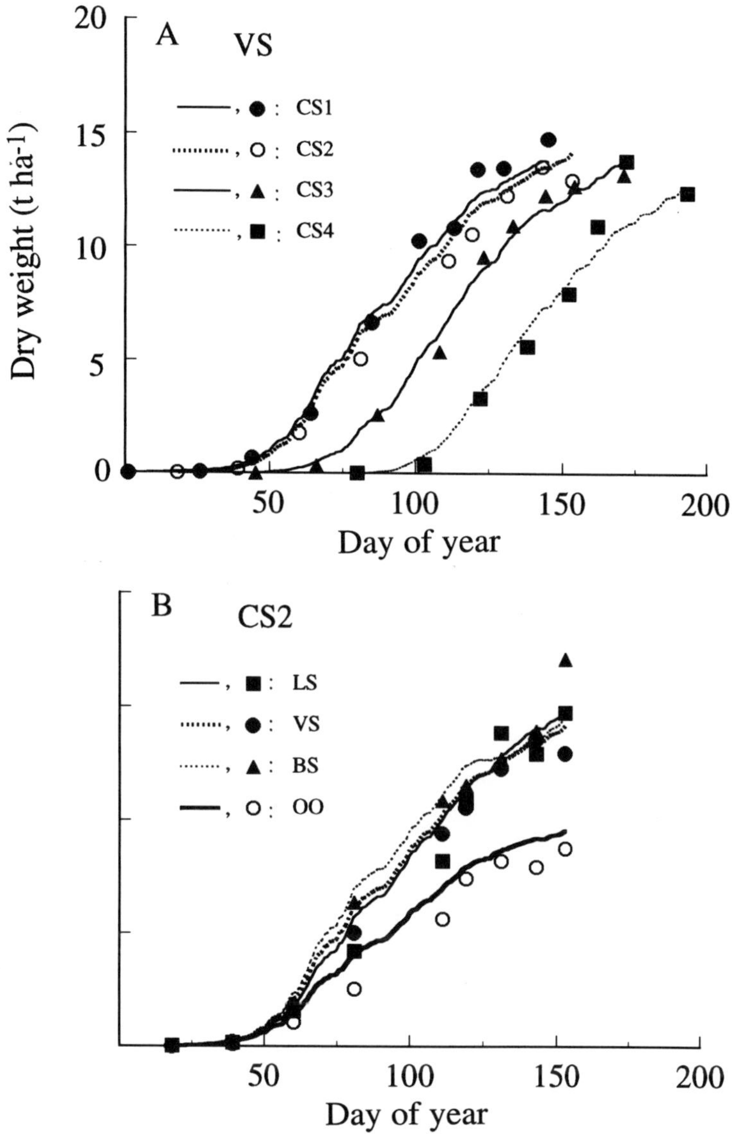

Figure 4. Simulated (lines) and measured (symbols) growth curves in dry weight (ton ha^{-1}); (A) for different cropping seasons (CS1, CS2, CS3 and CS4) in 1987 for the same N management (VS treatment); and (B) for different nitrogen managements (LS, VS, BS and OO) in the second cropping season in 1987 (CS2). For explanation of the codes see Tables 1 and 2.

seems to be necessary to accommodate the quite large *LRGR* reported for the seedling stage.

Although the largest value of leaf senescence rate (*LSSR*) and leaf destruction rate (*LSDR*) is chosen as the relative leaf death rate (*RDR*) in the model, *LSDR* was never

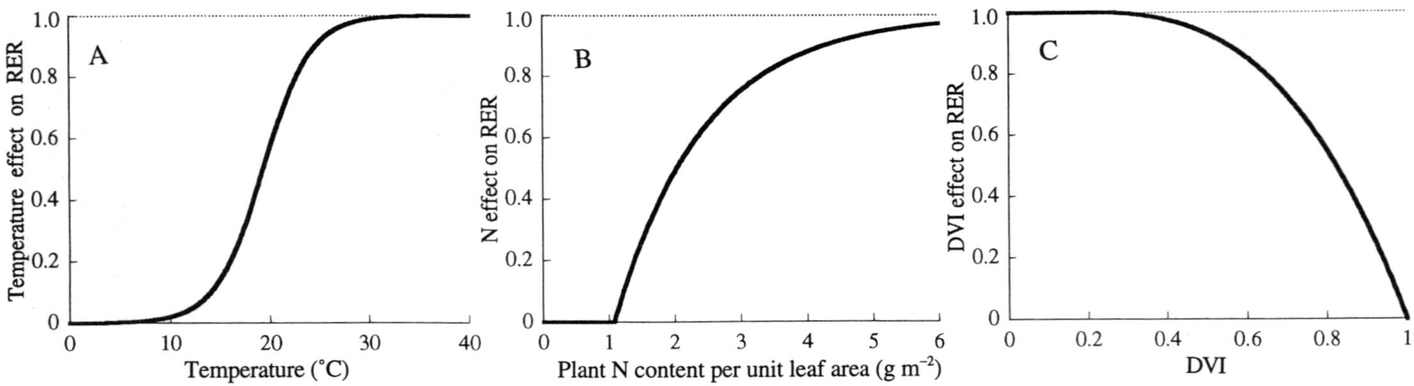

Figure 5. (A) Response curves of relative leaf expansion rate (*RER*) to temperature; (B) plant N content in the above-ground vegetative biomass per unit leaf area; and (C) to developmental index (*DVI*), estimated from fitting the model to measured rice growth curves.

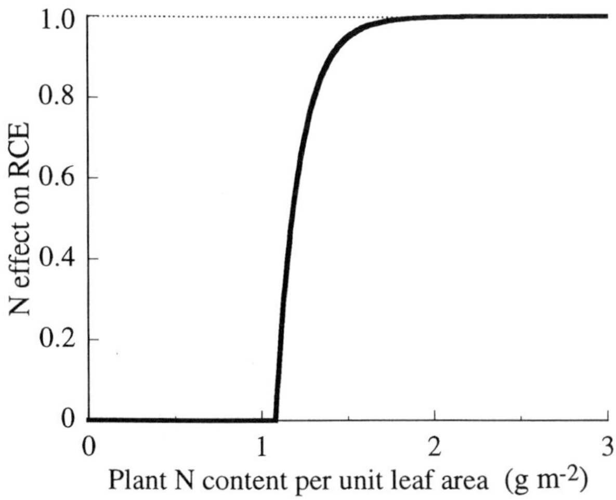

Figure 6. Response curve of radiation conversion efficiency (*RCE*) to plant N content in the above-ground vegetative biomass per unit leaf area, estimated from fitting the model to measured rice growth curves.

chosen before the late reproductive stage. Since no symptom of N deficiency was found in the field experiments, this relationship between *LSDR* and *LSSR* seems reasonable.

The maximum radiation conversion efficiency (*RCEmax*=1.81) estimated was slightly larger than the published values (Sinclair and Horie 1989; Horie 1987). It may result from the assumption that the light extinction coefficient (k_2) is fixed at 0.45 for all treatments, even though the rice canopy structures are considered to be different among treatments. The estimated maintenance respiration rate at 27°C (Rm_{27}=0.0076) was in the range of the values reported by Amthor (1980). The general shape of *RCE* response to plant N concentration (Figure 6) was similar to that reported by Sinclair and Horie (1989). The estimated value of 1.08 for critical plant N content per unit leaf area for *RCE* (*Nrc*=1.08), assumed equal to that for *RER* (*Nc*), is approximately equivalent to 0.26 to 0.45 $g\,m^{-2}$ of leaf N concentration on the leaf area basis. This range includes the critical leaf N concentration of 0.3 $g\,m^{-2}$, reported for rice by Sinclair and Horie (1989).

From the similarity between the estimated parameter values and published data, and the rationality in shapes of response curves and in relationships among parameters, all the estimated values for crop parameters are considered to be physiologically reasonable.

Conclusions

The model here presented successfully explained 32 sets of field data on the *LAI* growth and dry matter accumulation obtained under widely different environment conditions. Further, it has been clarified that the estimated values of crop parameters and the shapes of response curves of component growth functions to environments were physiologically reasonable and similar to published data. These results suggested that, in future, this model can be combined with an N-uptake model for simulating rice growth and yield in relation to soil-plant N dynamics.

References

Amthor J S (1984) The role of maintenance respiration in plant growth. Plant Cell Environ. 7:561–569.

Horie T (1987) A model for evaluating climatic productivity and water balance of irrigated rice and its application to Southeast Asia. Southeast Asian Studies 25:62–74.

Horie T, Nakagawa H (1990) Modeling and prediction of developmental process in rice. 1. Structure and method of parameter estimation of a model for simulating developmental process toward heading. Jpn. J. Crop Sci. 59:687–695.

Horie T, Sakuratani T (1985) Studies on crop-weather relationship model in rice. 1. Relation between absorbed solar radiation by crop and dry matter production. J. Agric. Met. 40:331–342.

Horie T, De Wit C T, Goudriaan J, Bensink J (1979) A formal template for development of cucumber in its vegetative stage. Proc. Koninklijke Nederlandse Akademie voor Wetenschappen (KNAW, Amsterdam, The Netherlands), Series C, Vol 82(4):433–479.

Horie T, Yajima M, Nakagawa H (1992) Yield forecasting. Agric. Syst. 40:211–236.

McCree K J (1970) An equation for the rate of respiration of white clover plants grown under controlled conditions. Pages 221–229 in Prediction and measurements of photosynthetic productivity. Pudoc, Wageningen, The Netherlands.

Monsi M, Saeki T (1953) Über den Lichtfaktor in den Pflanzengesellschaften und seine Bedeutung für die Stoffproduktion. Jap. J. Bot. 14:1–22 (in German).

Monteith J F (1977) Climate and the efficiency of crop production in Britain. Philosoph. Trans. R. Soc. London B 281:277–294.

Murata Y, Togari Y (1975) Summary of data. Pages 9–20 in Murata Y (Ed.) Crop productivity and solar energy utilization in various climates in Japan. Japanese committee for the International Biological Program – JIBP synthesis Vol. 11, Univ. of Tokyo Press, Japan.

Nakagawa H, Horie T (1987) Studies on modeling and prediction of developmental processes in rice. 3. Varietal differences of developmental characteristics. Jpn. J. Crop Sci. 56(2):189–190.

Penning de Vries F W T, Brunsting A H M, Van Laar H H (1974) Products, requirements and efficiency of biosynthesis: A quantitative approach. J. Theor. Biol. 45:339–377.

Shibles R M, Weber C R (1966) Interception of solar radiation and dry matter production by soybean planting pattern. Crop Sci. 6:55–59.

Sinclair T R, De Wit C T (1976) Analysis of the carbon and nitrogen limitations to soybean yield. Agron. J. 68:319–324.

Sinclair T R, Horie T (1989) Leaf nitrogen, photosynthesis, and crop radiation use efficiency: A review. Crop Sci. 29:90–98.

Yoshida S (1973) Effects of temperature on growth of the rice plant (*Oryza sativa* L.) in a controlled environment. Soil Sci. Plant Nutr. 19:299–310.

Influence of split application of nitrogen on foliar N content, photosynthesis, dry matter production and yield in short- and medium-duration rice cultivars

C. VIJAYALAKSHMI and M. NAGARAJAN

Department of Crop Physiology, Tamil Nadu Agricultural University, Coimbatore 641 003, Tamil Nadu, India

Key words: split application of N, foliar N content, photosynthesis, dry matter accumulation, grain yield

Abstract

This investigation was aimed at finding the facts and efficiencies of split application of N at different phenological phases of the rice crop. Physiological and plant factors contributing to efficiency of N utilization were also examined. Field experiments were conducted from 1992 to 1994 at Tamil Nadu Agricultural University, Coimbatore, India. N-uptake was highest between active tillering and panicle initiation. Proper apportioning of N between active tillering and panicle initiation enhance recovery of applied fertilizer N. The foliar N content and N-uptake increased by application of N at active tillering and panicle initiation, rather than by application of higher amounts of N at planting. Withholding N at active tillering or panicle initiation reduced grain yield through reduced spikelet formation. Increased foliar N content improved photosynthetic efficiency through higher amounts of chlorophyll and RuBP carboxylase, resulting in increased canopy photosynthesis. The highest grain yield was achieved by application of 85% to 100% of the total N applied of 120 kg ha^{-1} between active tillering and panicle initiation.

Introduction

In Asia, where 90% of the world's rice is grown and consumed, nearly 2.8 billion people derive 35–65% of the calories only from rice (Swaminathan 1989). By the year 2010, Indian rice production should increase up to 130 million tons in order to meet the needs of the population (Siddiq 1993). At this juncture, it is therefore imperative to investigate the opportunities to raise the level of rice production. Among nutrients which are essential for the rice crop, nitrogen (N) is the most important one in rice soils. The low average recovery of fertilizer N (i.e. 30–40%) suggests that there is considerable scope for improvement in N fertilizer efficiency. Application of N at critical growth stages would minimize losses and increase the recovery of N (Puckridge et al. 1992). Because higher fertilizer use and intensification of cropping patterns, together with varietal improvement for increasing rice production, has reached a plateau, a physiological approach to improve N-uptake and dry matter production through photosynthesis and efficient translocation of carbohydrates to grains should now be exploited to increase rice productivity. Thiyagarajan et al. (1994) emphasized that to achieve a target yield of 8 ton ha^{-1}, the leaf N concentration has to

M.J. Kropff et al. *(eds.): Applications of Systems Approaches at the Field Level, 285–294.*

Table 1. Details of the N split applications.

Treatments	Percentage of N applied at				Notation
	Basal	Active tillering (AT)	Panicle initiation (PI)	Flowering (FL)	
T1	50*	25*	25*	0	(50:25:25:0)
T2	25*	25*	25*	25*	(25:25:25:25)
T3	25	0	50	25	(25:0:50:25)
T4	25	50	0	25	(25:50:0:25)
T5	25	50	25	0	(25:50:25:0)
T6	15	0	60	25	(15:0:60:25)
T7	15*	60*	0	25*	(15:60:0:25)
T8	15*	60*	25*	0	(15:60:25:0)
T9	0	25*	50*	25*	(0:25:50:25)
T10	0	50	50	0	(0:50:50:0)

*^{15}N (5% atom. excess) labelled urea.

be maintained above a specified level which changes with time. Sink size (panicle and spikelet number) enhancement by late season N application, through reduced spikelet degeneration, was reported by Dingkuhn et al. (1992a). Rao (1987) reported that, in general, split application of granular urea was more efficient in terms of yields and N recovery as compared to other methods of application, namely deep placement of urea super granules and two sources of slow release N fertilizers such as sulphur coated urea and lac-coated urea.

Material and methods

Field experiments were conducted at Paddy Breeding Station, Tamil Nadu Agricultural University, in the soil belonging to fine, mixed, isohyperthermic, Vertic Ustropepts (Inceptisol). Two high-yielding short-duration cultivars, ASD 18 and ADT 36, and two medium-duration cultivars, Co 43 and ADT 38, were selected for this study. Phosphorus and potassium were applied at a rate of $60\,kg\,P_2O_5\,ha^{-1}$ and $60\,kg\,K_2O\,ha^{-1}$ respectively. Nitrogen in the form of urea was applied at a rate of $120\,kg\,N\,ha^{-1}$. But the total quantity of urea was given in split doses at different phenological stages of the crop. Details of the N applications are presented in Table 1. Since application of N at critical growth periods, active tillering and panicle initiation would minimize the losses and increase the recovery of N when compared with basal application, the phenophases transplanting (basal), active tillering and panicle initiation were chosen for collecting samples. Sink size (panicle and spikelet number) enhancement by N application at flowering was reported by Dingkuhn et al. (1992a). So, N was applied at flowering also and the plant samples were collected at active tillering, panicle initiation, flowering and harvest.

The total N content was estimated by the Bremner method as quoted by Black (1965) (micro-Kjeldahl distillation method) and expressed in terms of the percent content on dry weight basis. The N-uptake by the crop was calculated by multiplying the N content and total dry matter.

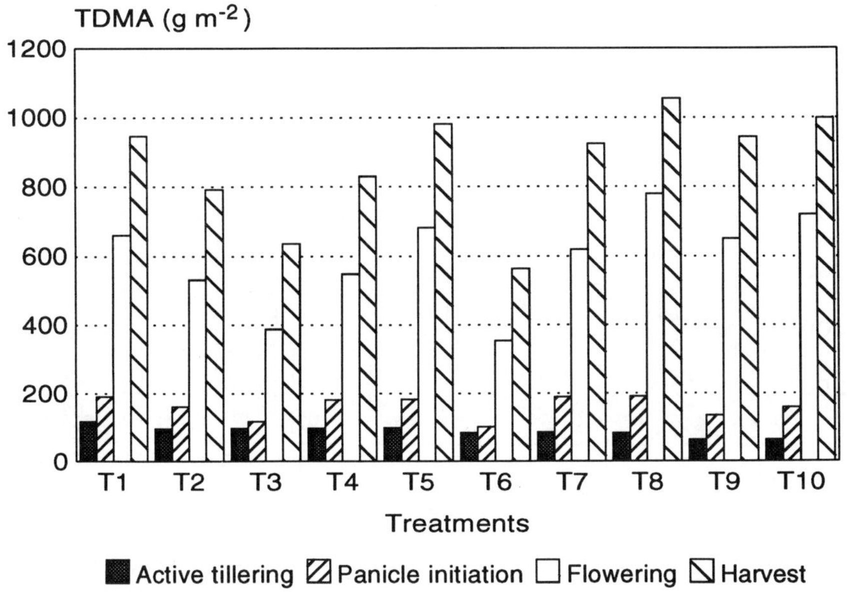

Figure 1. Total dry matter accumulation (TDMA, g m^{-2}) (short-duration cultivars).

Results and discussion

Total dry matter accumulation

Akita (1989) observed that the increased yield of the newly developed high-yielding variety Akenohoshi, which has genes from IR varieties, was mainly attributed to the increased sink capacity (spikelet number×grain weight) due to increased biomass production. Rao et al. (1994) reported that application of N at panicle initiation increased the dry weights of leaf, stem, root and also their N content and total N-uptake. Akita (1989) reported that high biomass production was closely associated with high yield.

Among the treatments, highest total dry matter of 1055 g m^{-2} was observed at harvest stage in treatment T8 in which N was applied in the ratio of 15:60:25:0 at planting, active tillering, panicle initiation and flowering respectively in short-duration rice cultivars. T8 was followed by the treatment T10, which received N in the ratio of 0:50:50:0 at planting, active tillering, panicle initiation and flowering recorded 998 g m^{-2}, while T1 (50:25:25:0) recorded 946 g m^{-2}. The total dry matter accumulation was lowest in treatments T3 and T6 (636 and 563 g m^{-2} respectively) and these two treatments received 0 N at active tillering (Figure 1). Reduction in total dry matter accumulation was also noticed in treatments T4 and T7 in which no N was applied at panicle initiation (830 and 924 g m^{-2}).

In the medium-duration cultivars, the highest total dry matter of 1174 g m^{-2} was recorded in T10 (0:50:50:0) followed by T8 (1159 g m^{-2}) which received N in the ratio of 15:60:25:0 at planting, active tillering, panicle initiation and flowering respec-

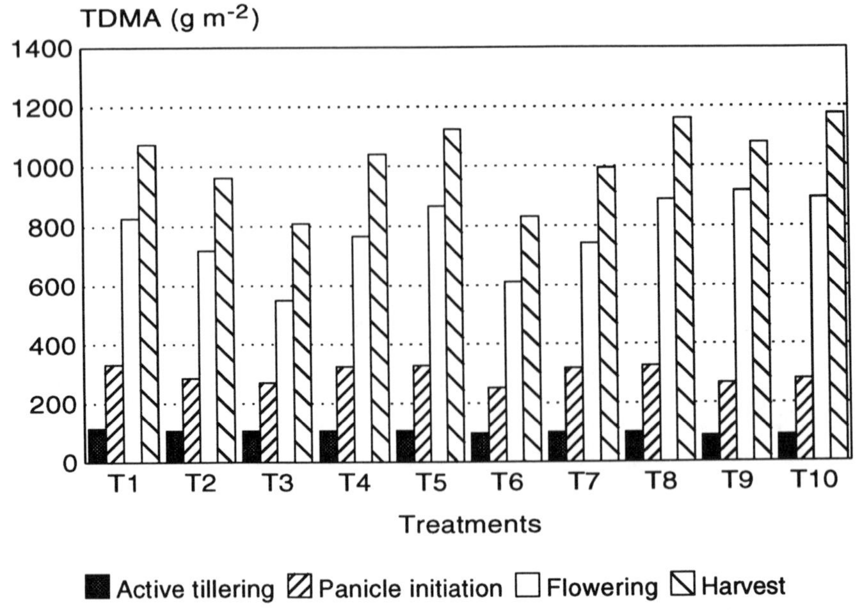

Figure 2. Total dry matter accumulation (TDMA, $g\,m^{-2}$) (medium-duration cultivars).

tively. Treatments T3 and T6 followed the same trend as that of short-duration cultivars (Figure 2).

Significant positive correlation existed between grain yield and total dry matter accumulation at panicle initiation (0.739**), flowering (0.984**) and harvest (0.994**). The correlation studies showed that the total dry matter accumulation at active tillering had no influence on grain yield probably because of transplanting shock and restricted root growth (Natarajan 1989).

Chlorophyll

Rao et al. (1986) reported that chlorophyll content plays a major role in controlling grain growth rate and grainfilling process, since nearly 74% of grain carbohydrate was translocated from photosynthesis after flowering (Cock and Yoshida 1972). A higher chlorophyll content in the reproductive phase may also be an indicator of high net photosynthetic rate and high grain yield (Liu 1980). A linear relationship between N and chlorophyll content in rice leaves was reported by Ishizuka (1976).

The total chlorophyll content in the short-duration cultivars at different phenological stages of crop growth revealed that treatments T7 and T8 recorded highest values (2.20 and 2.14 $mg\,g^{-1}$ respectively) at panicle initiation since these two treatments received 60% of total N at active tillering. At flowering, the highest chlorophyll content (2.82 $mg\,g^{-1}$) was estimated in T10 (0:50:50:0) (Figure 3). Medium-duration cultivars showed the same trend as that of short-duration cultivars. The treatments T8 (15:60:25:0) and T10 (0:50:50:0) showed that the chlorophyll content was above

Table 2. Influence of N split applications on soluble protein content (mg g^{-1}) (short-duration cultivars).

Treatment	Stages			
	Active tillering	Panicle initiation	Flowering	Harvest
T1	9.30	16.63	13.42	4.11
T2	7.08	14.96	13.02	6.32
T3	7.27	12.90	11.08	5.62
T4	7.97	16.80	12.46	5.45
T5	7.36	17.72	14.66	4.60
T6	6.19	12.35	10.40	5.72
T7	5.76	18.09	13.32	5.93
T8	6.03	18.33	15.08	4.80
T9	4.99	13.50	13.94	5.91
T10	5.12	14.65	14.15	5.09
	T×V	T×S	V×S	
CD (0.05)	0.01	0.02	0.01	

Table 3. Influence of N split application on soluble protein content (mg g^{-1}) (medium-duration cultivars).

Treatment	Stages			
	Active tillering	Panicle initiation	Flowering	Harvest
T1	8.95	17.63	14.22	5.06
T2	7.93	15.89	13.46	6.85
T3	7.42	13.63	11.94	6.21
T4	7.58	17.63	13.27	6.45
T5	7.47	18.10	15.00	5.70
T6	6.41	13.31	11.16	6.62
T7	6.48	18.47	14.29	6.81
T8	6.07	18.53	15.77	6.11
T9	5.68	14.38	14.07	6.44
T10	5.84	14.86	14.61	6.05
	T×V	T×S	V×S	
CD (0.05)	0.01	0.02	0.01	

2.0 mg g^{-1}, both at panicle initiation and flowering (data not shown). Significant positive correlation existed between grain yield and chlorophyll content at panicle initiation (0.682**).

Photosynthetic efficiency

The relationship between N content and photosynthetic efficiency has been well documented (Tsunoda 1984). Ingram et al. (1991) found that the photosynthetic rate correlates with grain yield, with the strongest correlations observed at panicle initiation and flowering. Dingkuhn et al. (1992b) also stated that the photosynthetic rate showed a strong correlation with grain yield. Makino et al. (1983) reported a direct correlation between soluble protein and carbon-fixing enzyme RuBP-carboxylase. Dey and Rao (1989) stated that a reduction in photosynthesis was due to the combination of reduction in N content per unit leaf area (69%) and stomatal conductance (31%).

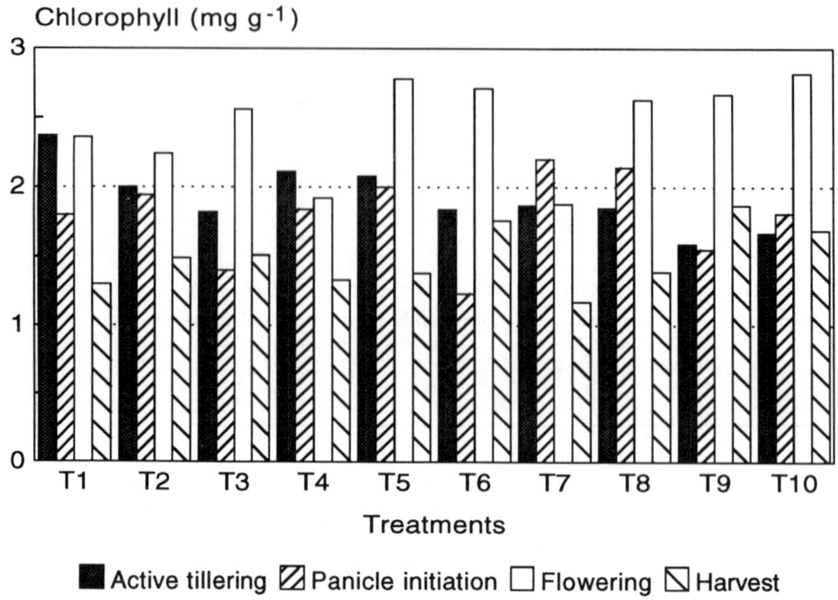

Figure 3. Influence of N split applications on chlorophyll content (mg g^{-1}) (short-duration cultivars).

The soluble protein content of short-duration cultivars estimated at all s'ages of crop growth showed that T8 (15:60:25:0) exhibited a higher value, both at panicle initiation and flowering. In these treatments, 85% of total N was applied at active tillering and panicle initiation. The treatments T3 and T6 recorded lowest values (12.90 and 12.35 mg g^{-1}) at panicle initiation. In these treatments, no N was applied at active tillering. Even though treatments T3 and T6 received 50 and 60% of total N respectively, at panicle initiation the deficit of N at active tillering was not compensated (Table 2). The same trend was observed in medium-duration cultivars. A significant positive correlation between grain yield and soluble protein content at panicle initiation (0.686**) and flowering (0.967**) was found (Table 3).

Foliar N-uptake

Dash et al. (1994) reported a higher N-uptake at panicle initiation by the rice cultivar IR36 when N was supplied at active tillering. Dingkuhn et al. (1992b) explained that the photosynthetic rate depends on foliar N concentration during the ripening phase. Makino et al. (1984) stated that the rate of leaf photosynthesis was promoted by the increase in leaf N content. Chau and Bhargava (1993) reported that the photosynthetic rate and assimilate production depend on foliar N concentration and the same resulted in an increased sink size (panicle and spikelet number).

The foliar content varied depending upon the phenological stage of the crop and the amount of N applied at previous stages. The data on the foliar N taken up at active tillering revealed that the highest N-uptake of 1.70 and 1.63 g m^{-2} was recorded in

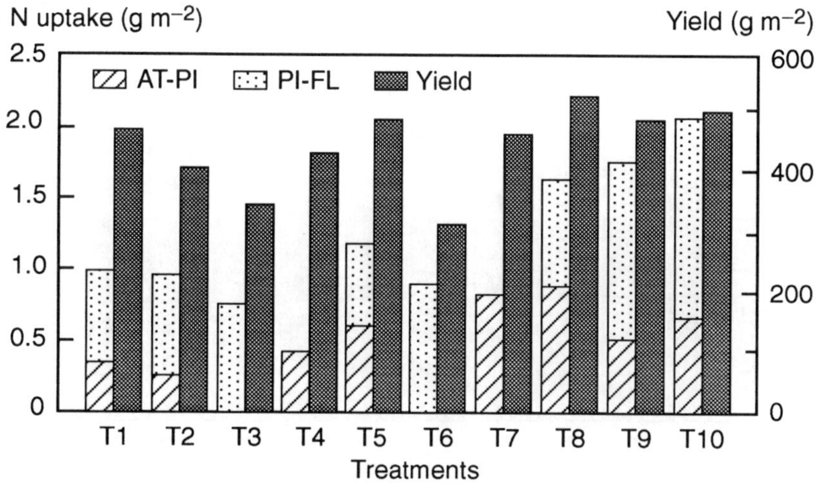

Figure 4. Influence of N split applications on leaf N-uptake and grain yield (g m^{-2}) (short-duration cultivars). AT = active tillering; PI = panicle initiation; FL = flowering.

treatments T7 (15:60:0:25) and T8 (15:60:25:0) respectively (Figures 4 and 5). At flowering, the highest N-uptake (2.55 g m^{-2}) was recorded in T10 (0:50:50:0) followed by T8 (2.36 g m^{-2}). The amount of N taken up during active tillering to panicle initiation had a significant positive correlation (0.851**), while no such correlation existed between the amount of N-uptake between panicle initiation and flowering and grain yield.

Grain yield and yield components

Penning de Vries et al. (1988) reported that maximum grain yield was obtained in treatments where N was applied in splits at suitable growth stages. Daradjat et al. (1994) reported that N fertilizer applied after panicle initiation was not efficiently used for grain yield formation. Rao et al. (1994) reported that delay in N application beyond the active tillering drastically reduced grain yield by reducing the number of panicles. They further stated that application of N in both the vegetative and panicle development phases resulted in a higher number of panicles, with a large percentage of filled grains and, therefore, high yields. Yoshida (1972) reported that there was a good correlation between the number of spikelets per square metre and total N-uptake at heading. Akita (1989) examined the relationship between spikelet number and N absorbed before heading and found that high N-uptake before heading was always related to high sink formation. Yoshida and Parao (1976) emphasized that more than 80% of the variation in yield of short-duration varieties could be attributed to the spikelet number per square metre.

Treatments T8 (15:60:25:0) and T10 (0:50:50:0) recorded high numbers of spikelets when compared with T1 (50:25:25:0). The higher number of spikelets and filled grains

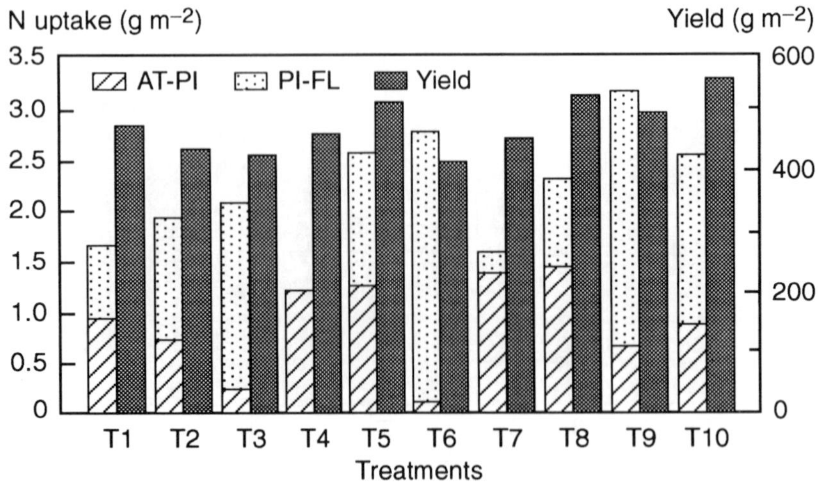

Figure 5. Influence of N split applications on leaf N-uptake and grain yield (g m^{-2}) (medium-duration cultivars). AT = active tillering; PI = panicle initiation; FL = flowering.

alone contributed to the yield improvement in T8 and T10, and it can be understood from their significant positive correlation with grain yield, i.e. 0.982** and 1.000** respectively. In short-duration cultivars, treatment T8 ranked first recording 11% yield increase over the treatment T1 (50:25:25:0). Treatment T8 was followed by T10 registering a 6% higher yield when compared with T1 in which 50% of the N was applied as a basal application. In medium-duration cultivars, T10 (50% of N at active tillering and 50% of N at panicle initiation) recorded the highest grain yield followed by T8 (15:60:25:0), with a yield increase of 13% and 10% respectively over T1 (Figure 4). This indicates that N applied at higher rates (85% and 100%) at active tillering and panicle initiation increased grain yield. Treatments T8 and T10 received only 15% and 0% of N as basal application, while T1 received 50% of the total N fertilizer as basal. This clearly indicates that N applied in larger quantities at transplanting was not superior to treatments which received lower levels, 15% or 0% of N as basal. A close observation of grain yield reveals the fact that treatments receiving maximum N from active tillering to panicle initiation, i.e. T8 and T10 (85% and 100% N respectively) performed better than T1. The reason for higher yield of the treatment T8 in which 85% of total N was applied at active tillering and panicle initiation in short-duration cultivars could be that the determination of panicle number and spikelet number occur almost simultaneously and N utilization is rapid and rather constant from active tillering till panicle initiation (Mamaril and Villapando 1984). But in medium-duration cultivars, two distinct periods of high N utilization are evident, at rapid tillering and the second peak of N utilization at panicle initiation. Hence, the N applied at the ratio of 50:50 at active tillering and panicle initiation helps to realize higher yields through higher N utilization than normal practice of applying 50% N as basal.

Grain yield in treatments T3 and T6 was reduced, with 27% and 34% respectively over T1 in the short-duration cultivars, while it was 11% and 13% in medium-duration cultivars. Both these treatments (T3 and T6) have not received any N at active tillering. The results of these two treatments support the earlier statement that major amounts of N are required between active tillering and panicle initiation. Even though T3 and T6 received 50% and 60% N respectively at panicle initiation, the N deficiency occurring at active tillering could not be compensated by N application at a later stage. The yield reduction in the medium-duration group when compared with short-duration cultivars in treatments T3 and T6 further supports the conclusion that N at active tillering is crucial for short-duration cultivars (Figure 5).

Conclusions

This investigation revealed that the quantity of N applied in the late vegetative (active tillering) and early reproductive phase (panicle initiation) increased foliar N content during the reproductive and ripening phase of crop growth. The higher foliar N-uptake between active tillering and panicle initiation increased the photosynthetic efficiency and, thereby, grain yield through higher dry matter accumulation. The N deficiency occurring at active tillering could not be compensated by N application at a later stage. The N applied at basal could not compensate for N deficiency in later phases.

References

Akita S (1989) Improving yield potential in tropical rice. Pages 41–71 in Progress in irrigated rice. International Rice Research Institute, P.O. Box 933, Manila, Philippines.

Black C A (1965) Methods of soil analysis. Vol. 2, Am. Soc. of Agron. no. 9, Madison, WI, USA.

Chau N M, Bhargava S C (1993) Physiological basis of higher productivity in rice. Indian J. Plant Physiol. 34(4):215–219.

Cock J H, Yoshida S (1972) Accumulation of ^{14}C-labelled carbohydrate before flowering and the subsequent redistribution and respiration in rice plant. Proc. Crop. Sci. Soc. Jpn. 41:226–234.

Daradjat A A, Utami P K, Fagi A M (1994) Nitrogen uptake and yield formation of rice cultivar IR64 under different urea application strategies on an alluvial ultisol in West Java. Pages 89–107 in Ten Berge H F M, Wopereis M C S, Shin J C (Eds.) Nitrogen economy of irrigated rice: Field and simulation studies. SARP Research Proceedings, International Rice Research Institute, P.O. Box 933, Manila, Philippines.

Dash R N, Rao K S, Nayak S K (1994) Growth, yield and uptake of nitrogen by irrigated rice as affected by timing of fertilizer application. Pages 1–13 in Ten Berge H F M, Wopereis M C S, Shin J C (Eds.) Nitrogen economy of irrigated rice: Field and simulation studies. SARP Research Proceedings, International Rice Research Institute, P.O. Box 933, Manila, Philippines.

Dey S K, Rao C N (1989) Influence of nutrient deficiency on photosynthesis and productivity in early rice varieties. Oryza 26:317–319.

Dingkuhn M, De Datta S K, Javellana C, Pamplona R, Schnier H F (1992a) Effect of late-season N fertilization on photosynthesis and yield of transplanted and direct-seeded tropical flooded rice. I. Growth dynamics. Field Crops Res. 28:223–234.

Dingkuhn M, DeDatta S K, Pamplona R, Javellana C, Schnier H F (1992b) Effect of late season N fertilization on photosynthesis and yield of transplanted and direct seeded tropical flooded rice. II. A canopy stratification study. Field Crops Res. 28:235–249.

Ingram K T, Dingkuhn M, Novero R P, Wijangco E J (1991) Growth and CO_2 assimilation of lowland rice in response to timing and method of N fertilization. Plant and Soil 132:113–125.

Ishizuka Y (1976) Nutrio-physiology of rice plant. Pages 76–92 in The fertility of paddy soils and fertilizer applications for rice. Food and Fertilizer Technology Centre for the Asian and Pacific Region, Taiwan, ROC.

Liu Z C (1980) A study on the photosynthetic characteristics of different types of rice. Scientia Agri. Sinica 3:3–10.

Makino A, Mae T, Ohira K (1983) Photosynthesis and ribulose 1,5-bisphosphate carboxylase in rice leaves. Plant Physiol. 73:1002–1007.

Makino A, Mae T, Ohira K (1984) Effect of nitrogen, phosphorus or potassium on the photosynthetic carboxylase content in the rice leaves during expansion. Soil Sci. Plant Nutr. 30:63–70.

Mamaril C P, Villapando R R (1984) Increasing the efficiency of nitrogen fertilizer on rice. Pages 1–30 in Food and Fertilizer Technology Centre, ASPAC. Extension Bulletin No. 216. Taipei, Taiwan, ROC.

Natarajan K (1989) Transformation and fertilizer nitrogen balance in flooded rice system using ^{15}N techniques. PhD thesis, Tamil Nadu Agricultural University, Coimbatore, India.

Penning de Vries F W T, Van Keulen H, Alagos J C (1988) Nitrogen redistribution and potential production in rice. Pages 513–520 in Sinha S K, Lane P V, Bhargava S C, Agrawal R K (Eds.) International congress plant physiology. Soc. Plant Physiol. Biochem., Water Technology Centre, IARI, New Delhi, India.

Puckridge D W, Wiergiveera A, Thongbai P, Sangtong P, Sattarsart A, Kongchum M, Runtan S (1992) Nitrogen uptake by deepwater rice during the preflood and flooded phases in relation to dry matter and grain production. Field Crops Res. 27(4):315–336.

Rao D L N (1987) Slow release urea fertilizer: Effect of flood water chemistry, ammonia volatilization and rice growth in an alkali soil. Fert. Res. 13:209–227.

Rao K S, Dash R N, Murty M V R (1994) The effect of time of nitrogen application on growth and performance of rice: A process study. Pages 14–30 in Ten Berge H F M, Wopereis M C S, Shin J C (Eds.) Nitrogen economy of irrigated rice: Field and simulation studies. SARP Research Proceedings, International Rice Research Institute, P.O. Box 933, Manila, Philippines.

Rao S P, Venkateswaralu B, Rao V S (1986) Studies on grain filling and grain growth rates of different rice varieties in relation to chlorophyll content. Indian J. Plant Physiol. 29(2):160–165.

Siddiq E A (1993) Rice production strategy for the 21st century. Oryza 30:186–196.

Swaminathan M S (1989) Role of rice in global food security. Oryza 26:1–2.

Thiyagarajan T M, Sivasamy R, Ten Berge H F M (1994) Time courses of leaf nitrogen concentration required to attain target yields in transplanted rice. Pages 56–69 in Ten Berge H F M, Wopereis M C S, Shin J C (Eds.) Nitrogen economy of irrigated rice: Field and simulation studies. SARP Research Proceedings, International Rice Research Institute, P.O. Box 933, Manila, Philippines.

Tsunoda S (1984) Adjustment of photosynthetic structures in three steps of rice evolution. Pages 89–115 in Tsunoda S, Takahashi T (Eds.) Biology of rice. Japan Sci. Soc. Press, Tokyo, Japan.

Yoshida S (1972) Physiological aspects of grain yield. Ann. Rev. Plant Physiol. 23:437–464.

Yoshida S, Parao F T (1976) Climatic influence on yield and yield components of lowland rice in the tropics. Pages 471–494 in Climate and rice. International Rice Research Institute, P.O. Box 933, Manila, Philippines.

Systems approaches to improve nitrogen management in rice

B. MISHRA

Department of Soil Science, Govind Ballabh Pant University of Agriculture and Technology, Pantnagar 263145, India

Key words: crop growth model, flooded soil, fertilizer efficiency, irrigated rice, lowland rice, nitrogen application

Abstract

Nitrogen is the most limiting nutrient in rice production. As the fertilizer N use efficiency in rice is low and N losses from flooded rice fields are high causing environmental pollution, efficient management of fertilizer N is desirable from both economic and environmental perspectives. A recent survey of farmers' fields in the rice-wheat belt of northern India revealed that the crop N-grain yield relationship was poor, reflecting inefficient management of fertilizer N by the farmers. Systems approaches for nitrogen management consider the dynamics of crop N demand and soil N supply including effective utilization of initial soil inorganic N, as well as the N that becomes available through mineralization of soil organic N during crop growth period. These are appropriately included in a simple crop growth model designed to simulate rice biomass production in relation to incident solar radiation and nitrogen supply from soil and fertilizer. The model is essentially based on the SARP model ORYZA_0 but has additional features for more realistic prediction of N availability from soil, fertilizer, manure and crop residues, and daily crop N-uptake. The model can be used as a tool to evaluate different N management strategies for rice in diverse soil-crop-weather conditions and a site-specific fertilizer application schedule for each situation can be formulated to optimize fertilizer N use in rice.

Introduction

Objective

Nitrogen is the most limiting resource for rice production in Asia. Though the use of N fertilizers has increased remarkably during the last two decades, fertilizer N efficiency in rice remains poor. Even in field experiments in which standard agronomic practices are followed, flooded rice generally recovers 20–40% of applied N, whereas upland crops normally recover 40–60% (Vlek and Byrnes 1986). The poor fertilizer N recovery efficiency in rice is largely due to loss of N from the soil-floodwater system following fertilizer application. The N loss from rice fields is not only an economic loss, but also a source of environmental pollution. Thus, judicious N management in rice is desirable, both from economic and environmental perspectives.

Technically, development of slow-release fertilizers for enhancing fertilizer N efficiency in rice looks promising but the products developed so far are quite expensive and uneconomical for rice farmers. Thus, the challenge of improving N use efficiency in rice has to be addressed by agronomic management options. This paper will focus

M.J. Kropff et al. *(eds.): Applications of Systems Approaches at the Field Level, 295–304.*

Table 1. Recovery of ^{15}N urea by rice.

Location	N applied ($kg\,ha^{-1}$)	Recovery of added N (%)	Reference
Cuttack	60	26.7	IAEA (1970)
Hyderabad	100	36.6	Rao and Shinde (1985)
Ludhiana	80	26.0	Katyal et al. (1985)
Pantnagar	60	38.5	Dhyani and Mishra (1993)
	120	37.2	

Table 2. Effects of timing of N application on fertilizer use efficiency in rice (*Source:* Dhyani and Mishra 1995).

Timing of N application ($60\,kg\,N\,ha^{-1}$)	Fertilizer use efficiency ($kg\,grain\,kg^{-1}\,N$ applied)	
	1987	1988
Transplanting	17.1	16.2
Tillering	21.4	21.5
Panicle initiation	21.8	21.7
Flowering	13.8	21.7

on the development of systems approaches and models to improve N management in rice.

Fertilizer N efficiency in rice

The term fertilizer N efficiency should be defined to make its meaning clear and unambiguous. Fertilizer N efficiency or N use efficiency may be expressed as agronomic efficiency which is the product of fertilizer N recovery efficiency (i.e. the increase in crop N per unit of applied N) and the physiological efficiency (i.e grain yield increase per unit increase in crop N). In rice, it is the fertilizer N recovery efficiency that is low and is considered to be a major constraint to achieving high yield per unit of added fertilizer N. The physiological efficiency of N in rice is greater than in high-protein bread wheat (Cassman et al. 1993).

The fertilizer N recovery efficiency in rice is not only low, but also varies with location, season, crop variety, soil type, weather and management (Table 1). Any factor that can influence N loss processes or plant root capacity for absorbing N has the potential to alter the fertilizer N recovery efficiency. Because of the involvement of a large number of factors relating to soil, crop, environment and management in N utilization by the crop, a systems approach is desirable to evolve appropriate N management strategies in rice. Systems approaches for enhancing the fertilizer N efficiency should consider the following:

(1) Defining daily crop N demand for optimum growth and desired yield.
(2) Quantifying soil N supply capacity including initial inorganic N present in soil at the time of rice transplanting and that which would become available daily through mineralization of organic N.

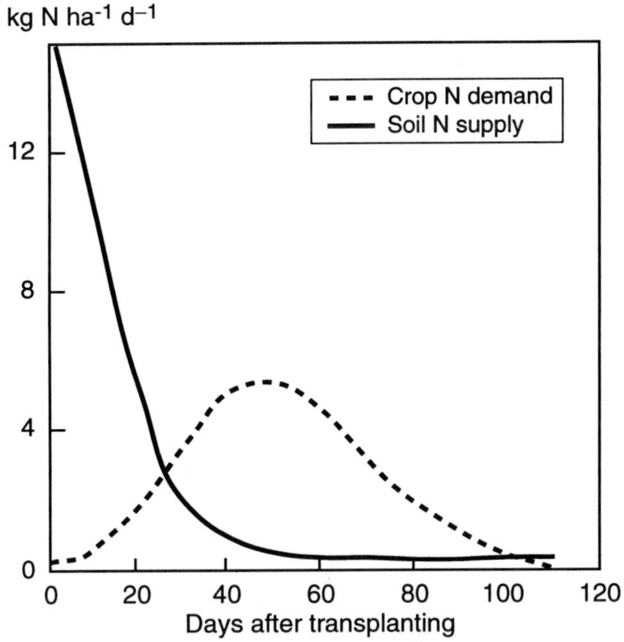

Figure 1. Dynamics of the rice crop N demand and the soil N supply (*Source:* Cassman et al. 1994).

(3) Estimating the recovery efficiency of fertilizer N applied at different times during pre-flowering and post-flowering stages.
(4) Calculating fertilizer N requirement to meet the crop N demand that is in excess of the soil N supply capacity (taking into account 1, 2 and 3 above).
(5) Adjusting the number of splits and timing of each fertilizer application for maximum agronomic efficiency and minimum N loss from the soil-floodwater system.

Crop N demand versus soil N supply

The rice crop has a characteristic N demand pattern (Figure 1). During early stages, crop N demand is low, but it increases rapidly during exponential growth phase attaining a peak around panicle initiation. The crop N demand decreases after flowering. The physiological efficiency for grain production of N absorbed by the crop at different stages is, however, not the same. In high production environments, the final grain production is sink-limited and the sink capacity is determined by the amount of N absorbed until panicle initiation (Yoshida 1981). Recovery efficiency of fertilizer N applied during the period between panicle initiation and flowering is high due to well developed root system, but its agronomic efficiency is relatively low due to sink limitation (Table 2). Optimum crop N demand for maximum fertilizer efficiency is, therefore, difficult to define accurately.

The pattern of soil N supply is just opposite to the pattern of crop N demand. It is highest when the crop is transplanted and decreases gradually attaining the lowest value after panicle initiation when the crop N demand is highest (Figure 1). It is noteworthy that in areas where the rice crop follows an upland crop (such as wheat in northern India), any amount of nitrate-N that might have accumulated in the soil before flooding the field for land preparation for rice is lost through leaching and denitrification. Accumulation of substantial quantities of nitrate-N in the soil before rice planting has been reported and growing a leguminous crop to capture this nitrate-N has been suggested (George et al. 1992). Practical difficulties involved in growing a crop during hot and dry summer months in northern India, however, restrict the adoption of this practice by farmers.

The need for fertilizer N application arises when soil N supply falls short of the crop N demand. Fertilizer N, however, does not stay for long in a flooded soil. The residence time of fertilizer N in the flooded soil is usually very short (1 – 2 weeks) due to its susceptibility to loss through ammonia volatilization and denitrification (Vlek and Byrnes 1986). Generally, a small fraction of the added fertilizer N is immobilized by soil microflora.

Status of current fertilizer recommendations

Presently, the recommendations for fertilizer N management that are made to rice farmers by agronomists and extension specialists do not account for the dynamic nature of the balance between crop N demand and soil N supply in the flooded rice ecosystem. For example, in India 50% of the fertilizer N is usually recommended at transplanting, though crop N demand is very low during crop establishment when the root system is not yet developed to capture the applied N. Moreover, soil N supply is usually sufficient to meet the crop N demand during 1–2 weeks after transplanting. This results in the loss of a major portion of the fertilizer N applied at transplanting through ammonia volatilization and denitrification. In a study using ^{15}N urea, Dhyani and Mishra (1993) observed that only 24.0–25.5% of the urea-N applied at transplanting could be recovered by the rice plants, whereas 37.6–39.4% was recovered when the same quantity was applied at panicle initiation.

Another criticism of the current fertilizer recommendation is its rigidity disregarding the differences in soil N supply and crop N demand balance from one location to another, and at the same location from one season to the other (Table 3).

Material and methods

Simulation model

The dynamics of crop N demand and soil N supply can be quantified by simulation models. For example, the ORYZA_0 model, developed by the SARP (Simulation and Systems Analysis for Rice Production) group, is a simple crop growth model that considers the dynamics of crop N demand and simulates biomass production in rice using leaf N status as the basis for intercepting solar radiation by the crop canopy

Table 3. Soil N supply capacity for rice.

Location	Season	Crop N without N fertilization ($kg\,ha^{-1}$)	Reference
Cuttack	Wet 1991	57.8	Rao et al. (1995)
	Dry 1992	95.8	
Delhi	Wet 1978	32.0*	Sharma and Prasad (1980)
Ludhiana	Wet 1983	44.0*	Bhandari et al. (1992)
Pantnagar	Wet 1980	40.7*	Chauhan and Mishra (1989)
	Wet 1987	55.4	Dhyani and Mishra (1995)
Ranchi	Wet 1987	66.2*	Mishra et al. (1994)
Thanjavur	Wet 1991–2	65.9	Sivasamy et al. (1994)
	Wet 1992–3	54.2	

*Excluding root N.

(Ten Berge et al. 1994). The model, however, considers an oversimplified and unrealistic situation of daily N application and a constant rate of soil N supply. This simple model was modified to account for a more realistic fertilizer application schedule. This, in turn, would make fertilizer recommendations more realistic and practical. The proposed model has the following additional features as compared to ORYZA_0 model:

- It accounts for the initial inorganic N present in the soil at the time of rice transplanting.
- Soil mineralization is expressed by first-order kinetics. Similarly, N mineralization from organic manures or crop residues can be accounted for.
- The entire N pool (soil + fertilizer) is subjected to losses by ammonia volatilization and denitrification (described empirically as a first-order reaction to maintain simplicity).
- Fertilizer mixing efficiency defines the distribution of the fertilizer between floodwater and soil.
- Urea hydrolysis is accounted for by using a first-order kinetics equation.
- Limitations to N uptake by rice roots are simplified and availability of soil N for actual uptake by roots is limited due to root accessibility factor to the same extent as fertilizer N.
- Photosynthetic efficiency of green leaf (N dependent) is assumed to remain constant throughout plant growth period irrespective of vegetative or reproductive phase.
- Actual dates of fertilizer N application are input to the model.
- Harvest index is calculated from a function based on observed data relating grain yield to crop biomass at maturity and cumulative solar radiation during reproductive phase (flowering to maturity).

The equation used for calculating daily dry matter production (crop growth rate) on the basis of leaf N status and incident solar radiation is essentially the same as described by Ten Berge et al. (1994) except the simplification mentioned above. It can be expressed as:

$$G = (p/R)(1 - e^{-ER/(pN_L)})$$

where

G = crop growth rate ($g\,m^{-2}\,d^{-1}$);
p = initial leaf N use coefficient (g dry matter g^{-1} leaf N d^{-1});
N_L = N content of the leaf canopy ($g\,m^{-2}$ ground surface);
E = initial solar radiation use coefficient (g dry matter MJ^{-1}); and
R = daily incident solar radiation ($MJ\,m^{-2}\,d^{-1}$).

For calculating the rate of soil N mineralization or rate of urea hydrolysis or rate of N loss, the following first-order kinetic equation is used:

$$-dS/dt = kS$$

with k being the first-order rate constant and S the state variable.

Evaluation experiment

From a field experiment conducted during the wet season of 1987 at the Crop Research Centre of G.B. Pant University of Agriculture and Technology, Pantnagar, India (Dhyani and Mishra 1995), 15 data sets representing different quantities (0, 60, 120, 180 and 240 kg N ha^{-1}) and different schedules of fertilizer (urea) application (at transplanting, maximum tillering, panicle initiation and flowering) were selected to evaluate the model with respect to N-uptake, biomass production and grain yield. Model parameters were determined by fitting the model to experimental data using an optimization procedure.

Results and discussion

Model evaluation

Total biomass of the rice crop recorded at maturity (panicle + stem + leaves + roots) in 15 treatments representing different quantities and schedules of fertilizer N application ranged from 9.21 ton ha^{-1} in the unfertilized plot to 17.36 ton ha^{-1} in the plot that received 240 kg N ha^{-1} (three splits). Corresponding values of biomass for the same treatments simulated by the modified model were 8.81 – 18.09 ton ha^{-1}. A good agreement between the simulated values and observed values of crop biomass was obtained ($r = 0.948$, Figure 2), with deviations ranging between −4.3% and 9.8%. Similarly, the model described the effect of the quantity and schedule of fertilizer N application on grain yield production of rice accurately ($r = 0.946$, Figure 3). Deviation of simulated grain yield from the observed data was minimal, from −5.19% to 5.91%. In spite of the simplicity of the model, N-uptake in these treatments was also reasonably well simulated ($r = 0.95$, Figure 4). The deviation of simulated N-uptake from the observed data was, however, higher (−13.45% to 15.66%) than that observed

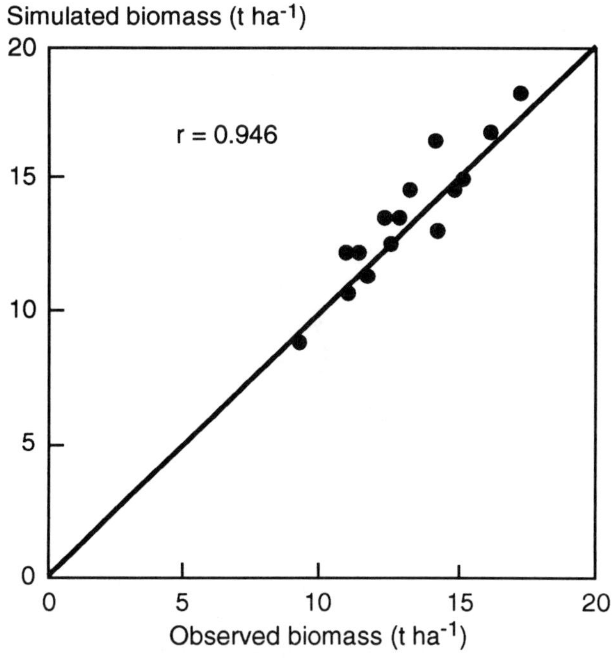

Figure 2. Observed versus simulated biomass of rice (ton ha^{-1}) at maturity.

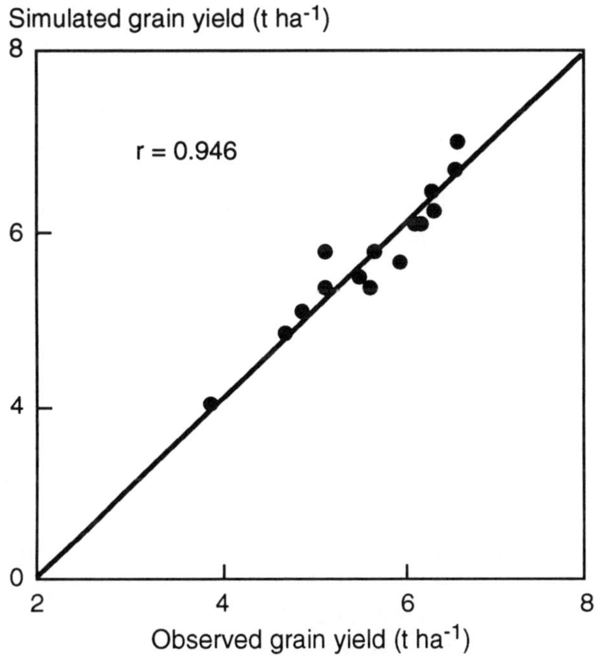

Figure 3. Observed versus simulated grain yield of rice (ton ha^{-1}).

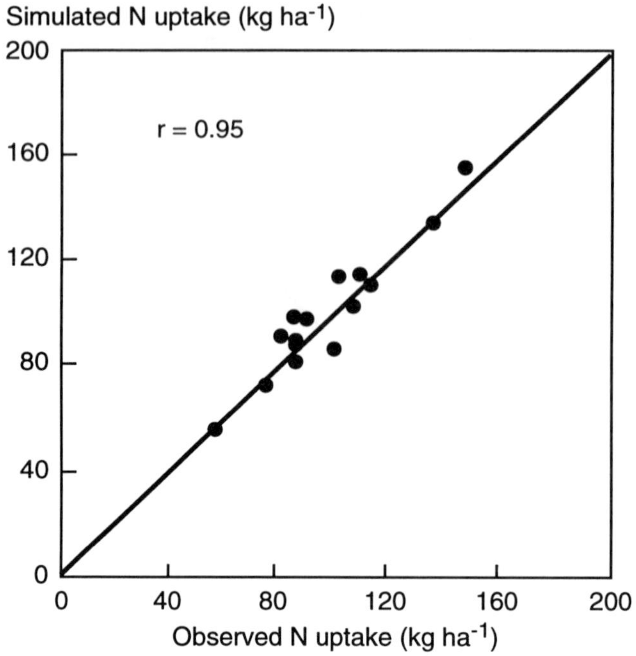

Figure 4. Observed versus simulated N-uptake by the crop at maturity (kg ha^{-1}).

in the case of biomass or grain yield. This is not unexpected in view of the simple empirical expression used for calculating N-uptake by the crop. Further requirements based on root distribution and its characteristics associated with N absorption would be needed for better prediction of N-uptake by rice (Mishra and Kirk 1994). The loss of N following fertilizer application is also predicted satisfactorily by the model as the amount of simulated N loss compared well with the measured data (Dhyani and Mishra 1993). Extensive testing of the model at several locations representing diverse soil and weather conditions would be desirable for evaluating its applicability.

Sensitivity analysis

Sensitivity analysis of the modified model with respect to initial conditions and input parameters was carried out to assess the model performance. Data given in Table 4 show the sensitivity of the model to initial soil organic N, with and without fertilizer N application.

As seen from the data given in Table 5, the model predicts that the current recommendation of three splits (50% basal + 25% at maximum tillering + 25% at panicle initiation) of fertilizer N is not the best N management option. Higher grain yield and better N use efficiency can be achieved by rescheduling the N application to either 25% basal + 50% at maximum tillering + 25% at panicle initiation or 25% basal + 37.5% at maximum tillering + 37.5% at panicle initiation, which will give

Table 4. Simulation results of initial soil organic N level on rice grain yield and crop N content, with and without N application.

Initial soil organic N ($kg\,ha^{-1}$)	No N application		$120\,kg\,N\,ha^{-1}$	
	Grain yield ($ton\,ha^{-1}$)	Crop N ($kg\,ha^{-1}$)	Grain yield ($ton\,ha^{-1}$)	Crop N ($kg\,ha^{-1}$)
100	2.31	25.2	5.54	82.5
200	3.14	39.5	5.86	95.2
300	4.00	53.4	6.15	107.8
400	4.59	67.3	6.38	120.2

Table 5. Simulation results of different N application schedules (applications: basal-tillering-panicle initiation) for $120\,kg\,N\,ha^{-1}$ on crop biomass, grain yield and crop N at maturity (initial soil organic N is $300\,kg\,ha^{-1}$).

N application schedule	Crop biomass ($ton\,ha^{-1}$)	Grain yield ($ton\,ha^{-1}$)	Crop N ($kg\,ha^{-1}$)
60-30-30*	14.69	6.07	108.0
30-60-30	15.43	6.27	115.9
30-30-60	15.27	6.23	120.8
40-40-40	15.21	6.21	115.5
30-45-45	15.45	6.28	119.0
45-45-30	15.08	6.18	112.0
45-30-45	15.04	6.17	114.9

*Standard agronomic recommendation.

$0.20\,ton\,ha^{-1}$ higher grain yield and 10.0% higher N-uptake than the standard agronomic recommendation at the N level of $120\,kg\,ha^{-1}$.

Conclusions

The simulation model based on systems approaches proposed here describes the dynamics of crop N demand and soil N supply and simulates biomass production, N-uptake and grain yield of rice satisfactorily. It can be used as a tool to evaluate different N management strategies, for rice in diverse soil-crop-weather conditions. Likewise, it can help formulating site-specific N application schedule for each situation to optimize fertilizer N use in rice.

References

Bhandari A L, Anil Sood, Sharma K N, Rana D S (1992) Integrated nutrient management in rice-wheat system. J. Indian Soc. Soil Sci. 40:342–347.

Cassman K G, Kropff M J, Gaunt J, Peng S (1993) Nitrogen use efficiency of rice reconsidered: What are the key constraints? Plant and Soil 155/156:359–362.

Cassman K G, Kropff M J, Yan Zhen-De (1994) A conceptual framework for nitrogen management of irrigated rice in high-yield environments. Pages 81–96 in Virmani S S (Ed.) Hybrid rice technology: New developments and future prospects. International Rice Research Institute, P.O. Box 933, Manila, Philippines.

Chauhan H S, Mishra B (1989) Fertilizer use efficiency of amended urea materials in flooded rice. J. Agric. Sci. (Camb.) 112:277–281.

Dhyani B P, Mishra B (1993) The fate of fertilizer nitrogen in a flooded rice field. J. Indian Soc. Soil Sci. 41:770–771.

Dhyani B P, Mishra B (1995) Scheduling nitrogen fertilization of rice on a Mollisol. Oryza 32 (in press).

George T, Ladha J K, Buresh R J, Garrity D P (1992) Managing native and legume fixed nitrogen in lowland rice based cropping systems. Plant and Soil 141:69–91.

IAEA - International Atomic Energy Agency (1970) Rice fertilization. Joint FAO/IAEA Division of Atomic Energy in Food and Agriculture. Technical Report Series 108, Vienna, Austria.

Katyal J C, Singh B, Vlek P L G, Craswell E T (1985) Fate and efficiency of nitrogen fertilizers applied to wetland rice. II. Punjab, India. Fert. Res. 6:269–290.

Mishra B, Kirk G J D (1994) A mechanistic model of N dynamics in flooded soil. Pages 293–301 in Ten Berge H F M, Wopereis M C S, Shin J C (Eds.) Nitrogen economy of irrigated rice: Field and simulation studies. SARP Research Proceedings, International Rice Research Institute, P.O. Box 933, Manila, Philippines.

Mishra P C, Sarkar A K, Mishra B (1994) Relative efficiency of different nitrogen sources in transplanted rice Chotanagpur region. J. Indian Soc. Soil Sci. 42:333–335.

Rao K S, Dash R N, Ten Berge H F M (1995) Numerical optimization of N fertilizer application to irrigated rice at Cuttack (India) in wet and dry seasons. Pages 9–25 in Thiyagarajan T M, Ten Berge H F M, Wopereis M C S (Eds.) Nitrogen management studies in irrigated rice. SARP Research Proceedings, International Rice Research Institute, P.O. Box 933, Manila, Philippines.

Rao K V, Shinde J E (1985) Transformation of labelled urea and its modified forms in tropical wetland rice culture. Plant and Soil 85:413–421.

Sharma S N, Prasad R (1980) Relative efficiency of urea, nitrification inhibitor treated urea and sulphur coated urea for rice. Indian J. Agron. 25:403–409.

Sivasamy R, Thiyagarajan T M, Ten Berge H F M, (1994) Nitrogen and rice: Uptake and recovery of applied nitrogen. Pages 31–55 in Ten Berge H F M, Wopereis M C S, Shin J C (Eds.) Nitrogen economy of irrigated rice. SARP Research Proceedings, International Rice Research Institute, P.O. Box 933, Manila, Philippines.

Ten Berge H F M, Wopereis M C S, Riethoven J J M, Drenth H (1994) Description of ORYZA_0 modules (version 2). Pages 43–55 in Drenth H, Ten Berge H F M, Riethoven J J M (Eds.) ORYZA simulation modules for potential and nitrogen-limited production. SARP Research Proceedings, International Rice Research Institute, P.O. Box 933, Manila, Philippines.

Vlek P L G, Byrnes B H (1986) The efficiency and loss of fertilizer N in lowland rice. Fert. Res. 9: 131–147.

Yoshida S (1981) Fundamentals of rice crop science. International Rice Research Institute, P.O. Box 933, Manila, Philippines. 269 p.

Use of simulation models to optimize fungicide use for managing tropical rice blast disease

S.B. CALVERO and P.S. TENG

Entomology and Plant Pathology Division, International Rice Research Institute (IRRI), P.O. Box 933, 1099 Manila, Philippines

Key words: disease management, rice blast, simulation, net return, fungicides, spray strategies, BLASTSIM.2, ORYZA1

Abstract

A modified blast simulation model, BLASTSIM.2 with a fungicide submodule named FUNGICID, was coupled to the rice crop growth model ORYZA1 to evaluate the 92 spray strategies intended for managing rice blast disease. These strategies were evaluated using IR72 crop parameter data taken from 1991–92 dry and wet season experiments at the IRRI experimental farm, and using as disease parameter inputs, 100 initial lesion numbers per hill and 0.2% initial severity, occurring either as an early (27 DAS) or late epidemic (41 DAS). Simulated values of weight of storage organs and above-ground biomass, estimated yield loss, area under disease progress curve (AUDPC), and net return from fungicide use served as criteria to identify optimum strategies at four disease-planting scenarios: early disease-dry season planting (early-dry), early-wet, late-dry, and late-wet. Using cluster analysis, three spray strategy groups (SSG) were obtained in all disease-planting scenarios for both years, except in 1991 for a late epidemic in wet season with two SSG. Principal component analysis characterized the three SSGs as optimum strategies for group 1, moderately favourable for group 2, and not favourable for group 3. Real-time field validation of the coupled model with the different spray strategies and improvement of FUNGICID in BLASTSIM.2 should be done to further assess the simulated blast management tactics.

Introduction

Wherever rice is grown, blast is one disease capable of causing tremendous yield reductions over a wide area in both lowland and upland rice ecosystems (Teng 1994). The sexual (*Magnaporthe grisea* Barr.) and the asexual forms (*Pyricularia grisea* (Cooke) Sacc.) of the fungus are both pathogenic and potentially destructive (Ou 1985). The sexual form is predominantly occurring in temperate environments, while the asexual form occurs in tropical environments. The severity of the disease in certain production situations is exacerbated by two factors – prevailing environmental conditions and crop management practices (Teng et al. 1991). This explains why some rice production areas are blast hot spots, while others, seemingly similar, are less affected by the disease.

A large amount of information on blast disease management is available from literature. Most of it deals with the epidemiology of blast, rice cultivar distribution relative to the pathogen's population dynamics, and crop production profiles of several rice areas. However, there still exist knowledge gaps that have to be filled to develop effective management of this disease. Teng (1994) proposes synthesizing knowledge-based information to generate a so-called 'blast disease management toolkit' which

M.J. Kropff et al. *(eds.): Applications of Systems Approaches at the Field Level, 305–320.*

is intended for use by both farmers and decision makers. This toolkit has four types of tools: knowledge, physical, communication, and policy tools that must have discernible epidemiological implications, so that any disease strategy will be sustainable (Teng 1994). Part of IRRI's blast research aims at the realization of this toolkit (P.S. Teng, IRRI, pers. commun.).

An important component of the knowledge tool is systems analysis which the use of simulation models falls into. This paper tackles three computer simulation models developed at IRRI: ORYZA1, a rice crop growth model (Kropff et al. 1994); BLASTSIM.2, a model simulating tropical rice-leaf blast pathosystem (Teng and Calvero 1991); and FUNGICID, a model that estimates fungicide persistence in a rice canopy and fungicide efficacy on blast (S.B. Calvero and P.S. Teng 1995, IRRI, unpubl. data). The three programs were coupled together to mimic actual crop production situations where blast occurs spontaneously, and with fungicide application as one option for crop management practices.

There is a need for modelling of production scenarios in order to have a better understanding of how a rice crop and a *P. grisea* isolate interact in natural systems. This understanding, in turn, can serve to explore how blast can be managed effectively, especially using fungicides. Another concern of this paper is for the coupled models to be part of the blast toolkit and for them to be applied in a more practical disease management setting.

Material and methods

Background on the ORYZA1 model

ORYZA1 is a model for rice growth and production in an irrigated lowland ecosystem (Kropff et al. 1994). This model calculates crop production under optimal growth conditions, light, temperature, and varietal characteristics that determine phenological, morphological, and physiological processes on a daily basis (Kropff et al. 1994). As shown in Figure 1, rates of dry matter production of the plant organs, leaf area development, and phenological development are calculated and integrated over time to generate dry matter production of the crop in a given growing season. The dry matter produced is partitioned among the various plant organs (i.e. leaves, stems, panicles and roots).

An important feature in ORYZA1 is the simulation of leaf area growth which is described to be affected by the canopy. Kropff et al. (1994) suggested that in an open canopy increase in leaf area is driven by temperature, while in a closed canopy increase in leaf area is relative to the increase in leaf weight. Integrating the daily growth rates of the organs and leaf area results in an increase in dry matter during the growing season (Kropff et al. 1994).

Background on the BLASTSIM.2 model

BLASTSIM.2 was developed to mimic the tropical leaf blast pathosystem (Teng and Calvero 1991). It is a mechanistic model that simulates in a deterministic mode the

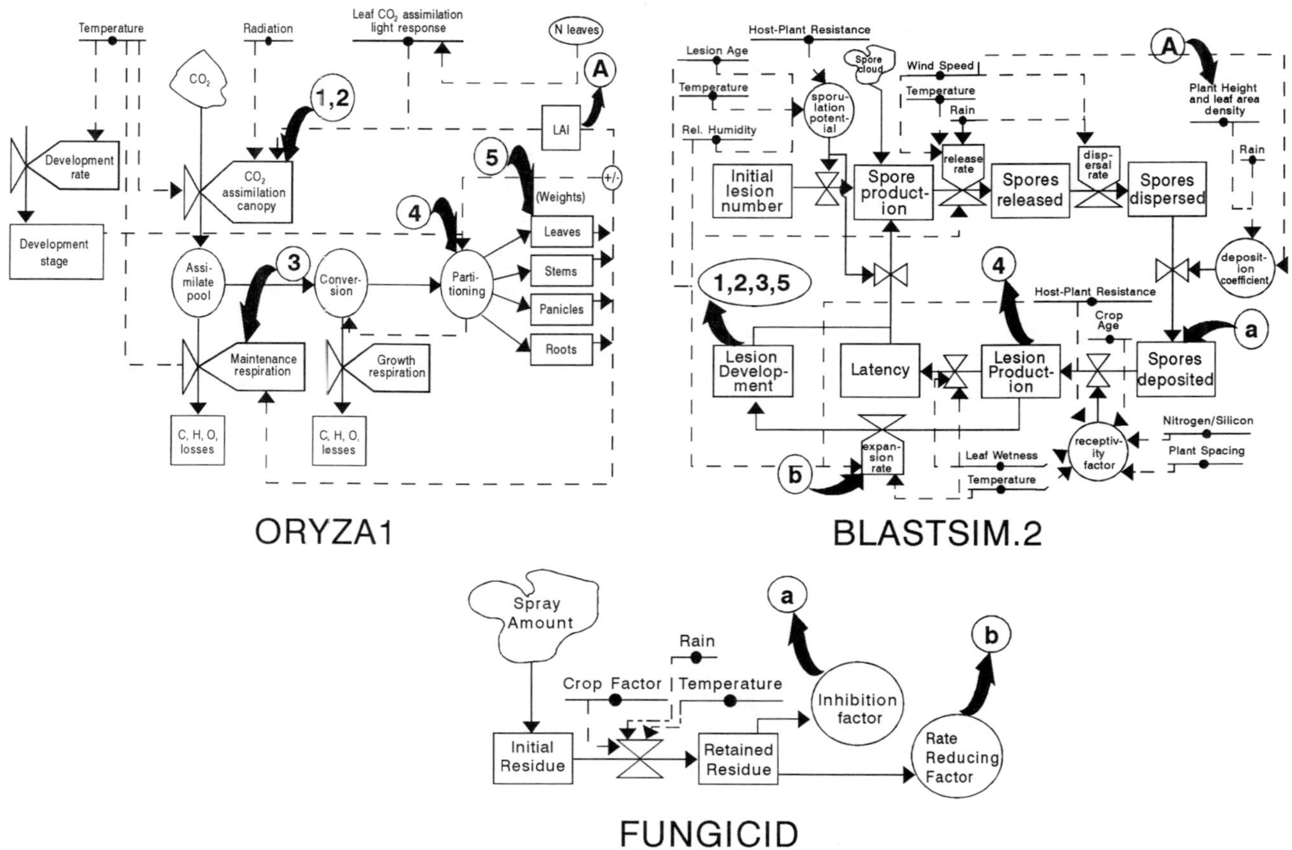

Figure 1. Relational diagram of the coupled models. Numbers, upper case, and lower case letters in circles indicate coupling points of BLASTSIM.2 to ORYZA1, ORYZA1 to BLASTSIM.2, and FUNGICID to BLASTSIM.2 respectively.

different stages of the pathogen's life cycle: sporulation, spore release, spore deposition, infection, latency, lesion formation and development (Figure 1). From these stages, the severity of the disease at a given day is calculated from lesion development expressed as lesion size.

BLASTSIM.2 requires data sets of daily weather variables (i.e. maximum and minimum temperature, rainfall, wind speed, relative humidity, maximum and minimum dewpoint temperatures) of a particular site, agronomic characteristics of a cultivar (i.e. crop height, leaf area, leaf width, cuticular resistance), and some edaphic factors (i.e. evaporation surface of the soil). Values of initial amount of disease (or disease severity) and number of lesions contributing to this initial amount, and the time this amount of disease is observed, are supplied to the model to initiate the simulation run.

One major feature of the model and an improvement from earlier disease simulators is that BLASTSIM.2 incorporates a dynamic simulation of leaf wetness duration through the DEWFOR model, developed by Luo and Goudriaan (1991). DEWFOR simulates dew formation induced by energy heat transfer (Campbell 1986). Leaf wetness duration is being used by BLASTSIM.2 as a driving parameter in leaf blast development.

Preliminary validation of BLASTSIM.2 under nursery and field conditions had been done with two blast-susceptible cultivars, IR50 and IR72 (Calvero and Teng 1992), and with a new rice plant type (Pangga 1995). To date, model validation is being conducted in Thailand under lowland conditions (S.B. Calvero, IRRI, unpubl. data).

Background on the FUNGICID model

FUNGICID was developed to quantify the efficacy of fungicides on rice blast disease and their persistence (retention) in a rice canopy. The model estimates fungicide persistence in the crop canopy through the amount of residue retained from initial application to its half-life. In the model, half-life of fungicide was set at 7.5 days (Bruhn and Fry 1982). Retention of fungicide in the crop canopy is driven by rainfall and temperature. Similarly, fungicide persistence is dependent on the initial residue retained in the canopy immediately after spray.

Redistribution of fungicide within a crop canopy is calculated in four canopy layers. In the model, the redistribution of fungicide is considered as a matrix table, $[k]_t$ (equation (1)), of four rows $(1, \ldots, i, \ldots, 4)$ and four columns $(1, \ldots, j, \ldots, 4)$ which represent the four canopy layers. Each cell in the matrix represents the proportion of fungicide $k_t(i,j)$ that remains on layer i from layer j after day t:

$$[k]_t = \begin{bmatrix} k_t(1,1) & 0 & 0 & 0 \\ k_t(2,1) & k_t(2,2) & 0 & 0 \\ k_t(3,1) & k_t(3,2) & k_t(3,3) & 0 \\ k_t(4,1) & k_t(4,2) & k_t(4,3) & k_t(4,4) \end{bmatrix} \tag{1}$$

Basically, the proportion retained on the topmost layer $k_t(1,1)$ is a function of rainfall, temperature, and cultivar which is calculated (Nokes and Young 1992) as:

$$k_t(1,1) = g_t \times d_t \times cv$$

where g_t and d_t are coefficients [$\in(0,1)$] that quantify the effects of rainfall and temperature respectively on the removal of fungicide, and cv is a coefficient [$\in(0,1)$] that corrects for cultivar effects. Equations describing g_t and d_t are shown in the Bruhn and Fry (1982) and Nokes and Young (1992) papers and will not be described here. A value of 1.0 was used for cv in the FUNGICID model.

The diagonal elements of the matrix [$k_t(i,i)$] which represent the proportion of fungicide retained in lower canopy layers i are estimated as:

$$k_t(i,i) = k_t(1,1)^{0.471} \quad i = 2, 3, 4$$

where 0.471 is the multiplication factor (Bruhn and Fry 1982) to describe fungicide loss in the lower canopy layers.

The non-diagonal elements in the matrix [$k_t(i,j)$] which represent the proportion of fungicide redeposited from upper canopy layers j onto the lower layers i are calculated as:

$$k_t(i,j) = [1 - k_t(j,j)] \times z_{(i-j)} \quad i > j$$

where $z_{(i-j)}$ describes the redistribution of fungicide on the foliage due to rainfall (Nokes and Young 1992). In FUNGICID, the values 0.05, 0.2, and 0.2 for $z_{(i-j)}$ were assigned for $(i-j)$ equal to 1, 2, and 3 respectively.

Coupling ORYZA1, BLASTSIM.2 and FUNGICID

Three types of coupling were made in the study; BLASTSIM.2 to ORYZA1, ORYZA1 to BLASTSIM.2, and FUNGICID to BLASTSIM.2.

There were five coupling points in the first coupling type (BLASTSIM.2 to ORYZA1) to represent the effect of disease on the crop physiology (Bastiaans 1993) (Figure 1): reduction in light use efficiency (coupling point 1), reduction in CO_2 assimilation at light saturation (coupling point 2), increase in maintenance respiration (coupling point 3), pathogen uptake of assimilates (coupling point 4), and increase in disease-induced leaf senescence (coupling point 5). The BLASTSIM.2-simulated values of severity and number of sporulating lesions in the rice canopy at a given day were used as parameters to couple the blast model to ORYZA1.

Reductions in CO_2 assimilation at light saturation ($AMAX$, $\mathrm{kg\,CO_2\,ha^{-1}\,leaf\,h^{-1}}$) and light use efficiency (EFF, $\mathrm{kg\,CO_2\,ha^{-1}\,leaf\,h^{-1}\,(J\,m^{-2}\,s^{-1})^{-1}}$) were estimated according to Bastiaans (1993):

$$AMAX = AMAX \times (1 - DISP)^{3.0}$$

$$EFF = EFF \times (1 - DISP)^{3.0}$$

where $DISP$ is the proportion of blast disease [$\in(0,1)$] and the value 3.0 is a proportionality factor (dimensionless) that expresses the ratio between measured reduction in photosynthetic rate due to a single lesion and the expected reduction assuming that only the lesion is photosynthetically inactive (Bastiaans 1993). The increase in maintenance respiration rate of the crop (*RMCR*, kg CH_2O ha^{-1} d^{-1}) was estimated (Bastiaans 1993) as:

$$RMCR = RMCR \times (1 - DISP) + 4.54 \times RMCR \times DISP$$

where the value 4.54 (dimensionless) is the ratio between the respiration of a lesion and that of an identical area of healthy leaf tissue. Bastiaans (1993) suggested that carbohydrate withdrawal for spore production was estimated at 200 kg CO_2 ha^{-1} (sporulating lesions)$^{-1}$ d^{-1}. In calculating such a withdrawal, a pathogen uptake of the daily total gross CO_2 assimilation of the crop (*PUDTGA*, kg CO_2 ha^{-1} d^{-1}) was estimated based on the area covered with blast lesions. The expression,

$$DTGA = DTGA - PUDTGA$$

estimates the reduced CO_2 assimilation (*DTGA*, kg CO_2 ha^{-1} d^{-1}) as a function of the pathogen uptake.

The increase in disease-induced leaf senescence was estimated by generating empirical regression equations that relate proportion of blast disease (*DISP*, independent variable) with the relative death rate of leaf tissues (*RDR*, d^{-1}) as the dependent variable. The empirical equations are described as:

$$RDR = 0.0000193 + 0.242 \times DTSP \quad 0 < DISP \leq 0.0004 \tag{2}$$

$$RDR = 0.447 \times DISP \quad DISP > 0.0004 \tag{3}$$

where the value 0.0000193 in equation (2) expresses the relative leaf death rate when *DISP* equals 0 (i.e. healthy leaf tissue). The values 0.242 and 0.447 in equations (2) and (3) describe the amount of increase in *RDR* when *DISP* value increases. A value of *RDR* was used to reduce the weight of green leaves (*WVLG*, kg ha^{-1}) by increasing the amount of leaves lost (*LLV*, kg leaf ha^{-1} d^{-1}) from the system.

For the second type of coupling (ORYZA1 to BLASTSIM.2), the simulated leaf area index from ORYZA1 (Figure 1; coupling point A) served as feedback variable to affect the subsequent disease progression in BLASTSIM.2. Leaf area index is also one of the driving variables for estimating dew period in the blast model. A change in the leaf area index resulted in a change in dew duration which subsequently affected disease development.

In the third coupling type (FUNGICID to BLASTSIM.2), it was assumed that fungicides affect *P. grisea* in two ways (Figure 1): reduction in lesion production via a reduction in the number of germinating spores and reduction in lesion expansion. The first way (Figure 1; coupling point a) was expressed as follows:

$$SPODEP = SPODEP \times (1 - PINHIB_t)$$

where *SPODEP* (number $hill^{-1}$) is the amount of *P. grisea* spores deposited on the leaves and $PINHIB_t$ is the proportion [$\in(1,0)$] of spores inhibited by a fungicide dose at day *t*. $PINHIB_t$ is expressed as:

$$PINHIB_t = 1/(\exp(FEFF_t) + 1)$$

where *FEFF* (dimensionless) is the fungicide efficacy factor calculated based on the concentration of fungicide residue on a foliage at day t ($RESD$, $\mu g\,cm^{-2}$). $FEFF$ is calculated as:

$$FEFF_t = -4 \times (\log_e(RESD)/2.303) - 5$$

where $\log_e$ is the natural logarithmic function.

The second way of coupling FUNGICID to BLASTSIM.2 (Figure 1; coupling point b) used the lesion expansion rate-reducing factors (MF) of fungicides as defined by Ishiguro (1994). The MF values were further translated as proportion of uninhibited lesion development, XMF (dimensionless) as driven by the age of the crop when fungicide was applied (AGE, days after sowing), and the fungicide application date from the time of initial application ($APDATE$, days):

$$XMF_t = MF \times AGE - APDATE$$

The rate of lesion expansion (RLX, $cm^2\,d^{-1}$) was reduced with the expression:

$$RLX = RLX \times (1 - XMF_t).$$

Spray strategies

A total of 92 spray strategies were used in the simulation study. These strategies were generated based on fungicide application, type of fungicide used and the sequence of application (e.g. benomyl first, then edifenphos), and fungicide dosage (Figure 2) as the components for blast management. Healthy crop (representing optimum strategy) and unmanaged diseased crop (least favourable strategy) were also included in the spray strategies.

Seven possible application times were considered with a minimum of one spray and a maximum of four sprays (Figure 2). The first spray was always applied either a week before or a week after initial disease symptom is expected to become visible. In the simulation study, benomyl and edifenphos were used as the two fungicide types. Although benomyl has a systemic mode of action, both fungicides were assigned here to have a protectant mode of action. FUNGICID to date has not been modified to include the mode of action of systemic fungicides. The lesion expansion rate reducing factors of these fungicides were determined by Ishiguro (1994). In the simulated scenarios, these fungicides were either applied singly or in combination. If applied in combination, the application sequence of the fungicide (Figure 2) distinguished a specific spray strategy. A total of 18 chemical sequences for single and combination applications were generated for the study. Three fungicide application dosages were used: high ($1.26\,kg\,ai\,ha^{-1}$), medium ($0.938\,kg\,ai\,ha^{-1}$), and low ($0.625\,kg\,ai\,ha^{-1}$).

Simulation

In simulating disease development with BLASTSIM.2, two initial disease observations periods (epidemic type) representing early and late disease onsets were used

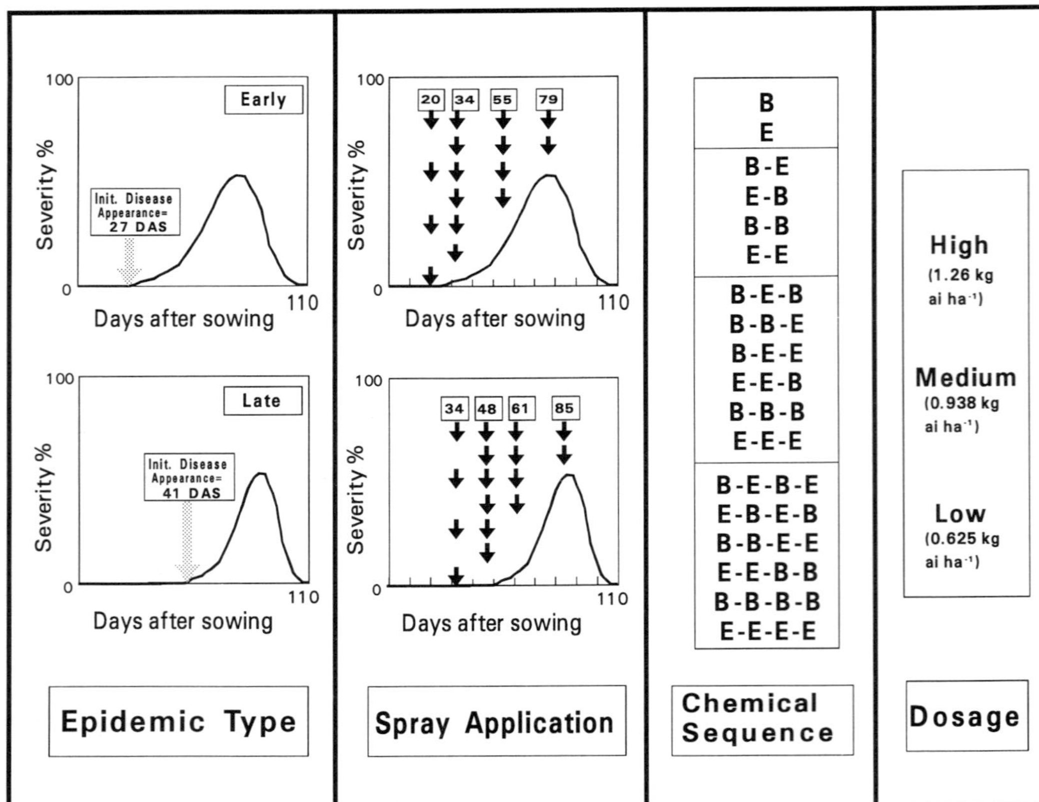

Figure 2. Criteria used in generating the 92 spray strategies. DAS=days after sowing, B=benomyl, and E=edifenphos. Black arrows and numbers in boxes in Spray Application criterion indicate fungicide spray and the time of spray application in DAS respectively.

(Figure 2). Early and late disease onsets were set at 27 and 41 days after sowing (DAS) respectively. For both disease onsets, the following input values were followed: 100 initial lesions $hill^{-1}$ and 0.2% disease severity. The receptivity factor of IR72 infected by *P. grisea* isolate PO6-6 (Calvero et al. 1992) was used in simulation. A nitrogen fertilization rate and plant spacing of $225\,kg\,N\,ha^{-1}$ and 20 cm respectively were also specified in the blast model.

Crop parameters (e.g. plant height, leaf area, etc.) of a rice cultivar, IR72 served as data inputs to both BLASTSIM.2 and ORYZA1 models. However, BLASTSIM.2 used the simulated leaf area index from ORYZA1 as one of the crop parameter inputs. The 1991 and 1992 weather data from the IRRI wetland station (14° 13′N; 121° 15′E) were used in both models.

Leaf blast progression and crop yield were simulated for the 92 spray strategies at four disease-planting scenarios during 1991 and 1992: early disease onset (27 DAS)-dry season sowing (on 4 January), early disease onset-wet season sowing (on 1 July), late disease onset (41 DAS)-dry season sowing, and late disease onset-wet season sowing.

Statistical analysis

For each spray strategy, five simulated outputs were determined which served as the dependent variables in the analysis: the weight of storage organ (WSO), the weight of above-ground biomass (WAG), the area under the disease progress curve (AUDPC), the estimated yield loss, and the net return from fungicide applications. In particular, net return was estimated as the difference between output cost and input cost. The return from the harvest derived from the price of rough rice (total yield) at PHP (Philippine Peso; \$1 US ≈ 26 Pesos) 3500 per 1000 kg (Rola and Pingali 1993) served as the output cost. Input cost included only the market prices of the fungicides (E. Nuevo, IRRI, pers. commun.) which are PHP 111.70 per 100 g and PHP 606 per litre for benomyl and edifenphos respectively.

The data set for the analysis was a matrix of 92 spray strategies and the five simulated outputs as rows and columns respectively. One data set was generated for each combination of the crop growing season (dry or wet), epidemic type (early or late), and year of observation (1991 or 1992). In order to determine the favourability of a spray strategy in managing rice blast, the cluster analysis using the PC-ORD system (McCune 1993) was used to determine spray strategy groupings (SSG) in each data set. The nature of the fungicides was not included in clustering as both fungicides were considered as protectants. An agglomerative clustering method using Ward's criterion (Greig-Smith 1983) was utilized to define SSG with relative Euclidean distance. Discriminant analysis in PC-SAS (SAS Institute 1988) was used to assess the membership composition in each SSG.

Principal component analysis (PCA) with variance/covariance as the cross-products matrix characterized the spray strategies in each SSG into either optimum, moderately favourable, and not favourable. Prior to PCA, data were normalized using equation

(4) (Greig-Smith 1983) to give the simulated outputs equal weights in the analysis:

$$b = x_{ij} / (\Sigma x_j^2)^{1/2} \tag{4}$$

where b is the normalized value and x_{ij} is the untransformed (actual) value of a simulated output at *i*th row and *j*th column of the matrix data set. Group characterization was determined by correlating the principal component scores with the values of the simulated outputs which was also graphically represented by the graphical overlay tool in PC-ORD as diamond figures in the ordination graph (Figure 3). Spray strategies in large diamond figures corresponded to high values for a simulated output.

Optimum spray strategies were identified by comparing the five simulated outputs of spray strategy 1 (healthy crop with no fungicide application and no disease) with that of the other members of SSG1. Comparison was done using Student's *t*-test. Spray strategies that gave the smallest differences ($Pr > t$ near unity) with spray strategy 1 were considered similar to it.

Results and discussion

Coupling of the models

Figure 1 showed the three coupling types: BLASTSIM.2 to ORYZA1, ORYZA1 to BLASTSIM.2, and FUNGICID to BLASTSIM.2. The five coupling points of blast effects on rice growth (coupling BLASTSIM.2 to ORYZA1) were found appropriate to explain the interaction of *P. grisea* with a rice host (Bastiaans 1993). When the host physiological variables were affected with blast, in general, a corresponding reduction in yield components was observed. However, when blast occurred at an early host vegetative stage, some of the simulated weights of storage organs (WSO) and above-ground biomass (WAG) were higher in a diseased crop than in a healthy crop. This could be explained by two reasons: the compensation effect brought about by a reduction in leaf area, and the insensitivity of certain crop growth stages to blast infection. The former reason was also exemplified by a simulation study conducted by Elings et al. (1994) with the BLIGHT model (a modified ORYZA1 model with bacterial blight and sheath blight disease modules incorporated). In their results, all inoculated plants gave an average of 5% increase in simulated WAG compared with that of a healthy crop. Similarly, a corresponding increase in simulated dry weights of leaves, stems, and panicles was observed in a diseased crop compared to a healthy crop. Plant compensation was also found in the wheat *Pseudocercosporella herpotrichoides* pathosystem (Murray and Bruehl 1986), where even at severe levels of disease, some healthy tillers remained which contributed to some yield but not to a total yield loss.

Identifying optimum spray strategy

Three spray strategy groups (SSG) were generated using cluster analysis for each disease-planting scenario, except in 1991 for a late disease onset (41 DAS) during a

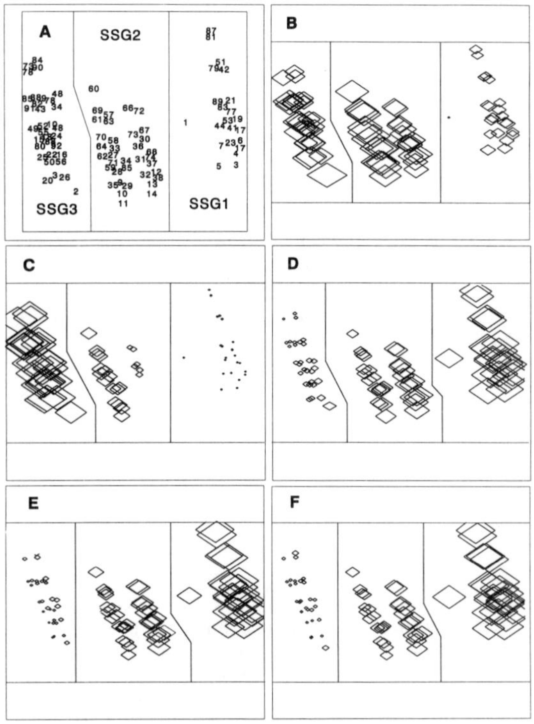

Figure 3. Ordination of 92 spray strategies for an early disease onset-dry season sowing scenario during 1991 (A) on the first two principal components showing the separation of these strategies to three spray strategy groups (SSG). Magnitude of principal component scores is also presented by the size of the diamond figures when overlaid with (B) area under disease progress curve, (C) yield loss, (D) net return, (E) weight of storage organs, and (F) weight of above-ground biomass.

wet season sowing (on 1 July) with only two SSG. Healthy crop (assigned as spray strategy 1) was always included in the first spray strategy group. This spray strategy group was initially described to include spray strategies that are optimum for blast management.

PCA revealed that only two principal components were sufficient enough to explain the variations that existed in all the data sets analysed. On average, the first two principal components explained 96% and 97% of the variations in the 1991 and 1992 data sets respectively (Table 1). In the 1991 data sets, the first principal component (PC1) was found to explain the direct effects of blast on rice yield variables, while the second principal component (PC2) indicated the economics of fungicide use. In the 1992 data sets, PC1 and PC2 indicated the economics of fungicide use and the direct effects of blast on rice yield respectively. Characterization of spray strategy groups was, therefore, based only on the first two principal components.

The characteristic of the first spray strategy group was further described by the graphical representation of correlation of the principal component scores with the five

Table 1. Principal component values, variances, and correlation on the first two principal component axes of five simulated outputs used as criteria in characterizing 92 spray strategies. PC1 and PC2 are principal components 1 and 2 respectively; numbers in parentheses are correlation coefficients of attributes on principal component values.

Attributes*	1991		1992	
	PC1	PC2	PC1	PC2
Early-dry				
WSO	−0.463 (−0.98)	0.226 (0.14)	−0.496 (−0.97)	0.116 (0.11)
WAG	−0.456 (−0.97)	0.367 (0.24)	−0.433 (−0.85)	0.513 (0.50)
AUDPC	−0.422 (−0.90)	−0.669 (−0.43)	−0.383 (−0.75)	−0.654 (−0.64)
YLOSS	−0.446 (−0.95)	−0.417 (−0.27)	−0.433 (−0.87)	−0.427 (−0.42)
RETURN	−0.449 (0.95)	−0.439 (−0.28)	0.474 (0.93)	−0.337 (−0.33)
Cum. var. %**	90.153	98.338	77.098	96.350
Early-wet				
WSO	0.552 (0.91)	−0.227 (−0.30)	−0.460 (−0.94)	0.351 (0.28)
WAG	0.534 (0.93)	−0.191 (−0.25)	−0.457 (−0.94)	0.382 (0.31)
AUDPC	−0.249 (−0.44)	−0.627 (−0.88)	−0.412 (−0.85)	−0.624 (−0.50)
YLOSS	−0.302 (−0.53)	−0.630 (−0.82)	−0.435 (−0.89)	−0.503 (−0.41)
RETURN	−0.538 (−0.94)	0.255 (0.33)	0.465 (0.97)	−0.298 (−0.24)
Cum. var. %**	61.272	95.239	84.457	97.578
Late-dry				
WSO	0.557 (0.99)	−0.143 (−0.17)	−0.499 (−0.99)	0.030 (0.03)
WAG	0.556 (0.98)	−0.153 (−0.18)	−0.498 (−0.99)	0.052 (0.05)
AUDPC	−0.148 (−0.26)	−0.710 (−0.86)	0.499 (0.99)	−0.043 (−0.04)
YLOSS	−0.227 (−0.40)	−0.653 (−0.79)	0.499 (0.99)	−0.030 (−0.03)
RETURN	−0.554 (−0.98)	0.160 (0.19)	0.077 (0.16)	0.997 (0.99)
Cum. var. %**	62.496	91.487	80.241	99.885
Late-wet				
WSO	0.575 (0.99)	−0.039 (−0.05)	−0.568 (−0.99)	0.108 (0.14)
WAG	0.567 (0.98)	−0.126 (−0.17)	−0.569 (−0.99)	0.100 (0.13)
AUDPC	−0.117 (−0.20)	−0.696 (−0.96)	−0.126 (−0.22)	−0.696 (−0.89)
YLOSS	−0.103 (−0.18)	−0.699 (−0.96)	−0.122 (−0.21)	−0.697 (−0.89)
RETURN	−0.569 (−0.99)	0.104 (−0.14)	0.569 (0.99)	0.096 (−0.12)
Cum. var. %**	60.210	98.098	60.792	93.368

*WSO = weight of storage organ ($kg\,ha^{-1}$); WAG = weight of above-ground biomass ($kg\,ha^{-1}$); AUDPC = area under disease progress curve; YLOSS = yield loss ($kg\,ha^{-1}$); RETURN = net return from fungicide application ($PHP\,ha^{-1}$).
**Cum. var. = cumulative variance.

simulated outputs (i.e. weight of storage organ, weight of above-ground biomass, area under the disease progress curve, yield loss, and net return). As shown in the example in Figure 3, spray strategies under the first group gave large diamond figures for weights of storage organ, above-ground biomass, and net return. The same spray strategies, however, gave small diamond figures for area under disease progress curve and yield loss. The size of the diamond figures actually represents the magnitude of the component scores of the different strategies for a particular simulated output. Spray strategy group 3 gave high (large-sized diamonds) component score values for area under disease progress curve and yield loss and low (small-sized diamonds) values for the yield components (weights of storage organ and above-ground biomass) and net return (see Figure 3 as an example). Spray strategy group 3 can be charac-

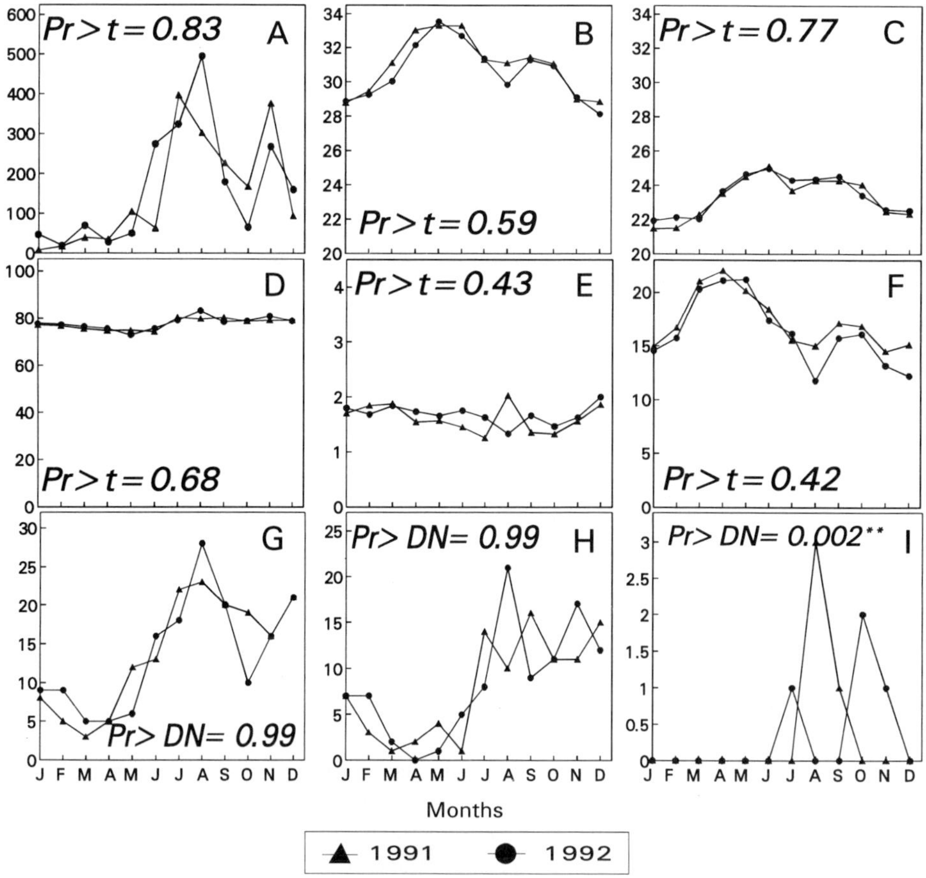

Figure 4. (A) Total rainfall in mm, (B) maximum temperature in °C, (C) minimum temperature in °C, (D) relative humidity (RH) in %, (E) wind speed in $m\,s^{-1}$, (F) solar radiation in $MJ\,m^{-2}$, (G) number of days that are wet, (H) number of days with RH > 80%, and (I) number of days with wind speed $>3.4\,m\,s^{-1}$ at IRRI during 1991 and 1992. $Pr>T$ and $Pr>DN$ are probabilities that are equal or exceeding the critical t-value and the maximum absolute deviation between the two cumulative distribution functions respectively. Probabilities followed with two asterisks (**) mean that weather between 1991 and 1992 is significantly different.

terized as the group that includes the least favourable strategies as unmanaged diseased crop (spray strategy 2) belongs to this group. Spray strategy group 2 has large component score for area under disease progress curve that is comparable with the third group, but has larger component scores on net return and yield components than the third group. Spray strategy group 2 can be characterized as the group consisting of moderately favourable to unfavourable strategies.

Table 2 shows the optimum spray strategies for 1991 and 1992. In 1991, the optimum spray strategy was found to be two sprays using edifenphos for the first application before blast symptoms are expected to appear and then benomyl two weeks

Table 2. Optimum spray strategies for blast disease management during 1991 and 1992. For application sequence, E-B means that edifenphos (E) is used during first application then benomyl (B) during second application.

Year	Spray strategy
1991	– 2 sprays – First application a week before an initial blast symptom is expected to appear; second application two weeks after first application – Application sequence: E-B – Dosage per fungicide: 0.938 kg ai ha^{-1}
1992	– 2 sprays – First application a week before an initial blast symptom is expected to appear; second application two weeks after first application – Application sequence: B-B – Dosage per fungicide: 0.625 kg ai ha^{-1}

Table 3. Economics (PHP ha^{-1}) of fungicide application based on the optimum spray strategies during 1991 and 1992.

Criteria	1991	1992
Net return from spraying	31 280.47	34 654.88
Net return with no spraying (diseased crop)	23 609.42	31 656.80
Increase in net return (%)*	24.5	8.0

*Calculated as: (Net return from spraying – Net return with no spraying)/Net return from spraying × 100.

after the first application. The rate of fungicide that is optimum for blast management was found at 0.938 kg ai ha^{-1}. In 1992, the optimum strategy was also two sprays using benomyl for both applications to be done as in 1991. The rate of fungicide optimum for blast management during 1992 was found to be at 0.625 kg ai ha^{-1}.

Considering the economics of fungicide applications, the spray strategies found optimum in 1991 and 1992 gave increases in net return of 24.5% and 8% respectively, if a diseased crop has not been sprayed (Table 3).

The difference in optimum strategies during 1991 and 1992 could be attributed to a difference in weather in those years (Figure 4), particularly with a weather variable pertaining to the number of days, with wind speed above 3.4 m s^{-1} (with $Pr > DN = 0.002$; DN is the maximum absolute deviation between two cumulative distribution functions). Although a majority of the weather variables gave almost similar values for 1991 and 1992, the difference in weather in August was also suspected to have a large effect on the favourability of certain spray strategies during the wet season. Nigg et al. (1977) and Bruhn and Fry (1982) observed that proper timing of fungicide is basically weather-dependent and weather variables, such as wind speed and rainfall, have been found to affect the efficacy of protectant fungicides.

In general, simulation results showed that two sprays using both edifenphos and benomyl at 0.938 kg ai ha^{-1} was found to be a favourable strategy in managing blast, if applied a week before blast symptoms are expected to appear or a day after initial symptoms have been detected in the plants.

Conclusions

Couplings of BLASTSIM.2, ORYZA1, and FUNGICID models have been shown to be very useful in exploring the potentialities of various spray strategies in managing blast. The coupled models described the actual production situation with blast as the one affecting yield, and fungicide as an option for managing the disease and preventing yield loss. Simulation suggests that blast management can be obtained with limited sprays during a growing season, and that combining different fungicide types would be feasible. The coupled model also showed that a spray management scheme can be planting season-specific, and depends on when infection may occur. The use of simulation approaches is potentially capable of knowing the proper timing of fungicide application which aids in deciding what optimum strategy should be applied to effectively manage blast. However, the coupled models should be validated in the field and FUNGICID modified to include the mode of action of a systemic fungicide to make simulation more realistic, dynamic, and applicable to actual production systems.

References

Bastiaans L (1993) Understanding yield reductions in rice due to leaf blast. PhD thesis Wageningen Agricultural University, The Netherlands. 127 p.

Bruhn J A, Fry W E (1982) A mathematical model of the spatial and temporal dynamics of chlorothalonil residues on potato foliage. Phytopathology 72:1306–1312.

Calvero S B, Teng P S (1992) Validation of BLASTSIM.2 model at the IRRI blast nursery and Cavinti, Laguna, Philippines. Intern. Rice Res. Newsl. 17:20–21.

Calvero S B, Guico E M, Teng P S (1992) Quantifying rice-leaf blast (Bl) genetic relationship via the receptivity factor. Intern. Rice Res. Newsl. 17:9.

Campbell G S (1986) An introduction to environmental biophysics. Springer-Verlag Inc., New York, USA. 159 p.

Elings A, Rossing W A H, Teng P S (1994) Structure and development of BLIGHT, a model to simulate the effects of bacterial leaf blight and sheath blight on rice. Pages 9–30 in Elings A, Rubia E G (Eds.) Analysis of damage mechanisms by pests and diseases and their effects on rice yield. SARP Research Proceedings, International Rice Research Institute, P.O. Box 933, Manila, Philippines.

Greig-Smith P (1983) Quantitative plant ecology, 3rd Edition. Blackwell Scientific Publications, Oxford, UK. 359 p.

Ishiguro K (1994) Using simulation models to explore better strategies for the management of blast disease in temperate rice pathosystems. Pages 435–450 in Zeigler R S, Leong S A, Teng P S (Eds.) Rice blast disease. CAB International, Wallingford, UK.

Kropff M J, Van Laar H H, Matthews R B (1994) ORYZA1: An ecological model for irrigated rice production. SARP Research Proceedings, International Rice Research Institute, P.O. Box 933, Manila, Philippines. 110 p.

Luo W, Goudriaan J (1991) Leaf wetness in rice crops caused by dew formation: A simulation study. Pages 320–327 in Penning de Vries F W T, Van Laar H H, Kropff M J (Eds.) Simulation and systems analysis for rice production (SARP). Pudoc, Wageningen, The Netherlands.

McCune B (1993) Multivariate analysis on the PC-ORD system. Oregon State University, Oregon, USA. 139 p.

Murray T D, Bruehl G W (1986) Effects of host resistance to *Pseudocercosporella herpotrichoides* and foot rot severity on yield and yield components in winter wheat. Plant Disease 70:851–856.

Nigg H N, Allen J C, Brooks R F (1977) Weather and pesticide residues. Proc. Int. Soc. Citricult. 2: 437–441.

Nokes S E, Young J H (1992) Predicting the persistence of efficacy of chlorothalonil on peanut leafspot. Trans. ASAE 35:1699–1708.

Ou S H (1985) Rice diseases. Commonwealth Mycological Institute, Kew, UK. 380 p.

Pangga I B (1995) Blast development on the new rice plant type in relation to canopy structure, microclimate, and crop management practices. MSc thesis University of the Philippines at Los Baños, Philippines. 163 p.

Rola A C, Pingali P L (1993) Pesticides, rice productivity, and farmer's health: An economic assessment. International Rice Research Institute, P.O. Box 933, Manila, Philippines. 100 p.

SAS Institute Inc (1988) SAS/STAT user's guide, release 6.03 edition. Cary, NC, USA.

Teng P S (1994) The epidemiological basis for blast management. Pages 409–434 in Zeigler R S, Leong S A, Teng P S (Eds.) Rice blast disease. CAB International, Wallingford, UK.

Teng P S, Calvero S B (1991) Computer simulation of tropical rice-leaf blast pathosystem using BLASTSIM.2. Phytopathology 81:1143.

Teng P S, Klein-Gebbinck H W, Pinnschmidt H O (1991) An analysis of the blast pathosystem to guide modeling and forecasting. Pages 1–30 in Rice blast modeling and forecasting. International Rice Research Institute, P.O. Box 933, Manila, Philippines.

Yield gap analysis of rainfed lowland systems to guide rice crop and pest management

H.O. PINNSCHMIDT[1], V. CHAMARERK[2], N. CABULISAN[3], F. DELA PEÑA[3], N.D. LONG[4], S. SAVARY[5], H.W. KLEIN-GEBBINCK[6] and P.S. TENG[5]

[1]*Tropeninstitut, Abt. Phytopathologie und Angew. Entomologie, Justus-Liebig-Universität Gießen, Bismarckstr. 16, 35390 Gießen, Germany*

[2]*Ubon Rice Research Centre, P.O. Box 65, Ubon Ratchathani 34000, Thailand*

[3]*ISDA, Mariano Marcos State University, College of Agriculture and Forestry, Batac, Ilocos Norte, Philippines*

[4]*Plant Protection Department, University of Agriculture and Forestry, Thu Duc, Ho Chi Minh City, Vietnam*

[5]*Entomology and Plant Pathology Division, IRRI, P.O. Box 933, 1099 Manila, Philippines*

[6]*Department of Agriculture, Food, and Nutritional Sciences, 4–10 Agriculture and Forestry Building, University of Alberta, Edmonton, Canada*

Key words: yield gap analysis, attainable yield level, pest problems, yield prediction, modelling, crop management, pest management, multiple regression analysis, principal component analysis, canonical correspondence analysis

Abstract

Yield gaps, and factors causing them, are of prime importance for agricultural production. There are too many yield-constraining factors in the variable rainfed lowland rice agroecosystems to be studied in full factorial experiments. An empirical-exploratory attempt is presented to characterize yield gaps and attainable yield levels as related to agroecosystem variables and to adjust yield expectations and identify research and management needs. Data on crop and pest management practices, soil conditions, weather, crop performance, and biotic and abiotic stresses were collected in over 600 plots in farmers' rainfed lowland rice-fields in northwest Luzon (Philippines), northeast Thailand, and the Mekeong River delta (south Vietnam) from 1992 to 1994. The CERES-Rice simulation model was used to estimate weather and nitrogen (N) limited attainable yield levels, while a simple empirical model was used to estimate yield trends based on fertilizer N and soil organic matter. Results of simulation runs indicated that the weather-adjusted yield gaps, i.e. the deviations of the observed yields from the weather-limited simulated yields, averaged about 35% in the Philippines, 45% in Vietnam, and 55% in Thailand. They were mainly due to N limitation in Thailand, where soil-N and fertilizer use is low, while in the other two countries they were mainly related to other constraints. In multiple regression analyses, terms related to soil carbon content and/or amount of N fertilizer captured the main yield trend well. The remaining yield variation could partly be explained by interactions between these main terms and severity levels of diseases and pest damage, water stress, and other variables related to soil conditions, crop and pest management, and weather. Factor and canonical correspondence analyses gave further insights into links among actual yield, estimated attainable yield levels and corresponding yield gaps, and other agroecosystem variables. The approach is useful for quantifying attainable yield levels and yield gaps at various constraint levels and leads to a better understanding of complex relationships between agroecosystem variables, improved yield expectations, and a better understanding of the role of yield-determining factors. This can be instrumental for prioritizing research on yield-limiting factors and guiding crop and pest management decisions.

M.J. Kropff et al. *(eds.): Applications of Systems Approaches at the Field Level, 321–338.*

Introduction

Many crop and most pest management efforts strive for reducing yield gaps to have crops exploit their full yielding potential at acceptable costs. Gomez and Gomez (1984) define yield gap (G) as the difference between the yield of a new technology (P) and farmers' yield (Q): $G=P-Q$. While this definition is largely orientated towards generation and transfer of agrotechnology, a broader definition of yield gap is useful for our purpose: a yield gap (y) is the difference between an attainable – even if hypothetical – yield level (Y_a) and the actual yield (Y), whether observed in a farmer's field or in an experiment: $y=Y_a-Y$. It is affected by various constraints and limitations, such as cultivar characteristics, cropping practices, weather and soil conditions, and stresses due to pests, diseases, and inadequate water supply. It is these limitations and constraints, as well as factors defining the attainable yield level, that one has to understand and manage if high actual yields are to be obtained. In a highly variable environment such as rainfed lowland, there are many possible causes for yield gaps. For example, water, soil, and weather related problems, together with problems imposed by insect pests, diseases, and weeds, may all constrain yields of rainfed rice in eastern India, depending on location, year, and season (Widawsky and O'Toole 1990). However, if all of these constraints are studied using full factorial designs, very large experiments would have to be conducted. In this paper, we present an empirical-exploratory attempt to study yield gaps and factors that cause them, using data from surveys and experiments carried out in farmers' rainfed lowland ricefields and using modelling techniques to estimate yield trends. While our broader research was initially aimed at obtaining baseline data on cropping conditions, pest problems, and pest management in rainfed lowland rice areas, the objectives of this paper were (i) to characterize yield gaps and attainable yield levels as related to agroecosystem variables with an emphasis on biotic stresses and (ii) to examine whether this information can be used to adjust yield expectations and identify management needs.

Material and methods

Data were obtained in holistic surveys and integrated crop protection trials conducted in farmers' fields at various rainfed lowland sites in northwestern Luzon (Philippines), northeastern Thailand, and the Mekeong River delta (south Vietnam) from 1992 to 1994. Details of this work have been outlined previously (Pinnschmidt et al. 1994a,b,c, 1995b). Four to 10 farmers' fields were randomly selected at each of 4–6 sites in Thailand and Vietnam in 1992, 1993, and 1994, and at 6 sites in the Philippines in 1993 and 1994. Fields were usually located within up to 5 km distance of a weather station. In some fields, only one farmer's practice plot was selected, while in others several plots receiving no fungicide and/or insecticide application, or 5–10 applications, or farmers' practices were established. We arrived at a total of 204 plots distributed among 102 fields in the Philippines, 201 plots within 93 fields in Thailand, and 233 plots in 99 fields in Vietnam. The total sample size was 638 plots within 294 fields, distributed among countries, sites, and years.

The plots were surveyed intensively during 5–6 field visits and a wide range of data was collected. Before planting, the depth of the top soil layer was determined and soil samples were taken and analysed for carbon content, texture, pH, CEC, available P, and exchangeable K. Farmers were interviewed about timing, kind, and intensity of crop and pest management practices that include the choice of cultivar, fertilizer application, weeding measures, pesticide use, and crop rotation. Crop phenological stage of Zadoks et al. (1974), tiller number, plant height, and percentage ground cover were determined 4–6 times during each cropping season. Similarly, visual examinations of biotic problems, as well as abiotic problems, were also determined during those times. Yield was determined using three 1 m^2 harvest samples per plot. Daily data on sunshine duration, minimum and maximum temperature, and rainfall were obtained from the weather stations. From these data, variables were derived to serve as potential explanators for observed yields and yield gaps (Table 1).

Different methods were used to obtain various indicators for attainable yield levels at various constraint levels for each plot. The CERES-Rice simulation model Version 2.1 (Singh et al. 1994; Alocilja and Ritchie 1988) was run using actual data on weather and planting practices to estimate the weather-limited attainable yield [*SY*(*W*) = *S*imulated *Y*ield, *W*eather-limited]. Solar radiation, an essential model input, was estimated from sunshine hours using an approach suggested by Mani and Rangarajan (1982). The deviation of the observed yield from *SY*(*W*) represents the weather-adjusted yield gap [*sy*(*W*)]. By including actual data on N and organic fertilizer input, soil pH, and soil organic matter, simulation runs were adjusted for N effects to obtain estimates for the attainable weather- plus N-limited yield [*SY*(*W,N*) = *S*imulated *Y*ield, *W*eather- plus *N*-limited]. The difference between *SY*(*W,N*) and the observed yield represents the weather- plus N-adjusted yield gap [*sy*(*W,N*)]. Other variables were assumed to be constant and genetic coefficients of rice variety IR72 were used in all simulation runs. Top soil carbon content was divided by 10 to estimate top soil N content. A single-equation empirical model (Neue 1985) was then used to estimate N-limited attainable yields [*EY*(*N*) = *E*mpirically estimated *Y*ield, *N*-limited] based on a term that considers the top soil N content in combination with the depth of the top soil layer [*EY*(*Nt*) = *E*mpirically estimated *Y*ield, *t*op soil layer *N*-limited], a term considering the fertilizer N [*EY*(*Nf*) = *E*mpirically estimated *Y*ield, *f*ertilizer *N*-limited], and an interaction term of these. The empirical model does not account for weather effects. If the yield estimates of the empirical model were negative, as happened in rare cases, they were set to zero. The deviation of the observed yield from *EY*(*N*) represents the N-adjusted (and weather-unadjusted) yield gap [*ey*(*N*)]. More details are given in Table 1.

About 150 primary variables were derived from the raw data to characterize various aspects of the agroecosystem. Only those that are relevant in the context of this paper are shown in Table 1. Multiple regression analyses were conducted using log or untransformed estimated yield gaps and observed yields as dependent variables and log and root transformed and/or untransformed variables with their interaction terms as independent variables. A stepwise regression procedure was used considering only variables significant at $p = 0.001$. Table 1 provides a list of root transformations

Table 1. List of variables and their acronyms, computations, and transformations used in this paper.

Non-stress-related variables:	
CARBON	Percent C-content of top soil layer
CARBON5	C-content of top soil layer 1.07–1.36%
CCOVER	Percent crop cover d^{-1}
CEC	meq exchangeable cations per 100 g soil of top soil layer (AAM method)
CLAY	Percent clay content of top soil layer
CPYEAR	No. crops per year (rice and non-rice crops)
CROTAT	Crop rotation, 0 = not practised, 1 = practised in all cases
DSEED	Planting method, 1 = direct seeding, 0 = other
EY(N)	Soil and fertilizer N limited empirically estimated attainable yield ('target yield', Neue 1985; ton ha^{-1} at 14% moisture) $= 0.05\ NAPPe - 0.000225\ NAPPe^2 + 0.05\ NAPPe\ Nt + (2.4152\ Nt - 6.0882 Nt^2)\ TLAYER$; see below for *NAPPe*, *Nt*, and *TLAYER*
ey(N)	*N* adjusted empirically estimated yield gap (%) $= [1 - YIELD/EY(N)] \times 100$
EY(Nf)	Fertilizer N limited empirically estimated attainable yield (ton ha^{-1} at 14% moisture) $= 0.05\ NAPPe - 0.000225\ NAPPe^2$
EY(Nt)	Top soil N limited empirically estimated attainable yield (ton ha^{-1} at 14% moisture) $= (2.4152 Nt - 6.0882 Nt^2)\ TLAYER$
FALLOW	Fallow period (from harvesting of previous crop until planting of current crop) in months
GDURAT	Growth duration (from seeding until harvesting) in days
KAPP	K_2O applied per season (kg ha^{-1})
MGMT	Pest management intensity = no. pesticide (fungicide, bactericide, insecticide, herbicide) applications
NAPP	N applied per season (kg ha^{-1})
NAPPe	Efficient $NAPP = 1.193\ NAPP^{0.9}$ (Neue 1985)
Nt	N content of the top soil layer (%), estimated based on C content = *CARBON/10* (after Neue 1985)
NONAPP	Number of N applications per season
NOHAPP	Number of herbicide applications per season
NOHWEED	Number of hand weedings per season
PAPP	P_2O_5 applied per season (kg ha^{-1})
PAVAIL	ppm available P of top soil layer (Olsen method)
PH	pH of top soil layer (measured in H_2O)
PH5	pH of top soil layer 5.5–6.6
SAND	Sand content of top soil layer (%)
SILT	Silt content of top soil layer (%)
SY(W)	Weather limited attainable yield simulated by CERES-Rice (ton ha^{-1} at 14% moisture)
sy(W)	Weather adjusted simulated yield gap (%) $= [1 - YIELD/SY(W)] \times 100$
SY(W,N)	Weather and N-limited attainable yield simulated by CERES-Rice (ton ha^{-1} at 14% moisture)
sy(W,N)	Weather and N adjusted simulated yield gap (%) $= [1 - YIELD/SY(W,N)] \times 100$
TLAYER	Top soil layer depth (cm)
TMAXG	Average daily max. temperature during generative phase (heading – late dough, °C)
TMIN	Average daily minimum temperature during whole season (°C)
VYIND	Varietal yielding index = actual yield/max. yield observed for a particular variety
YIELD	Observed yield (ton ha^{-1} at 14% moisture)
Stress-related variables	
B	Maximum % bird damage on panicles
BLB	Mean % bacterial leaf blight severity per day[1]
BLS	Mean % bacterial leaf streak severity per day[1]
DF	Mean % other defoliation per day[1]
DH	Cumulative adjusted % rate of stem borer dead heart damage[1]
DP	Maximum % dirty panicle severity
FET	Mean % iron toxicity per day[1]

Continued

Table 1. Continued.

FR	Maximum % foot rot severity
FSM	Maximum % false smut severity
GDROUGHT	Drought stress during generative phase (heading – late dough), 1 = occurred, 0 = did not occur
HDAM	Cumulative adjusted % rate of disease and/or damage affecting the whole hill (tungro, etc.)[2]
LB	Mean % leaf blast severity per day[1]
LCUT	Mean % defoliation per day[1]
LDIS	Mean % disease severity per day due to leaf diseases[1]
LINS	Mean % damage severity per day due to leaf insects: whorl maggot, thrips, sucking insects[1]
LDG	Maximum % lodging
LF	Mean % leaf folder damage per day[1]
LS	Mean % leaf scald severity per day[1]
NBS	Mean % narrow brown leaf spot severity per day[1]
OAD	Mean % other area damage per day[1]
ODAM	Mean % damage due to other (unspecified) biotic stresses per day[1]
OLD	Mean % other leaf disease severity per day[1]
OPD	Maximum % other panicle damage severity
OTD	Cumulative adjusted % rate of other tiller damage[2]
PB	Maximum % panicle blast severity
PCUT	Maximum % cut and/or mechanically damaged panicles
PDIS	Maximum % disease severity due to panicle diseases
PINS	Maximum % damage severity due to sucking insects on panicles
RDAM	Mean % disease and/or insect damage on roots per day[1]
RSV	Cumulative adjusted % rate of ragged stunt severity[2].
SFLOOD	Submergence stress during seedling phase, 1 = occurred, 0 = did not occur
SHR	Mean % sheath rot severity per day[1]
SIDL	Mean % sucking insect damage on leaves per day[1]
SIDS	Mean % sucking insect damage on sheaths per day[1]
TCUT	Cumulative adjusted % rate of cut tillers[2]
TDIS	Cumulative adjusted % rate of disease severity due to tiller diseases (ragged stunt, stem rot, tungro, others)[2]
TDRT	Total drought stress = sum of occurrence of drought stress during seedling, vegetative, and generative phase (0 – 3)
TINS	Cumulative adjusted % rate of tiller insect damage (stem borers)[2]
VDIS	Mean % severity d^{-1} of diseases on vegetative plant parts (leaf blades and leaf sheaths)[1]
VDROUGHT	Drought stress during vegetative phase (beginning of tillering – end of booting), 1 = occurred, 0 = did not occur
VINS	Mean % damage severity d^{-1} due to non-cutting insects on vegetative plant parts (leaf blades and leaf sheaths)[1]
WCOVER	Mean % weed cover per day[1]
WH	Cumulative adjusted % rate of stem borer white heads[2]
WM	Mean % whorl maggot damage per day[1]

Variable transformations[3]:
*CARBON***0.5, *CEC***0.5, *GDURAT***0.75, *NAPP***0.75, *KAPP***0.50, *NONAPP***0.75, *NOIAPP***0.333, *NOFAPP***0.333, *NODISIN***0.333, *NOHAPP***0.5, *NOPEST***0.5, *NOHWEED***0.5, *PAVAIL***0.2, *PH***0.75, *POTASS***0.5, *VYIND***0.75

[1]Computation of mean % damage/day (y): $y = \{\mathrm{sum}[x_{t2} + x_{t1})/2(t_2 - t_1)]\}/(t_h - t_e)$ where x = % disease or pests damage level, t = time in days after seeding, t_h = harvesting time, and t_e = time of emergence.

[2]Computation of cumulative damage rates of diseases and other pest damage types affecting the whole tiller (y): $y = \mathrm{sum}(x_{t2} - x_{t1})$ where x_{t1} and x_{t2} = % tillers affected by t_1 and t_2 respectively, and $x_{t2} > x_{t1}$. Adjustment for severity level: $x = x\{1 - [1 - (t/2)/GDURAT]^{**}[1/\exp(-1/((1 - x/100) + 0.000001))]\}$.

[3]All stress-related variables were transformed as $x' = x^{**}0.25$ except for water stress related variables that remained untransformed. (**means raised to the power of the value that follows.)

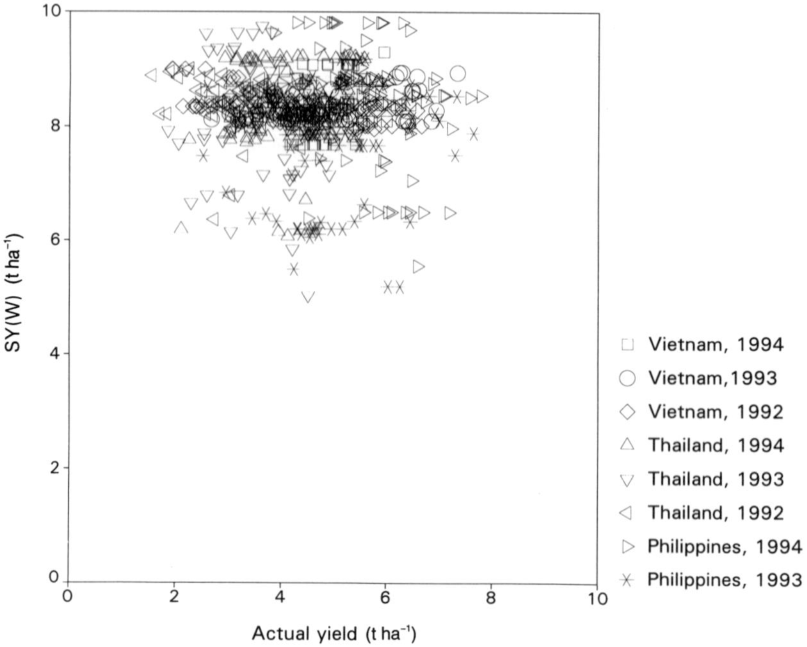

Figure 1. Simulated weather-limited attainable yields [*SY*(*W*)] versus actual yields by country and year. Simulation outputs of CERES-Rice Version 2.1 running on genetic coefficients of cv IR72, actual weather data, and actual planting practices. $R^2 = 0.001^{ns}$ across countries.

used for the respective variables. For analyses of monetary incomes and losses, the attainable income per hectare (US\$ ha^{-1}) was computed as: attainable yield (ton ha^{-1}) multiplied by price of brown rice (US\$ ton^{-1}). Monetary losses per hectare (US\$ ha^{-1}) was computed as: attainable income (US\$ ha^{-1}) multiplied by simulated yield gap (%) divided by 100. Based on farmers' interviews, the average price of brown rice (US\$ ton^{-1}) was assumed to be 230 in the Philippines, 160 in Thailand, and 100 in Vietnam. Data were further subjected to factor analyses (e.g. Hair et al. 1992) and canonical correspondence analyses (Ter Braak 1986; Dale et al. 1992) to explore and characterize the correlation structure of variables. Factor extraction was based on the principal component method assuming a minimum eigenvalue of 1 as criterion and factors were orthogonally rotated according to Kaiser's varimax criterion. Prior to conducting canonical correspondence analyses, the interval-scaled data for yield and attainable yield estimates were converted into ordinal scaled data with five classes per variable.

Results

The simulated weather-limited attainable yields [*SY*(*W*)] were comparably high across countries and did not show any correlation with observed yields (Figure 1). The yield

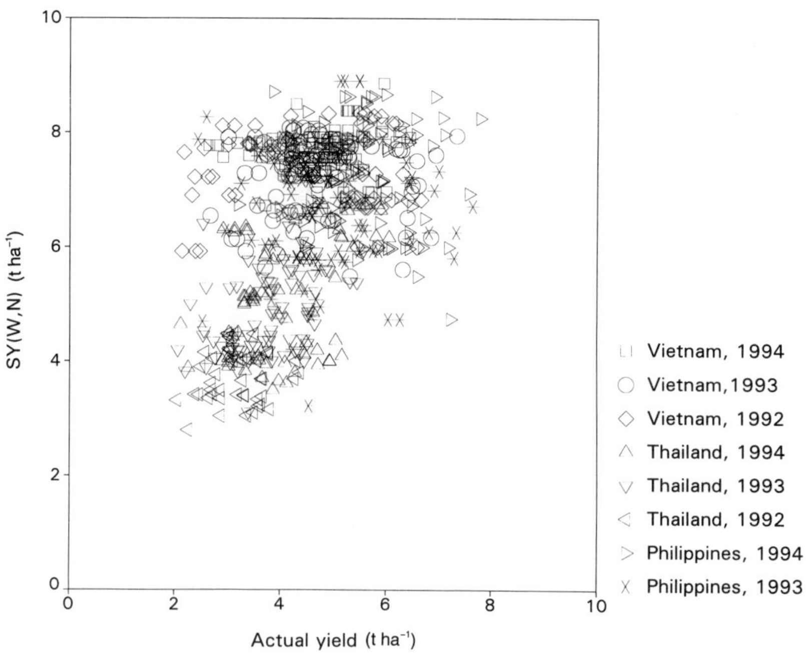

Figure 2. Simulated weather- and N-limited attainable yields [*SY*(*W*,*N*)] versus actual yields by country and year. Simulation outputs of CERES-Rice Version 2.1 running on genetic coefficients of cv IR72, actual weather data, and actual N management, planting practices, soil organic matter, soil pH, and estimated N residues of previous crops. $R^2=0.005^{ns}$ across countries.

gap, the deviation of the observed yield from the 1:1 line, averages about 55% of *SY*(*W*) in Thailand where actual yields are low, 35% in the Philippines, and 45% in Vietnam. Adjusting the simulations for N effects resulted in estimates of *SY*(*W*,*N*) considerably closer to the observations, reflecting (especially in Thailand) low attainable yield levels due to N limitation (Figure 2). The average yield gap remaining after adjusting for N effects was about 20% in Thailand, 25% in the Philippines, and 40% in Vietnam. An economical analysis of these results is presented in Figure 3. Average estimated yield gaps and attainable income levels were plotted along with calculated iso-curves of monetary losses. The isoloss curves and numbers along them denote equal monetary losses at different combinations of yield gaps and attainable income levels. The average monetary loss corresponding to an average income and yield gap level depicted by a data point in Figure 3 equals the value of the isoloss curve that would cut through that data point. This reveals that in the Philippines, where attainable income levels and average yield gaps are high, average monetary losses are also high. According to simulation results shown in Figures 1 and 2, they are mainly not related to N management and soil carbon, but to other limiting factors. In Thailand, although the weather-adjusted attainable income levels are intermediate, monetary losses are highest because the corresponding yield gaps are very high. Furthermore, N limitation seems to dominate yield gaps in Thailand. Attainable income levels are

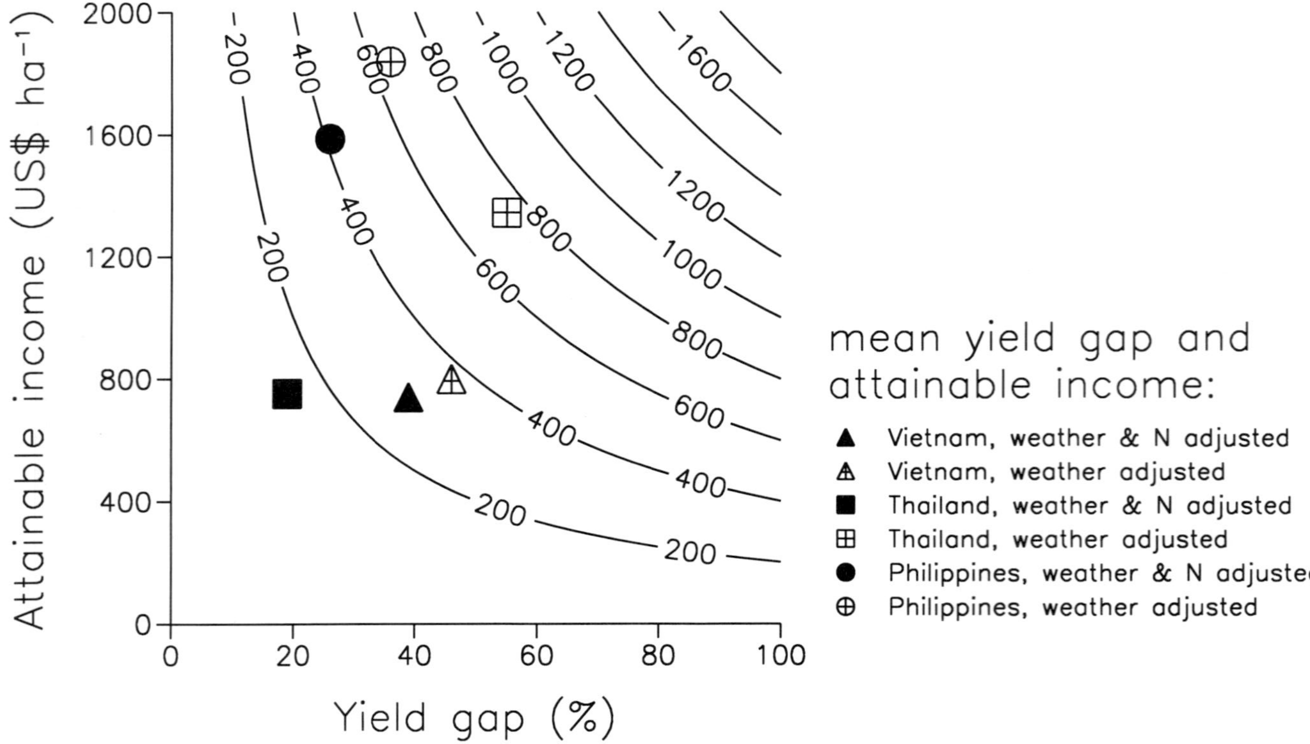

Figure 3. Mean weather and weather- and N-adjusted simulated yield gaps versus simulated attainable income levels by country as an overlay plot with iso-curves of monetary losses. Isoloss curves and numbers along them denote equal monetary losses (in US$) at different combinations of yield gaps and attainable income levels. Based on results presented in Figures 1 and 2. Yield gap (%) computed as: $\{[1 - (\text{actual yield, ton ha}^{-1})/(\text{simulated attainable yield, ton ha}^{-1})] \times 100\}$. Isoloss (US$) computed as: (% simulated yield gap/100)×(simulated attainable yield, ton ha^{-1})×(price per ton of brown rice, US$). Attainable income (US$) computed as: (attainable yield, ton ha^{-1})×(price per ton of brown rice, US$). Average price per ton of brown rice (US$): 230 (Philippines), 160 (Thailand), and 100 (Vietnam).

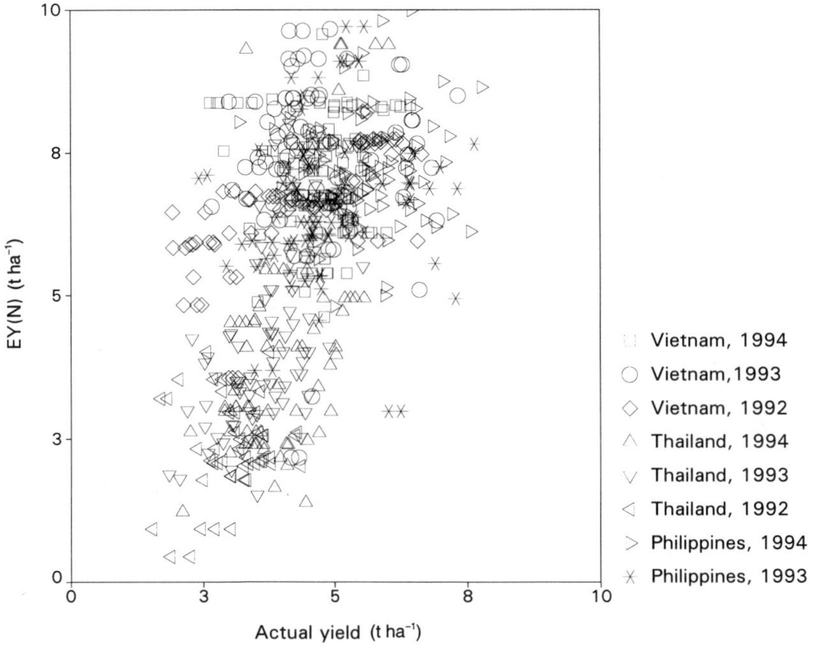

Figure 4. Empirically estimated N-limited attainable yields [*EY*(*N*)] versus actual yields by country and year. *EY*(*N*) was computed by a simple static equation based on soil C content (*CARBON*, %), depth of top soil layer (*TLAYER*, cm), and total N fertilizer input (*NAPP*, kg ha^{-1}) (*Source:* Neue 1985). $R^2=0.3^{***}$ across countries.

lowest in Vietnam. Despite high yield gaps in this country, the absolute monetary losses are relatively low compared to the other two countries.

The empirical N-driven model tends to underestimate the low yields in Thailand. In the Philippines and Vietnam, the model indicates average yield gaps similar to those estimated by the weather- plus N-adjusted simulation runs of CERES-Rice. Correlation of actual yields with model estimates was higher for the empirical model than for CERES-Rice (Figure 4).

Significant negative correlations were found between N fertilizer application, C- and clay content, cation-exchange capacity, and pH as independent variables and the total yield gap as represented by weather-adjusted simulated yields gaps [*sy*(*W*)] as dependent variable (not shown here). The highest negative correlation with *sy*(*W*), however, is displayed by the attainable N-limited yield level estimated by the empirical model [*EY*(*N*)] that reflects the yield efficient total nitrogen of the root zone (Figure 5).

The average yield differences between intensively protected and unprotected plots were generally low. In 1992, when fungicides and insecticides were applied, yield differences were negligible in Thailand and about 11% in Vietnam. In 1993, when fungicides were applied solely, yield differences were negligible in the Philippines and about 5% in Thailand and 6% in Vietnam (Table 2). Treatment effects on yield were significant but the effects on disease and pest damage severity levels were usually

Table 2. Mean percentage yield differences between intensively protected and unprotected plots. Standard deviations are in parentheses. Reference yield per field is the yield of the intensively protected plot.

1992 Insecticide and fungicide protection			1993 Fungicide protection		
Philippines	Thailand	Vietnam	Philippines	Thailand	Vietnam
no data	−0.9	11.4	0.4	5.1	6.0
	(20.9)	(9.6)	(13.2)	(14.4)	(7.8)
	n=20	n=23	n=17	n=15	n=12

Table 3. Mean percentage disease and pest damage severity levels across years. Standard deviations are in parentheses. See Table 1 for definition and computation of severity levels.

	Philippines	Thailand	Vietnam
Diseases on leaf blades, leaf collars,	1.5	1.3	17.8
and leaf sheaths (*VDIS*)	(3.0)	(2.5)	(10.6)
Insect damage on leaf blades, leaf collars,	0.3	0.1	4.1
and leaf sheaths (*VINS*)	(1.1)	(0.4)	(4.2)
Defoliation (*LCUT*)	0.2	0.2	2.9
	(0.4)	(0.2)	(3.7)
Panicle diseases (*PDIS*)	1.3	1.6	19.6
	(3.2)	(4.9)	(16.9)
Panicle damage due to sucking insects (*PINS*)	0.2	0.0	17.4
	(0.1)	(0.0)	(28.9)
Diseases affecting the whole tiller (*TDIS*)	1.1	0.7	12.5
	(7.5)	(2.5)	(17.9)
Tiller damage due to insects (*TINS*)	0.4	0.7	3.8
	(1.0)	(1.4)	(6.3)
Mechanical panicle damage (*PCUT*)	0.2	0.0	0.2
	(0.1)	(0.1)	(0.8)
Cut tiller damage (*TCUT*)	0.0	0.6	2.1
	(0.3)	(4.6)	(3.7)
Diseases and damage on roots (*RDAM*)	0.0	0.0	15.0
	(0.0)	(0.1)	(22.9)
Other damage, unspecified (*ODAM*)	2.3	0.3	1.1
	(7.9)	(1.3)	(4.2)

not significant (not shown here, see Pinnschmidt et al. 1995b). Mean disease and pest damage severity levels observed across years regardless of the treatment are shown in Table 3. The figures for Vietnam are about ten times higher than those for the Philippines and Thailand indicating very heterogeneous data across countries. However, in all countries severity levels of diseases were higher than severity levels of insect damage with respect to any plant organ. If it is assumed that pest-induced yield effects are proportional to any of the figures presented in Table 3, the corresponding pest-induced yield gaps would total about 10% in the Philippines and Thailand. In Vietnam, they would amount to about 90% which is not very realistic. This indicates that pest severity levels were overestimated in this country. The standard deviations shown in Table 3 are usually much higher than the means and reflect the very high variability of pest infestation levels across plots.

Log_e-transformed and untransformed yield gaps and actual yields were subjected to

Table 4. Results of multiple regression analyses of observed yields on agroecosystem variables including estimated attainable yield levels. See Table 1 for description of variables. Variables that went in the equations, their regression coefficients (*b*), and their standardized regression coefficients (*Beta*) are shown. *T* values of all regression coefficients and *F* values of all equations are significant at $p>0.001$. All variables in the equations that were not transformed to the natural log were transformed as indicated in Table 1.

All countries $Y=\ln(YIELD)$			Philippines $Y=\ln(YIELD)$			Thailand $Y=YIELD$			Vietnam $Y=\ln(YIELD)$		
Independent variables	*b*	*Beta*	Independent variables	*b*	*Beta*	Independent variables	*b*	*Beta*	Independent variables	*b*	*Beta*
ln[*EY(N)*]	0.22	0.4	ln(*PINS*+1.1)	−1.35	−0.5	*SY(W,N×TLAYER*	0.00	0.4	ln(*CARBON5*+1.1)	−0.05	−0.2
ln(*FET*+1.1)	−0.14	−0.1	ln(*TMIN*+1.1)	2.23	0.3	*EY(N)×PAVAIL*	0.06	0.3	ln(*LB*+1.1)	−0.07	−0.2
ln(*WM*+1.1)	−0.10	−0.1	(1-*EY(Nf)*/10)×*PINS*	1.27	0.6	*VDROUGHT×PDIS*	−0.54	−0.2	*SY(W,N)×VYIND*	−0.08	−0.2
ln(*LB*+1.1)	−0.11	−0.2	*EY(Nt)×NOHAPP*	0.05	0.6	*PH5*	0.49	0.3	(1−*EY(N)*/10)		
ln(*TMAXG*+1.1)	1.23	0.1	*VDROUGHT×LINS*	−0.17	−0.2	*OAD*	−0.50	−0.2	×*VDROUGHT*	−0.73	−0.3
VYIND×CLAY	0.00	0.2	*NAPP×PAVAIL*	0.00	0.2	Constant	−3.35		*EY(Nf)×PAVAIL*	0.08	0.4
VDROUGHT×PDIS	−0.13	−0.2	*KAPP×NOHAPP*	−0.03	−0.4				(1−*CARBON*/3)×*LINS*	−0.28	−0.3
CROTAT	0.06	0.2	Constant	−5.86					Constant	−1.69	
CARBON5	−0.06	−0.1									
PAVAIL	0.11	0.2									
LDG	0.03	0.1									
Constant	−3.40										
adj. *R* squared=0.59, *n*=637			adj. *R* squared=0.48, *n*=203			adj. *R* squared=0.55, *n*=200			adj. *R* squared=0.59, *n*=232		
Fitting actual yields (*YIELD*) to estimated yields (*Yp*) after back-transformation:											
YIELD=0.98×exp(*Yp*)+1.24 adj. *R* squared=0.56			*YIELD*=0.96×exp(*Yp*)+1.32 adj. *R* squared=0.45			(yield was untransformed)			*YIELD*=0.99×exp(*Yp*)+1.20 adj. *R* squared=0.54		

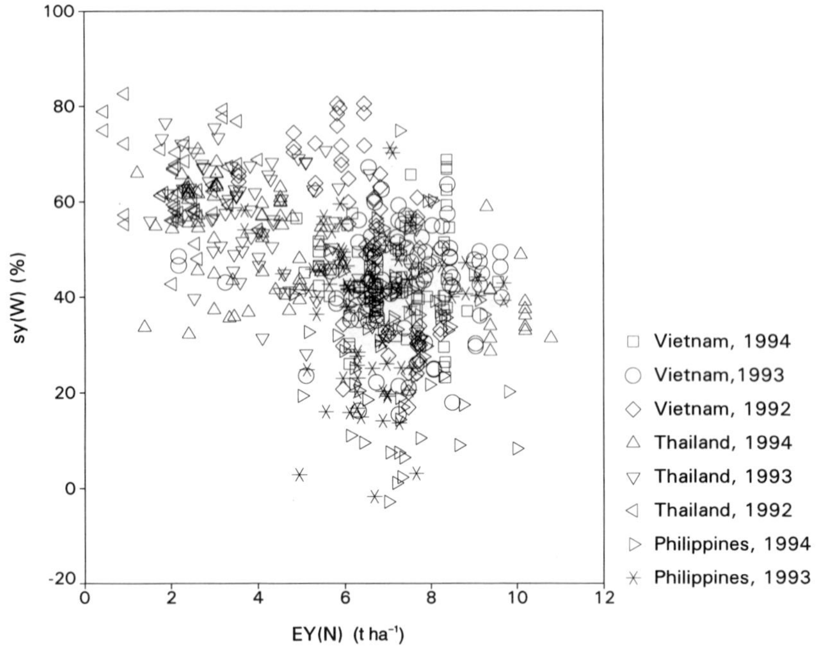

Figure 5. Simulated attainable yield gap adjusted for weather limitations [*sy*(*W*)] versus empirically estimated N limited attainable yield [*EY*(*N*)] by country and year. Note that *EY*(*N*) can simply be interpreted as the yield efficient total nitrogen of the root zone scaled as tons of grain yield per ha.

multivariate regression analyses using $\log_e$-transformed, root-transformed, and untransformed independent variables and their interaction terms. The equations presented in Table 4 remained as the most plausible regression models to estimate yield. In most cases, they are based on $\log_e$-transformed dependent variables. Log_e-transformed as well as root- and untransformed variables and interaction terms went in the equations as independent variables. The soil C- and/or the fertilizer N-limited empirically estimated attainable yield [*EY*(*N*), *EY*(*Nt*), *EY*(*Nf*)] and/or the weather and N-limited simulated attainable yield [*SY*(*W,N*)] are always included as main terms in any of these equations. High Beta values indicate their high relative importance. Pest severity (e.g. *WM*), pest management (*NOHAPP*), drought stress (*VDROUGHT*), and soil terms (e.g. *TLAYER*) often remain in the equations as interaction terms, depending on the country. To estimate actual yields after back-transformation of log-transformed data, the adjustments indicated in the lower portion of Table 4 are needed. Figure 6 shows a scatterplot of the yields estimated by the regression versus the actual yields after back-transformation and adjustments, indicating a slightly increasing variance with increasing estimated value which is due to the log-transformation.

Excerpts of the results of factor analyses are presented in Table 5. The table shows only factors and/or variables that displayed distinct relationships with yield in the factor analyses or in the regression analyses presented in Table 4. Factors extracted by the analyses are indicated in the table by either of the root names *SITEF*, *YIELDF*,

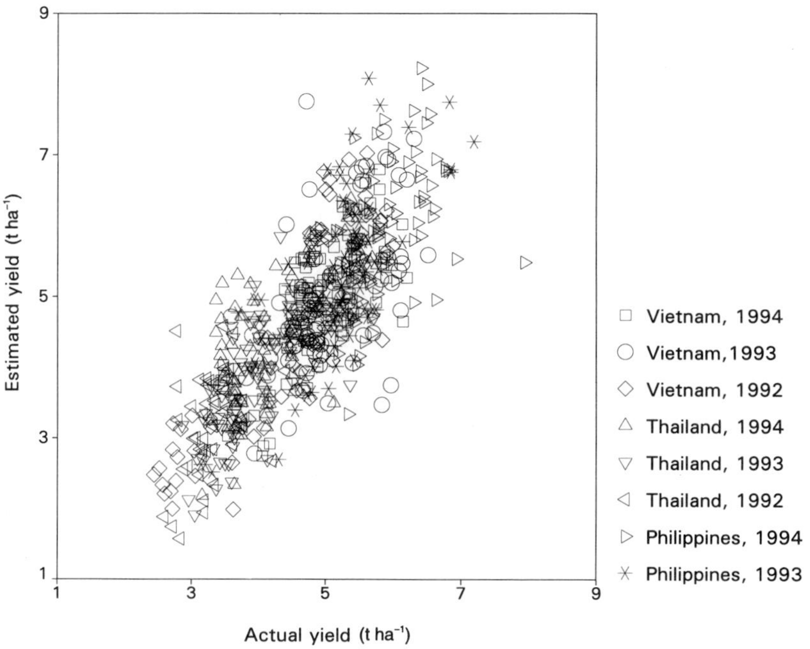

Figure 6. Yield estimates resulting from multiple regression analyses versus actual yields by country and year. The regression analyses were based on $\log_e$-transformed yield data. For more details see text.

PESTF, or *HOLF*. Factors with the root name *SITEF* were extracted from variables representing cropping practices and soil and site conditions. *YIELDF* denominates factors extracted from variables representing actual yield and estimated yield gaps. *PESTF* denominates factors extracted from variables representing disease and pest damage severities. *HOLF* denominates 'holistic' factors extracted from *SITEF*, *YIELDF*, and *PESTF* factors. Below each factor, the variables with the highest loadings for that factor and their factor loadings are listed. These variables can be thought of as varying along the respective common factor and composing it. Actual yield (*YIELD*) and some or all yield gap parameters vary along the common factor *YIELDF1* in all countries. The factor loading of *YIELD* is always negative, while the factor loadings of the yield gap parameters e.g. [*sy*(*W*), *ey*(*N*)] are positive. This indicates that actual yields decrease and yield gaps increase as values of factor *YIELDF1* increase and vice versa. Furthermore, results indicate how actual yields, estimated yield gaps, and variables that remained in the regression equations presented in Table 4 relate to other agroecosystem variables. For example, the holistic factor *HOLF1* of the data set from Thailand (right side of the middle part of the table) is composed of high positive loadings of site factor *SITEF3*, pest factor *PESTF1*, and yield factor *YIELDF2* and a high negative loading of yield factor *YIELDF1*. Looking at the original variables that these factors are composed of and taking into account the signs of the factor loadings, it can be concluded that in Thailand increasing yields correspond to decreasing yield gaps, decreasing weed cover (*WCOVER*), increasing

Table 5. Excerpts of the results of factor analyses of actual yields, attainable yield estimates, estimated yield gaps, and other agroecosystem variables, showing some of the extracted factors, the variables with the highest factor loadings, and the factor loadings.

Philippines

SITEF2*		SITEF3		SITEF5		SITEF8		SITEF9		YIELDF1		PESTF1		PESTF3		PESTF10		PESTF11		HOLF1	
NAPP	0.8	EY(Nt)	0.8	SILT	−0.8	NOHWEED	−0.6	VYIND	−0.7	YIELD	−0.9	DP	0.7	SIDS	0.9	OLD	0.7	NBS	0.6	SITEF8	0.8
EY(Nf)	0.8	TLAYER	0.7	VDROUGHT	−0.5	LDG	0.6	PAVAIL	0.6	sy(W)	0.8	PB	0.7	SIDL	0.9	LS	0.5	BLS	0.4	PESTF11	0.7
NONAPP	0.6	EY(N)	0.7	CARBON	0.5	GDROUGHT	0.5			sy(W,N)	0.8	OPD	0.6			WM	−0.4	WCOVER	−0.4	YIELDF1	−0.6
KAPP	0.6					NOHAPP	0.5			ey(N)	0.7	B	0.5								
PAPP	0.6											PINS	0.5								
SY(W,N)	0.5											OTD	0.4								
												SHR	0.4								

Thailand

SITEF3		SITEF4		SITEF6		SITEF8		YIELDF1		YIELDF2		PESTF1		PESTF3		PESTF8		PESTF10		HOLF1	
TLAYER	0.9	CPYEAR	0.8	GDROUGHT	0.7	VDROUGHT	0.7	YIELD	−0.9	ey(N)	0.8	DF	0.7	OAD	0.9	B	0.6	PB	0.7	SITEF3	0.8
EY(Nt)	0.8	FALLOW	−0.8	DSEED	0.6	CROTAT	−0.4	sy(W)	0.9	sy(W,N)	0.7	LF	0.7	ODAM	0.8	DP	0.6	FSM	0.6	PESTF1	0.7
EY(N)	0.8	PAVAIL	0.6	PH	0.4							WM	0.6	WH	0.4	BLB	−0.5			YIELDF2	0.6
LDG	0.7	SILT	0.5									DH	0.5							YIELDF1	−0.6
SY(W,N)	0.6											WCOVER	−0.4								

Vietnam

SITEF2		SITEF3		SITEF4		SITEF5		SITEF8		YIELDF1		PESTF5		PESTF7		PESTF9		PESTF10		HOLF4	
NAPP	0.8	CPYEAR	−0.7	PH	−0.7	EY(Nt)	0.8	PAVAIL	−0.7	YIELD	−0.9	RSV	0.7	WCOVER	0.7	LF	−0.6	SIDL	0.8	SITEF8	0.8
EY(Nf)	0.8	FALLOW	0.7	CARBON	0.6	EY(N)	0.7	SFLOOD	−0.5	sy(W)	0.9	LB	0.5	WH	−0.5	OPD	0.5			PESTF9	−0.6
NONAPP	0.7	GDURAT	0.7	PAPP	0.6	CEC	−0.5	VDROUGHT	0.5	sy(W,N)	0.8			OLD	0.4	FR	0.5			YIELDF1	0.4
SY(W,N)	0.6	NOHWEED	−0.4	KAPP	0.6	LDG	0.4			ey(N)	0.8			WM	−0.3						
		VYIND	0.4																		

*SITEF, YIELDF, PESTF, and HOLF are factor root names. Factors with the root name SITEF were extracted from variables representing cropping practices and soil and site conditions. YIELDF denominates factors extracted from actual yield and variables representing estimated yield gaps. PESTF denominates factors extracted from variables representing disease and pest damage severities. HOLF denominates 'holistic' factors extracted from SITEF, YIELDF, and PESTF factors. Only factors containing high loading variables related to yield (according to results of factor analyses and/or multivariate regression analyses presented in Table 4) and only variables having the highest factor loadings for a factor are shown. See Table 1 for description of variables.

defoliation (*DF*), leaf folder (*LF*), whorl maggot (*WM*), and dead heart (*DH*) damage, increasing top soil layer depths (*TLAYER*), increasing weather and/or N-limited attainable yield levels estimated by various methods [*EY*(*Nt*), *EY*(*N*), *SY*(*W,N*)], and increasing lodging (*LDG*). For the Philippines data, it shows that panicle insect damage (*PINS*) that went into the regression equation presented in Table 4 represents a whole spectrum of diseases and pest damage types such as dirty panicle (*DP*), panicle blast (*PB*), other panicle diseases (*OPD*), bird damage (*B*), other (unspecified) tiller damage, and sheath rot because it forms the common factor *PESTF1* with these.

Results of canonical correspondence analyses are presented for the Philippines data set from 1993 in Figure 7A and for the Vietnamese data set from 1992 and 1993 in Figure 7B. The figure is a biplot where the X- and Y-axes represent two dimensions that were extracted from the original variables. They can be interpreted in a similar way as principal components or factors extracted by factor analytic procedures. A variable name in the figure that is located at the tip of an arrow denominates an environmental variable, e.g. soil pH (*PH*) or applied fertilizer N (*NAPP*). The location of the tip of the respective arrow depicts the strength of the relationship between the environmental variable and an axis. The angle between arrows or between an arrow and an axis is an inverse measure of their correlation. For example, the X-axis of Figure 7A is closely related to the sand content of the soil (*SAND*), while the Y-axis is closely related to tiller insect damage (*TINS*). Variable names in Figure 7 that are numbered depict the category centroids of ordinal scaled data for yield and attainable yield estimates. For example, Yield 1 indicates the centroid of the lowest yield class and Yield 5 represents the centroid of the highest yield class. Yields 2, 3, and 4 depict intermediate yield classes. The strength of the relationship between an environmental variable and a category centroid is indicated by the length of the centroid's right angle projection onto the arrow or its extension. In the Philippines data set, there is a clear path of increasing yield (Yields 1–4) along the canopy cover (*CCOV*), total severity level of leaf diseases (*LDIS*) and total drought (*TDRT*) arrows. Very low yields (Yield 1) are associated with high severity levels of leaf diseases, high total drought stress, and sandy soils. High yields (Yield 4) can be found in the opposite direction in association with intensive pest management (*MGMT*) inputs, deep top soil layers (*TLYR*), high cation exchange capacity (*CEC*), high N fertilizer application (*NAPP*), and high crop cover (*CCOV*). Intermediate yield levels (Yield 2 and Yield 3) are located in between, thus being associated with conditions that are intermediate with respect to the above-mentioned variables. The Vietnamese data (Figure 7B) are more difficult to interpret because consecutive yield classes do not follow a straight path. However, actual yields (Yields 1–5) increase with carbon content (*CARB*), canopy cover (*CCOV*), number of nitrogen applications (*NONP*), and P and K input (*PAPP, KAPP*), while they decrease with total drought (*TDRT*).

Discussion

Yield gaps are imposed on crops by various constraints, especially in environments as variable as rainfed lowlands. Yield gaps directly translate into reduced food supplies

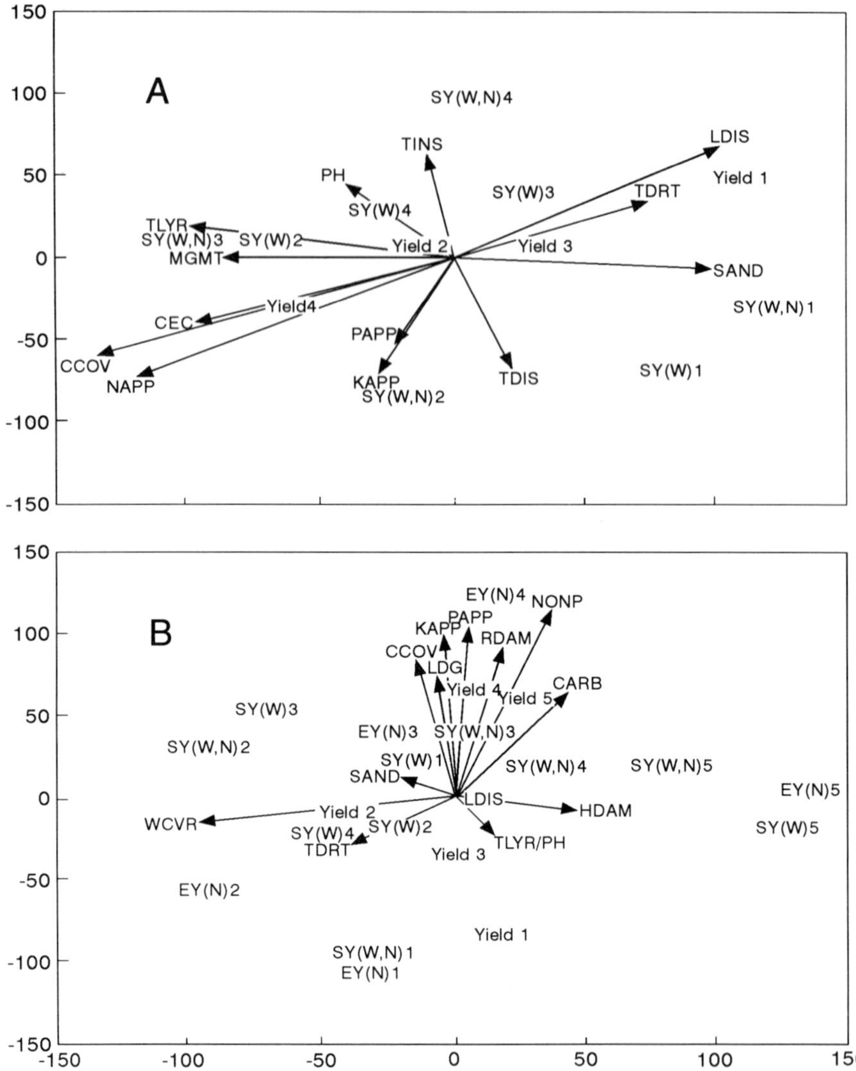

Figure 7. Results of canonical correspondence analyses (A) of some of the Philippine data from 1993 and (B) of some of the Vietnamese data from 1992 and 1993. *CARB=CARBON, CCOV=CCOVER, NONP=NONAPP, TLYR=TLAYER, WCVR=WCOVER*. For complete variable description see Table 1; for more details, see text.

and monetary losses. Many crop and practically all pest management efforts aim at closing yield gaps in order to exploit the full yield potential of crops at acceptable costs. Quantification of yield gaps at given constraint levels requires quantification of attainable 'reference' yields that correspond to those constraint levels. For example, the quantification of yield gaps induced by pests and N deficiency requires yield data of crops that are completely free of pest stress and N deficiency respectively. Such attainable reference yields are often difficult to obtain under field conditions. Simulation models can be valuable tools to estimate reference yields. Weather-limited yield estimates produced by dynamic simulation using CERES-Rice were similar across countries and plausible if compared to data from the literature (Penning de Vries 1993). Adjusting these estimates further by considering N limitations in the simulation compared favourably with yield estimates derived from a simple yet robust single-equation static model (Neue 1985) that considers soil N content, top soil layer depth, and efficient fertilizer N only. Yield gap estimates obtained by relating actual yields to simulated yields adjusted for various constraints may eventually serve as a basis for developing management decision support. Our results indicated major yield gaps due to low N supply in most of the fields in Thailand, resulting from low soil organic matter and low N fertilizer input. In contrast, yield gaps in the Philippines and in Vietnam where N supply is high, were mainly due to factors other than N limitation. This type of information can help in setting priorities with respect to studying and managing yield-limiting factors.

Yield gaps caused by varietal characteristics, water stress, pest damage, nutrient deficiencies not related to N, soil problems, and others, could not be addressed in our simulation studies due to a lack of adequate input data and/or a lack of adequate algorithms and routines of the simulation model. Genetic coefficients describing the yield-relevant characteristics of each cultivar would have been needed to include varietal effects in our yield gap analysis and soil hydrological data would have been needed to simulate water stress and its yield effects. Yield limitations imposed by soil problems, such as P and K nutrient deficiencies, cannot be simulated with the current versions of rice simulation models. Work on pest damage routines is under way to simulate pest-induced yield effects (e.g. Pinnschmidt and Teng 1994; Pinnschmidt et al. 1995a).

While dynamic simulation appears to be the most promising way to quantitatively disentangle yield effects caused by various factors, we supplemented the current capabilities of rice simulation models with a regression approach. Results underline the importance of soil organic matter – as a source for mineralizable soil N (Neue 1985) – and fertilizer N in setting yield trends and filling the major yield gap. Variables representing intensity levels and kind of pest damage, pest management, cropping practices, availability of soil nutrients, and water stress interacted with these trend terms. Positive regression terms can be interpreted as stabilizing yield levels, while negative terms can be interpreted as inducing yield gaps. The overall results led to an improved yield estimation, compared to yield estimates derived from simulation and static empirical models alone. However, to interpret the results in an agro-ecosystem context, we have to learn about interrelationships among yield predictors and other variables. Factor analysis (e.g. Hair et al. 1992) represents a well-established approach to study such interrelationships, but it is limited to interval or at least ordinal scaled variables and can only extract linear relationships. An alternative approach, although not multivariate in a strict sense, uses correspondence analysis (Savary et al. 1994). Canonical correspondence analysis (Ter Braak 1986; Dale et al. 1992) is a truly multivariate technique similar to correspondence analysis and greatly facilitates the visualization of intercorrelation structures in complex systems. It can help us understand in a holistic manner how yield gaps, yield trends, actual yields, and other agroecosystem variables relate to each other. Insights into these links need to be considered to guide research and management decisions.

References

Alocilja E C, Ritchie J T (1988) Upland rice simulation and its use in multicriteria optimization. IBSNAT Research Report Series 01, University of Hawaii, HI, USA. 95 p.

Dale M R T, Thomas A G, John E A (1992) Environmental factors including management practices as correlates of weed community composition in spring seeded crops. Can. J. Bot. 70:1931–1939.

Gomez K A, Gomez A A (1984) Statistical procedures for agricultural research. John Wiley and Sons, New York, USA. 680 p.

Hair J F, Anderson R E, Tatham R L, Black W C (1992) Multivariate data analysis. Macmillan Publishing Co., New York, USA. 544 p.

Mani A, Rangarajan S (1982) Solar radiation over India. Allied Publishers Private Ltd., New Delhi, India. 646 p.

Neue H-U (1985) Organic matter dynamics in wetland soils. Pages 109–122 in Wetland soils: Characterization, classification and utilization. International Rice Research Institute, P.O. Box 933, Manila, Philippines.

Penning de Vries F W T (1993) Rice production and climate change. Pages 175–192 in Penning de Vries F W T, Teng P S, Metselaar K (Eds.) System approaches for agricultural development. Kluwer Academic Publishers, Dordrecht, The Netherlands.

Pinnschmidt H O, Teng P S (1994) Advances in modelling multiple insect disease-weed effects on rice and implications for research. Pages 101–128 in Teng P S, Heong K L, Moody K (Eds.) Rice pest science and management. Selected papers from the International Rice Research Conference. International Rice Research Institute, P.O. Box 933, Manila, Philippines.

Pinnschmidt H O, Long N D, Mekwatanakarn P, Viet T T, Don L D, Teng P S, Dobermann A (1994a) Relationships between soil properties, crop and pest management practices, pest intensity, and crop performance in rainfed lowland rice. Intern. Rice Res. Notes 19(2):23–25.

Pinnschmidt H O, Long N D, Viet T T, Don L D, Teng P S (1994b) Characterization of pests, pest-losses, and production patterns in rainfed lowland rice of the Mekong Delta. Pages 223–242 in Denning G L and Xuan V-T (Eds.) Vietnam and IRRI: A partnership in rice research. IRRI, Manila, Philippines and Ministry of Agric. and Food Ind., Hanoi, Vietnam.

Pinnschmidt H O, Mekwatanakarn P, Long N D, Teng P S, Gaunt J L, Neue H-U (1994c) Empirical estimates of yield and pest potentials of farmer's rainfed lowland ricefields. Intern. Rice Res. Notes 19(2):30–31.

Pinnschmidt H O, Batchelor W D, Teng P S (1995a) Simulation of multiple species pest damage on rice. Agric. Syst. 48:193–222.

Pinnschmidt H O, Long N D, Mekwatanakarn P, Chamarerk V, Khumma S, Viet T T, Don L D, Abrina S, dela Peña F, Cabulisan N, Teng P S (1995b) Pests and pest losses as related to production situations in farmers' fields in rainfed lowland rice areas of Southeast Asia. Pages 555–586 in Fragile lives in fragile ecosystems. International Rice Research Institute, P.O. Box 933, Manila, Philippines.

Savary S, Elazegui F A, Moody K, Litsinger J A, Teng P S (1994) Characterization of rice cropping practices and multiple pest systems in the Philippines. Agric. Syst. 46:385–408.

Singh U, Ritchie J T, Godwin D C (1994) A user's guide to CERES-Rice–V2.10. International Fertilizer Development Center, P.O. Box 2040, Muscle Shoals, AL 35662, USA. 131 p.

Ter Braak C J F (1986) Canonical correspondence analysis: A new eigenvector technique for multivariate direct gradient analysis. Ecology 67(5):1167–1179.

Widawsky D A, O'Toole J C (1990) Prioritizing the rice biotechnology research agenda for Eastern India. Research Report of the Rockefeller Foundation, New York, USA. 86 p.

Zadoks J C, Chang T T, Konzak C F (1974) A decimal code for the growth stages of cereals. Weed Res. 14:415–421.

Quantification of the effects of bacterial blight disease on rice crop growth and grain yield

P.R. REDDY
Central Rice Research Institute, Cuttack 753006, Orissa, India

Key words: rice, bacterial leaf blight, crop losses, simulation modelling

Abstract
Experimental and simulation studies on the effect of epidemics of bacterial leaf blight in rice showed that at high levels of nitrogen fertilization, crop damage in terms of reduction in grain yield and total biomass was greater with epidemics initiated at early and late tillering compared to epidemics initiated at flag leaf appearance. Higher levels of N application favoured the production of larger amounts of green leaf area and disease epidemics. A disease onset at early and late tillering caused more damage than an infection at flag leaf appearance. In a second experiment, the onset time of disease was fixed but duration of the epidemics was varied with three levels of bacterial blight disease epidemics at different N levels. The results on total dry matter production and grain yield indicated stronger reductions in the plots where the disease was maintained throughout crop growth followed by inoculations up to late and early tillering. These differences were more pronounced at higher levels of N application.

Introduction

Bacterial leaf blight (BLB) disease of rice, caused by *Xanthomonas campestris pv. oryzae* is one of the major foliage diseases that limits increase in rice production. Disease severity, crop development stage at which the initial infection occurs, and the rate of subsequent disease spread determine yield loss. Increased crop N content has a positive effect on primary infection and subsequent disease spread (Reddy et al. 1979a; Mohanty 1981). A susceptible cultivar grown at high levels of N has an increased probability of early initial infection and a faster spread of the disease (Reddy 1994). Empirical damage functions were used to describe and estimate yield losses as an effect of single or multiple pests (Pinnschmidt et al. 1995). Yield loss studies by Ten Have and Kauffman (1972) and Reddy et al. (1979a,b) have quantified the relations between N application rate, disease severity, season and grain yield. This has resulted in a qualitative understanding of the host-pathogen-environment interactions and in recommendations for an improved disease management. However, validity of such recommendations is limited, as they are strongly influenced by disease onset, disease spreading rate, farm management practices, environmental conditions and their interactions. Physiologically based simulation models can be applied to quantify the effect of damage mechanisms and analyse their effects on crop growth and yield of rice. Actual data on crop and disease characteristics obtained through field experiments conducted during the wet season of 1993 at the Central Rice Research Institute, Cuttack, India, were used as default input into the BLIGHT model, developed by

M.J. Kropff et al. *(eds.): Applications of Systems Approaches at the Field Level, 339–348.*

Elings and Rubia (1994). The model was calibrated by Reddy (1994). Attempts are being made for model application through generation of disease scenarios and an effect of N×disease interaction. Therefore, the objectives of the research reported here are to understand the effects of different disease epidemics, and the interactions with nitrogen.

Material and methods

Field experiments

Two field experiments were conducted to quantify the effect of various epidemics of bacterial leaf blight (BLB) on rice growth and yield. The first was designed to determine the effect of the onset time of BLB epidemics on rice. The second experiment was designed to determine the effect of the duration of BLB epidemics.

The two field experiments were conducted during the dry season (January to May) of 1995 at the Central Rice Research Institute, Cuttack, India. Cultivar Annada was sown for both experiments on 2 January, and 29 day old seedlings were transplanted on 31 January to a well-puddled field in 28 plots. Single plots measured 1.8 m×4.2 m and were separated by an open space of 2 m. One plant per hill was planted, hill distance was 15 cm×15 cm and plant density was 444 000 plants ha^{-1}. Nitrogen fertilizer was applied in different doses for four N treatments as shown in Table 1. Treatments were allocated to plots in a randomized block design. Out of 28 plots, four (one plot for each N level) were not inoculated and treated as completely healthy plots. The remaining 24 plots were divided into 12 plots each for two experiments.

Experiment 1. Different epidemics of bacterial blight were created in individual plots at different levels of N application to study the impact of the timing of disease initiation. These epidemics were initiated at three different crop development stages, at early tillering – early onset (62 DAS); at late tillering – mid-onset (78 DAS); and at flag leaf appearance – late onset (86 DAS). The plots inoculated at early tillering were repeatedly inoculated (4 times) to maintain bacterial blight throughout the crop growth period. Inoculations were made thrice in the plots at late tillering and single inoculations in the plots at flag leaf appearance.

Experiment 2. In this experiment, epidemics were created in individual plots at different N levels by inoculating all the 12 plots at the same time, i.e. at the early tillering stage of crop growth. Maintenance of the epidemics ended in a phased manner, (i) throughout crop growth, (ii) up to late tillering and (iii) only for early tillering, to quantify the progress of natural disease spread for different periods on the crop and its impact on production. The plots, meant to maintain the disease for the entire growth period, were inoculated repeatedly (4 times). The disease was maintained up to late tillering by inoculating thrice, and twice inoculated for early tillering epidemic after which disease progress was maintained by natural spread.

Table 1. Amount and timing of N application in field experiments conducted during the dry season 1995, Cuttack, India.

Split no.	Days after transplanting	Date of application	Total amount of nitrogen application ($kg\,N\,ha^{-1}$)			
			65	145	200	250
1	10	10 Feb.	20	40	50	50
2	21	21 Feb.	15	35	50	50
3	28	1 Mar.	15	35	50	50
4	38	11 Mar.	15	35	50	50
5	55	28 Mar.	–	–	–	50

In both experiments the rice plants were inoculated using the clipping method (Kauffman et al. 1973). Crop and disease characteristics were observed regularly on 5 plants in 3 adjacent rows, starting at 20 days after transplanting, on days 50, 64, 71, 79, 87, 102, 113 and 123 for Experiment 1 and on days 50, 64, 73, 80, 89, 107, 117 and 123 for Experiment 2. Observed crop characteristics included leaf area and dry weights of leaves, stems and panicles. The canopy was initially considered as a single leaf layer and was later split in two and three layers. In each layer, healthy, diseased and dead leaves were separated. Diseased leaf area is considered to be the total area of all leaves that are partially covered by lesions. Disease severity of the diseased leaves was visually estimated and expressed as the ratio of diseased leaf area to total leaf area. Anthesis and physiological maturity of the crop of both the experiments were reached on 4 April and 2 May respectively, and the crop was harvested on 4 May.

Simulation model

The BLIGHT model (Elings and Rubia 1994), which is based on the ORYZA1 model (Kropff et al. 1994), was used to analyse the effects of bacterial blight disease on total dry matter production and grain yield. It incorporates damage mechanisms of diseased leaf area due to bacterial blight and sheath blight of rice and simulates crop growth under different disease conditions during one season. Effects of the disease on crop growth processes comprise the effects on the characteristics of the light response curve, e.g. CO_2 assimilation rate at light saturation, initial light use efficiency and dark respiration, which were assumed to be affected proportionally to disease severity. Effects of the disease on crop development are disregarded, however; effects on green leaf and stem area, dry matter partitioning, leaf N content and relative senescence rate are given in the input data file. Other plant characteristics have to be introduced as forcing functions, such as fractions healthy, and diseased leaf area and its severity, N content per unit leaf weight and specific leaf weight of healthy and diseased leaves.

The crop characteristics for rice cultivar Annada used earlier (Reddy 1994) for model calibration were also used for scenario studies. Disease levels were introduced into the model in two ways. First, the total leaf area was divided into either healthy

Table 2. Fraction diseased leaf area and severity in different disease onsets in simulated crop growth of rice cultivar Annada in the dry season of 1991. Development stage: 0 = germination; 1 = flowering; 2 = physiological maturity.

Treatment	Leaf area	Development stage								
		0.39	0.49	0.61	0.75	0.88	1.00	1.27	1.60	2.00
		Day of year								
		60	68	76	85	94	102	110	120	132
Early-onset										
Fraction disease	Layer 1	0	0	0.05	0.15	0.25	0.25	0.35	0.50	0.50
	Layer 2	0	0	0.15	0.25	0.25	0.35	0.40	0.50	0.55
	Layer 3	0	0	0	0.15	0.25	0.45	0.65	0.75	0.75
Severity	Layer 1	0	0	0.07	0.10	0.15	0.25	0.30	0.35	0.35
	Layer 2	0	0	0.05	0.15	0.20	0.25	0.25	0.35	0.35
	Layer 3	0	0	0	0.10	0.25	0.25	0.30	0.45	0.65
Mid-onset										
Fraction disease	Layer 2	0	0	0.15	0.25	0.25	0.35	0.40	0.50	0.55
	Layer 3	0	0	0	0.15	0.25	0.45	0.65	0.75	0.75
Severity	Layer 2	0	0	0.05	0.15	0.20	0.25	0.25	0.35	0.35
	Layer 3	0	0	0	0.10	0.25	0.25	0.30	0.45	0.65
Late-onset										
Fraction disease	Layer 3	0	0	0	0.15	0.25	0.45	0.65	0.65	0.75
Severity	Layer 3	0	0	0	0.10	0.25	0.25	0.30	0.45	0.65

(*H*) or diseased leaf area (*D*), for each of three canopy layers. For each canopy layer, fraction diseased leaf area (*FDSL*) was specified as:

$$FDSL = D/(D+H)$$

Second, diseased leaves in each canopy layer were visually assessed for percentage severity and a mean disease severity value for blight was calculated for each layer. Leaf area was simulated by the model. A value of 0.006 $(°Cd)^{-1}$ was used for relative growth rate of leaf area (RGRL). Development rates of vegetative (DVRV) and reproductive (DVRR) phases used in the model were 0.000719 $(°Cd)^{-1}$ and 0.001522 $(°Cd)^{-1}$ for cultivar Annada, calculated from its observed phenology. A value of 0.3 was used for the fraction of carbohydrates allocated to the grain from stem reserves.

Disease scenarios were created for three infection times, e.g. early tillering (early-onset), late tillering (mid-onset) and flag leaf appearance (late-onset). Yield reductions were studied at various levels of leaf N content, i.e. 0.06, 0.04, 0.02 and 0.01 $g\,g^{-1}$. For each level of leaf N content, different disease epidemics were assumed because of the fact that higher N application results in early infection and fast spread of the disease.

Simulation of early disease onset in dry season was initiated at early tillering at a fraction of diseased leaf area in the lower canopy layer (layer 1) of 0.05. This fraction increased up to 0.50 at maturity (Table 2). Later, the second and third leaf layers are infected, which results in 0.55 to 0.75 of the leaf area infected at harvest, respectively. Simulation of the mid-onset disease was initiated at late tillering at a fraction of diseased leaf area in the second canopy layer of 0.15. The bottom layer was assumed to remain healthy. Disease spread to the top canopy layer resulted in a fraction diseased leaf area of 0.75 at maturity. Simulation of the late-onset disease was initiated at flowering in the top canopy layer. Fraction diseased leaf area reached 0.75, with a severity of 0.65 (Table 2).

Results

Disease scenarios induced in the field at different levels of N application

Experiment 1: Early tillering epidemic – early onset. At high levels of N application (200 and 250 kg N ha^{-1}), the fraction diseased leaf area increased from 0.20 (shortly after disease onset) to 0.75 (at maturity) in the lower leaf layer and from 0.29 to 0.90 in the middle layer, while it increased from 0.46 to 0.75 in the top leaf layer. Progress of the disease was rapid, leading to maximum severity levels of 85–90% in all leaf layers. Application of 145 kg N ha^{-1} resulted in a fraction diseased leaf area which increased from 0.20 to 0.75 in the bottom leaf layer. It increased from 0.29 to 0.70 and 0.60 to 0.75 in the middle and top leaf layers respectively. Disease severity was between 75% and 90% in all leaf layers. At 65 kg N ha^{-1}, the fraction diseased leaf area increased from 0.26 to 0.65 in the bottom leaf layer, while it increased from 0.20 to 0.75 in the middle and from 0.42 to 0.50 in the top leaf layers. Disease severity at maturity in all layers was between 50% and 75%.

Experiment 1: Late tillering epidemic – mid-onset. Disease development at the level of 200 to 250 kg N ha^{-1} varied between 0.50 and 0.85 fraction diseased leaf area at crop maturity in different leaf layers, and between 75% and 85% of disease severity. Similarly, there was a steady progress at 145 kg N ha^{-1} with 0.50 to 0.65 fraction diseased leaf area at crop maturity. Maximum disease severity was also in the range 65–85%. At 65 kg N ha^{-1}, the fraction diseased leaf area increased from 0.20–0.65 to 0.22–0.55 in lower and middle leaf layers respectively. It was in the range 0.50–0.55 in top leaf layer. Severity of the disease gradually increased between 40% and 55% in different leaf layers at crop maturity.

Experiment 1: Epidemic at flag leaf appearance – late onset. At high levels of N application, the amount of diseased leaf area in the middle and top leaf layers was higher than in the bottom layers. The fraction diseased leaf area increased from 0.43 to 0.65 in the middle and from 0.48 to 0.85 in the top leaf layer. Similarly, the disease severity increased from 65% to 75% and from 75% to 85% in the middle and top leaf layers respectively. Inoculation at flag leaf appearance, in general, resulted in leaves which were diseased only for the last 30 days of crop growth period. In the plots applied with 65 kg N ha^{-1}, the maximum diseased leaf area fraction was 0.50 at maturity in the top leaf layer. At 145 kg N ha^{-1}, disease caused more damage in the middle and top leaf layers in terms of fraction diseased leaf area. Severity of the disease has been to the extent of 55–65% in the top leaf layer.

Table 3. Observed weights of storage organs (*WSO*, ton ha^{-1}) and total dry matter production (*WAG*, ton ha^{-1}) of healthy and different inoculated treatments of rice cultivar Annada, Cuttack, DS 1995.

Disease onset	N application (kg N ha^{-1})							
	250		200		145		65	
	WSO	WAG	WSO	WAG	WSO	WAG	WSO	WAG
Experiment 1								
Healthy	6.9	11.9	6.5	11.3	5.6	9.7	5.2	9.3
Early-onset	4.5	9.8	5.0	9.5	4.5	8.6	4.9	9.0
Mid-onset	5.3	9.7	4.9	8.9	4.6	8.5	4.7	8.2
Late onset	6.5	10.7	5.7	9.7	5.0	9.0	4.6	8.2
Experiment 2								
Healthy	6.9	11.9	6.5	11.3	5.6	9.7	5.2	9.3
Entire crop growth	4.4	8.2	5.5	9.2	4.8	8.3	3.6	7.3
Up to late tillering	5.2	9.9	5.1	9.8	5.2	8.8	4.4	8.6
Up to early tillering	5.7	10.7	5.7	10.4	5.2	8.8	4.4	8.0

Table 4. Simulated weights of storage organs (*WSO*, ton ha^{-1}) and total dry matter production (*WAG*, ton ha^{-1}) of healthy and different treatments of disease onsets of rice cultivar Annada.

Disease onset	Leaf nitrogen (g g^{-1})							
	0.06		0.04		0.02		0.01	
	WSO	WAG	WSO	WAG	WSO	WAG	WSO	WAG
Healthy	7.9	11.8	4.5	7.4	2.5	4.3	2.3	3.7
Early-onset	5.0	8.5	4.0	7.0	–	–	–	–
Mid-onset	5.0	8.5	4.2	7.1	2.4	4.2	–	–
Late onset	5.2	8.7	4.3	7.2	2.4	4.1	2.2	3.6

In general, it was observed that progress of the disease was fast at high levels of N application. However, at crop maturity fraction diseased leaf area did not differ in lower leaf layers between all the N treatments. A strong variation in the amount of diseased leaf area was observed in the inoculations made at late tillering and flag leaf appearance at low levels of N application.

Experiment 2: Inoculations for the entire crop growth period. In the plots where high N levels (200 and 250 kg N ha^{-1}) were applied, the progress of the disease was fast, with a fraction of diseased leaf area from 0.15 to 1.0 in the lower layer and from 0.11 to 0.85 and 0.35 and 0.75 in the middle and top leaf layers respectively. Similarly, disease severity ranged between 75% and 85% at crop maturity. There was moderate disease progress in the plots where 145 kg N ha^{-1} and 65 kg N ha^{-1} were applied. Diseased leaf area fraction ranged between 0.19 and 0.65 at crop maturity in different leaf layers. Severity was, however, high in the diseased leaves.

Experiment 2: Inoculations up to late tillering. Disease progress was high at higher levels of N application, with a fraction diseased leaf area which increased from 0.12 to 1.0 in lower and middle leaf layers. It increased from 0.15 to 0.65 in the top leaf layer. At 145 kg N ha^{-1} and 65 kg N ha^{-1}, the fraction diseased leaf area was high (increase from 0.17 to 1.0) in the lower leaf layer and less in the middle and top layers. Disease severity of infected leaves varied between 50% and 75% at crop maturity.

Experiment 2: Inoculations only at early tillering. At high levels of N application, fraction diseased leaf area was high in the lower leaf layer. The disease spread moderately to the middle and, to a less extent, to the top leaf layer. On the other hand, at lower levels of N application only a moderate disease spread was observed in the lower layers.

Observed effects of disease scenarios

Experiment 1. Disease initiation at early growth stages of the crop caused more damage followed by mid- and late-disease onsets (Table 3). The grain yield of the cultivar Annada was 6.9 ton ha^{-1} at 250 kg N ha^{-1}. It was reduced to 4.5, 5.3 and 6.5 ton ha^{-1} under the disease pressure at early, mid and late stages of crop growth respectively. Yield was 6.5 ton ha^{-1} at 200 kg N ha^{-1}. It was 5 ton ha^{-1} at early, 4.9 ton ha^{-1} at mid and 5.7 ton ha^{-1} at late disease onsets. At 145 kg N ha^{-1}, grain yield was 5.6 ton ha^{-1}. It was reduced to 4.5, 4.6 and 5 ton ha^{-1} at early, mid and late disease onsets respectively. The grain yield was 5.2 ton ha^{-1} at the level of 65 kg N ha^{-1}. It was 4.9, 4.7 and 4.6 ton ha^{-1} at different disease onsets (Table 3).

Maximum amount of total dry matter production for the cultivar Annada was 11.9 ton ha^{-1} in dry season when the crop was grown at 250 kg N ha^{-1} applied in the field. It was reduced to 9.8, 9.7 and 10.7 ton ha^{-1} at early-, mid- and late-disease onsets. See Table 3 for total dry matter production data at applications of 200, 145 and 65 kg N ha^{-1} for the different disease onsets.

The influence of different levels of disease epidemics on the amount of grain and total biomass indicates that early and mid onset of the disease and its continuous progress has greater effect on the crop than the late onset of the disease.

Experiment 2. The grain yield obtained from the healthy treatment of the cultivar Annada was 6.9 ton ha^{-1} at 250 kg N ha^{-1}. The yield was reduced to 4.4 ton ha^{-1} under the disease pressure throughout the crop growth period. It was 5.2 and 5.7 ton ha^{-1} when the inoculations were up to late tillering and early tillering respectively. At 200 kg N ha^{-1}, the healthy crop yielded 6.5 ton ha^{-1}. It was reduced to 5.5, 5.1 and 5.7 ton ha^{-1} when the disease was induced for the entire crop growth and up to late tillering and early tillering. The grain yield obtained from the healthy crop at 145 kg N ha^{-1} was 5.6 ton ha^{-1}. When the disease was induced for the entire crop growth and up to late tillering and early tillering, the yields were reduced to 4.8, 5.2 and 5.2 ton ha^{-1}. At 65 kg N ha^{-1}, the grain yield was 5.2 ton ha^{-1}. Under the three different disease epidemics, the yields were reduced to 3.6, 4.4 and 4.4 ton ha^{-1} (Table 3).

Total dry matter production at 250 kg N ha^{-1} was 11.9 ton ha^{-1}. It was reduced to 8.2, 9.9 and 10.7 ton ha^{-1} when the disease was induced for the entire crop growth, up to late tillering and early tillering respectively. See Table 3 for total dry matter production data at applications of 200, 145 and 65 kg N ha^{-1} for the disease pressure at varying periods.

In general, higher reductions in grain yield and total dry matter production were observed at increased levels of N application, when the disease pressure was maintained for the entire crop growth period. When the disease was allowed to spread naturally as a consequence of few artificial inoculations up to late tillering and early tillering stages, the effect on grain yield and total dry matter production was less marked.

Simulated disease scenarios

Nitrogen effect. Higher levels of leaf N generally increased simulated growth and yield of rice (Table 4). Cultivar Annada grown in the dry season gave a simulated total dry matter production of 11.8 ton ha^{-1} at a leaf N content of 0.06 g g^{-1} and it gradually decreased to 3.7 ton ha^{-1} at a leaf N content of 0.01 g g^{-1}. Simulated grain yield of Annada in dry season at a leaf N content of 0.06 g g^{-1} was 7.9 ton ha^{-1}. Reduction in leaf N content caused reduction of grain yield to 2.3 ton ha^{-1} at a leaf N content of 0.01 g g^{-1}.

Effect of disease build-up. Disease initiation at early growth stages of the crop caused most damage, followed by mid- and late-onsets. Early, mid- and late disease onsets reduced simulated total dry matter to 8.5, 8.5 and 8.7 ton ha^{-1} respectively (Table 4).

Simulated grain yield decreased from 7.9 to 5.0, 5.0 and 5.2 ton ha^{-1} due to early, mid and late disease onsets respectively. At the 0.04 g g^{-1} level, simulated yield of a healthy crop was 4.5 ton ha^{-1}. It was reduced to 4.0, 4.2 and 4.3 ton ha^{-1} under different disease pressures. In general, yield reductions were lower at low N levels (Table 4).

Discussion

The two field experiments were designed mainly to maintain the disease epidemics on the crop for different durations. Occurrence of varying levels of incidence and severity of the disease is common under farmers' field conditions. Thus, the information generated on the effects of different levels of disease pressures for different durations on the crop may become a useful source of guidance for rice cultivation. Earlier, researchers have also used successive inoculations to stimulate disease increase (Gregory et al. 1978). Similarly, long- and short-duration epidemics were also created by manipulating dates of inoculation to estimate yield loss due to stem rust of wheat (Romig and Calpouzos 1970) and *Septoria glycine* on soybean cultivars (Young and Ross 1978).

Observed effects of the disease in the first experiment indicate that reductions in grain yield and total dry matter were higher in early and late tillering inoculations at higher levels of N application. Though the disease was for a longer duration on the crop inoculated at early tillering, the difference in reduction between early and late tillering inoculations was negligible. This might be due to leaf area compensation mechanisms (Teng 1988; Reddy et al. 1994). The yield reductions were less marked in the plots inoculated at flag leaf appearance since the crop was affected for a short duration only. The crop grown at lower N levels was less affected due to slow disease progress in accordance with findings of Reddy et al. (1979a,b).

The results of the second experiment showed that losses were smaller in plots where the crop was exposed to natural disease spread after initial inoculations. This indicates that losses are higher when the disease progresses continuously on the crop especially grown at higher levels of N application. Higher levels of N favour increased leaf area formation providing more substrate for the disease and faster epidemics. Increased leaf area damage leads to increased losses. Grain yield and total dry matter productions are generally lower at low N application. Low N levels slow down the disease progress. Thus, in infectious environments, it would be more appropriate to grow the crop at moderate N levels.

The results showed highest yield losses at high N levels with early disease onset. With a decrease in N level, disease development was delayed and severity was also reduced. Therefore, yield losses were lower at low N levels. With decreasing leaf N content, total dry matter production was reduced, so was the effect of the disease.

Simulated results on the effect of different disease scenarios on grain yield and total dry matter production indicated higher reductions at early, mid and late disease onsets in the crop grown with 0.06 $g\,g^{-1}$ of leaf N than with leaf N levels of 0.04, 0.02 and 0.01 $g\,g^{-1}$. At the lower levels of leaf N, the effect of the disease in reducing the yield and total biomass was negligible (Table 4). On the other hand, the data collected from the field experiments showed that reductions in grain yield and total dry matter production were gradual with a decrease in the amount of N application in the field. Early and mid disease onsets caused significant reductions when compared with late onset of the disease at all levels of N application. However, two trends are clear from both simulation studies and field observations: the disease caused larger reductions at higher levels of N, and early and mid onset of bacterial leaf blight disease caused larger reductions than late onset at flag leaf appearance.

Reddy et al. (1994) reported that the effect of cumulative absorbed radiation on the crop depends upon the amount of green leaf area. The green leaf area reduces under varying levels of disease pressure. Different epidemic levels caused reduction in the photosynthetically active green leaf area at different growth stages of the crop reflecting in the reduction of total dry matter production.

Simulated and observed results indicate a general trend of reductions due to the disease which, in turn, is controlled by N supply to the crop. Reddy and Rubia (1995) have shown satisfactory model performance in terms of reproducing observed experimental results notwithstanding slight underestimations in some of the epidemic treatments. More insight in effects of N supply on leaf area development and disease epidemics is required for reliably simulating damage by leaf blight at different infection regimes.

Acknowledgements

The author is extremely thankful to Dr. R.S. Parods, Director-General, and Dr. E.A. Siddiq, Deputy Director-General, Indian Council of Agricultural Research, Krishi Bhavan, New Delhi, India, for the deputation granted to visit the International Rice Research Institute, Manila, Philippines, for a period of seven weeks to complete the work on simulation of the effect of bacterial blight disease on crop growth and yield of rice. Also, thanks are due to Dr. K.C. Mathur, Director, Central Rice Research Institute, Cuttack, India, for his encouragement in the studies.

Thanks are due to Dr. S.R. Obien, Director of the Philippine Rice Research Institute, Muñoz, Nueva Ecija, for providing all the required facilities besides taking care of personal well-being during the stay at PhilRice. Thanks are due to Dr. R.K. Singh and Mr. Noor of IRRI liaison office, New Delhia, India, for facilitating the visit to the Philippines.

References

Elings A, Rubia E G (1994) Models for the crop-pathosystems: Rice-bacterial leaf blight, sheath blight and stem borer. SARP, Crop Protection Theme: Pest management application programme, March 1994. International Rice Research Institute, P.O. Box 933, Manila, Philippines. 102 p.

Gregory L V, Ayers J E, Nelson R R (1978) Predicting yield losses in corn from southern corn leaf blight. Phytopath. 68:517–521.

Kauffman H E, Reddy A P K, Hsieh S P Y, Merca S D (1973) An improved technique for evaluating resistance of rice varieties to *Xanthomonas oryzae*. Plant Dis. Reptr. 57:537–541.

Kropff M J, Van Laar H H, Matthews R B (Eds.) (1994) ORYZA1: An ecophysiological model for irrigated rice production. SARP Research Proceedings, International Rice Research Institute, P.O. Box 933, Manila, Philippines. 110 p.

Mohanty S K (1981) Influence of host nutrition on the development of bacterial blight disease of rice. PhD thesis, Uktal University, Bhubaneswar, Orissa, India.

Pinnschmidt H O, Batchelor W D, Teng P S (1995) Simulation of multiple species pest damage in rice using CERES-Rice. Agric. Syst. 48:193–222.

Reddy A P K, Katyal J C, Rouse D I, Mackenzie D R (1979a) Relationship between nitrogen fertilization, bacterial leaf blight severity and yield of rice. Phytopath. 69:970–973.

Reddy A P K, Mackenzie D R, Rouse D I, Rao A V (1979b) Relationship of bacterial leaf blight severity to grain yield of rice. Phytopath. 69:967–969.

Reddy P R (1994) Simulation of the effect of bacterial blight disease on crop growth and yield of rice. Simulation and Systems Analysis for Rice Production, Internal Publication, Central Rice Research Institute, Cuttack, Orrisa, India. 46 p.

Reddy P R, Rubia E G (1995) Quantification and simulation of the effect of bacterial blight disease on rice crop growth and grain yield. SARP Report, International Rice Research Institute, P.O. Box 933, Manila, Philippines. 49 p.

Reddy P R, Nayak S K, Elings A (1994) Yield reduction due to different levels of bacterial blight disease of rice. Pages 31–42 in Elings A, Rubia E G (Eds.) Analysis of damage mechanisms by pests and diseases and their effects on rice yield. SARP Research Proceedings, International Rice Research Institute, P.O. Box 933, Manila, Philippines.

Romig R W, Calpouzos L (1970) The relationship between stem rust and loss of spring wheat. Phytopath. 60:1801–1805.

Teng P S (1988) Pest and pest-loss models. Agrotechn. Transfer 4:1–10.

Ten Have H, Kauffman H E (1972) Effect of nitrogen and spacing on bacterial leaf blight of rice. Indian Farming 21(10);8–13.

Young L D, Ross J P (1978) Brown spot development and yield response of soybean inoculated with *Septoria glycine* at various growth stages. Phytopath. 88:8–11.

Better biological control by a combination of experimentation and modelling

J.C. VAN LENTEREN and H.J.W. VAN ROERMUND
Departments of Entomology and Theoretical Production Ecology, C.T. de Wit Graduate School Production Ecology, Wageningen Agricultural University, P.O. Box 8031, 6700 EH Wageningen, The Netherlands

Key words: biological control, stochastic model, Monte Carlo simulation, individual-based population model, *Trialeurodes vaporariorum*, *Bemisia tabaci*, *Encarsia formosa*

Abstract
Greenhouse whitefly (*Trialeurodes vaporariorum*) and sweet potato whitefly (*Bemisia tabaci*) are very common, highly polyphagous pest insects worldwide. Biological control of greenhouse whitefly with the parasitoid *Encarsia formosa* has been applied with great commercial success during the past twenty years, while natural enemies for the sweet potato whitefly are now evaluated. Modelling has played a role in the process of selecting and improving the efficacy of releases of natural enemies, but often biologically unrealistic simplifications were part of these models which strongly limited their predictive value. We have developed a model which is unique in that it is individual-based and simulates the local searching and parasitization behaviour of individual parasitoids in a whitefly-infested crop. The model includes stochasticity and spatial structure based on location coordinates of plants and leaves. This model comprises several submodels for the parasitoid's foraging behaviour, the whitefly and parasitoid population development, the spatial distribution of whitefly and parasitoid within and between plants in the crop, and for leaf production. With the model we can simulate temporal and spatial dynamics of pest and natural enemy. The model will allow us to explain why the parasitoid can control whiteflies on some crops and not on others in large commercial greenhouses, to improve introduction schemes of parasitoids for crops where control was difficult, to predict effects of changes in cropping practices (e.g. greenhouse climate, choice of cultivars) on the reliability of biological control and, finally, to develop criteria for the selection of natural enemies.

Introduction

Out of the 1200 described whitefly species, only some twenty species may be considered as potential pests. Until recently, most research in control of whiteflies was directed towards the greenhouse whitefly (*Trialeurodes vaporariorum* (Westwood)) (Van Lenteren and Woets 1988). However, since the mid-1980s, whiteflies of the genus *Bemisia* created problems of such a large scale both in the field and in large, commercial greenhouses that concentration of research shifted to this pest (Gerling 1990). In natural ecosystems and agroecosystems where pesticides are not (or selectively) used, usually an array of natural enemies keeps the number of whiteflies at very low numbers: predators, parasitoids and pathogens all take their toll. Work in two cropping systems – tomatoes with *T. vaporariorum* pests in the 1960s in California (E.R. Oatman 1992, pers. commun.), and cotton with *Bemisia tabaci* (Gennadius) during the period 1925–1992 in Sudan (B. Munir 1992, pers. commun.) – has shown

M.J. Kropff et al. *(eds.): Applications of Systems Approaches at the Field Level, 349–363.*

that whiteflies can be kept under perfect *natural control*. When pesticides are applied, natural enemies are decimated and whitefly pests are the result. Furthermore, changes in crop rotation, shortening of fallow periods, and concurrent or overlapping growth of whitefly-sensitive crops may result in such a high and continuous whitefly pressure that natural enemies are not capable to sufficiently reduce whitefly numbers.

If natural control is insufficient, inoculative or inundative releases with natural enemies can be made. Commercial biological control of greenhouse whitefly through releases of the parasitoid *Encarsia formosa* Gahan is used at present on about 5000 ha of greenhouse crops and in most countries with an important greenhouse industry (Van Lenteren 1995). Although other types of natural enemies of greenhouse whitefly are known (predators and entomopathogenic fungi; Gerling 1990), it is mainly introductions of parasitoids that have led to economically feasible control. Biological control of *Bemisia* species is still problematic and a worldwide search for effective natural enemies is going on currently. Much of this research is opportunistic and evaluation of the control potential of natural enemies is often a purely empirical process. Neither the farmer nor the scientist is much helped by such an approach, because many projects are terminated prematurely if success is not obtained quickly and, thus, scientific insight does not evolve. Therefore, our biological control approach is based on obtaining insight in the functioning of natural enemies in agroecosystems and to develop criteria for a scientifically sound selection of natural enemies (Vet and Dicke 1992; Van Lenteren 1993). That such an approach does not have to result in a slowdown of practical biological control programmes, which is often claimed by biological control practitioners, is illustrated by the successes we obtained in greenhouses where during the past 25 years some 20 natural enemy species have been evaluated which are commercially used for pest control nowadays (Van Lenteren 1995). Modelling has always formed an essential part of our approach (e.g. Van Roermund 1995).

One of the problems we faced with biological control of whiteflies with *E. formosa* was that in some key greenhouse crops (e.g. tomato) results were very reliable, while on other important crops (e.g. cucumber) control often failed. Another, more recent problem is that *E. formosa* does not seem to be a suitable parasitoid for the control of *Bemisia* and new natural enemies need to be evaluated. We started a long-term project in which modelling and experimentation are integrated, with the aim to obtain quantitative understanding of this tritrophic system to explain failure or success of biological control. Because of the multitude of relationships between trophic levels, the most important quantitative factors can only be traced after integration of all relevant processes. Systems analysis and simulation are powerful tools for this purpose. This approach bridges the gap between knowledge at the individual level and understanding at the population level.

In this paper, we summarize work of the first phase, which was limited to the tritrophic system: tomato – greenhouse whitefly – *E. formosa*. During the next phase, we will, based on newly collected data, add different crops to the model, adapt it for *Bemisia* spp. and include other natural enemies. The model will allow us (i) to explain why the parasitoid can control whiteflies on some crops and not on others in large commercial greenhouses, (ii) to improve introduction schemes of parasitoids for crops where control was difficult, (iii) to predict effects of changes in cropping practices (e.g. greenhouse climate, choice of cultivars) on the reliability of biological control and (iv) to develop criteria for the selection of natural enemies.

Methodology

Data collection from literature

When we started with the first phase of this project, many data were already available about development of tomato crops and life-history parameters of the two insect species (immature development rate, immature mortality, adult longevity, sex ratio, fecundity and oviposition frequency) (for whitefly reviewed in Hulspas-Jordaan and Van Lenteren 1989). Furthermore, extensive studies had been done on the whitefly's host-plant preference and performance, on selection of feeding and oviposition sites and on spatial distribution patterns (reviewed in Van Lenteren and Noldus 1990). Also, comprehensive work had been done on biology and foraging behaviour of the parasitoid when confined to small experimental arenas (reviewed in Noldus and Van Lenteren 1990). A critical analysis of all this earlier work resulted in a selection of data on life-history parameters of greenhouse whitefly and *E. formosa*, and with these data the relationships between life-history parameters and temperature were estimated by non-linear regression (Van Roermund and Van Lenteren 1992a,b).

Time allocation of the parasitoid on tomato leaflets: behavioural observations

As we intended to develop a biologically realistic model (see discussion for reasoning) based on individual behaviour of parasitoids, we needed information on the time allocation with respect to searching for and oviposition in hosts by the parasitoids on tomato leaflets. Leaflets form the basic unit for host searching and parasitization, and the time spent on these leaflets plus the type of activities performed, largely determines the outcome of the model. Information was collected, gaps in knowledge identified, and a number of experiments were run to obtain lacking information. The experiments consisted of directly observing parasitoid behaviour in a number of different search situations, like clean leaves, leaves with low/high host densities, leaves with unparasitized/parasitized hosts, etc. This work resulted in information about the parasitoid's walking speed, walking activity and host handling behaviour, and the tendencies to depart from leaves related to the quality and number of hosts on the leaves (Van Roermund and Van Lenteren 1995a,b; Van Roermund et al. 1994).

Foraging behaviour of the parasitoid on plants: quantification through simulation

The behavioural data were used as input in a stochastic Monte Carlo simulation model of the parasitoid's foraging behaviour. Earlier research made clear that *E. formosa* does not use volatile or visual cues to locate infested plants. Once on a plant, they are searching at random. After finding a host, host remains or honeydew, the searching pattern does not change, but the parasitoid does stay much longer on such a leaflet (Noldus and Van Lenteren 1990; Van Roermund et al. 1994).

The complexity of the model was increased step by step: the searching behaviour was first simulated for a small experimental arena, then for a tomato leaflet and, finally, for a tomato plant. For each level of complexity, the model was first validated with independent experimental data before the next step was made. The model on foraging on the plant level enables quantification of the parasitization process of *E. formosa* from the moment of landing on the plant until departure from that plant (Van Roermund et al. 1996a,b,c).

Population dynamics of pest and natural enemy in a crop: how biocontrol works

The final part of this project consisted of designing a model which simulates the population dynamics of the whitefly–parasitoid interaction in a commercial tomato crop at realistic and varying temperature conditions. This model comprises several submodels, one of which is the model of the parasitoid's foraging behaviour on tomato leaflets, other submodels concern the pest and parasitoid population development, the spatial distribution of whitefly and parasitoid in the canopy, the dispersion of adult whiteflies and parasitoids from leaf to leaf, and finally a tomato leaf production model. For a comprehensive description of the models, the reader is referred to Van Roermund (1995). A simplified scheme is given in Figure 1. Information on the most important submodels follows below.

Development of whitefly and parasitoid was modelled by using a boxcar train procedure (Goudriaan and Van Roermund 1993), in which the successive developmental stages and variation in development within each stage is taken into account.

The whitefly population model is an explanatory state-variable model. Whitefly stages are represented by boxcar trains. The development rate of each stage, the adult reproduction and the longevity are affected by temperature, hence temperature is used as the driving variable. The time step of the model was chosen to be 0.05 d (1.2 h), which is about one-tenth of the smallest time coefficient (residence time of individuals in one boxcar). Such a small time step is needed for accurate numerical integration.

The parasitoid population model is similar to the whitefly population model, with adjustments for specific *E. formosa* characteristics. The main adaptations were a 100% female sex ratio (the parasitoid is thelytokous), different development rates and mortality of parasitoid immatures depending on the whitefly stage they parasitized and a separate submodel for oviposition based on the foraging behaviour of the parasitoid. Besides temperature, the daily light period appeared to be also an important driving variable for the parasitoid, as the parasitoid only searches for hosts during day time. The time step of this model was also chosen as 0.05 d (1.2 h).

The final model combines both population models and simulates the population dynamics of whitefly and parasitoid during a growing season of a crop. Whitefly and parasitoid disperse in the canopy almost exclusively by flying from leaflet to leaflet,

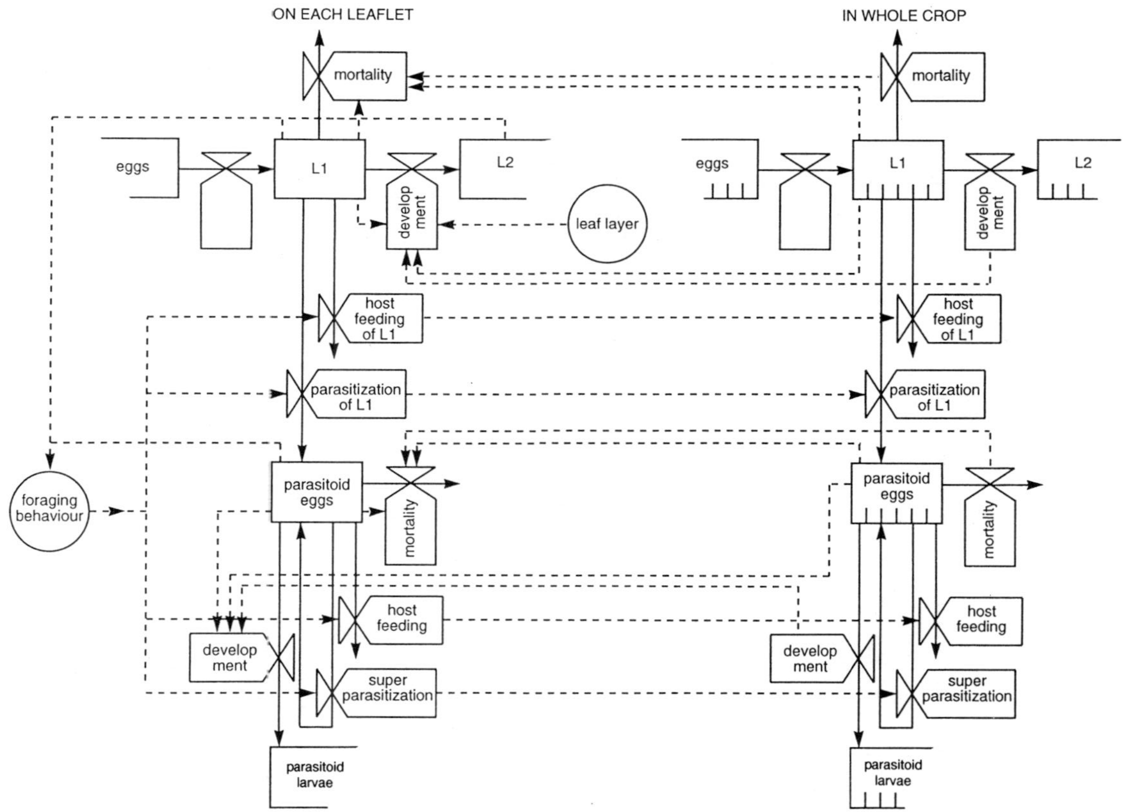

Figure 1. Relational diagram for the population growth of greenhouse whitefly and *Encarsia formosa*, with details on the interaction between host and parasitoid; the state variables or integrals (insect numbers) are presented within rectangles, the rates of change within valve symbols, auxiliary variables within circles; the flow of material is indicated by solid lines + arrows, the flow of information by dotted lines + arrows; small bars in a rectangle indicate series of integrals (boxcar train); L1, L2... are the first, second, etc. larval stages of whitefly.

walking to other leaves is hardly observed. Therefore, the leaflet is chosen as the spatial unit. In the model, the canopy is divided into plants with a certain number of leaves, and leaf senescence and leaf emergence is simulated. The coordinates of each plant, leaf and leaflet represent their position in the canopy. The interaction between both populations in the greenhouse is achieved by the local foraging behaviour of the parasitoid adult on leaves (Figure 1). The stochastic submodel on the foraging behaviour of the parasitoid determines ultimately the outcome of events in the whitefly and parasitoid populations. New parasitoids are generated after parasitizations, while the whitefly population decreases in size as result of parasitizations. Host feeding by parasitizations kills additional whiteflies. The time step of the submodel on foraging is only 0.00005 d (4.32 s) which is smaller than the smallest time coefficient in the model (antennal rejection of host which lasts about 15 s).

The foraging efficiency of the parasitoid is strongly affected by host density and distribution, which are both included in the model (vertical distribution within the plant and horizontal distribution between plants). The model simulates the spatial distribution of all whitefly and parasitoid stages in the canopy, based on flight behaviour of individual adults, oviposition, and immature development.

Input data

The model is initialized by introduction of a certain number of adult whiteflies and parasitoid pupae in the crop. Number of plants, initial numbers, locations and dates of releases of whiteflies and parasitoids have to be defined. Variable, actual temperature, and daylight duration for each day are the driving variables.

Validation

Validation data for the whitefly population model were available as six independent population counts on tomato grown in small greenhouses with 10–15 plants, in which 100–300 one day old whitefly females were released, and where the numbers of emerged whitefly pupae were counted after 40, 60 and 80 d; temperature was regularly measured during these experiments (Joosten and Elings 1985).

Validation for the combined whitefly-parasitoid population dynamics model was possible with data from an experiment in a commercial, large greenhouse with 18 000 tomato plants where absolute counts of white (unparasitized) and black (parasitized) pupae were made during a 112-day period (Eggenkamp-Rotteveel Mansveld et al. 1982a,b).

Results

Validation

The simulated population increase of greenhouse whitefly in the absence of parasitoids agrees well with the observations; an example is given in Figure 2. This result can

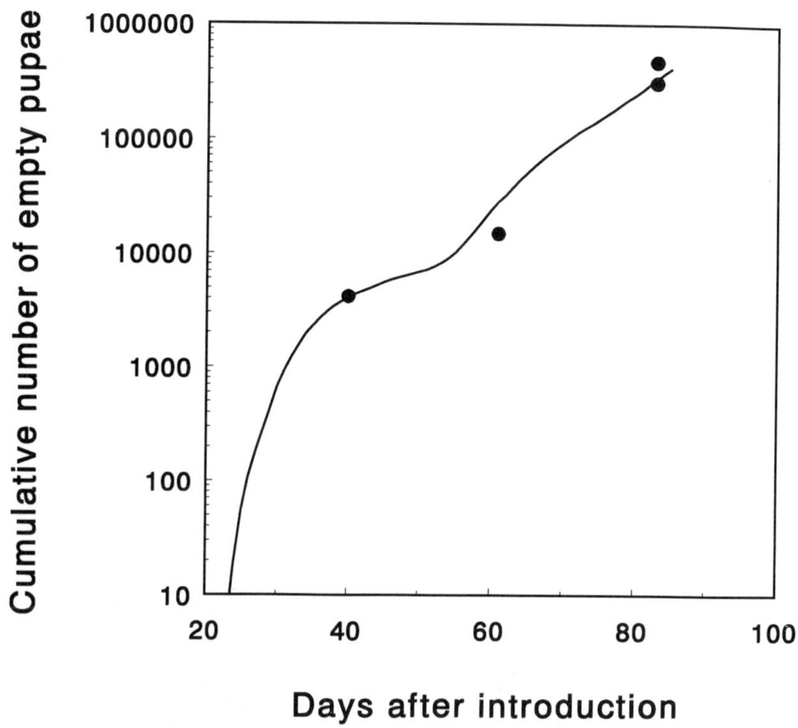

Figure 2. Simulated (line) and observed (dots) numbers of empty whitefly pupae in a small greenhouse with tomato plants (on day 1, 100 whitefly females were released on 10 tomato plants, whitefly pupal counts made on circa 40, 60, and 80 days after release).

be explained to a large extent by the accurate estimates of the life-history parameters, which are based on many experiments at a wide range of temperatures (Van Roermund and Van Lenteren 1992a).

The final model was validated with population counts from an experiment in a commercial tomato crop where parasitoids were introduced by the producer of natural enemies. For initial validation, we used one of the large whitefly patches in the greenhouse comprising 1344 plants. The number of white and black pupae was simulated for a 112-day period – the same period during which absolute counts were made in the greenhouse – with and without introductions of parasitoids (Figure 3). At the end of the simulation, the whitefly population without parasitoid introduction was more than 300 times larger than when parasitoids were released.

The total number of prepupae/pupae for five simulation runs for an area representing 1/20 of the greenhouse area is given in Figure 4. The simulated numbers of prepupae/pupae agree well with the observations; the variation between the simulations is large which is due to the low initial whitefly density. Also percentage visual parasitism (Figure 5) shows good agreement with observed percentages. Apparently, the foraging behaviour as well as the life-history of *E. formosa* were correctly modelled.

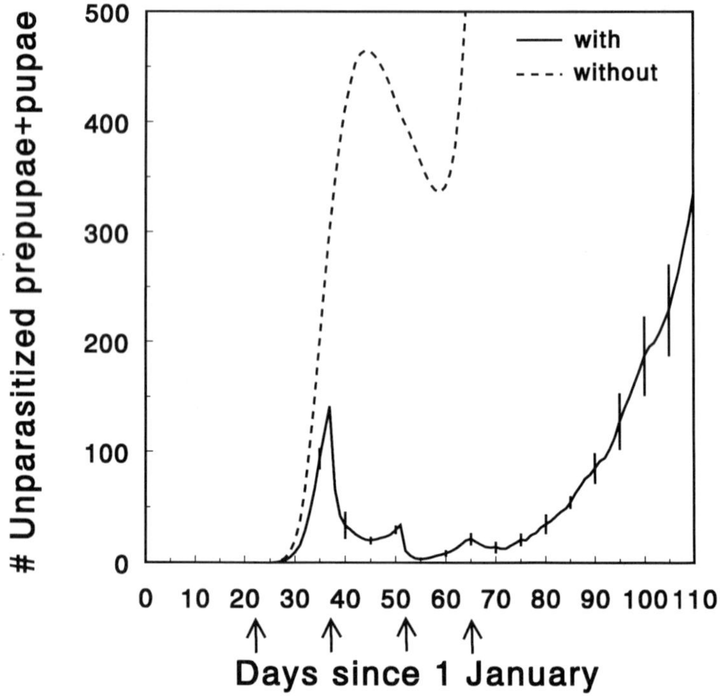

Figure 3. Simulated numbers of unparasitized whitefly prepupae/pupae in a large patch (1344 plants) in a commercial greenhouse, with and without parasitoid releases; bars represent standard deviations (5 replicates), arrows indicate releases of parasitoids.

Sensitivity analysis

For a sensitivity analysis of the final model, the initial whitefly density from one of the commercial greenhouse patches (rows of 25 plants, 25 plants per row=625 plants) was taken. The timing and number of released parasitoids was done according to commercial conditions: parasitoids were released as black pupae on 25 cards evenly distributed over the crop. Simulations were done 5 times per situation. Attention was focused on the increase or decrease of the whitefly 'pressure' in the crop (cumulative whitefly population during the first 100 d). With the detailed and powerful model now available, we could determine the effect of variations in many input parameters. The initial sensitivity analysis consisted of a change of 25% in the value of one particular parameter compared to the 'standard run', for a total of 32 input parameters. The most important results are discussed below (see Van Roermund (1995) for a comprehensive discussion).

Effect of variation in whitefly parameters (life history and others). The following parameters of whitefly were tested: immature development rate, daily oviposition, longevity, immature mortality, degree of variation in stage duration, sex ratio, initial population size, average flight distance from plant to plant, and average residence

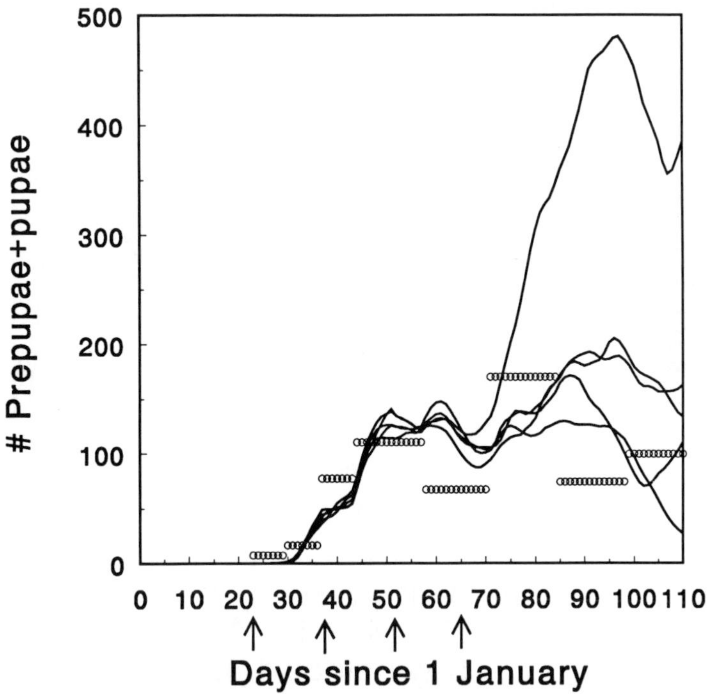

Figure 4. Simulated (lines; 5 replicates) and observed (dots) total numbers of white and black prepupae/pupae on 900 plants (1/20 of total greenhouse), with parasitoid releases; arrows indicate releases.

time on leaf of dead hosts and empty pupal cases. For whiteflies, only a change of the whitefly developmental rate significantly influenced the whitefly population: a 25% decrease in development rate resulted in a 57% decrease of cumulative whitefly numbers.

Effect of variation in parasitoid parameters (life history and others). The following parameters of the parasitoid were tested: immature development rate, longevity, immature mortality, degree of variation in stage duration, initial population size, and average flight distance from plant to plant. For the parasitoid, variation in two parameters resulted in significant changes in the whitefly population: longevity and number of released parasitoids. The strongest influence was from a 25% increase in longevity, which reduced the whitefly population by 36%.

Effect of variation in foraging parameters of the parasitoid. Fourteen parameters related to the parasitoid's foraging behaviour were tested in the sensitivity analysis. Variation in five of these resulted in a significant decrease of the whitefly population: walking activity (48%), walking speed (43%), width of searching path (33%), probability of oviposition in host (30%), search time on lower (infested) leaf side compared

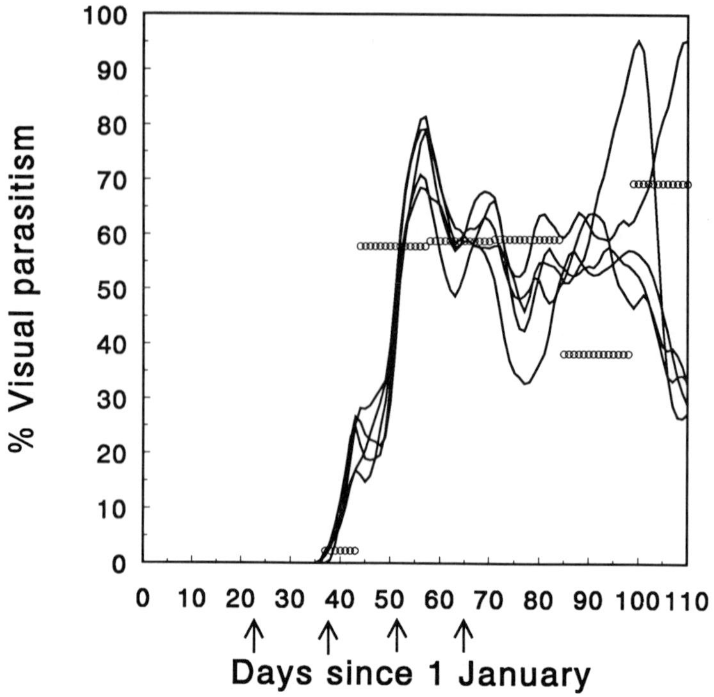

Figure 5. Simulated (lines; 5 replicates) and observed (dots) percentages of visual parasitism (percent black pupae) on 900 plants, with parasitoid releases; arrows indicate releases.

to search time on upper side (23%). The percentages between brackets give the whitefly population reduction. In addition, variation of the host size (a 25% increase in size of host stages) resulted in an increased encounter probability, and a 26% decrease in the whitefly population.

The observed variation of the parasitoid's giving-up time (GUT is the time until departure from a leaf after landing, or after the last host encounter), strongly influences whitefly population reduction. First, the minimum value of the giving-up time is important. When it is too short, parasitoids depart before encountering hosts and whitefly populations are not sufficiently reduced, even though time would be saved on the majority of leaves which are uninfested. Then, the existence of an arrestment effect appeared important, as longer searching after contact with a host leads to increased parasitism of other hosts on that leaf. Finally, the observed variation in GUTs appeared essential for a long-lasting relationship between pest and parasitoid. When variation in GUT was excluded, the whitefly population became less stable and nearly became extinct.

Another important point to stress here is that at low pest densities – and that is where the grower aims at with biological control – a high rate of reproduction of the parasitoid is of limited value. In the literature, too much attention is paid to the maximum reproduction capacity of natural enemies, instead of searching efficiency.

Effect of plant growth and leaf size. When leaflet size is decreased by 25%, the probability increases that whiteflies are found and the whitefly population decreased by 57%.

Recently emerged whitefly adults migrate to young leaves in the top of the plant to feed and oviposit. Slower leaf production results in a longer stay and more ovipositions of whitefly adults on a particular leaflet, which leads to a more aggregated host distribution. In such situations, whiteflies are suppressed to much lower numbers: a higher aggregation increases the probability of being found by the parasitoid, and results in stronger parasitoid arrestment. A 25% reduction in leaf production rate results in a 40% reduction of the whitefly population.

Effect of variation in the driving variables temperature and daylength. When the temperature was taken on average 1°C lower, the whitefly population was reduced by 30% when no parasitoids were released. In the presence of parasitoids, the whitefly population size was hardly affected by a 1°C decrease in temperature.

A reduction in daylength of 1 h resulted in a whitefly population increase of 19%, simply because parasitoids do not search in darkness and, therefore, parasitize fewer whitefly.

Effect of variation in parasitoid releases (number, location and timing). The location of release sites and the number of released parasitoids were both important factors for the reduction of the whitefly population. When all parasitoids were released in the centre of the plot where the first adults were observed, instead of being released at regular spatial intervals, the whitefly population was reduced by 63%. When the number of released parasitoids was increased by 25%, the whitefly population decreased by 30%.

If all parasitoids are released at once, timing of release is very important to obtain a sufficient reduction of the whitefly population, and in general single introductions result in unstable whitefly populations when compared with multiple releases. Timing of multiple releases (4 times with a two-week interval which is the procedure followed by the natural enemy company) is less important, although very early releases resulted in less stable populations than later releases.

Discussion

The cardinal idea behind theoretical explanations for successful biological control has long been that efficient natural enemies operate by creating a stable pest-enemy equilibrium at low densities and that aggregation of natural enemies at patches with high host or prey densities is the critical feature that results in stability (Waage and Hassell 1982). The central role of a low, stable pest equilibrium has recently been challenged and different explanations have been put forward (see e.g. Murdoch 1990). The present analysis indicates that the parasitoid does not create a stable pest-enemy equilibrium during the growing season of a crop, but that it decreases the whitefly's population growth sufficiently to prevent it from causing yield losses. With the Dutch practice of releasing parasitoids in 4 waves with a two-week interval, whiteflies are suppressed initially to very low densities which do not increase above the economic threshold density during the growing season, rather than creating a stable pest-enemy equilibrium at low densities.

Most of the models on population dynamics of parasitoid–host or predator–prey relationships developed so far use experimentally determined functional response curves as input (see e.g. Hassell 1986). Functional response curves describe the relationship between the host density and number of hosts attacked per natural enemy per unit of time. Such functional response curves are generally based on experiments on leaves or in small arenas, and then extrapolated to the crop level by deriving rates of parasitism from the average host density in the crop. This implicitly assumes that the observed relationship for small units like leaves is also valid at higher spatial levels, which is rather unrealistic, particularly when hosts show a strongly clustered distribution in the crop (which is the case for whitefly) and when functional response curves are non-linear (which is the case for many natural enemies, including *E. formosa*).

The model described in this paper links the population dynamics of whitefly and the parasitoid through simulation of the foraging behaviour of individual parasitoids in a crop where whiteflies are present in common distribution patterns and age classes. Therefore, this model does not have the inherent problems of functional response-based models. Individual-based models like this one are a necessity when local interactions and stochasticity are important (Judson 1994).

Another problem with many models of population dynamics is that they treat insects usually as identical, synchronized individuals. In our model, the so-called boxcar train technique for simulation of development was used, which is able to handle all possible development stages simultaneously, as well as the variation in developmental time for each stage (Goudriaan and Van Roermund 1993). As a result, development of whitefly and parasitoid could be simulated very realistically.

With the model, characteristics of the crops were identified that strongly influence the feasibility of biological control, like the effect of leaf surface, leaf structure, and leaf size on the foraging efficiency of the parasitoid. It is particularly the combined effect of these factors that can be evaluated with this model for different crops or plant varieties. The submodel of plant growth is still simple. We aim to include a more realistic tomato growth model in the future, if sensitivity analysis indicates this is of importance.

Also the effect of differences in cultivars on the rate of pest development and the effect of biological control could be evaluated. Simulation results indicate, for example, that breeding cultivars for resistance against whitefly is much more profitable when it concentrates on an increase of the immature developmental time of whitefly instead of aiming at an increase of immature mortality or reduced fecundity.

In addition, the model pointed at weak points in previous evaluation programmes of natural enemies. In biological control research, parasitoids are usually tested in small-scale experiments at high host densities before introduction in the field (Waage 1990). As a result, maximum daily oviposition of parasitoids is measured, whereas for field performance at low pest densities effective host searching is the most important process. With the model, we were able to identify a number of parasitoid characteristics that can easily be measured in the laboratory and provide important information about its searching efficiency, such as walking speed, walking activity, probability of host acceptance and the parasitoid's arrestment effect when searching on clean and infested leaves. The model thus assists in identification of the characteristics which compose an efficient natural enemy for various types of biological control. As a result, selection of natural enemies can be drastically improved.

Three approaches for the selection of natural enemies emerge from the literature: (i) evaluation based on individual attributes of natural enemies, (ii) evaluation based on integration of individual attributes and (iii) evaluation based on ecosystem studies.

In the evaluation based on *individual attributes* of natural enemies (called the reductionist approach by Waage 1990), agents are selected on the basis of particular biological attributes or life-history characteristics (e.g. duration of development, fecundity, searching efficiency). Theory dissects natural enemies into simple sets of characters, which can be viewed and compared independently. This approach is no longer popular, although it is still used.

In the evaluation based on *integration of individual attributes*, one comes up with a composite picture of the pest reduction potential of the natural enemy (e.g. Van Lenteren 1986). When carefully applied, this method has proved to be of value, and the model we developed for the tritrophic system *Encarsia*-whitefly-crop (and influenced by the environment) is an example of such an evaluation.

The evaluation based on *ecosystem studies* (named the holistic approach by Waage 1990) proceeds from the theoretical notion of how natural enemies fit into the broad ecology of the pest and its other mortality factors. Here, community concepts predominate, expressed in arguments for density-specific agent complexes, multiple introductions, and filling 'empty' natural enemy niches. This approach is not often applied, but strongly supported by some biocontrol workers (e.g. Ehler 1990). Although it is scientifically attractive, it is not applicable yet, and we feel that it will take many more years before it may be of use, contrary to what e.g. Ehler (1990) suggests.

Currently, there are good evaluation criteria available to allow for a choice between useless and potentially promising natural enemies. Such a choice prevents research on and introduction of inefficient natural enemies. With the gradual improvement of evaluation criteria and a further integration of criteria, ranking among the promising natural enemies will be possible. Systems analysis and simulation are important tools for such an integration, but go beyond the mere comparison of parasitoids, as they also take characteristics of the pest, the crop and the environment into account. In such a complete framework, and with models including sufficient biological detail, the role of different natural enemies can be analysed.

Acknowledgements

This is article number XLVI in the series, The parasite–host relationship between *Encarsia formosa* (Hymenoptera: Aphelinidae) and *Trialeurodes vaporariorum*

(Homoptera: Aleyrodidae). The second author was financially supported by The Netherlands Technology Foundation (NWO-STW).

References

Eggenkamp-Rotteveel Mansveld M H, Ellenbroek F J M, Van Lenteren J C (1982a) The parasite-host relationship between *Encarsia formosa* (Hymenoptera: Aphelinidae) and *Trialeurodes vaporariorum* (Homoptera: Aleyrodidae). XII. Population dynamics of parasite and host in a large, commercial glasshouse and test of the parasite-introduction method used in The Netherlands. Part I. J. Appl. Entomol. 93:113–130.

Eggenkamp-Rotteveel Mansveld M H, Ellenbroek F J M, Van Lenteren J C (1982b) The parasite-host relationship between *Encarsia formosa* (Hymenoptera: Aphelinidae) and *Trialeurodes vaporariorum* (Homoptera: Aleyrodidae). XII. Population dynamics of parasite and host in a large, commercial glasshouse and test of the parasite-introduction method used in The Netherlands. Part II. J. Appl. Entomol. 93:258–279.

Ehler L E (1990) Introduction strategies in biological control of insects. Pages 111–134 in Mackauer M, Ehler L E, Roland J (Eds.) Critical issues in biological control. Intercept Ltd., Andover, UK.

Gerling D (Ed.) (1990) Whiteflies: Their bionomics, pest status and management. Intercept Ltd., Andover, UK. 348 p.

Goudriaan J, Van Roermund H J W (1993) Modelling of ageing, development, delays and dispersion. Pages 89–126 in Leffelaar P A (Ed.) On systems analysis and simulation of ecological processes. Kluwer Academic Publishers, Dordrecht, The Netherlands.

Hassell M P (1986) Parasitoids and population regulation. Pages 201–224 in Waage J K, Greathead D J (Eds.) Insect parasitoids. Academic Press, London, UK.

Hulspas-Jordaan P M, Van Lenteren J C (1989) The parasite–host relationship between *Encarsia formosa* (Hymenoptera: Aphelinidae) and *Trialeurodes vaporariorum* (Homoptera: Aleyrodidae). XXX. Modelling population growth of greenhouse whitefly on tomato. Pages 1–54 in Wageningen Agricultural University Papers 89, Wageningen, The Netherlands.

Joosten J, Elings A (1985) Onderzoek naar de toetsmethoden voor resistentie tegen kaswittevlieg (*Trialeurodes vaporariorum* (Westwood)) in tomaat (*Lycopersicon esculentum* Mill). MSc thesis, Wageningen Agricultural University, The Netherlands. 62 p.

Judson O P (1994) The rise of the individual-based model in ecology. Trends in Ecology and Evolution 9:9–14.

Murdoch W W (1990) The relevance of pest-enemy models to biological control. Pages 1–24 in Mackauer M, Ehler L E, Roland J (Eds.) Critical issues in biological control. Intercept Ltd., Andover, UK.

Noldus L P J J, Van Lenteren J C (1990) Host aggregation and parasitoid behaviour: Biological control in a closed system. Pages 229–262 in Mackauer M, Ehler L E, Roland J (Eds.) Critical issues in biological control. Intercept Ltd., Andover, UK.

Van Lenteren J C (1986) Parasitoids in the greenhouse: Successes with seasonal inoculative release systems. Pages 342–374 in Waage J K, Greathead D J (Eds.) Insect parasitoids. Academic Press, London, UK.

Van Lenteren J C (1993) Parasites and predators play a paramount role in pest management. Pages 68–81 in Lumsden R D, Vaughn J L (Eds.) Pest management: Biologically based technologies. American Chemical Society, Washington DC, USA.

Van Lenteren J C (1995) Integrated pest management in protected crops. Pages 311–343 in Dent D (Ed.) Integrated pest management. Chapman and Hall, London, UK.

Van Lenteren J C, Noldus L P J J (1990) Whitefly–plant relationships: Behavioural and ecological aspects. Pages 47–89 in Gerling D (Ed.) Whiteflies: Their bionomics, pest status and management. Intercept Ltd., Andover, UK.

Van Lenteren J C, Woets J (1988) Biological and integrated control in greenhouses. Annu. Rev. Entomol. 33:239–269.

Van Roermund H J W (1995) Understanding biological control of greenhouse whitefly with the parasitoid *Encarsia formosa*: From individual behaviour to population dynamics. PhD thesis, Wageningen Agricultural University, The Netherlands. 243 p.

Van Roermund H J W, Van Lenteren J C (1992a) The parasite–host relationship between *Encarsia formosa* (Hymenoptera: Aphelinidae) and *Trialeurodes vaporariorum* (Homoptera: Aleyrodidae). XXXIV. Life-history parameters of the greenhouse whitefly, *Trialeurodes vaporariorum* as a function of host plant and temperature. Pages 1–102 in Wageningen Agricultural University Papers 92.3, Wageningen, The Netherlands.

Van Roermund H J W, Van Lenteren J C (1992b) The parasite–host relationship between *Encarsia formosa* (Hymenoptera: Aphelinidae) and *Trialeurodes vaporariorum* (Homoptera: Aleyrodidae). XXXV. Life-history parameters of the greenhouse whitefly parasitoid *Encarsia formosa* as a function of host stage and temperature. Pages 103–147 in Wageningen Agricultural University Papers 92.3, Wageningen, The Netherlands.

Van Roermund H J W, Van Lenteren J C (1995a) Foraging behaviour of the whitefly parasitoid *Encarsia formosa* on tomato leaflets. Entomol. Exp. Appl. 76:313–324.

Van Roermund H J W, Van Lenteren J C (1995b) Residence times of the whitefly parasitoid *Encarsia formosa* Gahan (Hym. Aphelinidae) on tomato leaflets. J. Appl. Entomol. 119:465–471.

Van Roermund H J W, Hemerik L, Van Lenteren J C (1994) The influence of intra-patch experiences and temperature on the time allocation of the whitefly parasitoid *Encarsia formosa*. J. Insect Behav. 7: 483–501.

Van Roermund H J W, Van Lenteren J C, Rabbinge R (1996a) Analysis of the foraging behaviour of the whitefly parasitoid *Encarsia formosa* in an experimental arena: A simulation study. J. of Insect Behav. (in press).

Van Roermund H J W, Van Lenteren J C, Rabbinge R (1996b) Analysis of the foraging behaviour of the whitefly parasitoid *Encarsia formosa* on a leaf: A simulation study. Biological Control, Theory and Application in Pest Management (in press).

Van Roermund H J W, Van Lenteren J C, Rabbinge R (1996c) Analysis of the foraging behaviour of the whitefly parasitoid *Encarsia formosa* on a plant: A simulation study (submitted).

Vet L E M, Dicke M (1992) Ecology of infochemical use by natural enemies in a tritrophic context. Annu. Rev. Entomol. 37:141–172.

Waage J K (1990) Biological theory and the selection of biological control agents. Pages 135–157 in Mackauer M, Ehler L E, Roland J (Eds.) Critical issues in biological control. Intercept Ltd., Andover, UK.

Waage J K, Hassell M P (1982) Parasitoids as biological control agents – a fundamental approach. Parasitology 84:241–268.

Quantitative evaluation of growth and yield of rice plants infested with rice planthoppers

T. WATANABE[1], L.T. FABELLAR[2], L.P. ALMAZAN[2], E.G. RUBIA[2,3], K.L. HEONG[2] and K. SOGAWA[1,4]

[1]*Kyushu National Agricultural Experiment Station, Nishigoshi, Kumamoto 861-11, Japan*
[2]*International Rice Research Institute, P.O. Box 933, 1099 Manila, Philippines*
[3]*Philippine Rice Research Institute, Maligaya, Muñoz, Nueva Ecija 3119, Philippines*
[4]*Japan International Research Centre for Agricultural Sciences, 1–2 Ohwashi, Tsukuba, Ibaraki 305, Japan*

Key words: rice, rice growth model, brown planthopper, *Nilaparvata lugens*, white-backed planthopper, *Sogatella furcifera*, Integrated Pest Management

Abstract
Characteristics of the infestation of, and damage to, rice plants by the rice planthopper, the brown planthopper (BPH) and the white-backed planthopper (WBPH) in Japan were reviewed and a quantitative analysis and growth models were used to evaluate the feeding effect on rice growth and yield. At the vegetative stage, WBPH feeding reduced the shoot dry weight about 4 times more than the total dry matter removed by insect feeding. WBPH feeding has a direct effect on biomass production by removing phloem sap, and some indirect effects by reducing some physiological processes. BPH feeding also has direct and indirect effects, and causes severe damage to rice plants after flowering. Field experiment suggested that the translocation rate of carbohydrates from shoot to panicle at the maturing stage was not affected by BPH feeding as long as the carbohydrates that remained in the reserves pool was sufficient for grain growth. Photosynthetic reduction was observed in infested plants by using stable isotope-labelled carbon dioxide. The direct effect (removal of the phloem sap) and the indirect effect (reduction in leaf area) were introduced into the rice growth model. This model could describe not only the reduction in grain weight, but also the plant death that is referred to as 'hopper burn'. The simulation outputs were influenced by fluctuations in weather conditions and the infestation period, though the infestation density was the same.

Introduction

The brown planthopper (BPH), *Nilaparvata lugens* Stål, and the white-backed planthopper (WBPH), *Sogatella furcifera* Horváth, are the most important insect pests of rice in Asia. These planthoppers selectively feed on phloem sap of the rice plants (Sogawa 1982). They are known to migrate long distances from the tropical zone to the temperate zone (Kisimoto 1976). In the rainy season (Baiu), June and July, they invade Japan from China when the low-level jet stream associated with the Baiu front stretches from China to Japan (Seino et al. 1987). After immigration to the paddy fields, both planthopper populations have three generations during the rice growing season.

The mean density of BPH immigrants is 0.01 to 0.1 per hill in the Kyushu district, southwestern Japan, which receives more immigrants than the other regions (Kisimoto 1979). In spite of low initial densities, a steep and steady population growth of BPH

M.J. Kropff et al. *(eds.): Applications of Systems Approaches at the Field Level, 365–382.*

Table 1. Rates of dry matter consumption (mg d^{-1} $insect^{-1}$) by fourth and fifth instar nymphs and adult females of the brown planthopper (BPH) and white-backed planthopper (WBPH) that feed on phloem of rice plants (*Source:* Sogawa 1992).

Fractions	BPH		WBPH	
	Nymphs	Adults	Nymphs	Adults
Honey dew	0.54	1.54	0.28	0.65
Insect biomass	0.10	0.31	0.06	0.07
Respiration	0.04	0.14	0.03	0.06
Total	0.68	1.99	0.37	0.78

results in high density in the third generation (Kuno 1968), causing 'hopper burn' (Kisimoto 1965). Although immigrants of both planthopper species occur simultaneously, the initial densities of WBPH are ten to 100 times higher than those of the BPH (Kuno 1968; Watanabe 1994). WBPH invades the whole area of Japan and its immigrant densities decrease linearly with the distance from the west coast of Kyushu (Kisimoto 1979). Peak density is reached in the first or second generation (Kuno 1968; Watanabe 1994).

Annual fluctuations in the field populations of planthoppers were analysed using records of collections by a light trap over a 40-year period (Watanabe et al. 1994a). WBPH immigrant densities have increased since the mid-1970s. The variances in the densities of BPH immigrants and successive generations have also been increasing since the early 1980s. Accurate monitoring of population growth and damage to the rice crop are necessary for the integrated management of rice planthoppers. Compared with the progress in understanding the population dynamics and monitoring systems of the rice planthoppers, a little information is available on their feeding effects on rice growth and yield.

In this paper, we (i) describe the characteristics of the planthopper infestation and damage to rice plants, (ii) give quantitative analysis of growth and yield of the infested rice plants, and (iii) present the results of simulations of the effect of BPH feeding on grain yield and death of rice plants.

Characteristics of the planthopper infestation and damage to rice plants

It is helpful to know the feeding dynamics of the planthoppers to quantify the relationship between the infestation rate and plant growth, or to couple them to the rice growth model. The rate of honeydew excretion increased exponentially as planthopper nymphs developed and was closely related to the increase in the dry weight of the planthoppers (Kenmore 1980). Sogawa (1992) quantified the relationships between planthopper feeding dynamics and their developmental stage, crop age and temperature. Honeydew excretion was maintained at a high level by female adults but this was much lower in males. Thus, the population of late instar nymphs and adult female planthoppers acts as a major extra sink for the rice crop. BPH drained 2 to 3 times more dry matter from the rice plant than WBPH (Table 1). Eighty percent of the total

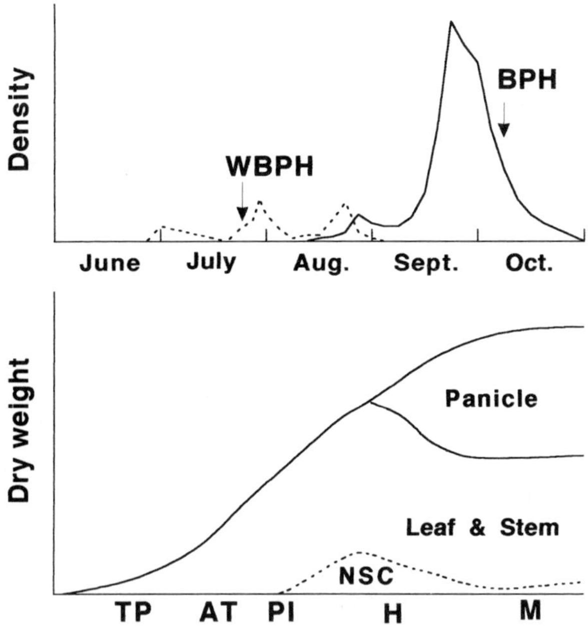

Figure 1. Typical population dynamics of BPH and WBPH, and the rice growth pattern in the Kyushu district, southwestern Japan. TP, Transplanting; AT, Active tillering; PI, Panicle initiation; H, Heading; M, Maturity; NSC, Non-structural carbohydrate.

dry matter removed from the phloem was directly excreted in the form of honeydew and only 10–15% was assimilated into insect biomass (Table 1). The rate of honeydew excretion by both planthopper species varied significantly with crop age, leaf nitrogen concentration and temperature (Sogawa 1992). The maximum rate was reached at two weeks after transplanting, was maintained up to maximum tillering and declined until booting (WBPH), or flowering (BPH) (Sogawa 1992).

The population dynamics and infestation patterns of both planthoppers in the Kyushu district are shown with a typical rice growth pattern in Figure 1. The transplanting date in Japan varies from early April to mid-July, and the immigrant peak of planthoppers usually occurs in late June to early July. Hopper burn usually occurs by BPH after flowering of the rice plants because of the high density in the third generation. Even if the planthoppers do not cause such severe damage, their infestations still result in a significant decrease in rice yield (Sogawa and Cheng 1979). BPH infestation causes some metabolic changes, such as an increase in free amino acids and a decrease in soluble protein in leaves (Sogawa 1971; Cagampang et al. 1974). BPH nymphs and adults usually stay and feed on the lower sheaths near the water surface (Kuno 1968).

Compared with BPH, WBPH has a higher immigrant density and an earlier peak generation. The type of damage to rice plants varied widely with infestation behaviour and the infestation period of the WBPH (Table 2). WBPH immigrants prefer young

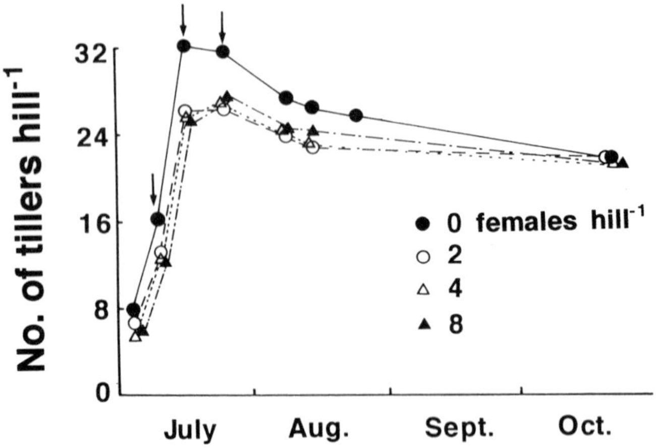

Figure 2. Fluctuations in the number of tillers at each infestation level. Arrows indicate the date when the F-values were significantly large ($p<0.05$) by single-factor ANOVA. Infestation period was 26 June to 3 July (8 to 15 DAT) (*Source:* Watanabe et al. 1994b).

Table 2. Type of infestation by WBPH and its damage to rice plants in Japan.

Rice growth stage	Type of infestation	Type of damage
A Transplanting to early tillering	Oviposition to the leaf sheath	Discoloration of the leaf sheath Reduction in the growth rate
B Tillering	Phloem sap feeding	Reduction in plant height and tillering
C Panicle initiation to booting	Phloem sap feeding	Reduction in plant height Reduction in number of spikelets
D Booting to flowering	Phloem sap feeding	Brown-coloured spikelets Black-cracked rice
E Flowering to maturity	Phloem sap feeding	Reduction of grain maturity Sooty mould Death of plant

rice plants and oviposit on the leaf sheath. Rice plants that suffer intensive oviposition show discoloration of the leaf sheath. During the vegetative stage, sap sucking by the nymphs and adults of the following generation reduces the plant biomass and number of spikelets (Naba 1992; Watanabe and Sogawa 1994). When WBPH infest rice plants, they aggregate to the flag-leaf sheath at booting, and to the panicles after flowering, which induces discoloured spikelets and cracked rice (Noda 1986). Although the main infestation of WBPH occurs during the vegetative stage on Kyushu Island, plant death (such as hopper burn) occurs as a result of intensive WBPH infestation in the northern district of Japan (Matsumura 1991).

Table 3. Coefficients of correlation between yields and density of WBPH (*Source:* Watanabe and Sogawa 1994a).

	Density of WBPH at 2nd generation	
	Mean density	Peak density
Yield ($g\,m^{-2}$)	0.632*	0.732*
Number of panicles (m^{-2})	0.252	0.022
Number of spikelets ($panicle^{-1}$)	0.348	0.583
Number of spikelets (m^{-2})	0.583*	0.741**

*$p<0.05$.
**$p<0.01$.

Analysis of growth and yield of infested rice plants

White-backed planthopper (WBPH)

The effects of the period and intensity of WBPH infestation on the growth and yield of rice plants were investigated by field-cage experiments (Watanabe and Sogawa 1994a; Watanabe et al. 1994b). Rice seedlings were transplanted in late June. Infestation levels were manipulated by changing the planthopper densities. Insects were released at early and maximum tillering. Insecticide spray plots were prepared for each treatment to evaluate separately the effects of oviposition damage by female immigrants (infestation type A in Table 2) and feeding damage by subsequent progeny (infestation type B). When WBPHs were released at early tillering, intensive oviposition on the leaf sheath significantly reduced the plant height and number of tillers of infested rice plants until maximum tillering. These growth differences between infested and uninfested plants disappeared gradually towards flowering (Figure 2). Watanabe et al. (1994b) observed these recovery processes by measuring the spectral reflectivity of the rice plant canopy using a portable spectroradiometer (450–1150 nm wavelength). The flowering date was delayed at the early infested plots (Watanabe and Sogawa 1994a). Female oviposition on the leaf sheath at maximum tillering did not cause a reduction in the number of tillers. No correlation was observed between the initial density of the released WBPH and yield, although there was a significant correlation between the yields and yield component (number of spikelets) and the density of WBPH in the second generation, which occurs during panicle initiation (Table 3). Based on our results, biomass reduction at early tillering due to intensive oviposition by WBPH would recover during the vegetative stage if phloem sap feeding could be prevented from tillering to panicle initiation.

Further greenhouse experiments were conducted to quantify the relationship between feeding intensity and biomass reduction in the rice plant (Watanabe and Sogawa 1994b). Rice plants at active tillering and panicle initiation were used. WBPHs were allowed to feed during the nymphal period and were removed when the adult stage emerged. The leaf area and shoot dry weight decreased during infestation by the nymphs. A linear relationship was found between the total dry weight of WBPH and the loss of shoot dry weight (Figure 3). Total dry matter consumption by sap feeding on rice plants during the period from the first instar nymph stage to

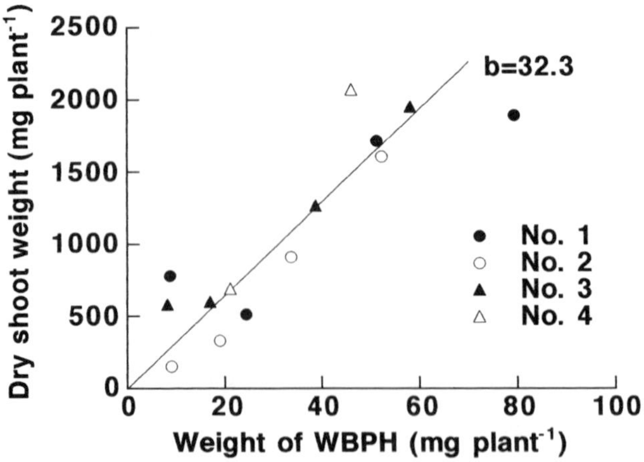

Figure 3. Relationship between total dry weight of WBPH and loss of dry shoot weight. Numbers 1–4: Experiments under different weather conditions (*Source:* Watanabe and Sogawa 1994b).

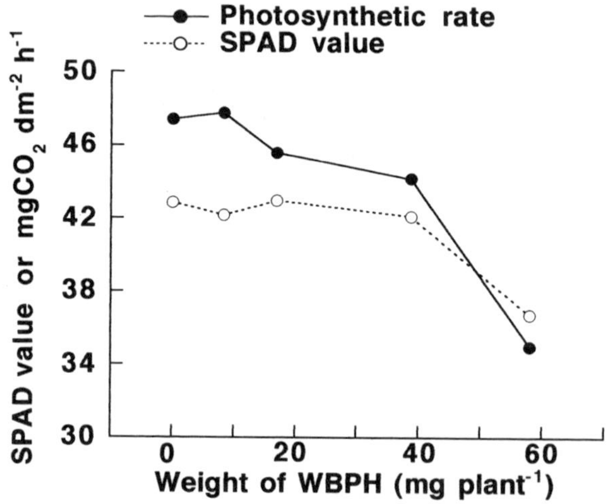

Figure 4. Relationships between total dry weight of WBPH and photosynthetic rate and leaf colour value of the top leaf at panicle initiation. Photosynthetic rate was measured using a portable photosynthesis system (LICOR LI6200), and leaf colour was measured using a Minolta SPAD 502 chlorophyll meter (*Source:* Watanabe and Sogawa 1994b).

the adult stage was about 7 mg per 1 mg dry weight of WBPH (Sogawa 1992). The shoot dry weight decreased 32.3 mg per mg insect (Figure 3). WBPH feeding reduced the shoot dry weight about 4 times more than the total dry matter removed by insect feeding. This result suggests that the dry shoot weight of infested rice plants is reduced not only by direct feeding, but also by indirect effects. Phloem sap feeding by WBPH reduced both the chlorophyll content and the photosynthetic rate of leaves (Figure 4). The reduction in leaf area and photosynthetic rate probably work together to reduce the daily production of photosynthates.

Brown planthopper (BPH)

The rice plant does not produce further tillers or leaves after flowering. A part of the photosynthates are consumed in respiration, and the remainder accumulates in a 'reserves' pool in the stem. The translocation of carbohydrates from the reserves pool to the panicle, rather than production of new plant parts, starts just after flowering. This pool is constructed from non-structural carbohydrates (NSC) that accumulate before flowering and are assimilated after flowering (Figure 1). We supposed that the damage from feeding on phloem sap after flowering was also divided into a direct effect (drainage of photosynthates), and three indirect effects: translocation blockage of photosynthates (Kenmore 1980), a reduction in the photosynthetic rate and an increase in maintenance respiration.

Field experiments were conducted to determine the relationship between the feeding intensity of BPH and the grain yield (Watanabe 1994). At the panicle initiation, BPH adults were released at several densities to the paddy field in order to cause feeding damage by the next generation, which emerges after flowering. Panicle dry weight, shoot (all plant parts above the ground) dry weight and insect density were measured at two- to three-week intervals. Both the yield and growth rate of the shoots and panicles decreased linearly when the BPH incidence (density $\times$ observation interval) increased (Figure 5). The reduction in the panicle growth rate was lower than that of the shoot growth rate.

Kenmore (1980) showed that phloem translocation was blocked by the feeding of BPH on young rice plants. If the translocation blockage of photosynthates occurred by BPH feeding on mature plants, then photosynthates would accumulate in the reserves pool, and panicle growth rate would be lower than the shoot growth rate. From our field experiment, we assumed that the translocation rate at the maturing stage was not affected by BPH feeding as long as the carbohydrate remaining in the reserves pool was sufficient for grain growth.

Effects of phloem feeding by BPH on the rice canopy photosynthesis were measured by using the stable isotope ^{13}C (Watanabe and Kitagawa, in preparation). Rice plants were planted in a pot of $0.02\,m^2$. Infestation started at booting. One hundred large nymphs (third to fourth instar) were released in each pot, and were allowed to feed for 10 days. After the infestation period, each pot was put into a closed acryl chamber and $^{13}CO_2$ was fed under clear sky conditions. After feeding, each leaf was measured for area, oven-dried and weighed. The amount of ^{13}C assimilated by each leaf position was determined. The areas of the fourth and fifth leaves of infested plants were reduced. BPH feeding caused the death of the lowest leaf. There were no significant differences in the areas of the first to third leaves between the infested plants and control. The excess ^{13}C abundance per leaf area was used as an index of photosynthetic activity of the leaves. These values decreased gradually from the top to the lower leaves in both treatments. The rate of reduction of the ^{13}C abundance in the infested plants was higher than that in the control (Table 4). The ^{13}C abundance in the third to fourth leaves was significantly different between the two treatments.

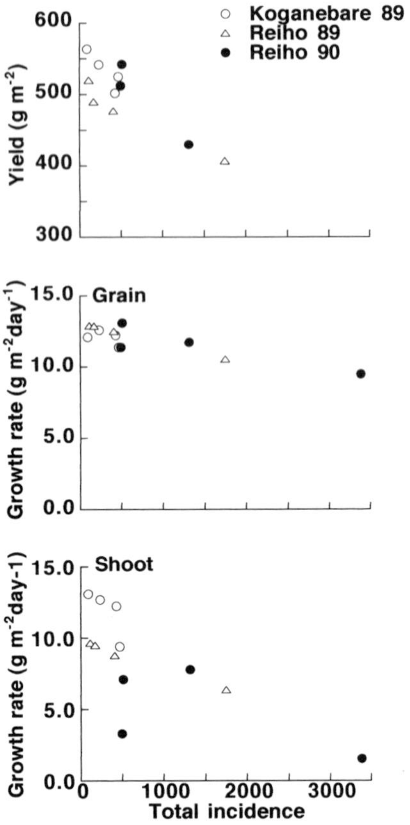

Figure 5. Relationships between the incidence of BPH (density×observation interval) and grain yield and daily growth rate of grain and shoot dry weight in the field experiment (*Source:* Watanabe 1994).

Table 4. Effects of phloem sap feeding by BPH on the rice canopy photosynthetic rate and leaf area.

Leaf position	Photosynthetic rate	Leaf area
1	0.85*	0.99*
2	0.78	1.05
3	0.59	1.06
4	0.43	0.81
5	0.39	0.62
Total	0.72	0.92

*Each value was calculated as the ratio between the infested plants and control.

Simulation of the effects of BPH feeding on grain growth and death of rice plants

Rice growth models have been constructed by several researchers (Iwaki 1975; Angus and Zandstra 1980; Horie 1987; Alocilja and Ritchie 1988; Penning de Vries et al. 1989; Graf et al. 1990). Plant growth models are useful not only for quantitative estimates of yield, but also for analysing the physiological processes that result from plant damage by insect pests and diseases. Rice growth models have been used for evaluating feeding damage by BPH (Kenmore 1980), the leaf folders (Benigno et al. 1988; Graf et al. 1992; De Jong and Daamen 1992) and the stem borer (Rubia and Penning de Vries 1990). Kenmore (1980) coupled the effect of BPH feeding with the rice model (Angus and Zandstra 1980), and calculated how rice grain yield was affected by solar radiation, BPH density and the rice stage at which infestation began. However, his model did not incorporate the indirect effects of BPH feeding or describe the process of hopper burn. Our objectives were to (i) simulate and understand the effect of BPH feeding on grain growth and death of rice plants, and (ii) evaluate the effects of fluctuations on the infestation period and meteorological conditions on yield.

Models

To simulate the effect of BPH feeding on the rate of translocation and on the rate of accumulation of carbohydrate in grain, we modified the MACROS-L1D model of Penning de Vries et al. (1989). In the vegetative phase of rice plants, new biomass, which is produced in leaves, is allocated to leaves, stems and roots to develop new organs. There is no quantitative information on changes in assimilate partitioning in rice plants that have been attacked by insect pests in the vegetative stage. Therefore, it is difficult to simulate the growth change in rice plants that are infested before flowering. In addition, the third generation of BPH (with the highest density) usually surges after flowering. Therefore, in our model, the simulation starts at flowering. To evaluate the contributions of direct and indirect effects on grain growth separately, two models were formulated: a basic model in which only the direct effect of BPH feeding (carbohydrate reduction) is introduced, and an extended model which included both the direct and an indirect effect (reduction of leaf area).

Basic model (model 1). Canopy gross photosynthesis ($PCGC$, $\mathrm{kg\,CO_2\,ha^{-1}\,h^{-1}}$) is calculated in the subroutine FUPHOT (Penning de Vries et al. 1989) using the maximum rate of photosynthesis of single leaves ($PLMX$, $\mathrm{kg\,CO_2\,ha^{-1}\,h^{-1}}$), initial light use efficiency ($PLEA$, $\mathrm{kg\,CO_2\,ha^{-1}\,h^{-1}\,(J\,m^{-2}\,s^{-1})^{-1}}$), leaf area index ($LAI$, $\mathrm{m^2\,m^{-2}}$), and the photosynthetically active radiation (PAR, $\mathrm{J\,m^{-2}\,s^{-1}}$):

$$PCGC = PLMX \times (1.0 - \exp(-PLEA \times PAR / PLMX)$$

PLMX and *PLEA* depend on temperature.

At flowering, the green leaf area reaches a maximum (*LAImax*) and then starts to decline, even under pest-free conditions. The rate of leaf area loss (*DLA*) was a function of development stage (*DVS*):

$$DLA = LAImax \times FL(DVS)$$

where *FL(DVS)* is a function of *DVS* that gives the relative loss rate of leaves.

Respiration was divided into two processes (McCree 1974): maintenance respiration in each part of the plant, and growth respiration of the grain.

Carbohydrate accumulation rates in the grains were determined by the daily demand of grains (*Gp*) and the translocatable carbohydrates from reserves pools (*Ga*) (Spitters et al. 1989):

$$RCAGR = \min(Gp, Ga)$$

where *Gp* is a function of *DVS*, temperature, and the ratio between the real grain weight (*WGR*) and the potential maximum grain weight (*GCAP*):

$$Gp = Gp_0 \times (1.0 - (WGR/GCAP)^3)$$

where Gp_0 is the maximum value of *Gp*. *Ga* is affected by the reserve carbohydrate level (*RESL*), which is calculated from the weight of reserve carbohydrates (*ARES*) and the weight of the shoots (*WSS*):

$$RESL = ARES/WSS$$

$$RSV = FR(RESL)$$

$$Ga = ARES \times RSV$$

The translocation rate (*RSV*), which is a function of *RESL*, starts to decrease gradually when *RESL* is lower than 0.10. When *RESL* is lower than 0.03, *RSV* decreases rapidly.

BPH infestation was defined as a consumption of carbohydrate in the reserves pools. BPH feeding rate was affected by the host plant age and temperature (Sogawa 1992). Honeydew excretion by the BPH on the rice plants before flowering was variable depending on the rice development stage and temperature. However, it became relatively constant after flowering. A constant sucking rate of 1 mg carbohydrates per mg BPH dry weight per day was assumed (*SRBP*). Therefore, the carbohydrate weight drained by BPH (*CDEMAB*) is calculated as follows:

$$CDEMAB = SRBP \times AWBP \times NUBP$$

where *AWBP* is the dry weight of *BPH* (g) and *NUBP* is the number of *BPH* (m^{-2}).

After flowering, the daily phenological development rate (*DVR*) can be calculated as a function of temperature (Horie and Nakagawa 1990). The simulation stops when the crop reaches full maturation ($DVS = 2.0$), or when the carbohydrate reserves are less than 0 for three consecutive days (Penning de Vries et al. 1989).

Extended model (model 2). As we previously mentioned, the change in translocation rate due to BPH infestation was very low. Furthermore, we have no data on the effect of BPH feeding on maintenance respiration. BPH nymphs and adults usually stay and feed on the lower sheaths near the water surface (Kuno 1968). The reduction in photosynthetic activity and leaf death started in the lower leaves because of BPH infestation (Table 4). This phenomenon was introduced in the simulation processes, a relational diagram is given in Figure 6. The canopy of host plants was derived into 5 leaf layers with similar leaf areas. The SUPHOL photosynthesis sub-

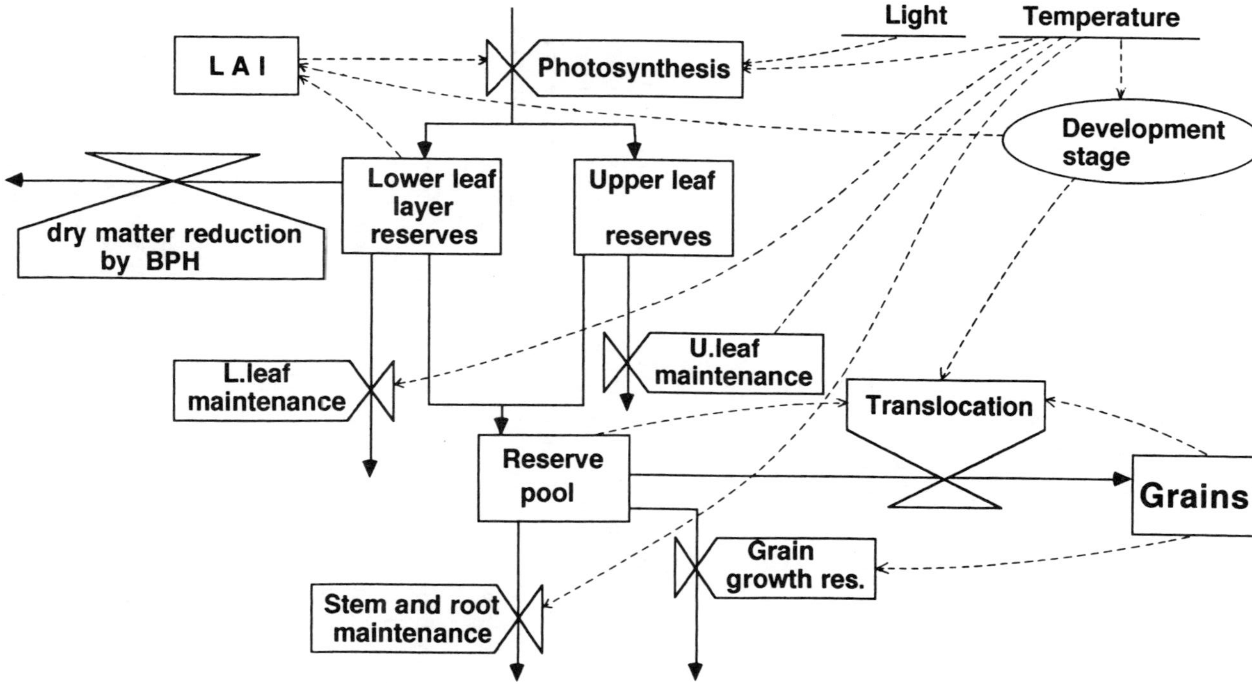

Figure 6. Relational diagram of a rice growth model that accounts for BPH feeding (model 2).

routine (Penning de Vries et al. 1989) was developed to calculate photosynthesis in each leaf layer. The SUPHOL subroutine requires the distribution of leaf angles and the active and dead leaf areas in each leaf layer. We used the erect leaf angle distribution in the first and second layer from the top and a spherical distribution in the third to fifth layers (Penning de Vries et al. 1989). The dead leaf area at the initial condition is 0. Each layer has a hypothetical carbohydrate pool, and the assimilates produced each day are preserved in the respective carbohydrate pools after subtracting the maintenance respiration of each layer.

BPH initially feed on the carbohydrate pool of the lowest layer. There is no translocation of carbohydrate from the other layers to the lowest one. If the BPH population ingests more carbohydrate than is assimilated in the leaf layer, the leaves in the layer die progressively. These processes also include a change in photosynthetic activity. When the entire lowest leaf layer dies, the BPH population moves to, and continues to feed on, the next leaf layer. The simulation stops when the leaf area index (*LAI*, $m^2 m^{-2}$) is lower than 1.2 (this value is arbitrarily chosen), or when both carbohydrate reserve pools and daily photosynthesis supplies are completely consumed.

Model simulation

Initial values for the simulation runs, such as values for *LAI* and top dry weight, were obtained from data collected for the *japonica* variety Reiho at flowering in 1989 in Kyushu National Agricultural Experiment Station, Chikugo, Fukuoka, Japan (33° 12′N; 130° 29′E). We also used the daily mean temperature and radiation, collected from the same station in 1989. Planting density was 20 hill m^{-2}.

Simulated grain weight at full maturation (yield = 569 g m^{-2}) was nearly the same as that observed in the field (571 g m^{-2}). The growth rate of grain in the simulation was affected by BPH feeding. In model 1, grain weight decreased linearly with the increase in BPH density (Figure 7A). However, the phenomenon of hopper burn was not simulated satisfactorily. We observed that hopper burn occurred when the mean density of BPH in the third generation increased to more than 200 per hill (Watanabe 1994). Kenmore (1980) also showed that hopper burn occurred at densities greater than 200 BPH late instar nymphs per hill. The simulation output of model 1 showed that the incorporation of a direct effect into the rice model was not sufficient to simulate the effect of BPH feeding on grain growth.

In model 2, the simulated grain growth under insect-free conditions was very similar to that in model 1 (Figure 7B). However, the simulation showed that the host plants died during the process of grainfilling, when the BPH density is 200 or more per hill. Model 2 could simulate not only the reduction in grain weight, but also plant death (hopper burn). The yield simulated by model 2 decreased more rapidly than did the yield described by model 1, when the mean density of BPH was greater than 100. This showed that the death rate of leaves was accelerated by insect feeding when the BPH density increased.

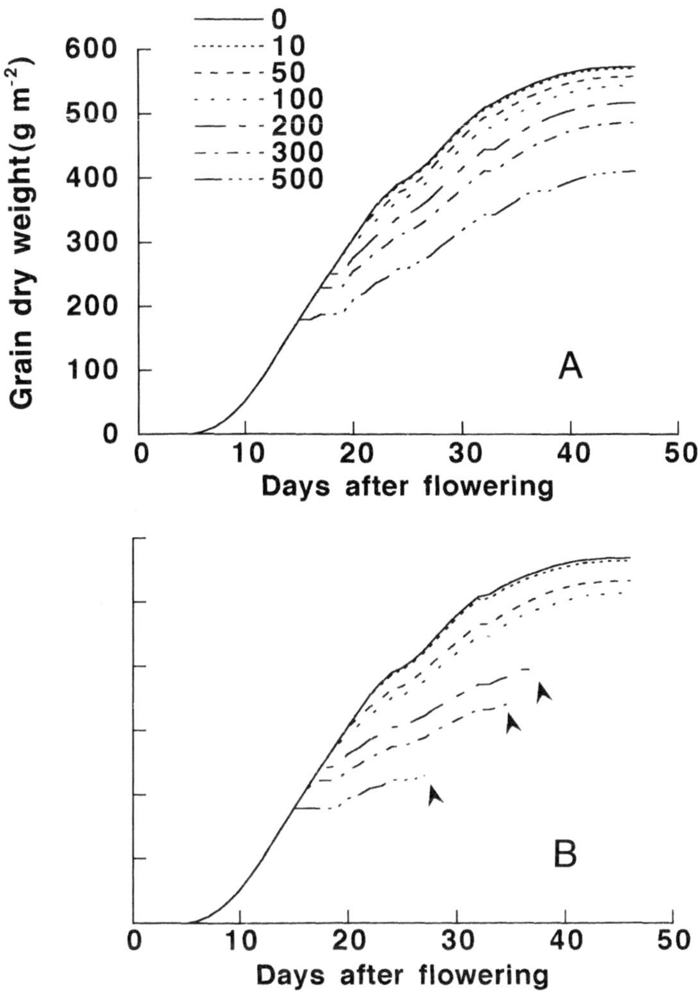

Figure 7. Effects of BPH feeding on grain yield in (A) model 1 and (B) model 2. Numbers indicate BPH density per hill. In model 2, arrows indicate that the simulation was stopped during grainfilling due to death of the rice crop when the BPH density increased to more than 200 per hill.

Model validation

We used model 2 to evaluate the effects of BPH infestation in field experiments. The field experiments were conducted at the International Rice Research Institute (Philippines) in the dry season of 1994 in order to quantify the growth response of rice plants infested with the BPH population. BPH were released to the rice plants 10 days before flowering (T1), at flowering (T2) and 10 days after flowering (T3) at a density of 10 pairs per hill. When the nymphs of the next generation reached the fifth instar, all nymphs were collected and the plants were sampled. It was found that

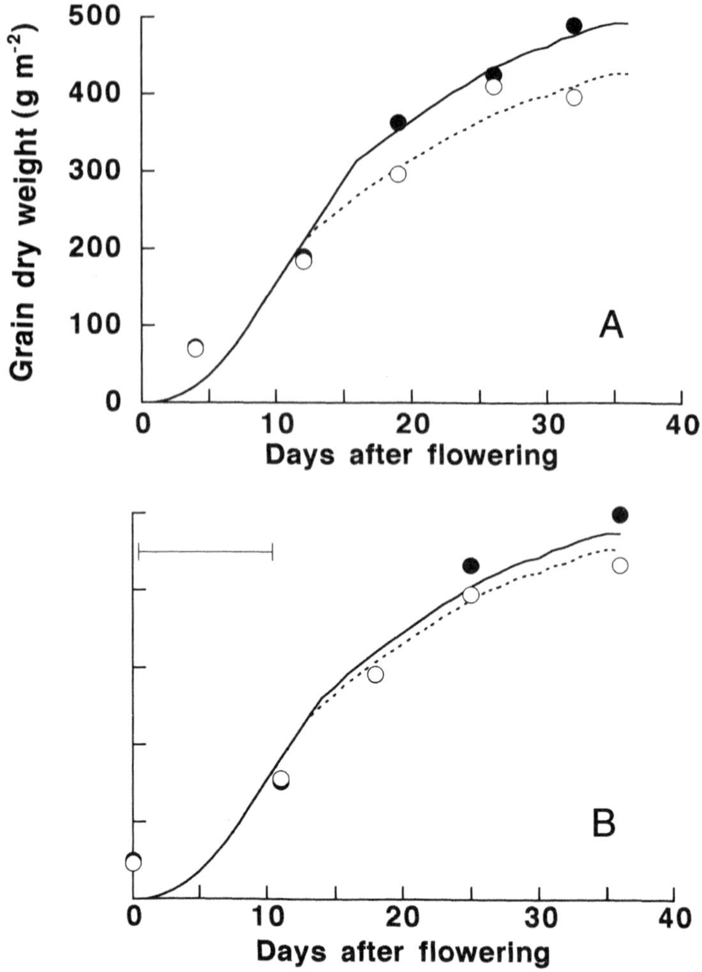

Figure 8. Simulated and observed grain growth. Model 2 was used for the simulation. Field experiment was done at IRRI in the dry season of 1994. (A) BPH infestation started 10 days before flowering and ended at flowering. (B) Infestation started at flowering. Horizontal bar indicates the infestation period (10 days). Circles are observed data, black: control, white: infested. Lines are simulated output, solid line: control, broken line: infested.

the later the infestation on the rice plants, the lower was the population growth rate of BPH. At the end of release period, the BPH biomasses were 200, 120 and 30mg per hill at T1, T2 and T3 respectively. The shoot dry weight and panicle dry weight of the T1 and T2 plants were reduced by BPH infestation. After the infestation period, the difference between the infested plants and control increased as the plants matured.

Our model does not simulate plant growth before flowering. Therefore, we used the *LAI* and shoot dry weight data that were observed at flowering in both the infested and control plants as the initial values for the simulation at T1, and ran the simulation

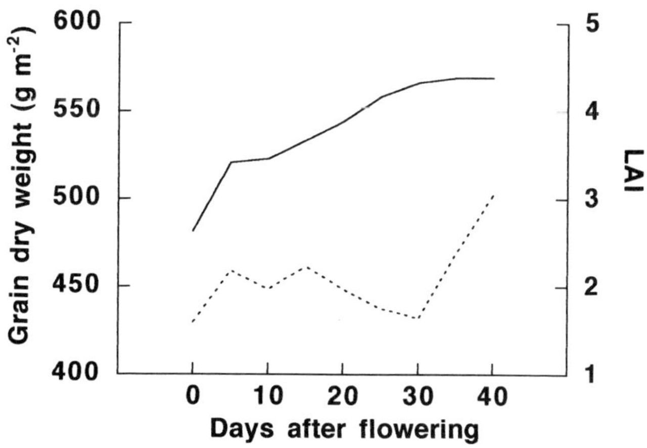

Figure 9. Effects of the length of BPH infestation on grain yield (solid line) and *LAI* (dashed line) at the end of the maturing period using model 2. In the model, 200 BPHs per hill attacked the plants continuously for 15 days. Initial infestation dates were changed at 5-day intervals.

under insect-free conditions after flowering (Figure 8A). At T2, we used both the initial plant parameter at flowering and the BPH biomass data for the simulation (Figure 8B). We did not simulate the effect of BPH infestation that was done at T3, because of the low final density.

The outputs of simulations for infestations at both T1 and T2 showed that model 2 could describe the grain growth in the field infested with BPH (Figure 8). The reduction in grain growth started after the infestation period and was caused by (i) a decrease in daily assimilates due to the reduction in *LAI*, and (ii) a decrease in carbohydrates in the reserves pool. The results of our simulation and field experiments seem to support our hypothesis that BPH feeding does not directly affect carbohydrate translocation.

Application of the model

Infestation period. The simulation was carried out by changing the starting date of the infestation by 5-day intervals using model 2 (Figure 9). Two hundred BPHs per hill attacked the plants continuously for 15 days. The *LAI* at full maturation (DVS=2.0) in each simulation except 35 days after flowering showed small differences, even though the starting date of the infestation was changed. However, the earlier the infestation started, the lower was the simulated grain yield. Since the *LAI* is the main driving force for the amount of daily photosynthesis, decreasing the *LAI* earlier resulted in a greater change in the total carbohydrate production.

Weather conditions. Simulations were run with model 2 for a simulation period of several years using natural weather conditions to evaluate the effect of weather on the simulation output. We used the same initial plant parameters and the actual tem-

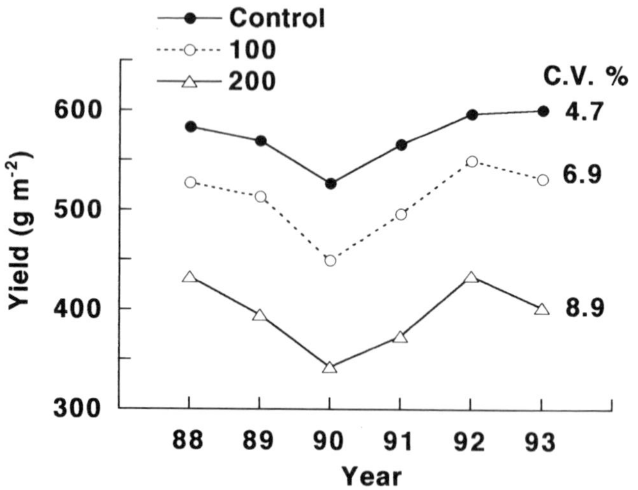

Figure 10. Fluctuations in the simulated grain yield of control and infested plants. Numerical values indicate the coefficient of variances of grain yield at each infestation level.

peratures and solar radiation from 1988 to 1993. Infestation rates of 0, 100 and 200 BPHs per hill were used in the model. Grain yields of both the control and infested plants fluctuated as a result of the weather conditions (Figure 10). The yield ratio (yield of infested plants/yield of control plants) also fluctuated (from 0.92 to 0.85 at 100 BPHs per hill and from 0.74 to 0.65 at 200 BPHs per hill). The yield and yield ratio depended on the solar radiation. The higher the temperature during the maturation period, the faster the plants developed. For instance, in 1990 when temperature and radiation during the maturation period were higher and lower respectively than in other years, photosynthetic products were fewer and the maturation period was shorter. As the infestation rate increased, the coefficient of variance of the yield increased.

Conclusions

The mechanisms and processes of plant damage caused by sap-sucking planthoppers have not been as well studied as those relating to tissue consumers, since there has been a lack of appropriate artificial methods for simulating sap feeding-type infestations. We conducted some experiments and developed simulation models, and revealed that plant growth and yield were affected not only by the direct effect (phloem sap feeding), but also by indirect effects such as the reductions in photosynthetic rate and leaf area. BPH, unlike leafhoppers, does not cause necrosis of plant cells or vascular bundles. The effect of translocation blockage of assimilates due to sap feeding on young rice plants reported by Kenmore (1980), was relatively small on grain growth after flowering. Some other indirect effects, such as the effects on maintenance respiration and nitrogen remobilization, have not yet been determined.

By using the crop growth model, we showed that variations in yield occurred with different infestation and climatic conditions, although the initial plant biomass and the planthopper infestation levels were stable. These approaches will improve our knowledge of compensation mechanisms, the interaction between plants and insects, and the effect of climatic change on crop yields.

Acknowledgements

The first author expresses his sincere thanks to Professor T. Horie, Kyoto University, for his valuable suggestions. This study was supported in part by a grant-in-aid from the Bio Cosmos Program of the Ministry of Agriculture, Forestry and Fisheries (BCP-96-I-B-4), and from the IRRI-Japan Shuttle Research Project of the International Rice Research Institute.

References

Alocilja E C, Ritchie J T (1988) Upland rice simulation and its use in multicriteria optimization. IBSNAT Research Report Series 01, University of Hawaii, USA. 95 p.

Angus J F, Zandstra H G (1980) Climatic factors and the modeling of rice growth and yield. Pages 189–199 in Proceedings of a symposium on the Agrometeorology of the rice crop. International Rice Research Institute, P.O. Box 933, Manila, Philippines.

Benigno E A, Shepard B M, Rubia E G, Arida G S, Penning de Vries F W T, Badong J P (1988) Simulation of rice leaf-folder population dynamics in lowland rice. IRRI Research Paper Series 135:1–8.

Cagampang G B, Pathak M D, Juliano B O (1974) Metabolic changes in the rice plant during infestation by the brown planthopper, *Nilaparvata lugens* Stål (Homoptera: Delphacidae). Appl. Entomol. Zool. 9:174–184.

De Jong P D, Daamen R A (1992) Simulation of yield loss by the rice leaffolder *Cnaphalocrocis medinalis* under different growing conditions. J. Plant Prot. Trop. 9:117–123.

Graf B, Rakotobe O, Zahner P, Delucchi V, Gutierrez A P (1990) A simulation model for the dynamics of rice growth and development: I. The carbon balance. Agric. Syst. 32:341–365.

Graf B, Lamb R, Heong K L, Fabellar L T (1992) A simulation model for the population dynamics of rice leaf-folders (Lepidoptera: Pyralidae) and their interactions with rice. J. Appl. Ecol. 29:558–570.

Horie T (1987) A model for evaluating climatic productivity and water balance of irrigated rice and its application to southeast Asia. Southeast Asian Studies 25:62–74.

Horie T, Nakagawa T (1990) Modeling and prediction of development process in rice. I. Structure and method of parameter estimation of a model for simulating developmental process toward heading. Jpn. J. Crop Sc. 59:687–695.

Iwaki H (1975) Computer simulation of vegetative growth of rice plants. Pages 105–121 in Crop productivity and solar energy in various climates in Japan. Japanese Committee for the International Biological Program – JIBP Synthesis Vol. 11, University of Tokyo, Japan.

Kenmore P (1980) Ecology and outbreaks of a tropical insect pest of the green revolution, the rice brown planthopper *Nilaparvata lugens* (Stål). PhD thesis, University of California, Berkeley, CA, USA. 226 p.

Kisimoto R (1965) Studies on the polymorphism and its role playing in the population growth of planthopper, *Nilaparvata lugens* Stål. Bull. Shikoku Agric. Exp. Stn. 13:1–106.

Kisimoto R (1976) Synoptic weather conditions inducing long-distance immigration of planthoppers, *Sogatella furcifera* Horváth and *Nilaparvata lugens* Stål. Ecol. Entomol. 1:95–109.

Kisimoto R (1979) Brown planthopper migration. Pages 113–124 in Brown planthopper: Threat to rice production in Asia. International Rice Research Institute, P.O. Box 933, Manila, Philippines.

Kuno E (1968) Studies on the population dynamics of rice leafhoppers in a paddy field. Bull. Kyushu Agric. Exp. Stn. 14:131–246.

McCree K J (1974) Equations for the rate of dark respiration of white clover and grain sorghum, as functions of dry weight, photosynthetic rate, and temperature. Crop Sci. 14:509–514.

Matsumura M (1991) Characteristics of recent population growth patterns of the white-backed planthopper, *Sogatella furcifera* Horváth in the Hokuriku district. Proc. Assoc. Pl. Prot. Hokuriku 39:47–50.

Naba K (1992) Loss assessment of feeding damage due to the white-backed planthopper, *Sogatella furcifera*

Horváth, on ordinary planting paddy in Horishima prefecture. Bulletin of Hiroshima Prefectural Agriculture Research Center 55:7–16.

Noda H (1986) Damage to ears of rice plants caused by the white-backed planthopper, *Sogatella furcifera* (Homoptera: Delphacidae). Appl. Entomol. Zool. 21:474–476.

Penning de Vries F W T, Jansen D M, Ten Berge H F M, Bakema A (1989) Simulation of ecophysiological processes of growth in several annual crops. Simulation Monographs, Pudoc, Wageningen, The Netherlands. 271 p.

Rubia E G, Penning de Vries F W T (1990) Simulation of yield reduction caused by stem borers in rice. J. Plant Prot. Trop. 7:87–102.

Seino H, Shiotsuki Y, Oya S, Hirai Y (1987) Prediction of long distance migration of rice planthoppers to Northern Kyushu considering low-level jet stream. J. Agric. Met. 43:203–208.

Sogawa K (1971) Effects of feeding of the brown planthopper on the components in the leaf blade of rice plants. Jpn. J. Appl. Entomol. Zool. 15:175–179.

Sogawa K (1982) The rice brown planthopper: Feeding physiology and host plant interactions. Ann. Rev. Entomol. 27:49–73.

Sogawa K (1992) Modeling of dry matter drain by the rice planthoppers on rice. Kyushu Agric. Res. 54: 104.

Sogawa K, Cheng C H (1979) Economic thresholds, nature of damage, and losses caused by the brown planthopper. Pages 125–142 in Brown planthopper: Threat to rice production in Asia. International Rice Research Institute, P.O. Box 933, Manila, Philippines.

Spitters C J T, Van Keulen H, Van Kraalingen D W G (1989) A simple and universal crop growth simulator: SUCROS87. Pages 145–181 in Rabbinge R, Ward S A, Van Laar H H (Eds.) Simulation and systems management in crop protection. Simulation Monographs, Pudoc, Wageningen, The Netherlands.

Watanabe T (1994) Population dynamics of long distance migratory rice planthoppers, *Nilaparvata lugens* Stål and *Sogatella furcifera* Horváth, and growth and yield analysis of rice plant infested with these planthoppers. PhD thesis, Kyoto University, Japan. 193 p.

Watanabe T, Sogawa K (1994a) Growth and yield analysis of rice plants infested with long-distance migratory rice planthoppers. I. Effects of period and intensity of the white-backed planthopper, *Sogatella furcifera* Horváth (Homoptera: Delphacidae), infestation on vegetative growth and yield. Japan J. Appl. Entomol. Zool. 38:153–160.

Watanabe T, Sogawa K (1994b) Growth and yield analysis of rice plants infested with long-distance migratory rice planthoppers. III. Quantitative evaluations of vegetative growth of rice plants infested white-backed planthopper, *Sogatella furcifera* Horváth (Homoptera: Delphacidae). Jpn. J. Appl. Entomol. Zool. 38:275–281.

Watanabe T, Sogawa K, Suzuki Y (1994a) Analysis of yearly fluctuations in the occurrence of migratory rice planthoppers, *Nilaparvata lugens* Stål and *Sogatella furcifera* Horváth, based on light-trap data in northern Kyushu. Jpn. J. Appl. Entomol. Zool. 38:7–15.

Watanabe T, Yamamoto H, Sogawa K (1994b) Growth and yield analysis of rice plants infested with long-distance migratory rice planthoppers. II. Measurement of recovery of vegetative growth of rice plants infested with the white-backed planthopper, *Sogatella furcifera* Horváth (Homoptera: Delphacidae), by spectral reflectivity. Jpn. J. Appl. Entomol. Zool. 38:169–175.

Addressing sustainability of rice-wheat systems: analysis of long-term experimentation and simulation

J. TIMSINA, U. SINGH[1] and Y. SINGH[2]

[1]*International Rice Research Institute, P.O. Box 933, 1099 Manila, Philippines*
[2]*G.B. Pant University of Agriculture and Technology, Pantnagar-263145, India*

Key words: CERES-Rice, CERES-Wheat, long-term experiments, simulation, sustainability, risk analysis

Abstract

The rice-wheat system dominant in the Indogangetic region of South Asia, is facing a number of complex problems, with severe implications on the sustainability of the system. Agronomic, soil, and historical weather data from a long-term experiment (LTE) at Pantnagar, India, were used for analysing yield trends, and in an attempt to explain sustainability related issues for the rice-wheat systems. Validated models (CERES-Rice and CERES-Wheat) were run in sequence using a Sequence Analysis Driver of DSSAT V3.0 for simulating long-term yields of rice and wheat for the LTE. Experimental data indicated that rice yields have declined, whereas wheat yields have increased during the same period. Long-term simulation indicated that, except for some outliers, the yield trends were similar to the trends obtained from the experiment. The Driver was then used in running the rice-wheat experiment for 20 years and 15 replications using generated weather data. The Sequential Analysis Program of DSSAT V3.0 was then used in analysing the trends of model outputs (yields, N, organic C) using percentiles and cumulative probability functions. Yield and soil fertility (total N, organic C) trends in context with sustainability of rice-wheat systems are explained using both long-term experimental and simulation results.

Introduction

In an earlier paper, it was reported that rice-wheat systems in the Indogangetic region have begun to run into sustainability problems, and that crop simulation modelling may be the potential tool in studying and quantifying the sustainability related issues in these systems (Timsina et al. 1995). The 'Wageningen' SUCROS and IBSNAT CERES-Rice and CERES-Wheat models were validated for Los Baños in the Philippines and Pantnagar in India, representing the humid tropics and subtropics respectively, using independent data sets from each place. The paper also demonstrated that CERES-Rice and CERES-Wheat can predict the grain yields satisfactorily under high and low soil fertility conditions (Timsina et al. 1995).

Sustainability of rice-wheat systems has recently been one of the major concerns among researchers, farmers, and policy makers in South Asia. While many long-term experiments (LTEs) are going on in India, Nepal, and Bangladesh, the data have just been piled and not yet analysed and interpreted. Due to differing soils, varying weather and changing management practices, the results from the LTEs vary from place to place and among fields within a place. We hypothesized that results of the LTE when used together with the simulation models could help explain the increase

M.J. Kropff et al. *(eds.): Applications of Systems Approaches at the Field Level, 383–397.*

or decline in the yields of the crop in sequence and, hence, could help studying and quantifying the sustainability issues in the systems. Hence, the objectives of the paper are: (i) to review relevant literature on LTEs and sustainability for conceptualizing and relating it to the rice-wheat system sustainability, (ii) to validate the CERES-rice-wheat model using data from one of the LTEs in India, and (iii) to demonstrate the use of the sequential model in analysing risk and/or sustainability in relation to the rice-wheat system under high and zero N fertilizer conditions.

Systems sustainability and rice-wheat systems

Concepts and measurements

Sustainable agricultural production systems should on the one hand meet the requirements of the farm household in terms of food, income, and leisure (Kruseman et al. 1993), and should on the other hand not endanger the productive capacity of the natural resource base to guarantee possible production in the future (Meerman et al. 1992). The two aspects are directly related, as the quality of the resource base determines its productive capacity, and the optimum amounts of growth factors are needed to attain target production levels.

Sustainability has been viewed differently by different people. Harrington (1993) provides various definitions and interpretations of and issues related to sustainability, and classification of sustainability problems in relation to agricultural systems. System function, or performance, can be quantified in many ways. In its widest sense, sustainability involves biology, ecology, economics, and sociocultural and political considerations. Economists are concerned with the conditions for profit maximization and the discounted value of future returns from current investment and technology choices. Ecologists look at the stability and resilience in terms of material and energy balances, interaction among trophic levels, regulating feedback mechanisms, and biodiversity. Agronomists seek to harness solar, soil, and water resources to optimize crop yields in relation to input requirements. Farmers are concerned with maximizing their return on labour, land, or capital depending on the socioeconomic and policy environments in which they operate. Lynam and Herdt (1988) suggested the total factor productivity (TFP, i.e. ratio of the total output value to the total cost of all inputs used to produce the crop) and the partial factor productivity (PFP, i.e. ratio of total output value to the input of interest), as the measures of system performance. Later, Cassman and Pingali (1995) reported that though TFP as the measure of sustainability may be more appropriate for monitoring and comparing system performance at the farm or regional level, the PFP can be equally as useful as the complete TFP index in monitoring productive trends in the LTEs. Monteith (1990) provides some techniques for quantifying sustainability for district level, as well as for field level analyses, and emphasizes that crop simulation modelling could be the short-cut way of quantifying sustainability. Singh and Thornton (1992) also provide some examples of using crop models for sustainability and environmental quality assessment.

System performance and soil quality

The dominant natural resources of an agroecosystem are soil and water, and both have large impact on the productivity. Soil quality can be assessed in terms of chemical, physical, and biological properties of soil that govern the supply and utilization of nutrients and water by the crops grown. Chemical properties relate to nutrient retention characteristics (CEC), nutrient supplying or buffering capacity, pH, salinity, and toxicities; the physical properties are such as size and continuity of pores, water retention traits, infiltration, aggregate stability, and bulk density; the biological properties could be such as nutrient mineralization, pathogenic/beneficial microbes, weeds, and nematodes (Cassman 1993). Any research on quantifying system performance or sustainability should be able to measure the changes in soil quality and their relationship with productivity over the long run. Soil organic carbon and nitrogen are two of the important soil quality parameters that need to be assessed for quantifying sustainability.

Long-term experiments and yield decline

The issues of sustainability as related to biophysical environments can best be understood by LTEs as they can expand our knowledge of the fundamental processes that influence the performance of cropping systems over longer periods. LTEs had been used by agricultural scientists for decades to look at trends in agricultural productivity. Several of such experiments now exist in the world and most of them focus on soil fertility and look at the depletion or build-up of soil nutrients over time. The most famous of these experiments is the Broadbalk winter wheat experiment at Rothamsted, England, which was initiated in 1843 (Jenkinson 1991). Yields of winter wheat in Broadbalk increased from 1852 to 1986 in all the fertilizer treatments (both chemical and organic). Soil organic carbon levels increased markedly in the plots receiving farmyard manure (FYM) and have still not reached equilibrium after nearly 150 years. Long-continued use of inorganic fertilizer containing N, P, K, and Mg increased soil organic carbon level but the increase was much less than that brought about by FYM.

In the USA alone, there are about 25 LTEs over 25 years old of which 12 are more than 50 years old. The Morrow plots in Urbana, Illinois (1876), the Sanborn field plots in Columbia, Missouri (1888), the Magdruder plots in Stillwater, Oklahoma (1892), and Old Rotation plots in Alabama (1896) are among the four oldest LTEs that provide useful information on the effect of crop rotation and organic and inorganic manuring on soil health and its relationship with productivity (Mitchell et al. 1991). There are variable results regarding increase or decrease of soil organic matter, organic N, and yields from several LTEs. For example, in the Morrow plots, the crop rotations plus appropriate fertilization not only produced the highest crop yields, but also maintained soil N and organic C at the highest levels. In the Sanborn field, continuous legumes, legumes system, or animal manures did not maintain soil productivity without supplemental additions of plant nutrients, and continuous wheat did not remain productive over time whether manure or commercial fertilizers were used. In the Old Rotation experiment, soil organic C in the plough layer ranged from 0.12% in the continuous cotton plots with no legume and no fertilizer N to 0.33% in plots in the 3-year rotation; and cropping system and winter legumes had a large influence on soil organic matter. In the Magdruder plots, soil organic matter declined rapidly during the first 35 years, maintained a lower rate of decline during the next 52 years, and then stabilized in the continuous wheat plots (Mitchell et al. 1991).

In all the LTE sites at the International Rice Research Institute (IRRI) and several other research stations in the Philippines, rice yields in the continuous rice cropping experiments declined both in high NPK input levels that were considered to be optimal at the outset of the experiment and in control plots without nutrient inputs (Cassman and Pingali 1995). Weather data summarized by Flinn et al. (1982) and Flinn and De Datta (1984) indicated that the yield decline was not associated with changes in solar radiation, and that the influence of viral diseases, typhoons, and bacterial leaf blight and leaf streak was isolated by statistical analysis, and yet still found a significant linear yield decline. The soil organic matter and total N in the 0–20 cm puddled top soil increased slightly since 1963 in control treatments without fertilizer N inputs, and there has been a significant enrichment in soil organic carbon and N in treatments that received high inputs of fertilizer N. Cassman and Pingali (1995) concluded that there were no differences in soil N supplying capacity between control and high N treatments and hypothesized that yields were limited by an insufficient N supply in the IRRI LTE.

In India, LTEs were started as early as 1885 with the establishment of the first permanent manurial experiment based on the Rothamsted model at Kanpur, Uttar Pradesh, and since then several LTEs involving several different cropping patterns have been set up at different places. The aim of all these LTEs was to evaluate the long-term effect of the organic and inorganic manuring on crop production and soil health (Nambiar 1994). Effects of long-term application of single nutrient and nutrient combinations with and without organic manures on soil fertility have been evaluated since 1972–73 under the All-India Coordinated Long-term Fertilizer Experiments by scientists from the Indian Agricultural Research Institute (Nambiar and Ghosh 1984). Manuring at optimal to superoptimal (100–150%) NPK doses over the 14-year period (1971–85) improved appreciably the soil organic carbon content of the plough layer (0–15 cm) on Ustochrepts but decreased it in Eutochrepts and Hapludols soils. FYM application further increased the soil organic carbon content, though with variable responses for variable soils. Application of N at optimal to superoptimal (100–150%) NPK doses over the 16-year period (1971–87) improved the level of available soil N on the Ustochrepts and Eutochrepts, and the incorporation of FYM (along with optimal NPK dose) enhanced the N build-up remarkably. The highest N build-up was observed with the combined application of optimal NPK dose and FYM both in case of Haplaquepts and Hapludolls, but the soil N declined on Tropaquepts even at superoptimal NPK (150%) dose as compared with the initial level (Nambiar 1994).

Using simulation models to analyse field experiments

The discussion for this section is largely based on Acock and Acock (1991). Most of the data used in the validation of crop simulation models and some of the data used in their development come from short-term field experiments (STEs). The STEs have the obvious advantage that the treatments are under the control of researchers making observations. However, they suffer from two problems: (i) there may be a carry over of fertilizer salts, herbicides, and other inputs from previous treatments; and (ii) continuous application of the treatments may cause changes in soil conditions which only become apparent after the treatments have been continued for many years. In LTEs, the unknown effects of previous treatments are gradually eliminated over time and some factors are constant over time.

In addition to use of field data in validating the models, it is possible to use simulation models to interpret the results of long-term research. In our earlier paper, we discussed two groups of simulation models, the Wageningen-originated SUCROS models and the IBSNAT CERES models, and emphasized crop simulation modelling as a potential tool for studying and quantifying sustainability issues such as those that exist in rice-wheat systems (Timsina et al. 1995). Models can be used to determine how much of the year-to-year variation in yield is attributable to changes in weather and soil conditions. Then hypotheses about possible causes of the remaining variation can be tested by programming them into models. Bell and Fischer (1994) used the CERES-Wheat model for estimating wheat yield gain as a function of weather conditions in the Yaqui Valley of northwestern Mexico. Small changes in the growing season temperature over the years appeared to be the key aspect of weather affecting yearly yield fluctuations. Reddy and Baker (1990) used the cotton model, GOSSYM, in the same way in a recent analysis of factors affecting cotton yields in the long-term cotton breeding trials in several states in the USA. They concluded that cotton lint yield decline in cotton growing states over the study period was not caused by any deterioration in the weather but could be due to factors other than weather.

The use of crop models to test hypotheses about the causes of variation in field data, and the use of field data for validating and improving models is an iterative process. Hypotheses about the causes of variation can be incorporated into crop models which are then validated against field data. This validation reveals flaws in the original hypothesis and leads to the proposal and testing of new hypotheses. Testing a hypothesis in a simulation model is much more rigorous than testing it as a concept only. Both experimentation and simulation are essential components to understand the phenomena and processes regarding crop growth and development.

Validation of the sequential model

LTE site and experimental yields

For the analysis of trends in rice and wheat yields, data sets from 1977–78 to 1993–94 cropping seasons from the LTE entitled, 'Long-range effect of continuous

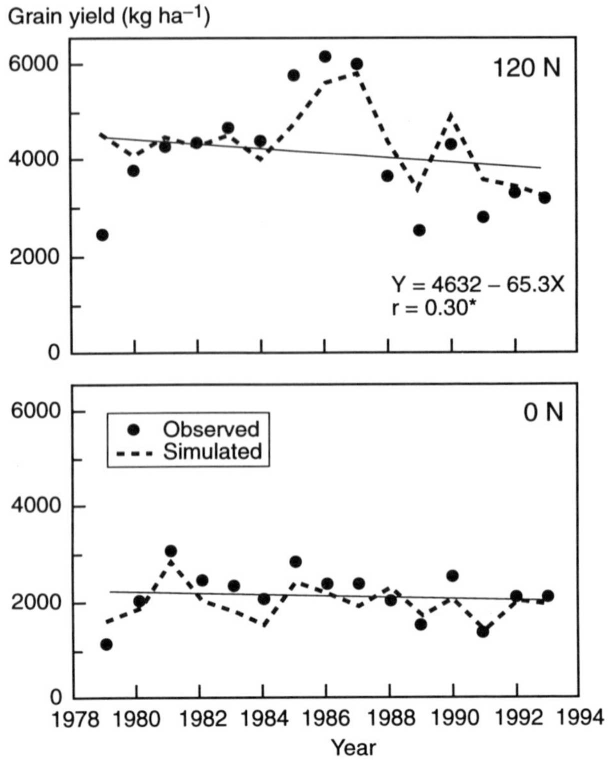

Figure 1. Long-term simulated and observed grain yields of rice at two nitrogen application levels for a long-term experiment at Pantnagar, India.

rice-wheat cropping and manuring on the soil fertility and yield stability' from Pantnagar was used (see Timsina et al. (1995) for soil characterization and treatment details on the LTE). The soil of the experimental site is moderately well to well-drained, pH ranging from 5.4 to 6.6, and base saturation ranging from 79 to 95% throughout the profile. It has moderate to high fertility with organic matter of 1.9% in 0–20 cm depth, 1.1% in 20–40 cm depth, and less than 1% in the lower depths (Despande et al. 1971). Soil samples were collected from 0–15 cm depth before the start of the LTE in 1977 and the results were as follows: total N, 0.104% N; available P, 26.8 kg ha^{-1}; exchangeable K, 178 kg ha^{-1}; exchangeable cations, 0.4 mhos cm^{-1}; organic C, 0.986%; sand, 28%; silt, 42%; clay, 29.7%. Total N, available P and exchangeable K were not analysed after each year of experimentation, but were analysed occasionally only.

In the same paper (Timsina et al. 1995), the yield trends of rice and wheat over 17 years using mean data (i.e. grain yields obtained from all levels of potassium and phosphorus in the different plots were averaged for 0, 60, 120, and 180 kg N ha^{-1} applications) were analysed. There was a clear trend of decreasing rice yields over

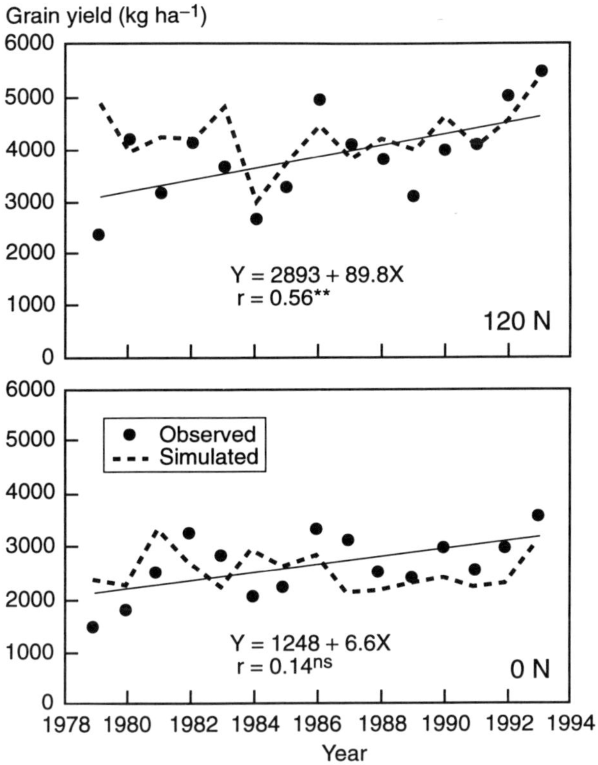

Figure 2. Long-term simulated and observed grain yields of wheat at two nitrogen application levels for a long-term experiment at Pantnagar, India.

the years for all treatments, especially since year 11, and the decrease was highest for control (N0) treatment ($r=-0.48$; $b=-74$) and lowest for 120 kg N treatment ($r=-0.3$; $b=-65.3$) (Figure 1). Wheat yields, on the other hand, consistently increased over the years, with the highest increase with 120 kg N treatment ($r=0.56$; $b=89.8$) and lowest for 0 kg N treatment ($r=0.14$; $b=6.6$) (Figure 2).

Careful examination of management practices revealed that transplanting dates of rice, sowing dates of wheat, and the rice and wheat varieties used were different during different years of experimentation. Rice variety Prasad was used in 1978, Jaya from 1979 to 1985, and Pant-4 from 1986 to 1993. Transplanting date of rice was late July in the earlier years, and it was late June in later years. In wheat, variety RR21 was used from 1978–79 to 1985–86, while UP2003 was used from 1986–87 to 1993–94. Sowing date of wheat also varied across years. The experiment was supposed to be conducted under irrigated condition, but the irrigation amount varied depending upon rainfall and water availability in the irrigation canals. As a result, crops suffered from drought stress in some years.

Model validation

Different rice and wheat varieties were used in different years and accordingly genetic coefficients for each variety were determined using the program GenCalc V3.0 of the DSSAT V3.0 (Hunt and Pararajasingham 1994). The genetic coefficients are calculated and discussed in the companion paper (Timsina et al. 1995). Since planting dates in each year differed, actual planting dates were inputted for validation purpose. Thus our inputs to the models were only transplanting/planting dates, variety, and real historical weather, and initial soil N. The latter was available only for few years, and so interpolation was done for those years where the data were not available. Validation of the CERES-Rice (Singh et al. 1993) and CERES-Wheat (Godwin et al. 1990) models using long-term data from the LTE showed that except for few years simulated yields were close to the experimental yields. In years 1, 3, 5, and 11, observed rice yields were much lower than the simulated ones with high N (Figure 1). The observed wheat yields were also much lower than the simulated ones in years 1, 11, and 13 (Figure 2). Simulated rice, as well as wheat yields, were generally underestimated with low N. Careful examination of the management practices revealed that there were occurrences of pests and diseases, as well as water shortage, in those years. Although some simulated yields were much above or below the actual yields, the models, in general, satisfactorily predicted the long-term trends of rice and wheat yields. Such a discrepancy is expected since we did not have data on many other soil related parameters which could influence simulation results. We concluded that if at least soil N and organic C were available for each year, our simulation results on long-term yields would be much closer to experimental results. Nevertheless, we were confident in the use of these models for analysing the risk associated with rice-wheat systems under high-N, as well as low-N, conditions.

Sequence analysis program

The Sequence Driver Program of DSSAT V3.0 is capable of running up to 10 treatments or sequences. The Analysis Program, however, is capable of analysing only one treatment at a time, but can analyse summary output files with a maximum of 1200 separate simulation runs (Thornton et al. 1994, 1995). Long-term weather data were generated from 15 years of actual weather data using the program WGEN, a statistical weather generator to perform probabilistic and stochastic analysis (Thornton et al. 1994).

Sustainability and risk analysis

To see whether decline in rice yields was due to adverse weather or some other factors, Jaya and Pant-4 rice varieties were transplanted on 18 July each year using actual weather data and identical management practices. Figure 3 shows that the potential yields of these varieties were higher by 500–2000 kg ha^{-1} than the yields for the 120 kg N ha^{-1} application, and that the yields were not declining. These results show

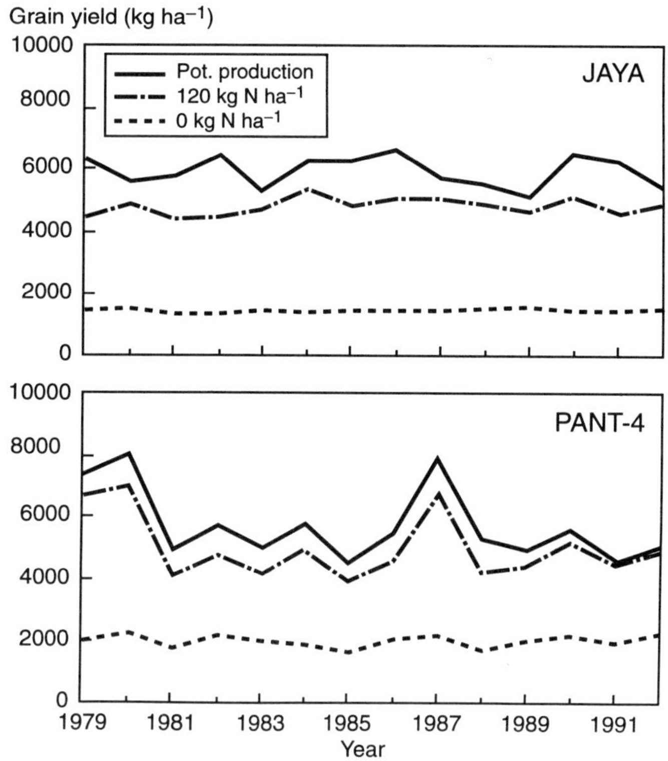

Figure 3. Simulated rice grain yields at potential production, 120 and 0 kg N ha^{-1} for 18 July plantings, Pantnagar, India.

that weather did not cause any decline in yields. In case of the RR21 and UP262 wheat varieties, the potential yields and yields limited by water and nutrients were almost similar, and again the yields were not declining (Figure 4). Results indicate that increase in wheat yields over years was probably not due to more favourable weather, but was possibly due to the use of UP262, a high-yielding variety in the later years. Thus we were confident that decline in rice and increase in wheat yields was not due to effects of weather.

A simulation experiment was performed with sequencing versions of CERES-Rice and wheat models to explain the yield decline in rice. The experiment, run as a 20-year sequence, was replicated 15 times. The experiment was made up of a repeating sequence of the rice–fallow–wheat–fallow models. In each year, rice was transplanted on 18 July and wheat was sown on 3 December. Crop management was the same in all seasons of the sequence; the varieties used were Jaya (rice) and RR21 (wheat). Automatic irrigation was used which applied water whenever needed. All rice and wheat residues after harvesting the grains were removed from the field. Simulations for high N (120 kg N ha^{-1}) and low N (0 kg N ha^{-1}) were conducted. The following paragraphs discuss and analyse model outputs such as grain yields, and

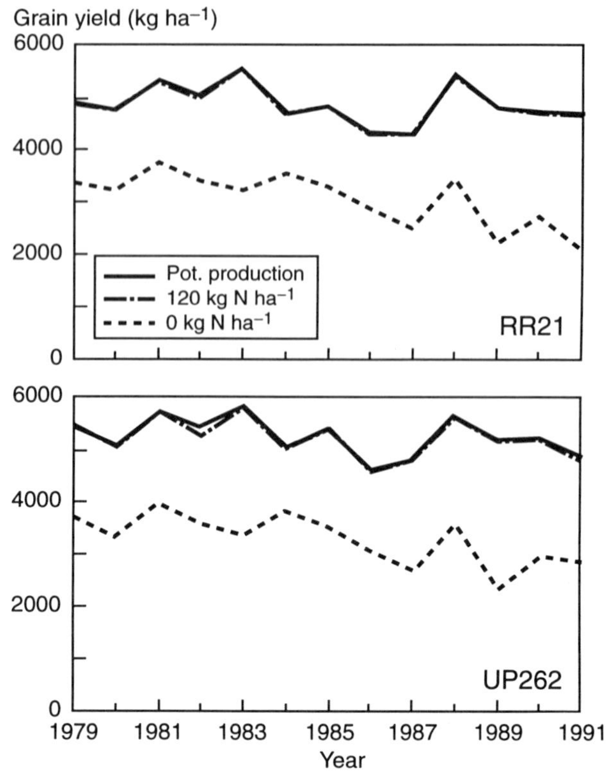

Figure 4. Simulated wheat grain yields at potential production, 120 and 0 kg N kg^{-1} for 3 December plantings, Pantnagar, India.

soil organic carbon and nitrogen using percentile plots and cumulative probability functions (CPFs) for high and zero-N fertilizer conditions.

Figures 5 and 6 show simulated yield distributions of rice and wheat for continuous rice-wheat system with 120 and 0 kg N ha^{-1} for 0th, 25th, 50th (the tiny circle), 75th, and 100th percentile. Simulated yields were most variable for 0th percentile for both 120 and 0 kg N ha^{-1} treatments. Yield trends were similar for each crop and each percentile.

Another way of looking at the variability in yields in terms of cumulative probability functions (CPFs) is presented in Figure 7 for the years 1979, 1982, 1985, 1988, and 1991. At all probability levels, rice yields were always higher in 1979 than in any other years. In all other 4 years, yields were little less than 4.2 ton ha^{-1} at 50% probability level; at lower probability level, yields were higher in 1991 than in other years. At probability levels higher than 50%, yields in all years were almost similar. For wheat, however, the scenario was different. Linear regressions of year on simulated yields of rice showed that the yields were declining, with the r^2 of 0.55, but no linear relationship was observed for wheat (data not shown).

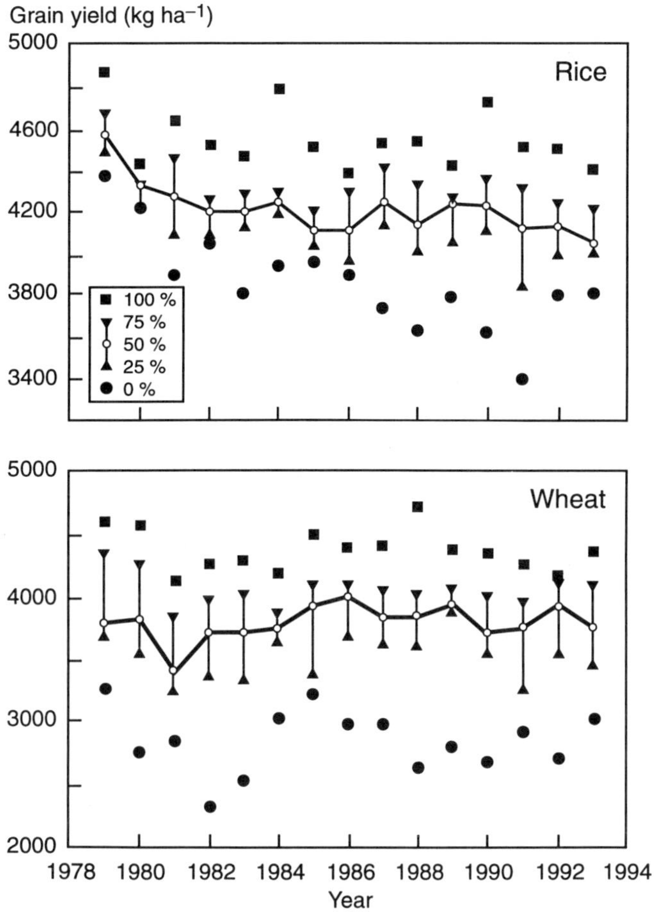

Figure 5. Simulated yield distributions of rice and wheat for continuous rice-wheat systems at 120 kg N ha^{-1} for 15 years at Pantnagar, India, replicated 15 times.

Regression of sequence year number on simulated soil organic N and C for 120 kg N ha^{-1} (Figure 8) illustrates a decline in soil organic N and C. This shows that the soil nutrients such as organic N and C were decreasing over time, causing decline in rice yields. This corroborates with the experimental results presented earlier, as well as with results from the Magdruder plots as reported by Mitchell et al. (1991), but contrasts with results from intensively cultivated irrigated rice-rice systems in which rice is continuously grown in submerged soil (Cassman et al. 1995; Cassman and Pingali 1995). In the Magdruder plots with wheat crop, soil organic matter declined for first 35 years, but in the LTEs with irrigated rice-rice systems, the total organic matter and total N did not decline, even in zero-N plots, yet yields declined and could be restored by increased N fertilizer addition rates.

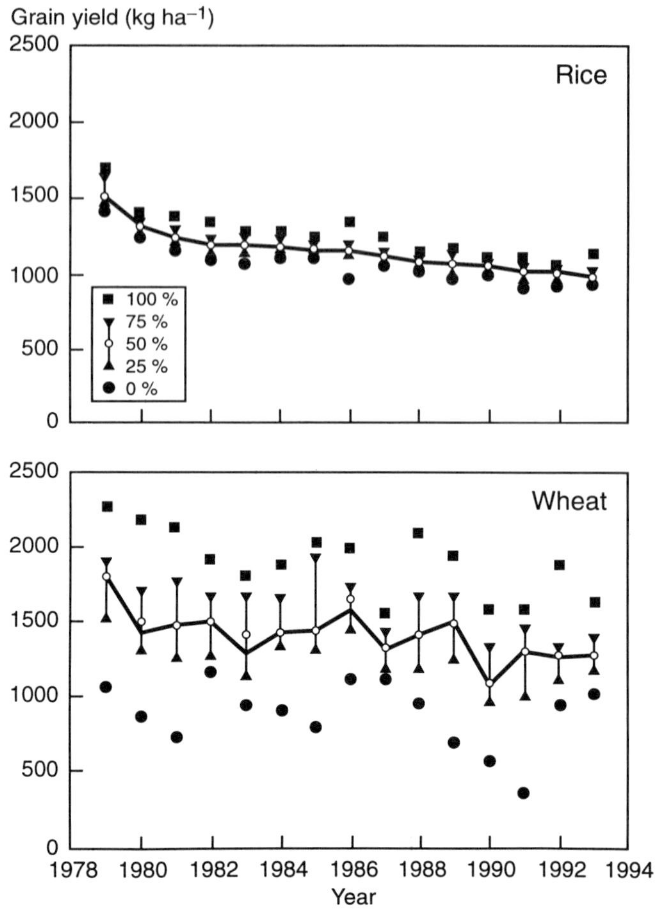

Figure 6. Simulated yield distributions for rice and wheat for continuous rice-wheat systems at $0\,kg\,N\,ha^{-1}$ for 15 years at Pantnagar, India, replicated 15 times.

The fact that simulated wheat yields showed an increasing trend was due to the simulated increase in N availability under aerobic conditions.

Conclusions

The results demonstrate the possibility and potentiality of using crop simulation models in explaining the yield trends in rice-wheat systems. Decreases in rice yields were probably associated with decline in organic carbon and nitrogen. These results differ from those of the LTEs on intensively cultivated rice-rice systems in the Philippines in which rice yields declined, despite the fact that total organic matter and total N did not decline.

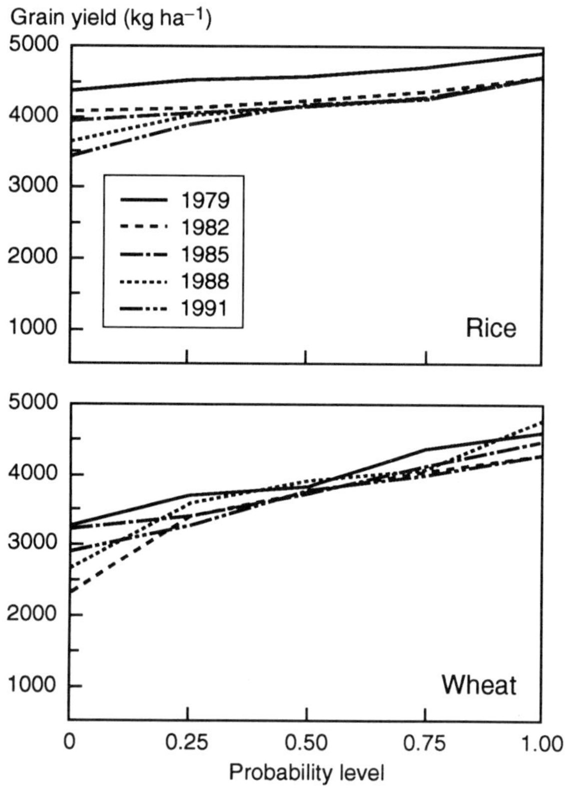

Figure 7. Cumulative probability functions for grain yields at 120 kg N ha^{-1} for different years at Pantnagar, India.

References

Acock B, Acock M C (1991) Potential for using long-term field research data to develop and validate crop simulators. Agron. J. 83:56–61.

Bell M A, Fischer R A (1994) Using yield prediction models to assess yield gains: A case study for wheat. Field Crops Res. 36:161–166.

Cassman K G (1993) The challenge of quantifying sustainability: Long-term agronomic experiments and farm monitoring. Pages 20–28 in Harrington L W, Hobbs P R, Cassaday K A (Eds.) Methods of measuring sustainability through farmer monitoring: Applications to the rice-wheat cropping pattern in South Asia. Proceedings of the Workshop, 6–9 May 1992, Kathmandu, Nepal. CIMMYT, IRRI, and NARC. CIMMYT, Mexico D.F., Mexico.

Cassman K G, Pingali P L (1995) Extrapolating trends from long-term experiments to farmers' fields: The case of irrigated rice systems in Asia. Pages 63–84 in Barnett V, Payne R, Steiner R (Eds.) Agricultural sustainability in economic, environmental and statistical considerations. John Wiley and Sons, London, UK.

Cassman K G, Steiner R, Johnston A E (1995) Long-term experiments and productivity indexes to evaluate the sustainability of cropping systems. Pages 231–244 in Barnett V, Payne R, Steiner R (Eds.) Agricultural sustainability in economic, environmental and statistical considerations. John Wiley and Sons, London, UK.

Despande S B, Fehrenvacher J B, Ray B W (1971) Mollisols of Tarai region of Uttar Pradesh, Northern India. 2. Genesis and classification. Geoderma 6:195–201.

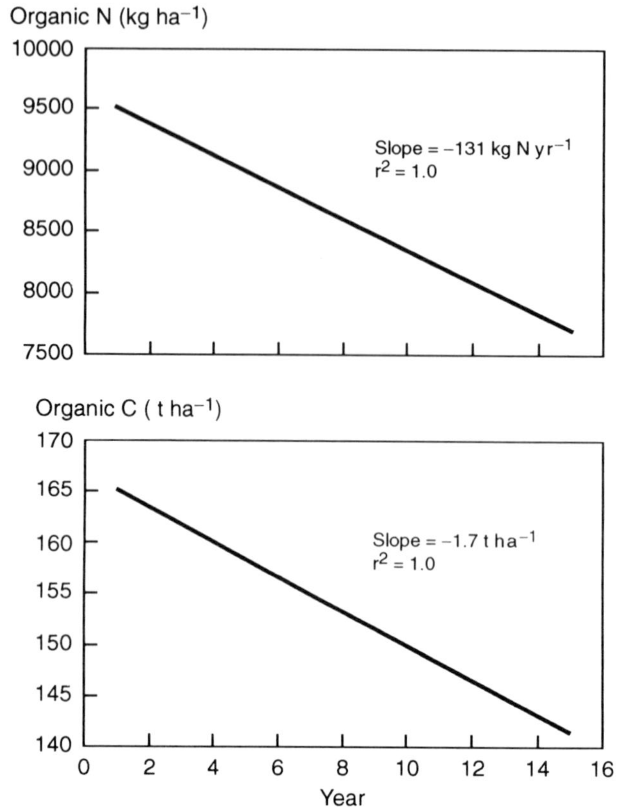

Figure 8. Regression of sequence year number on simulated soil organic N and C for 120 kg N under rice-wheat systems grown for 15 years at Pantnagar, India.

Flinn J C, De Datta S K (1984) Trends in irrigated rice yields under intensive cropping at Philippine research stations. Field Crops Res. 9:1–15.

Flinn J C, De Datta S K, Labadan E (1982) An analysis of long-term rice yields in a wetland soil. Field Crops Res. 5:201–216.

Godwin D C, Ritchie J T, Singh U, Hunt L A (1990) A user's guide to CERES-Wheat V2.10. Simulation manual. IFDC-SM-2. IFDC, Muscle Shoals, AL, USA. 94 p.

Harrington L W (1993) Interpreting and measuring sustainability: Issues and options. Pages 3–19 in Harrington L W, Hobbs P R, Cassaday K A (Eds.) Methods of measuring sustainability through farmer monitoring: Applications to the rice-wheat cropping pattern in South Asia. Proceedings of the Workshop, 6–9 May 1992, Kathmandu, Nepal. CIMMYT, IRRI, and NARC. CIMMYT, Mexico D.F., Mexico.

Hunt L A, Pararajasingham S (1994) GenCalc: Genotype coefficient calculator. Pages 201–233 in Tsuji G Y, Uehara G, Balas S (Eds.) IBSNAT, DSSAT V3. Vol. 3–4. University of Hawaii, Honolulu, HI, USA.

Jenkinson D S (1991) The Rothamsted long-term experiments: Are they still of use? Agron. J. 83(1): 2–11.

Kruseman G, Hengsdijk H, Ruben R (1993) Disentangling the concept of sustainability: Conceptual definitions, analytical framework and operational techniques in sustainable land use. DLV Report No. 2, AB-DLO, P.O. Box 14, Wageningen, The Netherlands. 52 p.

Lynam J K, Herdt R W (1988) Sense and sustainability: Sustainability as an objective in international agricultural research. Paper presented at CIP-Rockfeller Conference on Farmers and Food Systems. CIP, Lima, Peru.

Meerman F, Van De Ven G W J, Van Keulen H, De Ponti O M B (1992) Sustainable crop production and protection. Discussion Paper, Ministry of Agriculture, Nature Management and Fisheries, Department OSL, P.O. Box 20401, The Hague, The Netherlands. 48 p.

Mitchell C C, Westerman R L, Brown J R, Peck T R (1991) Overview of long-term agronomic research. Agron. J. 83(1):24–29.

Monteith J L (1990) Can sustainability be quantified? ICRISAT Conference Paper No. 538. ICRISAT, Andhra Pradesh, India. 16 p.

Nambiar K K M (1994) Soil fertility and crop productivity under long-term fertilizer use. ICAR, New Delhi, India. 150 p.

Nambiar K K M, Ghosh A B (1984) Highlights of research of long-term fertilizer experiments in India (1971–82). Long-term Fertilizer Experiments Research Bulletin No. 1, ICAR, New Delhi. 72 p.

Reddy V R, Baker D N (1990) Application of GOSSYM to analyse the effects of weather on cotton yields. Agric. Syst. 32(1):83–95.

Singh U, Ritchie J T, Godwin D C (1993) A user's guide to CERES-Rice V2.10. Simulation manual IFDC-SM-4. IFDC, Muscle Shoals, AL, USA. 131 p.

Singh U, Thornton P K (1992) Using crop models for sustainability and environmental quality assessment. Outlook on Agriculture 21(3):209–218.

Thornton P K, Wilkens P W, Hoogenboom G, Jones J W (1994) Sequence analysis user guide. Pages 67–136 in Tsuji G Y, Uehara G, Balas S (Eds.) DSSAT V3.0. University of Hawaii, Honolulu, HI, USA.

Thornton P K, Hoogenboom G, Wilkens P W, Bowen W T (1995) A computer program to analyse multiple-season crop model outputs. Agron. J. 87:131–136.

Timsina J, Singh U, Singh Y, Lansigan F P (1995) Addressing sustainability of rice-wheat systems: Testing and application of CERES and SUCROS models. Pages 633–656 in Proceedings of the International Rice Research Conference, 13–17 February 1995. International Rice Research Institute, P.O. Box 933, Manila, Philippines.

Systems approach in the design of soil and water conservation measures

L. STROOSNIJDER and P. KIEPE

Department of Irrigation and Soil and Water Conservation, Wageningen Agricultural University, Nieuwe Kanaal 11, 6709 PA Wageningen, The Netherlands

Key words: run-off, soil loss, soil and water conservation, modelling, systems approach, conservation tillage, agroforestry

Abstract

In this report, two examples are presented of systems approaches in the design of soil and water conservation measures for agricultural development. The first example deals with conservation tillage in the semi-arid tropics. This practice aims at a better infiltration of water into the soil that stimulates vegetative growth and nutrient uptake early in the growing season. In good rainfall years, this may enhance millet grain production but in dry years this conservation tillage may do more harm than good. In semi-arid regions, both water and nutrients are production-limiting factors. The most limiting factor depends on the rainfall distribution over the year, which is notoriously variable in semi-arid climates. Daily rainfall distributions and intensities of 35 years were used in a problem-driven combination model (using existing submodels) that describes crop growth and grain production, the water balance, crust formation and the nutrient balance. Nutrient availability is calculated on the basis of soil data and compared with crop requirements at maturity. Actual production is either the water- or the nutrient-limited output. Results show that conservation tillage without an increase in nutrient input has a marginal effect. Average grain production increases from 320 to 350 kg ha^{-1} and no improvement in the production risk is achieved. When nutrient limitations are eliminated, production increases from 320 to 470 kg ha^{-1} without conservation tillage and from 320 to 560 kg ha^{-1} with conservation tillage. However, the coefficient of variation (SD/AVG), thereby, increases from 40% to 65%. In the second example, the model SHIELD (Simulation of Hedgerow Intervention against Erosion and Land Degradation) was used for an optimization of the spacing of agroforestry hedges on sloping land in Kenya. Again, an existing crop growth model was used and adapted for this special use. Changes were made with respect to rainstorm intensity, infiltration rate, run-off, landform, mulch and crop cover, spatial arrangement in the crop-hedge agroforestry system and the vulnerability of the soil to erosion. The model computes run-off and soil loss, as well as crop production in competition with the hedge. Results are used to calculate optimal spacing of the hedges which depends on user-defined goals for crop production, use of hedge products and incentives for soil conservation.

Introduction

Due to land degradation caused by erosion, some areas will suffer from food shortages, while other areas will be subject to environmental disaster. In the last 45 years, already 20% of the land in the world has been abandoned because it suffered from erosion (Pimentel et al. 1995). Thus, erosion represents a serious time-bomb and may threaten political stability in the world. Erosion has both physical and chemical aspects. Until recently, physical aspects such as soil loss from agricultural fields and sediment load in streams received most of the attention. However, there is no water erosion without run-off. So, understanding and quantification of run-off should

M.J. Kropff et al. *(eds.): Applications of Systems Approaches at the Field Level, 399–411.*

precede that of erosion. Even less obvious, and hidden for a long time, are the amount of plant nutrients and other chemicals that get lost with run-off and in the eroded material. For the selection of appropriate conservation measures, for their technical design and for an (economic) appraisal, quantified knowledge about the effects of conservation on run-off and nutrient availability is needed (Stroosnijder 1994). This paper provides two examples of the systems approach. The first example (DUET) deals with the assessment of water conservation tillage in an environment where both water and nutrients can be limiting crop production. In the second example (SHIELD), simulation modelling is used to determine the optimum spacing between hedgerows in a soil and water conservation system.

Modelling the effect of water or nutrient conservation

Appraisal of the effect of water and nutrient conservation on crop production is essential for conservation planning, because run-off and erosion affect water and nutrient availability. When, for instance, water availability is the most limiting factor for crop production, the conservation of nutrients will not stimulate growth and hence such conservation will not be very effective. An appraisal implies a quantitative understanding of all processes determining crop growth under the local circumstances. Deterministic crop growth models have been developed with the aim to explain and understand crop growth as it can be observed at different scales of observation. So, modelling seems a relevant tool for the appraisal of nutrient conservation.

When looking for an appropriate model, one may choose an existing model or develop a new one. Since many models exist already, it must be possible to find a suitable existing model and adapt it to the specific user's and environment's needs. In choosing such an existing model, there are a few essential choices to be made: (1) the complexity of the model, (2) time step of modelling and (3) time horizon of the effect. The complexity of a model depends, among others, on whether one can assume one or more growth variables irrelevant. So-called potential growth models assume both water and nutrients amply available. This makes these factors irrelevant and water and nutrient submodels often do not even exist in these level-1 models. Level-2 models assume either water or nutrients amply available while the other factor is then part of the dynamic model. A 2A-model, for instance, assumes ample nutrient availability. Growth depends here, apart from the crop and climatic factors that determine potential growth, on the amount of available water. So, these models have to simulate the water balance, i.e. all gains and losses of water in the simulated system. A 2B-model does the same but then the other way around. It assumes ample water and simulates nutrient availability. Level-3 models take both the water and nutrient balances into account simultaneously and are obviously more complicated than level-2 or level-1 models.

There are very few level-3 models. Existing ones are difficult to manage due to their complexity, their data demand and the little attention that was paid to make these models user-friendly. Therefore, a simple method was used in this paper to overcome this problem. First, a level-2A model was used to estimate the so-called water-limited crop production. Then, with a 2B-level model, the nutrient-limited production was calculated separately. Actual production was then assumed to be the minimum of these two values. Obviously, this is only a rough estimate of the real production since synergistic effects between water and nutrients, as taken into account in level-3 models, can only be incorporated to a limited extent.

Each process, being part of the deterministic modelling of crop growth, has its optimal time step. In level-2 models, processes of the water balance have a much smaller optimum time step than crop growth processes. In such cases, parametric modelling (Stroosnijder 1982) is used to match water balance processes with crop growth processes, so that calculation time steps of one day can be used. However, it must always be checked whether crucial process information is not lost in this upscaling of the time frame. Models that are using time steps of one day often make use of computer languages that were specially developed for the simulation of continuous processes like growth.

Plant nutrients occur in soils in at least three phases: (A) readily (in the current season) available nutrients, (B) in organic matter and (C) in the mineral reserve. When, due to conservation measures, soil is retained that otherwise would have been eroded the beneficial effect on crop growth lasts for a long time because nutrients of all three phases are gained (compared to the without conservation case). This is especially so when the release of nutrients from organic matter is the most important source of readily available plant nutrients. Therefore, in the case of soil conservation, the appraisal of nutrient conservation must be calculated over a period of 15 years or more.

Modelling agroforestry systems

There are two major challenges in modelling agroforestry (Kiepe 1995b). First, agroforestry comprises a range of disciplines, which complicates the modelling effort for mono-disciplinary specialists. Second, agroforestry is a relatively new science where many processes are not yet fully understood. There are still many lacunas in agroforestry, notably in the tree/crop interface or the zone where both tree and crop are active. Agroforestry cannot draw on large data bases as agriculture can, with all its information on the physiology and growth characteristics of the major food crops and responses to constraints, like deficiencies in radiation, water or nutrients. This is not so for trees, where most process-based research has concentrated on forest stands rather than individual trees and where the soil is often treated as a single black box. Furthermore, a crop has often only one option to respond to environmental constraints, while a tree has several strategies and might even switch in its approach to tackle constraints over the years. For instance, to respond to high transpiration losses an annual crop closes its stomata. Apart from closing stomata too, a tree may also opt to expand its rooting system or to shed its leaves. To circumvent lacunas, the approach to modelling agroforestry systems can be semi-empirical until more is known about tree behaviour and the interaction with crops.

The agroforestry technology that shows great potential for soil conservation as well

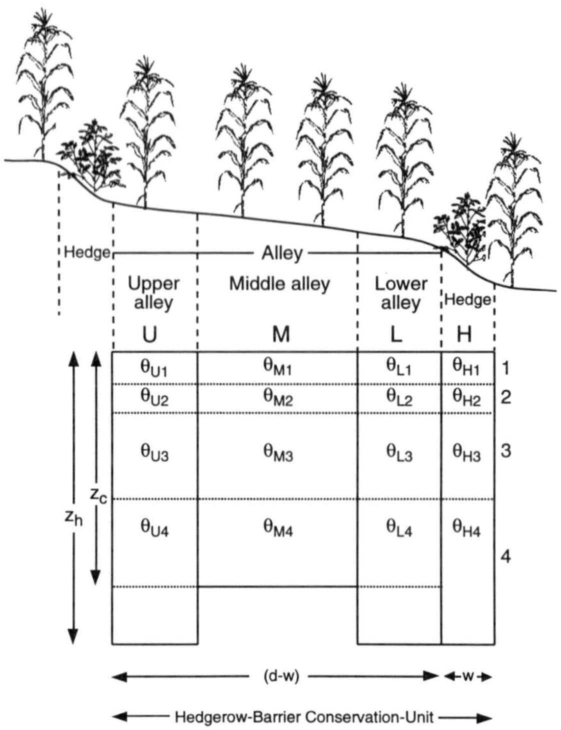

Figure 1. Schematic cross-section through a hedgerow-barrier conservation-unit divided into soil compartments, depicting the upper alley (U), middle alley (M), lower alley (L) and the hedgerow (H), 4 soil horizons (1, 2, 3 and 4), maximum rooting depth of the crop (z_c) and the hedgerow (z_h) and the soil water content θ of each compartment. This cross-section forms the basis of the SHIELD model (Simulation of Hedgerow Intervention against Erosion and Land Degradation).

as applicability to a wide range of environments is the hedgerow barrier technology. Compared with other agroforestry technologies, it has a simple geometry that is easy to describe mathematically, which makes it suitable as a starting-point for the modelling approach (Figure 1). Modelling is especially wanted for hedgerow barrier systems, because crucial information in its design, like the optimum spacing between hedgerows, is unknown due to the complexity of the processes involved. For instance, in soil conservation the ideal hedgerow spacing depends on a range of biological, edaphic and climatic factors. Even when the ideal spacing for soil conservation is established, it is not necessarily the same as the ideal spacing for plant production. Therefore, the optimum spacing of a hedgerow barrier system should be derived with respect to both plant production and soil conservation.

There are two reasons to link a hedgerow barrier model to an existing crop growth model. First, modelling soil and water conservation aspects of hedgerow barrier systems without considering the effect of hedgerows on crop production is imprudent. Water uptake by plants is an important component of the water balance and essential to compute the water content of the soil. In turn, the water content of the soil is essential to compute the vulnerability of the soil to erosion. Therefore, water uptake by hedgerow and the crop in the alley cannot be omitted. Secondly, it is important to estimate the yield of the crop because a land user will be reluctant to opt for a conservation technology if the economic benefits are negligible. Linking hedgerow-barrier system modelling to crop growth models is, therefore, a necessity. However, finding a compromise between conflicting time constants is a problem. For crop growth modelling, the choice of a daily time step is an obvious as well as a convenient one, because a day represents a natural unit consisting of complete cycles for processes that can be characterized distinctly by environmental parameters. Flow processes concerning the water balance usually need much smaller time constants. However, in practice large differences occur depending on the maximum gradients that occur. For infiltration, often time steps of seconds or minutes are needed, while internal drainage (redistribution) can sometimes be modelled on a weekly time-basis. The best compromise seems daily time steps because of the importance of plant growth related processes together with validated parametric modelling (Stroosnijder 1982) of the water balance, or a submodel with a different time step, e.g. SAWAH (Ten Berge et al. 1992).

Brief description of DUET

In this report, two examples will be briefly discussed. The first example (Stroosnijder et al. 1994) is a numerical computer model called DUET that simulates crop growth on a daily basis and was used for semi-arid regions where rainfall distribution over the year is notoriously variable. DUET is a problem-driven combination model (using existing validated submodels) that describes crop growth and grain production, the water balance, crust formation and the nutrient balance.

The water balance submodel is similar to SWATRE (Belmans et al. 1983); nutrient availability is calculated with QUEFTS (Janssen et al. 1990) on the basis of soil data and compared with crop requirements at maturity, resulting from the crop growth submodel of WOFOST (Van Diepen et al. 1989). Synergy between water and nutrients was taken into account to a limited extent. When water infiltrated deeper into the soil due to water conservation in the SWATRE submodel, roots were also allowed to penetrate deeper into the soil and could subsequently 'graze' more nutrients in the QUEFTS submodel. The model is written in FORTRAN and can be run on a PC. For a full description of DUET see Stroosnijder et al. (1994).

Examples of the use of DUET

DUET was used for the appraisal of various types and frequencies of conservation tillage on the growth of millet (*Pennisetum glaucum* L., R. Br.) in the West African Sahelian zone. Soil data of loamy sands from Niono in Mali (5° 45′ W; 14° 30′ N) were used and climate data of Burkina Faso. Conservation tillage aims at a better infiltration of water into the soil. This stimulates vegetative growth and nutrient

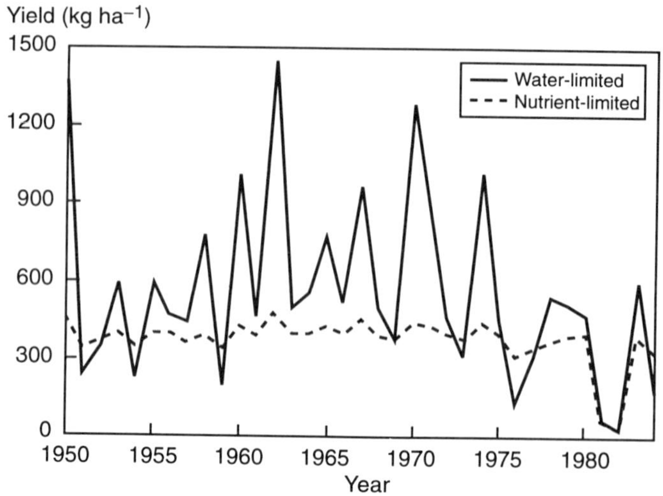

Figure 2. Water- and nutrient-limited grain millet production in Burkina Faso for 35 years (1950–85) with water conservation tillage practised every 15 days.

uptake early in the growing season. In good rainfall years, this may stimulate grain production but in dry years it is feared that such conservation tillage may do more harm than good. Actual production is either the water- or the nutrient-limited output, which were calculated for 35 years (1950–84, Figure 2).

Results show that conservation tillage without an increase in nutrient input has a marginal effect. Average grain production increases from 320 to 350 kg ha^{-1} and no improvement in the production risk is achieved. When nutrient limitations are eliminated, production increases from 320 to 470 kg ha^{-1} without conservation tillage and from 320 to 560 kg ha^{-1} with conservation tillage. However, the coefficient of variation (SD/AVG) thereby increases from 40% to 65%. The conclusion was that it would be better to use a conservation measure where water as well as nutrients are conserved simultaneously. There was no necessity to validate DUET, since the model uses validated submodels and was used for appraisals only. Although the predicted yields were low, they do not differ substantially from the actual yields of unfertilized sandy farmers' fields in the Sahel.

Brief description of SHIELD

The starting-point in the development of SHIELD was that the model should be linked to an existing crop growth model with feedback between crop growth and water availability. The model structure that was selected was SUCROS2 (Van Keulen et al. 1992), because it is mechanistic, clear and allows the user to make all necessary changes to adapt it for any required purpose. The tropical maize version of SUCROS2, called MAIZE2 (Stroosnijder 1989), became the basis of SHIELD. The last character of the names of both crop growth models indicates Production Level 2. Hence, SHIELD assumes an ample fertilized crop, free of pests and diseases. SHIELD was written in the simulation language FST (Fortran Simulation Translator) that generates a FORTRAN program and the corresponding data files (Rappoldt and Van Kraalingen 1996), which allows the user to work in a FORTRAN simulation environment. The applicability of the crop growth model was enhanced for use at higher elevations by changing the psychrometer constant in the Penman-Monteith combination equation into a function of altitude. Therefore, the potential evaporation became also altitude dependent. Necessary changes for modelling run-off and soil loss were a revised water balance and adding a parametric-modelled soil balance (Kiepe 1995b). The water balance now includes critical features like infiltration and moisture deficit of the topsoil, which determine Hortonian overland flow, hydraulic conductivity, which determines topsoil saturation overland flow, and surface depression storage. The soil balance was added to the water balance by parametric modelling of the sediment concentration in run-off. The spatial arrangement of crops and hedgerows, the effect of mulch and hedgerows on infiltration and the spatial uptake of water were added to compute the effect of mulch and hedgerows on run-off, soil loss and crop production. The hedgerow barrier system under consideration reduces run-off by increasing the infiltration by two mechanisms. First, the ability of mulch application to increase infiltration, which has been known for a long time (e.g. Mannering and Meyer 1963). Second, hedgerows planted on the contour as barriers increase the infiltration by 3 to 8 times (Kiepe 1995a). A reduction in run-off implies a reduction in soil loss. A number of relations was quantified during four years of field work in Machakos, Kenya. The most relevant equations are given below.

Effect of mulch on infiltration, run-off and soil loss

The effect of mulch on daily infiltration was described as (Kiepe 1995b):

$$I_m = (1 + a_m \sqrt{M}) I_c \tag{1}$$

where the infiltration in mulched soil (I_m in mm) is computed using a mulch-impact parameter (a_m in $\mathrm{ha^{1/2}\,ton^{-1/2}}$), the amount of mulch ($M$ in $\mathrm{ton\,ha^{-1}}$) and the infiltration in non-mulched soil (i.e. the control) (I_c in mm). Run-off from a mulched plot (R_m in mm) is calculated by:

$$R_m = P - I_m \tag{2}$$

where P (in mm) is the daily rainfall.

The effect of mulch on soil loss can be described as:

$$S_m = b_m \mathrm{M}^{-1} 0.01\, c R_m \qquad \text{for } M \geq b_m \tag{3}$$

where soil loss (S_m in $\mathrm{ton\,ha^{-1}}$) is a function of the barrier-effect parameter of mulch (b_m in $\mathrm{ton\,ha^{-1}}$), the amount of mulch applied (M), the sediment concentration (c in $\mathrm{kg\,m^{-3}}$) and run-off from the mulched compartment (R_m).

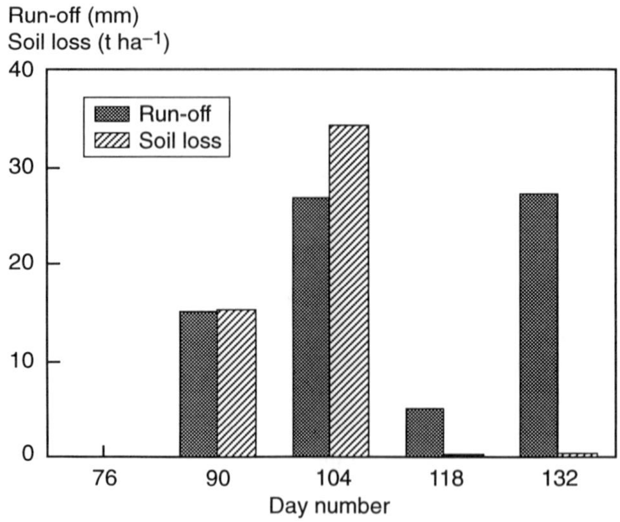

Figure 3. SHIELD output: the effect of run-off and soil loss from design storms on day numbers 76, 90, 104, 118 and 132. Results of day 104 are consistent with measured values.

Effect of hedgerows on infiltration, run-off and soil loss

The infiltration in hedgerow barrier systems without mulch application for a single-row hedge ($I_{h(1)+a}$) can be calculated according to:

$$I_{h(1)+a} = (1 + (a_h(d - w_1)w_1 d^{-2}))I_c \tag{4}$$

where the hedgerow-impact on infiltration (a_h), the distance between the hedgerows (d in m) and the width of the hedgerow (w_1 in m) are required. Run-off from a hedgerow plot ($R_{h(1)+a}$ in mm) is calculated by:

$$R_{h(1)+a} = P - I_{h(1)+a} \tag{5}$$

The infiltration in single-row hedgerow barrier systems with mulch application ($I_{h(1)+m}$) is calculated by:

$$I_{h(1)+m} = (1 + ((a_m \surd(d - w_1)d^{-1}M) + (a_h(d - w_1)w_1 d^{-2}(1 - a_m \surd d(d - w_1)^{-1}M))))I_c \tag{6}$$

Run-off from a hedgerow plot with mulch application ($R_{h(1)+m}$ in mm) is calculated by:

$$R_{h(1)+m} = P - I_{h(1)+m} \tag{7}$$

Equations similar to equations (4) and (6) were also developed for multiple row hedges. Soil loss of single hedgerow barrier systems without mulch application ($S_{h(1)+a}$ in ton ha^{-1}) can be calculated with:

$$S_{h(1)+a} = 2b_h \surd(d^{-1})\, 0.01\, cR_{h(1)+a} \tag{8}$$

where soil loss is a function of the barrier-effect parameter of hedgerows (b_h), the hedgerow distance (d), the sediment concentration of run-off from the control plot (c) and run-off from the hedgerow barrier system ($R_{h(1)+a}$).

Soil loss of single hedgerow barrier systems with mulch application ($S_{h(1)+m}$ in ton ha^{-1}) is computed by:

$$S_{h(1)+m} = 2 b_m \sqrt{(d^{-1} M^{-1})}\, 0.01\, c R_{h(1)+m} \qquad \text{for } M \geq 4 d^{-1} b_m^{\,2} \tag{9}$$

where soil loss is related to the hedgerow distance (d), run-off from the mulched hedgerows ($R_{h(1)+m}$) and c, b_m and M.

Sediment concentration in run-off

The proposed parametric modelling of the soil loss uses an estimate of the sediment concentration of run-off first (c in kg m^{-3}), followed by a multiplication with run-off and the hedgerow and mulch barrier parameters. Factors that influence the sediment concentration in SHIELD are: the energy of the storm (E_k in J), the vulnerability of the soil to erosion (v in mm), landform (LF in m$^{1/2}$) and total soil cover (F_T).

$$c = E_k\, v\, LF\, F_T^{-1} \tag{10}$$

Equation (10) is only valid for the specific units and variables as described in Appendix 1. The dimension found after multiplication of these four factors does not conform to the dimension of the sediment concentration (kg m^{-3}), which is a common problem in establishing a relation between erosive factors and soil loss (Wischmeier and Smith 1978).

Examples of the use of SHIELD

Usually, soil and water conservation engineers use the concept of a design storm for the dimensioning of conservation structures. A design storm is a storm that occurs with a predetermined frequency, e.g. once every ten years. In circumstances when the time of the storm is crucial because of the wetness of the soil profile, this concept fails. Therefore, the concept was modified into a design event, i.e. a design storm for a specific time of the year. Figure 3 presents run-off and soil loss from the same design storm falling on different dates on which the soil has different states of wetness. From Figure 3 can be concluded that the timing of the design storm makes a difference with respect to run-off and soil loss.

Conservation tillage is a form of soil and water conservation technology that is widely used. One can optimize the depth of tillage but also its frequency. Figure 4 shows the effect of six different tillage schemes on soil loss. No-till gives the highest run-off and soil loss. Primary and secondary tillage do not make much difference. In reality, no-till gives indeed the highest run-off, but soil loss decreases due to an increased biological activity in the soil. However, such changes in biological activity are not yet incorporated in SHIELD.

The optimum distance between conservation hedges depends on the limit that is set for the tolerable soil loss. If this is 2 ton ha^{-1}, this distance varies between less

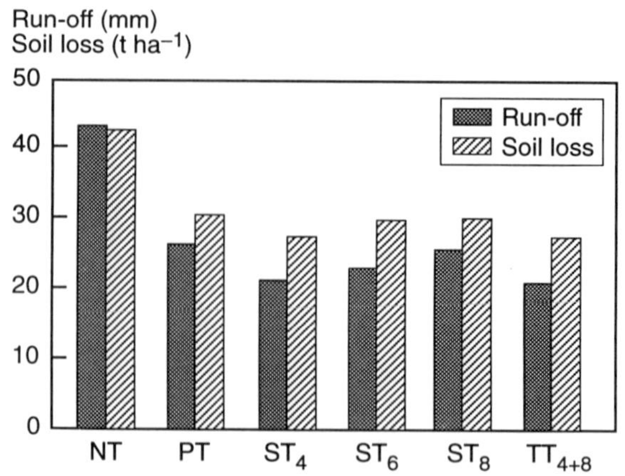

Figure 4. SHIELD output: the effect of tillage frequency (75 mm depth) on run-off and soil loss. No-till (NT), primary tillage before planting (PT), secondary tillage after 4 (ST_4), 6 (ST_6) and 8 (ST_8) weeks and tertiary tillage (TT_{4+8}) after 4 and 8 weeks (20-season average).

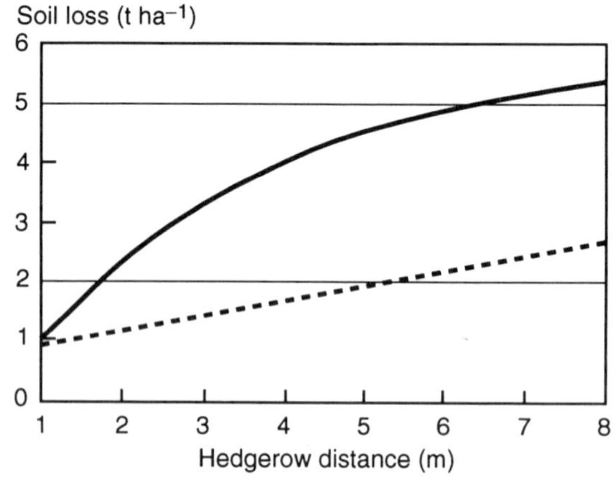

Figure 5. SHIELD output: the effect of the distance (m) between single-row hedges on soil loss (ton ha^{-1}, 20-season average) without mulch (solid line) and with mulch (dashed line) applied between the hedges.

than 2 m without mulch on the alley to 5 m with mulch application for a single hedge as can be seen in Figure 5.

Conclusions

Models like DUET and SHIELD allow the user to appraise the effect of the many conservation measures that soil and water conservation experts have in their toolbox. Where conservation technology is expensive and funds are limited, such an appraisal is a must. However, before measures can be implemented, the model needs to be validated with experimental data first, because previous technical failures have contributed to the bad image of soil and water conservation measures. Even without technical flaws, there are still enough cultural, social, economic, political and ethical problems to overcome before conservation measures become a real success. The latter is urgently needed, since land degradation is one of the most important threats to a sufficient food supply in the world. After appraisal, by taking the local agroecological conditions into account the users can also optimize the design of the measure that has been selected. It is a drawback that there are still no easy to use and adapt level-3 models. The present practice to run a level-2A model and a level-2B model and take the minimum or maximum values of the two outcomes does not fully credit possible synergetic effects between water and nutrient conservation.

References

Belmans C, Wesseling J G, Feddes R A (1983) Simulation of the water balance of a cropped soil: SWATRE. J. Hydrol. 63:271–286.

Hudson N (1986) Soil conservation. Revised reprint. Batsford, London, UK. 324 p.

Janssen B H, Guiking F C T, Van Eijk D, Smaling E M A, Van Reuler H (1990) A system for quantitative evaluation of the fertility of tropical soils (QUEFTS). Geoderma 46:299–318.

Kiepe P (1995a) Effect of *Cassia siamea* hedgerow barriers on soil physical properties. Geoderma 66: 113–120.

Kiepe P (1995b) No runoff, no soil loss: Soil and water conservation in hedgerow barrier systems. Tropical Resource Management Paper No. 10, Wageningen Agricultural University, The Netherlands. 156 p.

Mannering J V, Meyer L D (1963) Effects of various rates of surface mulch on infiltration and erosion. Soil Sci. Soc. Am. Proc. 27:84–86.

Pimentel D, Harvey C, Resosudarmo P, Sinclair K, Kurz D, McNair M, Crist S, Sphritz L, Fritton L, Saffouri R, Blair R (1995) Environmental and economic costs of soil erosion and conservation benefits. Science 267:1117–1123.

Rappoldt C, Van Kraalingen D W G (1996) The FORTRAN Simulation Translator; FST Version 2.0. Introduction and reference manual. Quantitative Approaches in Systems Analysis, Report No. 5, AB-DLO and C.T. De Wit Graduate School for Production Ecology, Wageningen, The Netherlands. 147 p.

Stroosnijder L (1982) Simulation of the soil water balance. Pages 175–193 in Penning de Vries F W T, Van Laar H H (Eds.) Simulation of plant growth and crop production. Simulation Monographs, Pudoc, Wageningen, The Netherlands.

Stroosnijder L (1989) Maize growth in a tropical environment. Part II. Rainfed production. Dept. Irrig. and Soil Water Conservation, Wageningen Agricultural University, The Netherlands. 65 p.

Stroosnijder L (1994) Population density, carrying capacity and agricultural production technology in the Sahel. Pages 3–24 in Reenberg A, Markussen B (Eds.) The Sahel. AAU Reports 32, Danish International Development Assistance (Danida), Denmark.

Stroosnijder L, Hoogmoed W B, Berkhout J A A (1994) Modelling effects of water conservation tillage in the semi-arid tropics. Pages 235–251 in Libbey J (Ed.) Bilan hydrique agricole et sécheresse en Afrique tropical. Eurotext, Paris, France.

Ten Berge H F M, Jansen D M, Rappoldt C, Stol W (1992) The soil water balance module SAWAH, description and user's guide. Simulation Reports CABO-TT 22, Department of Theoretical Production Ecology, Wageningen Agricultural University, The Netherlands. 78 p.

Van Diepen C A, Wolf J, Van Keulen H, Rappoldt C (1989) WOFOST: A simulation model of crop production. Soil Use and Management 5:16–24.

Van Keulen H, Goudriaan J, Stroosnijder L, Lantinga E A, Van Laar H H (1992) Crop growth model for water-limited conditions (SUCROS2). Pages 27–72 in Van Laar H H, Goudriaan J, Van Keulen H (Eds.) Simulation of crop growth for potential and water-limited production situations (as applied to spring wheat). Simulation Reports CABO-TT 27, Department of Theoretical Production Ecology, Wageningen Agricultural University, The Netherlands.

Wischmeier W H, Smith D D (1978) Predicting rainfall erosion losses, a guide to conservation planning. Agric. Handb. No. 537, USDA, Washington DC, USA. 58 p.

Appendix 1

Kinetic energy is defined as half the product of mass and velocity squared. The total mass of a rainstorm equals the amount of rainfall (A in mm), while the velocity equals the rainfall intensity (i in mm h^{-1}). The kinetic energy (E_k in J) of a storm is then:

$$E_k = 0.5 A i^2 \qquad \text{(A1)}$$

The vulnerability of the soil to erosion can be calculated as:

$$v = (\%\ \text{sand} + \%\ \text{silt})\ (\%\ \text{clay})^{-1} C_O^{-1} I_p \theta_i \theta_s^{-1} \qquad \text{(A2)}$$

Vulnerability to erosion depends on the cohesive forces that soil particles exert on each other. These forces depend on texture, soil organic matter content and the amount of water in the soil. The effect of organic matter on the resistance of soils to erosion is taken inversely related to the organic carbon content (C_O^{-1}). The cohesive forces between soil particles of the Lixisols in Machakos are weakened when water molecules are present between the particles, especially when the water content of the soil is above field capacity. Hence, the amount of water present in the soil profile is of crucial importance. The effect of water in the topsoil is calculated as the antecedent water content divided by the water content at saturation ($\theta_i \theta_s^{-1}$). Drainage of soil water after a rainstorm takes about two days in Machakos. Entrapped air can impede infiltration and can cause aggregates to 'explode' causing slaking and crusting. This was taken into account by adding the cumulative amount of water that infiltrated on the two days preceding a rainstorm (I_p in mm).

The combined effect on landform (LF in m$^{1/2}$) is:

$$LF = \phi^2 \sqrt{L} \qquad \text{(A3)}$$

Landform is defined by shape, slope angle and slope length. A uniform slope loses more soil than a concave, but less than a convex slope, and is usually taken as reference. The relation between slope angle (ϕ in %) and soil loss is exponential and for the tropics the value of 2 is appropriate. The effect of slope length on soil loss per unit area is equal to the square root of the length (L in m) (Hudson 1986).

The total soil cover ($F_{T(t)}$) is:

$$F_{T(t)} = F_{c(t)} + F_{m(t)} \qquad \text{(A4)}$$

The fraction of soil cover by a maize crop ($F_{c(t)}$) and leaf area index (LAI) were measured for a full season and appeared to be closely related:

$$F_{c(t)} = 0.30\, LAI \qquad \text{(A5)}$$

Correlation was high ($n = 10$, $R^2 = 0.966$), but is only valid in Machakos, Kenya, for this specific maize cultivar (*Zea mays*, cv. Katumani Composite B) and planting scheme (0.3 m × 0.9 m).

The fraction soil cover by mulch ($F_{m(t)}$) of cassia (*Cassia siamea*, Lam., recently changed into *Senna siamea*) was described by:

$$F_{m(t)} = F_{m(0)} k^{-t} \qquad \text{(A6)}$$

where $F_{m(t)}$ is the mulch cover fraction at time t (in days), $F_{m(0)}$ is the initial mulch cover fraction and k is the decay constant. Mulch of fresh tree prunings decays over time due to decomposition. The rate of decomposition depends on temperature and humidity, as well as on plant characteristics, notably the ratio between polyphenolics and nitrogen. A decay constant for cassia in Machakos of $k = 1.04$ provided a good correlation for six seasons of measurements ($n = 37$, $R^2 = 0.965$).

The conversion of the weight of fresh prunings applied as surface mulch to soil cover can be described by:

$$F_{m(0)} = 1 - \exp(-UM) \tag{A7}$$

where U is the specific leaf area of the mulch (ha ton^{-1}) and M is the mulch application rate (ton ha^{-1}). For cassia in Machakos a specific leaf area of 0.79 ha ton^{-1} was found for oven-dry leaves ($n = 10$, $R^2 = 0.997$).

Farming systems for sustainable agriculture and environmental quality

R.S. KANWAR[1], D.L. KARLEN[2], C. CAMBARDELLA[2] and T.B. MOORMAN[2]

[1]*Agricultural and Biosystems Engineering, Iowa State University, Ames, Iowa 50011, USA*
[2]*USDA-ARS, National Soil Tilth Laboratory, Ames, Iowa 50011, USA*

Key words: sustainability, farming systems, water quality

Abstract

Field studies were initiated in 1992 to investigate the impacts of eight integrated crop and chemical management systems on crop production and groundwater quality. The tillage and chemical management systems included the use of late spring NO_3-N test (LSNT) in a corn and soybean rotation respectively, with either no-till or chisel ploughing, and broadcast and banding of herbicides under no-till and chisel ploughing. All experimental plots were instrumented for sampling subsurface drainage water for quantity and quality. An automatic data collection system was used to collect water samples for water quality analyses during the 1993, 1994 and 1995 growing seasons from 36 field plots (size 0.4ha). Results of this study indicate that the use of LSNT resulted in equal or better yields and lower NO_3-N concentrations in the drainage water for both tillage practices. The LSNT resulted in NO_3-N concentrations of 9.3, 8.0 and 8.9 mg l^{-1} under no-till, and 11.1, 7.1 and 11.4 mg l^{-1} under chisel plough for 1993, 1994 and 1995 respectively. The highest corn yields were obtained with LSNT for both tillage systems in 1993, 1994 and 1995. Also, banding of herbicides resulted in lower herbicide losses to shallow groundwater. The overall results of this study show that alternative chemical and crop management systems are important to the preservation of environmental quality and increased crop productivity.

Introduction

Pesticides and nitrate from fertilizers and manure application have been detected in the surface and groundwater in many agricultural regions of the world. The current practices of fertilizer and pesticide application methods and rates are believed to be contributing significantly in the contamination of groundwater (Randall and Iragavarapu 1995; Milburn and Richards 1994). Groundwater pollution from agricultural activities is of particular concern in the USA because 50% of the population depend on groundwater for drinking water supplies (Hallberg 1989). The US Environmental Protection Agency (EPA 1990) reported that nitrate-nitrogen (NO_3-N) was the most commonly found agricultural pollutant found in more than 50% of rural and community drinking wells surveyed nationwide. If the potential for contamination is not reduced by developing and successfully demonstrating the innovative N and pesticide management practices, the potential for contamination will remain and could result in additional regulations.

Tillage, planting systems and chemical management systems that minimize the potential for chemical run-off and leaching losses to groundwater need to be developed. Several studies have shown that NO_3-N concentrations in the subsurface drainage

M.J. Kropff et al. *(eds.): Applications of Systems Approaches at the Field Level, 413–418.*

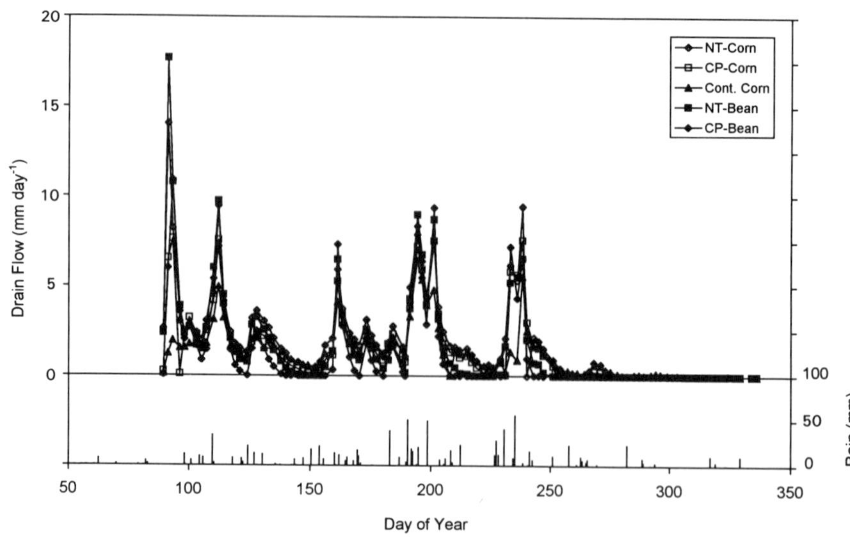

Figure 1. Subsurface drain flows from plots with five different production systems for 1993.

waters from existing agricultural production systems often exceed 10 mg l^{-1}, a drinking water standard set by the EPA (Baker and Johnson 1981; Bergstrom 1987; Jemison and Fox 1994; Kanwar et al. 1988, 1993, 1995; Kladivko et al. 1991).

The objectives of this study were to develop new N, herbicide and tillage management strategies to reduce the leaching of chemicals to shallow groundwater. Field experiments were conducted at the Iowa State University's Northeast Research Center near Nashua, Iowa, during the growing seasons of 1993, 1994 and 1995. This paper will summarize three years' data on water quality and crop yields and evaluate the effects of various practices on groundwater quality.

Material and methods

This study was conducted on a predominantly kenyon silty-clay loam soil with 3–4% organic matter. These soils have seasonally high watertables and benefit from subsurface drainage. Pre-Illinoian glacial till units of 60 m thickness overlie a carbonate aquifer used for water supply. However, in some areas, bedrock is almost at the surface.

The study site has 36 experimental plots (size 0.4 ha) with fully documented tillage and cropping records for the past sixteen years. The subsurface drainage system has been in place at this site for more than fifteen years. The subsurface drainage system was installed in 1979 into all of the 36 blocks (67 m×59 m). The drain lines were installed about 1.2 m deep at 28.5 m spacings. Each plot has one drain line passing through the middle of the plot, and there is a drain line at each of the two borders. The drain lines at the borders pick up any cross-contamination from the surrounding plots. There are a maximum of ten plots in a row and plot rows are separated by an uncultivated area of 9 m width. The middle drain lines of all the plots were intercepted and connected to individual sumps for measuring subsurface drainage flow and collecting water samples for chemical analyses. To monitor drain flow on a continuous basis, each drain sump has a 100 V effluent pump, water flow meter and an orifice tube to collect water samples for water quality analysis. The water flow meters were connected to dataloggers for recording drain flow monitoring data. For water quality sampling, an orifice tube was used to deliver about 0.2% of the drain water into a sampling bottle each time effluent is pumped from the sump. For NO_3-N analyses, subsurface drain water samples were collected three times a week and weekly samples were collected for pesticide analyses when drain lines were flowing.

Farming practices included 12 no-till (NT) and 18 chisel plough (CP) plots under corn-soybean rotation and six continuous corn (CC) chisel plough plots. Chisel plough plots were ploughed in the fall and field cultivated in the spring before planting. No-till crops were planted directly into stubble. All soybean plots were planted with a no-till drill. All corn plots were cultivated once for weed control, except in 1993 when wet conditions prevented cultivation. Two N fertilizer treatments were applied to CC plots: single, pre-plant urea-ammonium nitrate (UAN) application at 135 kg N ha^{-1} and fall manure application at approximately 135 kg N ha^{-1}. CP rotation corn received three N treatments: single, pre-plant UAN application at 112 kg N ha^{-1}, fall manure application at approximately 112 kg N ha^{-1} and 30 kg N ha^{-1} pre-plant application with additional UAN sidedressed based on late spring soil nitrate test (LSNT). NT rotation corn received the same N treatments as CP rotation corn without the manure treatment. CP plots with a pre-plant UAN application of 112 kg N ha^{-1} were also given banding and broadcast treatment of two herbicides (extrazine and alachlor at application rates of 2.8 and 2.5 ai kg ha^{-1} at the time of planting). In the banding treatment, typically half (or less) of the amount of herbicide is needed as herbicide is applied on the rows only and weed control is done through cultivation in between the rows. All management treatments were replicated three times. Soil and water samples were analysed for NO_3-N and herbicides at the USDA-ARS National Soil Tilth Laboratory in Ames, Iowa. For NO_3-N analysis, a weighed soil sample was mixed with 2N (normal) KCl (potassium chloride) solution. The mixture was shaken for one hour and then filtered. The resulting filtrate was analysed with a Lachet Model AE ion analyser. The herbicide samples were analysed using a gas chromatograph.

Results and discussion

Rainfall patterns were highly variable from 1993 to 1995. The year 1993 was an extremely wet year with a growing season rainfall of 1000 mm, compared to a normal precipitation of 750 mm for the study site. The extremely wet conditions caused subsurface drains to flow continuously during the entire growing season (Figure 1). In 1994 and 1995, the precipitation during the growing seasons was close to the normal precipitation of 750 mm and subsurface drains did not flow continuously during both growing seasons.

Table 1 gives the overall yearly average NO_3-N concentrations in the subsurface drainage water as a function of tillage, crop rotation and N-fertilizer management

Table 1. Yearly average NO_3-N concentrations in the subsurface drainage water as a function of tillage and N management practice.

Tillage	Rotation	N management practice	N application rate ($kg\,ha^{-1}$)			NO_3-N concentration ($mg\,l^{-1}$)		
			1993	1994	1995	1993	1994	1995
No-till	CS	Late spring test	160	147	190	9.3	8.0	8.9
No-till	CS	Single N	112	112	112	8.8	8.5	9.8
Chisel plough	CS	Late spring test	103	167	159	11.1	7.1	11.4
Chisel plough	CS	Single N	112	112	112	8.9	9.1	12.5
Chisel plough	CS	Swine manure	84	237	392	11.6	10.1	13.4
Chisel plough	CC	Single N	135	135	135	11.6	7.5	11.9
Chisel plough	CC	Swine manure	82	280	502	11.1	14.7	27.0
Chisel plough	Strip crop					7.0	5.8	6.5
No-till	Forage crop		0	0	0	6.0	3.9	1.6

Table 2. Corn yields as a function of tillage and N management practice.

Tillage	Rotation	N management practice	Corn yields ($ton\,ha^{-1}$)		
			1993	1994	1995
No-till	CS	Late spring test	6.5	7.3	6.3
No-till	CS	Single N	4.1	6.3	4.5
Chisel plough	CS	Late spring test	6.8	8.2	5.5
Chisel plough	CS	Single N	4.9	7.9	5.2
Chisel plough	CS	Swine manure	5.6	8.2	5.6
Chisel plough	CC	Single N	4.2	5.8	4.2
Chisel plough	CC	Swine manure	2.8	7.3	4.8

system for three years (1993–95). The data in Table 1 show that in all three years, NO_3-N concentrations for some of the alternative management systems (LSNT, and pre-plant liquid N and swine manure application for corn-soybean rotation under NT and CP) being evaluated have the potential for reducing the leaching of NO_3-N to shallow groundwater. The LSNT has resulted in NO_3-N concentrations of 9.3, 8.0 and 8.9 $mg\,l^{-1}$ under no-till system, and 11.1, 7.1 and 11.4 $mg\,l^{-1}$ under chisel plough conditions for 1993, 1994 and 1995 respectively. All these NO_3-N concentrations values were either below or close to the 10 $mg\,l^{-1}$ drinking water standard that is often used to evaluate different farming systems for water quality enhancement. The NO_3-N concentrations from manured plots were much higher under both corn-soybean rotation and continuous corn because of the difficulty in applying the desired N rate with swine manure. Data from Table 1 indicate clearly that the LSNT (even with higher total N rate) resulted in lower NO_3-N concentrations in the shallow groundwater compared to a single pre-plant application of 112 $kg\,ha^{-1}$.

Table 1 also shows the yearly average NO_3-N concentrations in the subsurface drainage water for continuous-corn production with UAN application of 135 $kg\,N\,ha^{-1}$ and fall manure applications. Achieving the desired N application rates with liquid swine manure continues to be one of the most challenging problems of this study. Our intended application rate for N from swine manure was 135 $kg\,ha^{-1}$ but laboratory

Table 3. Effect of banding and broadcast on herbicide loss with subsurface drainage water.

Tillage	Herbicide	Banding		Broadcast	
		Max. conc. (ppb)	Loss ($g\,ha^{-1}$)	Max. conc. (ppb)	Loss ($g\,ha^{-1}$)
Chisel plough	Atrazine	1.23	0.40	1.62	5.80
	Metolachlor	3.82	0.12	1.58	0.54
Chisel plough (manure)	Atrazine	1.15	0.30	1.24	3.30
	Metolachlor	3.01	0.29	1.91	1.60

analysis of manure indicated that actual N application rates to continuous corn plots were 82, 280 and 502 $kg\,N\,ha^{-1}$ for 1993, 1994 and 1995 respectively. Because of the difficulty in applying the intended amount of N with manure, NO_3-N concentrations in the subsurface drainage water from continuous corn manured plots were 11.1, 14.7 and 27.0 $mg\,l^{-1}$ which were much higher compared to plots that received 135 $kg\,N\,ha^{-1}$.

Table 2 gives corn yields for 1993, 1994 and 1995 as a function of tillage and N management practices under corn-soybean rotation. On the average, corn yields were quite low in 1993 and 1995 because of the extremely wet conditions in 1993 and hail damage in 1995. In spite of the wet conditions in 1993, the LSNT resulted in the highest yields for no-till and chisel plough conditions. Also, the LSNT resulted in higher yields in 1994 and 1995 when compared with a single N application of 112 $kg\,N\,ha^{-1}$ (Table 2). Manured plots resulted in overall higher yields in 1994 and 1995 for both continuous-corn and corn-soybean rotations (Table 2). Table 2 also shows the importance of rotations in enhancing corn yields as continuous corn plots resulted in lower yields compared to corn from rotation plots despite greater inputs of N (135 $kg\,ha^{-1}$).

Table 3 gives herbicide leaching losses to subsurface drainage water as affected by N management and banding versus broadcast practice. The effect of banding is highly significant in reducing the overall yearly atrazine and metolachlor concentrations in the subsurface drainage water. Data in Table 3 also indicate that banding of herbicides could reduce the herbicide leaching losses to less than one-tenth compared to the surface broadcast practice under both manured and non-manured plots. Also, these data indicate that atrazine leaching losses were much lower from manured plots compared to the non-manured plots. This shows that swine manure may have a positive effect in reducing the leaching of atrazine to groundwater because of two reasons. One reason could be the increased microbial activity in the manure plots which may increase the microbial degradation of atrazine making it less available for leaching to groundwater. The second reason could be the greater degree of adsorption of atrazine to organic matter supplied by swine manure which also makes atrazine less available for leaching.

Conclusions

Experiments were conducted to evaluate the effects of eight different farming systems on productivity and water quality. Three years (1993–95) of data from this study clearly indicate that lower NO_3-N concentrations in the shallow groundwater can be achieved by reducing N application rates to 112 kg ha^{-1}. Use of late-spring NO_3-N test (LSNT) and differential N fertilization rates based on LSNT resulted in the lowest NO_3-N concentrations in subsurface drainage water under both no-till and chisel plough treatments when compared with manure application or single N application rates of 112 kg ha^{-1}. Data on corn yields indicate that manured plots resulted in overall higher yields in 1994 and 1995, because of the increased levels of N received from swine manure applications. On average, no-till plots yielded lower yields and the use of LSNT resulted in better yields than a single application of 112 kg N ha^{-1}.

References

Baker J L, Johnson H P (1981) Nitrate-nitrogen in tile drainage as affected by fertilization. J. Environ. Qual. 10:519–522.

Bergstrom L (1987) Nitrate leaching and drainage from annual and perennial crops in tile-drained plots and lysimeters. J. Environ. Qual. 16:11–18.

EPA – Environmental Protection Agency (1990) National pesticide survey. Summary results of EPA's national survey of pesticides in drinking water wells. Draft 31 October 1990. US-EPA, Washington DC, USA.

Hallberg G R (1989) Pesticide pollution of groundwater in the humid United States. Agr. Ecosys. Environ. 26:299–367.

Jemison Jr J M, Fox R H (1994) Nitrate leaching from nitrogen fertilized and manured corn measured with zero tension pan lysimeters. J. Environ. Qual. 23:337–343.

Kanwar R S, Baker J L, Baker D G (1988) Tillage and split N-fertilization effects on subsurface drainage water quality and corn yield. Trans. ASAE 31:453–460.

Kanwar R S, Stoltenberg D E, Pfeiffer R, Karlen D L, Colvin T S, Simpkins W W (1993) Transport of nitrate and pesticides to shallow groundwater systems as affected by tillage and crop rotation practices. Pages 270–273 in Proceedings of National Conference on Agricultural Research to Protect Water Quality. 21–24 February 1993, Minneapolis, MI, USA.

Kanwar R S, Colvin T S, Karlen D L (1995) Tillage and crop rotation effects on drainage water quality. Proceedings of the National Conference on Water quality: Clean water – clean environment – 21st century. Publication 2, ASAE 95(3):163–166.

Kladivko E J, Van Scoyoc G E, Monke E J, Oates K M, Pask W (1991) Pesticide and nutrient movement into subsurface tile drains on a silt loam soil in Indiana. J. Environ. Qual. 20:264–271.

Milburn P, Richards J L (1994) Nitrate concentrations of subsurface drainage water from a corn field in southern New Brunswick. Canadian Agr. Engr. 36:69–77.

Randall G W, Iragavarapu T K (1995) Impact of long-term tillage systems for continuous corn on nitrate leaching to tile drainage. J. Environ. Qual. 24:360–366.

Harnessing crop research data to develop expert systems

K. MURALIDHARAN[1] and E.A. SIDDIQ[2]
[1]*Directorate of Rice Research, Hyderabad 500 300, India*
[2]*Indian Council of Agricultural Research, New Delhi 110 001, India*

Key words: coordinated experiments, crop research data, expert systems, rice

Abstract
Agricultural research and development in India is achieved mainly through all India coordinated crop improvement projects (AICCIPs). They are multi-institute, multi-disciplinary research and development projects of the Indian Council of Agricultural Research (ICAR). Eighteen projects based at strategically important locations collectively involve all agricultural universities and other institutions in the country for agricultural growth and sustainability. These projects test promising breeding materials, varieties, hybrids, composites and agronomic practices. In the process, improved high-yielding or superior genotypes, suited for different ecosystems are identified. Over 2000 varieties in several crop species have been released for commercial cultivation. The new technology developed through AICCIPs has enabled small farmers to cross the threshold of non-availability. The rise in on-farm employment opportunities and ancillary enterprises contributed to increases in real wages. AICCIPs represent a very large investment in time, resources and expertise on the part of many researchers, farmers, administrators and policy makers involved in agricultural development in India. The rationale for developing the AICCIP data base was that the provision of information with which to make decisions is a critical bottleneck in the development process. The availability of abundant data in AICCIPs, useful for increasing efficiency in management and production of most crop species, is demonstrated. Using AICCIPs' data base, suitable expert systems for crops could be developed to provide decision makers with the type of information they need to make better decisions. The attempts to organize an expert system on rice using data and knowledge gained from AICCIPs is described here for illustrative purposes.

Introduction

India is a vast country with diverse environmental conditions. There is a wide range of rainfall, temperature, soil type and abiotic stresses. For scientists around the country, access to information such as data on the available varieties and crop management held by the governmental agencies can be difficult to obtain. Crop, climate and weather patterns do not stop at state boundaries, but data about them often do. Attempts were made to create a new AICCIP data base to help break down some of these barriers. All data available from thousands of experiments conducted will be pooled in AICCIP data base for sharing crop data among all scientists, researchers and developmental staff at every district level. Although many state governments and coordinating units may be reluctant to share even the basic data, developing a programme that is of general interest would lead to liberalization of the kind of data circulated. One of the AICCIP data base's first formal activities is to compile an inventory of related research work already completed or under way in the various all India coordinated crop improvement projects throughout the country. Most of such information is

M.J. Kropff et al. *(eds.): Applications of Systems Approaches at the Field Level, 419–436.*

recorded as observations by hand, stored in notebooks and annual reports. Extensive experimental data has been gathered in different crop species covering agronomical, soil, physiological and protection aspects of crop production. The AICCIP data base also includes yield performance of varieties and trait details. This AICCIP data base will computerize these data and train technicians to use them. For the individual crop commodities, effort in the development of an appropriate expert system is illustrated here using rice as an example.

All India coordinated crop improvement projects (AICCIPs) network

The initiation of all India coordinated research projects by ICAR resulted in a planned agricultural research and development in the country for the following crops: rice, wheat, maize, sorghum, pearl millet, small millet, barley, soybean, potato, tobacco, cotton, sugarcane and jute. Most crop commodities received the desired technological support from the coordinated crop improvement projects. Today, ICAR is funding 18 such projects at the national level, covering most of the economically important crops. Presently, ICAR funds a total of 78 coordinated projects in India covering the disciplines of crop sciences, plant protection, horticulture, agronomy, soil science, engineering, animal sciences, fisheries, etc. The state governments bear 25% of the cost for their own locations or centres. These projects focus attention on major national or regional problems by involving the available scientific work force at different institutions, so that the suitable answers are found through their joint but complementary efforts (Paroda 1992). The projects help in joint programme planning, interdisciplinary and also interinstitutional interaction, and collective effort in testing the technologies at several locations before their release to farmers. The collective process aids in defining the goals and objectives clearly in view of the priority of the problems, and the most relevant approaches for experimentation. Through such projects, ICAR enforces effective monitoring of the needed technology support provided by the different centres in crop production.

AICCIPs and crop data generation

One of the ICAR coordinating unit's functions is to ensure proper data recording and reporting. After finalizing the technical programme in a workshop, the project coordinator has to ensure timely supply of seed material for different coordinated trials or experiments, arrange for the chemicals required for testing (such as fertilizers, growth hormones, herbicides, fungicides, insecticides, etc.), and also supply data books to different centres. Data books ensure proper uniformity in reporting the results, which are then compiled, examined and processed through the statistical unit of the coordinating cell. The reports have all auxiliary data to facilitate effective discussion in the workshop. The comparative performance of each centre with respect to conduct or non-conduct and success or failure of experiments is also given in a tabular form in the report from the point of view of monitoring. Concerned principal

Table 1. Experiments conducted during 1994 under all India coordinated rice improvement projects (*Source:* DRR 1996).

Discipline	Dry season	Wet season	Centres
Agronomy	9	162	41
Plant breeding	30	188	54
Entomology	36	56	38
Plant pathology	–	229	34
Plant physiology	–	12	12
Soil science	–	29	17
Hybrid rice	2	84	12
Total	77	760	–

and zonal coordinators give their input in report writing. Annual reports are made using this information. A very large number of trials are organized and conducted at many locations each year in different crop commodities. During 1994 in rice alone, 837 experiments were completed (Table 1).

Such all India testing of promising breeding material (varieties, hybrids, composites, agronomical practices and other input use) helps in identifying the most stable, high-yielding or superior genotypes suited for different agroclimatic conditions. In the process, varieties are identified and released not only for the zones or regions where they have been bred, but also for other regions as well. This helps in complementing the efforts of relatively upcoming centres where the breeding programme is not strong enough to cater to the needs of the specific region. Also, common testing helps in avoiding any possible bias towards the promotion of a breeder's own varieties. In the process, only the best materials are identified for release.

Methodology for testing and evaluation

Normally, it takes a minimum of three to four years to identify a promising variety (Figure 1). Cooperating centres nominate their promising material either for a National Screening Nursery (NSN), if the number is large, or directly for an Initial Evaluation Trial (IET). NSN and IET experiments are conducted for one year. In these, materials that show promise on the basis of either significant yield advantage or better disease or pest resistance, drought tolerance, quality attributes, etc. are promoted to the next stage trial (Advanced Variety Trials – AVT). No variety that is inferior in yield to the best check is promoted to AVT. Best entries are tested in AVT for two years (AVT I and AVT II). Simultaneously, agronomical data are generated on promising varieties and the reactions to diseases and pests are recorded under hot-spot conditions at selected centres where competent expertise is available. Also simultaneously, field data are generated through on-farm trials conducted either by the extension staff or by the scientists themselves.

On-farm research is a multidisciplinary process attempting to find solutions to the field problems of farmers, adopting farmer first and farmer last as an operational approach. The two types of on-farm research trials are usually: (i) formal replicated

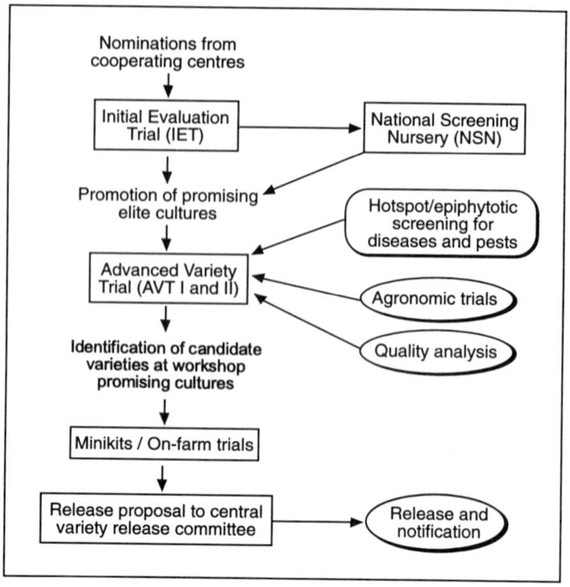

Figure 1. Procedure for varietal testing and release through all India coordinated crop improvement projects (AICCIP) (*Source:* Paroda 1992).

yield trials consisting of a group of 8–10 promising genotypes identified at the zonal or the national level and managed or supervised by the researchers at relatively few (3–6) locations; and (ii) unreplicated large plot (>0.10ha) observation trials consisting of one or two promising candidate genotypes, identified for release as recommended cultivars in addition to a local control and management by farmers using farm advisory. This trial is monitored by extension or research workers at many locations (20–500) depending on the envisaged recommended area for the cultivar. This is also known as a 'minikit' trial in India.

Multilocation data is gathered on both plant performance and site characteristics in order to study the genotype×environment interactions. Efforts are made to understand plant responses to specific environmental attributes, and determine the levels and growth stages of tolerance to specific stresses. The data also provide an opportunity to determine the key sites for different biotic and abiotic stresses.

To facilitate a uniform methodology in data collection, a written handout is made available to all scientists conducting trials. The experimental methodologies including input use to be adopted at all locations are described in full. Numerical scales and corresponding descriptors for evaluating various morphological, physiological, agronomical and stress tolerance traits are summarized in this instructional handout. The scoring methodologies are jointly developed by the scientists. Reporting of data of trials under coordinated projects is often not perfect. Some centres do not opt for conduct of certain experiments; at some others, the experiments are poorly conducted due to unforeseen problems. As a result of such deviations, exceptionally high coefficient of variance (CV %) or very low yield levels are recorded. Therefore, the data are scrutinized thoroughly on receipt by noting all the comments from the experimenters, as well as by statistical analyses.

Benefits and results from AICCIP

The AICCIPs are a unique mechanism in building a nation-wide cooperative interdisciplinary research network linking ACAR institutes with the state agricultural universities to focus attention on commodities of national importance. A national research grid was forged out of scattered, independent and duplicating efforts some of which competed with each other for resources and material. The innovative nature of the institutional change was highlighted by Paroda (1990), who made a deep critical analysis of the system. There does not appear to be any comparable instance of an institutional innovation resulting in such a high pay-off in production. The AICCIP concept was to pool together the national resources, technical and material, for increasing crop production. AICCIPs generated improved technology through interdisciplinary research, and evaluated the suitability, adaptability and transferability of technology to different ecological regions. They have proved to be the most effective instruments for the development and transfer of improved technology and for creating a cooperative research network in the country. India has the distinction of developing hybrids in pearlmillet, cotton, castor and pigeonpea for the first time in the world. Due to concerted coordinated research efforts, over 2000 improved varieties in different crop species have been notified for commercial cultivation. The new technologies that have been developed during this period have enabled many small farmers to cross the threshold of non-viability. The rise in employment opportunities on on-farm and ancillary enterprises has contributed to increases in real wages.

Additional available crop research data

Data on germplasm

The Indian region is an important centre of origin for more than 20 major agriculture and horticulture crops. These include rice, beans, cotton, sugarcane, citrus, mango, banana, yam, numerous medicinal and aromatic plants, several common vegetables and spices. India has a wide diversity in the available native germplasm (ICAR 1992–93). In fact, nearly 160 domesticated species of economic importance, 350 species of the wild relatives and 800 species of ethnobotanical interest are native to this region (NBPGR 1995). Such a diversity in species is an invaluable reservoir of genes needed by breeders for development of superior varieties. Recent advances in biotechnology and genetic engineering offer scope for selection and introduction of candidate genes from any germplasm into modern varieties. For this, it is essential to document the available sources and add this knowledge to the AICCIP data base for enabling access to the material needed by the breeder or other users.

Data on climate and weather

Over most parts, southwest monsoon rainfall is one order of magnitude more than in the rest of the year. Seasonal reversals of pressure gradients and winds make patterns of annual averages misleading. The gradual rise in temperature through spring to summer does not happen as with the onset of southwest monsoon. Temperatures drop sharply in June–July. The monsoon period of four months from June–September is out of step with three-monthly seasons. The usual classification into spring, summer, autumn and winter is not adopted. January–February period is called the 'winter period'. The hot weather period is from March–May. June to September is called the southwest monsoon period and October–December is the post-monsoon period (Rao 1981). The annual rainfall is mainly accounted by the southwest monsoon. In the post-monsoon period, some states like Assam receive high rainfall.

India has a large diversity in crop-growing environments: irrigated, rainfed lowland, rainfed upland and flood-prone. Since the major portion is rainfed, national production is strongly tied to the distribution of rainfall. In some of the states, erratic rainfall leads to drought during the vegetative period, but later in the season the crop may be damaged by submergence due to high rainfall. In the eastern states, damage due to flash floods can be high. The performance of southwest monsoon is of great significance to Indian farmers. The areas receiving >750 mm have a coefficient of variation of 30% or less. On average, the country receives 88 mm of rain; 80% of the total annual rainfall occurs during the four months from June to September. The long-range forecast made towards the end of May by the Indian Meteorology Department uses 16 regional and global land-ocean-atmosphere parameters.

A very large part of the Indian subcontinent experiences heavy clouding after the onset of monsoon and the period of sunshine is about a third of the maximum possible. In some areas, the solar radiation is less in June–September due to heavy monsoon clouds than in December–February. May is the month of maximum potential evaporation in many parts. The cloudiness during the southwest monsoon period substantially reduces the potential evaporation. Thus, the monsoon provides rains and conserves water by restricting evaporation. Computations have been made on the duration of water availability periods in different regions (Rao 1981; FAI 1986–87).

All the coordinated research centres have a weather station either under the Indian Meteorological Department and its Agromet division or ICAR or state government. Data collections are available for the entire AICCIP period at all these centres. This can be easily organized into a user-friendly data base for easy access.

Data on soil and related constraints

Soil is a vital but essentially non-renewable resource. It is fundamental for wealth creation through agriculture, forestry and an increasing number of natural products not traditionally associated with agriculture. The soil is the major component of the environment and is at the interface between the atmosphere and natural waters. It acts as a filter, or buffer, protecting the quality of water and air. Poor soil management can lead to pollution of both. The soil itself can be damaged by erosion, contamination, salinization or acidification. The distribution of different soil types in each and every district is well documented. Detailed knowledge on soil is already available based on hundreds of soil profiles from around the country (Murthy et al. 1982). There is a further scope for linking soil resource map data (NBSS 1995). Fertility status of the soils of India is also well documented periodically. Details are readily available on the nutrient deficiency in soils (Ghosh and Hassan 1976, 1980; FAI 1986–87, 1994–95). The levels of response to fertilizer application in the deficient areas are also well known.

Many experiments have been conducted under AICCIPs to generate knowledge on soil process and to develop strategies that reconcile environmental concern with the need to produce more food and other products. Additionally, the production-oriented surveys help to document and understand the soil-related constraints as illustrated using data on rice in Andhra Pradesh. Soils in Andhra Pradesh are low to medium in organic carbon, low in available phosphorus and medium to high in available potash. Deficiencies of zinc and phosphorus are widespread. High responses to applied nutrients particularly to phosphorus and zinc to cereals are observed. Zinc deficiency in rice was the production constraint during 1968, while calcium limited yields in 1972. Zinc deficiency is a serious problem in northern districts, Nagarjunasagar and Sriramsagar irrigation project areas. Potassium depletion was generally slow in black soils. But in red soils, depletion rate was very high when rice was grown continuously without potash fertilization. Besides nutrient deficiencies, other disorders such as sulphide injury have been recorded in some red and black soils of the Nagarjunasagar project left canal area. Response of rice to phosphorus ranged from 16 to 58 kg grain kg^{-1} nutrient in the left canal, and 27 to 44 kg grain kg^{-1} nutrient in right canal areas under Nagarjunasagar irrigation project. The response was about 14 to 66 kg grain kg^{-1} nutrient in soils under the Sriramasagar irrigation project. Rice responded to potash in Guntur, Cuddapah, Karimnagar, Nellore, Chittoor and Anantpur districts. The critical levels for phosphorus and potash in soil for rice are 17 and 178 kg ha^{-1} respectively. Rice responded to zinc in black soil regions of Nellore and Mahaboobnagar (DRR 1965–94).

Data on biotic constraints

Each year during the crop season, several surveys are made routinely to gather information on the biotic production constraints such as diseases, insect pests, rodents, etc. that cause considerable yield losses. Besides observations on the occurrence of biotic stresses, the intensity of the stress is also recorded. A detailed knowledge of the various factors that influence the population of the pest or disease causing organism would help in forecasting their outbreaks. Experimental data relevant to an understanding of the outbreaks are available from the many studies through AICCIPs. Temperature, relative humidity, rainfall and daylength are the most important climatic factors that affect pathogen populations. A particular set of these factors in a locality may have specific influence on the biology of a pest or disease-causing organism.

These factors can lead to an outbreak or a suppression of the pathogen in that locality. Any generalization made about the outbreak of a pathogen based on a set of climatic factors may not always be correct. However, certain clues may help in predicting a pest incidence. The type of AICCIP data available on most important pathogens of crop commodities is illustrated with the knowledge gained on stem borer (*Scirpophaga incertulas*) and sheath rot (*Sarocladium oryzae*).

*Knowledge on stem borer (*Scirpophaga incertulas*)*

High temperature and low rainfall can cause severe stem borer infestation in rice (Table 2). The peak emergence of stem borer populations, as indicated by light trap catches, was when maximum temperature ranged from 28°C to 36°C, minimum from 19°C to 23°C and relative humidity from 76% to 99%. Temperature above 38°C or below 15°C, and rainfall exceeding 158 mm per week, were unfavourable for the build-up of stem borer (Pasalu et al. 1996). Percentage incidence of dead heart and white ear was correlated negatively with rainfall and minimum temperature, and positively with maximum temperature. Activity of this pest was more from August to May in double cropped areas. In places having distinct generations, usually the first generation (smaller in size) appeared in the nursery or immediately after transplanting. The population increased in later broods that often caused serious damage. Therefore, the stem borer was more destructive in the late planted or second crop. During periods when there was no rice crop and the prevalent temperature was not optimum for larval development, the full grown larvae undergo diapause. But wherever two or more rice crops are grown in a year, the insect remains active year round. It undergoes only a temporary quiescent stage or weak diapause in the last larval instar, during the brief periods of non-availability of host plants.

The modern high-yielding rice varieties with luxuriant growth of foliage provide a thick crop canopy beneath which the microclimate becomes quite congenial for development and multiplication of stem borer. Evidence indicates that greater severity of stem borer is not due to any single factor, but due to a combination of management practices. These factors include continuous planting of the same variety over large areas, excessive use of nitrogenous fertilizers, assured irrigation, close plant spacing and indiscriminate use of insecticides.

*Knowledge on sheath rot (*Sarocladium oryzae*)*

Sheath rot disease markedly reduced rice productivity (Venkataraman and Muralidharan 1988) whenever panicle emergence was affected by climatic factors. The fluctuations in monsoon and the late rains, particularly from panicle initiation to booting stages in wet season, enable sheath rot disease to appear and cause damage to production in many states (Table 2). The intensity of sheath rot is usually severe in crops already infected with other biotic stresses such as rice tungro virus, bacterial blight or stem borer. Besides reducing grain yields, the disease affects grain quality (Ahmed et al. 1978). Sheath rot, considered to be a minor disease, has assumed greater

Table 2. Favourable or unfavourable meteorological conditions for occurrence, peak activity and biology of yellow stem borer (*Scirpophaga incertulas*). The outbreak conditions, perpetuation of pathogen and influence of cultural practices on sheath rot (*Sarocladium oryzae*) disease in rice (*Sources:* Pasalu et al. 1996; Venkataraman and Muralidharan 1988).

Yellow stem borer	
Favourable/unfavourable meteorological conditions	High temperature (28–36°C) Low rainfall Excess rainfall detrimental (> 158 mm/week) Below 19°C limits
Peak activity	August to May: abundant
Biology	Vegetative phase – dead hearts Reproductive phase – wheat ears Eggs near tips 50–80 eggs in egg mass Hatch in 5–8 days Emerge between 1900 and 2100 h Larvae crawl for 30 min Enters shoot More incidence in alkaline/acid soils
Sheath rot	
Outbreak conditions	Rain or drizzle Prolonged dew period at flowering High humidity (RH > 90%) Moderate temperature
Perpetuation	Seed-borne Left over infected material Straw piles Perfect stage (?) Collateral hosts
Cultural practices	Varietal susceptibility Poor panicle exsertion High N levels

importance in the recent years, particularly in Uttar Pradesh, Punjab, Karnataka, Bihar and West Bengal. *S. oryzae* is known to infect many other host species. The primary factor responsible for sheath rot outbreak is the availability of moisture at the site of infection. At flowering, occurrence of high relative humidity as a result of rain, drizzle or dew fall helps the pathogen establish and spread.

Pest monitoring and economic thresholds

In the ecology-oriented approach to pest management, monitoring plays an important role. Monitoring provides information that enables appropriate remedial measures to be taken in time to counter the severe incidence of a pest or disease and avert economic losses in grain yields. Pathogen populations must be maintained at levels below those causing economic injury (Table 3). The economic threshold for a pathogen indicates the pest density at which control action should be taken. The economic threshold includes cost of control, crop value, changing susceptibility of crop to pest or disease damage, changing density and composition of pathogen popu-

Table 3. Economic threshold levels as decision aids for intervention in management of pests and diseases in rice (*Source:* Pasalu et al. 1996; Muralidharan et al. 1990).

Threshold levels for insect pests	
Brown plant hopper	5–10 insects $hill^{-1}$
Green leaf hopper	2 insects $hill^{-1}$ in tungro endemic areas, 20 insects $hill^{-1}$ in other areas
Stem borer	5% dead hearts in veg. stage; 1 moth m^{-2} at flowering
Gall midge	1 gall m^{-2} in endemic area or 5% galls during tillering
Rice hispa	1 adult or 1 damaged leaf $hill^{-1}$
Leaf folder	1–2 freshly damaged leaves $hill^{-1}$
Thrips	5% plants with curled leaves
Whorl maggot	20% damaged hills
Case worm	1–2 cases $hill^{-1}$
Cut worm	1 larva $hill^{-1}$
Gundhi bug	5 or more bugs m^{-2}
Threshold levels for diseases	
Blast	Light (2–5% leaf area damaged) in nursery; 5% leaf area damaged at tillering and PI or 1–2 neck infections and 10% tillers affected
Sheath blight	10% or more affected tillers
Foot rot	Light in nursery
Stem rot	10% or more tillers affected
Brown spot and panicle discoloration	Light
Bacterial leaf blight	2–5% disease severity at tillering; 10% or more tillers affected at PI
Tungro	One affected hill m^{-2}

lations, etc. Careful monitoring of pathogen populations and their threshold values greatly helps to reduce the pesticide application.

Prediction using survey and threshold levels

Rice farmers face numerous disease problems that limit production. The data from surveys and basic research on the epidemiology of pathogens that limit production have resulted in a better understanding on the disease outbreaks. Analyses of disease problems provide clues on the nature of their occurrence and favourable factors associated with sudden outbreaks. Using the basic information on diseases, data from field observations are interpreted to develop a prediction system (Nagarajan and Muralidharan 1995).

In disease management, identification of the key pathogens in the given target area is the first step. The quantitative estimates on the contribution of each constraint to the total loss provide a rational base for arriving at a decision for intervention. Characterization of explanatory variables such as variety planted, yield, etc. help in detecting the patterns of relationships among variables. Within-field movement and dispersal gradients and large area movement or migration provide more information on the nature of the epidemic outbreak. Crop planning and deployment of resistant varieties, and rescheduling planting date, harvest date and sequence crops can effectively alter the recurrence of epidemics.

Accessing and using crop research data

The first coordinated crop improvement project was initiated in 1957 on maize. In the past four decades, the agricultural scene in India underwent profound changes. More land has been brought under the plough. New crops (soybean, sunflower) have been introduced. In some areas, traditional crops have been given up (bajra and groundnut in the Punjab). Consumer demand profiles are changing. At present, India has an exportable surplus of wheat, rice, castor oil, etc. AICCIPs responded so admirably to catalyse a change that ushered the Green Revolution. They can certainly respond again to changed emphasis on increasing further the production and ensuring sustainability. Productivity-related programmes dominated the research agenda of the crop improvement projects so far. Sustainability, competitiveness, environmental concerns, biomass and renewable energy sources and processing related attributes should additionally constitute the research agenda for the future crop improvement.

Data processing represents the first and foremost step in developing a computing model for use by researcher or extension worker from a rural area. The AICCIP data base will provide on-line access to weather data, choice of crop species for any specified agroecosystem, national data bank on varieties and donors used as parents in their development, crop production methods, crop protection practices, use of input such as nutrients, growth hormones and pesticides and their interactions and economics of production.

Assembling information, data and knowledge on rice

As a first step, attempts were made to assemble and analyse the information from AICCIP on rice. Over 30 years' data from the coordinated experiments (DRR 1965–94) were pooled on varieties, crop yield response to pest incidence, production-oriented constraints (DRR 1975–94) such as soil problems, variations in monsoon and prevalence of diseases and pests.

Variety data base

AICCIP on rice has resulted in the release of 474 improved varieties for commercial cultivation. For each of these released varieties, data on yield performance across locations and years are available. The performance of Ratna in the coordinated experiments (Table 4) is presented as an example. Data are also available on many other important characteristics of these varieties. They include: year of release, parents used and donor traits, breeding methodology adopted (pure-line selection, crossing, mutation, etc.), flowering and maturity duration, photosensitivity, tillering capacity, distinguishing morphological characters, resistance or susceptibility to biotic and abiotic stresses, and ecosystem and area suitable for commercial cultivation. A specimen data on Vikramarya, a high-yielding variety with resistance to green leaf hoppers, tungro virus and blast is cited as an example (Table 5).

Table 4. Performance of cv Ratna, an early duration high-yielding variety released for irrigated ecosystems in India (*Source:* DRR 1965–94).

Year	No. of locations tested	No. of genotypes in experiment	Mean grain yield (ton ha^{-1})	
			Experimental mean	Ratna
1976	49	26	3.79	4.01
1977	50	26	3.56	3.68
1978	57	16	3.83	3.80
1979	55	15	3.86	3.59
1980	55	15	3.90	4.14
1981	43	14	4.07	4.20
1986	35	30	4.26	4.25
1987	7	18	4.77	4.47
1988	7	18	4.57	4.97
1989	7	16	4.76	4.41
1991	20	49	4.87	4.57
1992	22	31	4.65	4.86
1993	24	21	4.72	4.50
1994	21	26	4.39	3.98

Genotype performance data analysis

The yield performance of rice genotypes, in the all India coordinated experiments (DRR 1975–94), was studied. These coordinated variety experiments were established for specific ecosystems and for specific crop maturity periods. Ecosystem-wise a comparison was made using the overall mean grain yields recorded in these experiments (Prasad et al. 1995). All the locations in rainfed upland, rainfed semi-deepwater and rainfed deepwater areas were included under favourable ecosystem for rice. Similarly, all the locations in rainfed shallow lowland and irrigated areas were included under favourable ecosystem. In these experiments under unfavourable ecosystem, the mean grain yield remained around 2–3 ton ha^{-1} irrespective of the new genotypes introduced over the years. However, under favourable ecosystem the mean grain yield of the genotypes reached beyond 4 ton ha^{-1} (Figure 2). These analyses further showed that genotypes in any of these experiments during the years 1974–94 gained very little in terms of productive capacity. Such analyses help alter the course of crop research.

Developing empirical relations

AICCIPs have been evaluating the commercially available growth hormones, pesticides and coated or modified fertilizers. Every experiment has been conducted at multilocations for several reasons. This rich data base can be used in a number of ways as illustrated in the following. The use of herbicides in the all India, coordinated herbicide evaluation experiments in rice from 1980 to 1992 resulted in different levels of weed infestation and rice yields. Weed dry weight, a continuous variable, was used as a precise measure of weed influence (Muralidharan et al. 1994). Because locational effects would be more uniform within an ecosystem, data sets from 16

Table 5. A record on cv Vikramarya in the rice varieties data base (*Source:* data base on rice variety, DRR, Hyderabad, India).

Variety name	Vikramarya
Released by Central and State:	Andhra Pradesh
Year of release	1986
IET (Initial Evaluation Trial) No.	7302
Designation	RP 1125-1598-1-4-3
Cross	RP 6-13×Ptb 2
Name of the releasing agency	SVRC: Andhra Pradesh
Name of persons responsible	Ramakrishnarao and Ghosh
Place of origin of the variety and the university	AICRIP, Hyderabad
Method of evolution of the variety	Pedigree
Breeding objectives	Resistance to tungro virus and medium maturity duration
Whether recommended by seminar/ conference/workshop/state seed subcommittee	Rice workshop
Target area/suitable ecosystem	Irrigated ecosystem Shallow lowland ecosystem (up to 30 cm water depth)
Description of the variety/hybrid	Medium maturity duration
Average plant height	120 cm
Maturity period (seed to seed)	135 days
Flowering duration	105 days
Photosensitivity	Photo-insensitive
Anthocyanin pigment	Absent
Pigmented parts	–
Other distinguishing morphological characters	–
Grain characters	Long bold
Grain quality	Non-scented
Grain colour at maturity	Straw coloured
Furrows on the grain	Absent
Kernel colour	Absent
Abdominal white	Absent
Hulling (%)	79
Milling (%)	69
Head rice recovery (%)	65
Alkali value	–
Volume expansion	–
Water uptake	–
Kernel elongation ratio	1.7
Amylose (%)	29
Protein (%)	9
100 grain weight	2.8 g
Reaction to pests and disease (resistance or tolerance or susceptibility)	Resistant to green leaf hopper, *Nephottetix* spp. Resistant to tungro virus
Area under adoption	Andhra Pradesh, Tamil Nadu and West Bengal

diverse experimental sites were grouped into four ecosystems: hills, irrigated, rainfed lowlands and rainfed uplands. The mean biomass of weeds was higher in the rainfed uplands and hills, but lower in the irrigated ecosystem and lowest in the rainfed lowlands. The degree of association between differences of weed weight over the control and the difference of grain yield over the control were highly significant except for the rainfed uplands. Using linear regression models, Muralidharan et al. (1994) calcu-

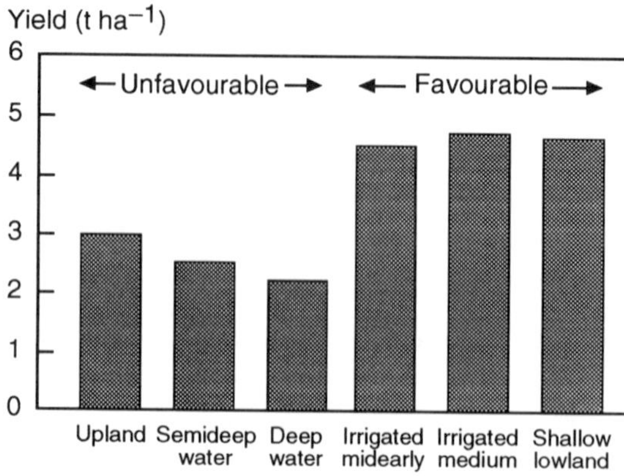

Figure 2. Overall mean performance of elite breeding lines evolved in India (*Source:* Prasad et al. 1995).

Table 6. Rice production output data for a given contiguous block with districts in the state of Uttar Pradesh (*Source:* production statistics data base, DRR, Hyderabad).

District	Triennium ending 1990–1991			Irrigated area (%)	Fertilizer consumption* ($kg\,ha^{-1}$)
	Area (ha)	Production (ton)	Productivity ($kg\,ha^{-1}$)		
Gorakhpur	123 523	181 058	1466	0.66	68.19
Deoria	252 761	473 522	1873	17.80	96.13
Basti	213 025	360 829	1694	0.02	51.45
Kheri	167 932	371 643	2213	40.26	86.72
Bahraich	238 433	336 811	1413	1.37	27.20
Gonda	287 058	400 269	1394	0.18	29.29

*For gross cropped area.

lated that for every gram decrease in weed dry weight, the approximate increase in grain yield is predicted to be: $3.85\,kg\,ha^{-1}$ in hills, $4.32\,kg\,ha^{-1}$ in irrigated and $3.61\,kg\,ha^{-1}$ in rainfed lowlands. The evaluation experiments with herbicides thus helped in deriving yield loss models for different ecosystems. Similar efforts have been made (Muralidharan et al. 1996) to predict yield losses from stem borer (*Scirpophaga incertulas*), gall midge (*Orseolia oryzae*), blast (*Magnaporthe grisea*) and sheath rot (*Sarocladium oryzae*).

Developing expert systems

Farmers regularly receive information on crop production and protection from extension personnel, mass media like radio, television and press, agro-input dealers, traders and neighbouring farmers. Mostly information is through spoken communication; occasionally a few visual and written communications reach farmers. The village agricultural functionaries are not always readily accessible or available to farmers when most needed. Therefore, the private traders who sell various agro-inputs or landlords inevitably serve as advisers to farmers on crop production

(Muralidharan et al. 1991). None of the information channels disseminate any information on the possible yield loss. In the tropical farming system, the extension functionaries need to assimilate, retain, recall and quickly communicate a vast amount of information required for farmers. The data generated from AICCIP experiments could be used to develop knowledge base and empirical relations. This would later help build expert systems on most important crop commodities. The development and use of a rule-based expert systems for crop production recommendations would reduce the need for such highly trained personnel. This would also eliminate the bottlenecks in the flow of information to farmers.

Expert systems are written in artificial intelligence programming languages which handle problems with subjective numerical data, rules of thumb and judgement of knowledge (Huggin et al. 1986). These systems are designed to capture knowledge and problem-solving logic employed by human experts in a given field (Runnels and Shortliffe 1987). The objective of the system is to provide accurate, readily accessible and quick decision support information in planning production, diagnosing constraints in production and selecting the suitable management option. The expert systems include easy to use, menu-driven screens and require just a personal computer without any programming knowledge or skill. Expert systems are an ideal tool in diagnosis, design and testing of improved systems.

The expert system on rice currently under development is an interactive, menu-driven program that allows recommendation on crop production planning and management in the electronic media form. Menu choices offer a facility whereby the user can add additional recommendations for a new production constraint not currently available on the data base. It is a rule-based program written in C++. Expert system is a menu-based program, minimizing the time required to learn it. All it takes to use the program is the ability to read English characters and numbers. All the data collected are transferred into electronic form using FOXPRO, a commercial DBMS program. The information is organized as per crop growth stages. The AICCIP data base is indexed. The expert system for rice requires an IBM PC or compatible personal computer with 640 KB RAM, a hard disk – minimum 80386 processor with MS-DOS version 6.22 or higher – to increase speed, access time and display pictures of crop damage. A suitable monitor is needed to display colour images to illustrate the constraints in production and management recommendations.

The rules that make up the knowledge base are structured as IF-THEN statements. The IF portion of the statement provides a premise that consists of certain parameters and their values. Parameters are variables or attributes that are important to the identification of the constraint in crop production or protection. The system is expected to serve as a computerized consultant. Users are prompted to provide input data to indicate district and area where a consultation on rice production is required. Users are then guided to indicate the choice of the inquiry from a series of menus. The menus require a single choice from a list of several possible responses. Also, in appropriate cases, the system offers other menus from which the user may choose all or none of the offered selections. Simple rules are provided in the knowledge base for identifying available choice of varieties and traits of donor germplasm used

Table 7. Query and output on technology for maximizing rice productivity in the wet season in Uttar Pradesh, India. Input districts: Gorakhpur, Deoria, Basti, Kheri, Bahraich, Gonda. Initially, three options appear: variety; soil, water and nutrient management; and optimum cultural management. On selection, details appear on the screen as shown in table columns; all the options can also be selected as in this table.

OUTPUT:		
1. VARIETY	2. SOIL, WATER AND NUTRIENT MANAGEMENT	3. OPTIMUM CULTURAL MANAGEMENT
	SELECT ONE OR ALL OPTIONS AND ENTER	
1	2	3
Choice available in mid-early maturing (100–110 days) varieties	*Incorporate basal N dose prior to last puddling	*Timely transplant, preferably by the first week of July
Govind, Saket 4, Vikas, Rasi, Sasyasree	*Improve drainage and ameliorate soil salinity/alkalinity through gypsum or pyrite application and by green manuring	*Adopt closer spacing of 20×10 cm with 2–3 seedlings per hill to ensure optimum population
Choice available in medium maturing (120–135 days) varieties	*Ensure adequate irrigation at critical crop growth stages such as panicle initiation to flowering	*Wherever random transplanting is practised maintain about 50 hills m^{-2}
Sonasali, Vikramarya, Pant Dhan 4, Sarjoo 52	*For good water management, level the field and maintain 3–5 cm water until the beginning of the ripening phase	*Transplant seedlings of optimum age (25–30 day old) at shallow depth of 3–4 cm
	*For effective split application of nitrogen, drain the field prior to top dressing and irrigate after 24 hours	*In case of late transplanting, adopt bunch plantings (5–7 seedlings $hill^{-1}$) with 40–45 day old seedlings at closer spacing of 15×10 cm
	*Adopt adequate and balanced use of fertilizers, wherever required	*Gap fill twice at 7 and 15 days after transplanting
	*Spray zinc sulphate (at 0.5%) along with urea (2%) twice: 10 and 20 days after sowing. Repeat spray 15 days after planting in Tarai area	*Adopt fill twice at 7 and 15 days after transplanting
	*Green manure to supplement fertilizer nitrogen	*Adopt line sowing and chemical weed control under direct sown conditions
	*Apply coated N fertilizers such as neem cake coated urea in soils with excessive percolation or apply N through urea in 3–4 or more splits	*Use herbicides for timely control of weeds. Apply butachlor or benthiocarb at 1.5 kg a.i. ha^{-1} at 5–7 days after transplanting. Alternatively, hand weed twice: 20 and 40 days after transplanting
		*Harvest crop at 20–24% grain moisture content for maximizing yield and milling recovery

as parents in these varieties, or for identifying and diagnosing physical, input, biotic or abiotic constraints. From the production-oriented constraints data base (DRR 1975–94), response to a query on the occurrence of a particular stress in any defined rice growing district in India can easily be obtained.

The program explains in the first two screens the background on its development and the main menu offers the user six choices: Input Data, Crop Details, Examine Rules, Add Rules, Index Databases and Quit. Considering the parameters entered, the system would come up with a recommendation and possible decision. It also informs that a recommendation and decision as finally displayed may not be the only solution. It then further prompts the user to enter parameters that would be required to make an intelligent recommendation. Once the final parameter is entered, expert system searches through the data base containing the various parameters to find out what recommendation is stored matching the parameters entered. The recommendations are then displayed on the screen. For example, by entering a contiguous block of districts as input data, either production statistics (Table 6) or recommended technology for adoption can be accessed (Table 7). Specific information can be obtained by selecting a particular option from the main menu. A 'help' function incorporated into the knowledge base provides descriptions of terms, site or field symptoms, single plant symptoms and field diagnostic procedures. One of the special features that distinguishes the rice expert system from the conventional program is its ability to address the problem and offer solutions when given incomplete data or unknown information. Sometimes key features essential for absolutely certain diagnosis of a constraint are not apparent, or detected from input data. Then colour images of the constraints are used to help the user to arrive at a conclusion. Efforts are being made to display colour slides to illustrate the crop damage from different biotic and abiotic stresses. They also help the user identify a disease or pest on the computer screen to enable a visual identification, and a reference relating it to the field situation.

More than one constraint often exists in a district in any farmer's field. In such cases, the computer stores the selections from different menus. The colour images also alert the users on the possibility of additional constraints. After shortlisting the constraints, suitable solutions are provided.

References

Ahmed N J, Kanniyan S, Venkatrao A (1978) First record of *Acrocylindrium* sheath rot disease of rice in India. Sci. Culture 75:544–545.

DRR – Directorate of Rice Research (1965–1994) Annual Report. Directorate of Rice Research (AICRIP), Hyderabad, India.

DRR (1975–94) Production-oriented survey. Directorate of Rice Research (AICRIP), Hyderabad, India.

DRR (1996) Research highlights 1994–95. Directorate of Rice Research, Hyderabad, India. 44 p.

FAI – Fertiliser Association of India (1986–87) Fertiliser statistics. Fertiliser Association of India, New Delhi. 390 p.

FAI (1994 –95) Fertiliser statistics. Fertiliser Association of India, New Delhi. 429 p.

Ghosh A B, Hassan R (1976) Available potassium status of Indian soils. Bull. Indian Soc. Soil Sci. 10: 1–15.

Ghosh A B, Hassan R (1980) Results of soil test carried out and data provided by soil testing laboratories. Fert. News 25:19–24.

GOI – Government of India (1996) Agricultural statistics at a glance. Department of Economics and Statistics, Department of Agriculture and Cooperation, Ministry of Agriculture, Government of India, New Delhi. 140 p.

Huggin I F, Barret J R, Jones D D (1986) Expert systems: Concepts and opportunities. Agric. Eng. 67: 21–23.

ICAR – Indian Council of Agricultural Research (1992–93) Annual Report. Indian Council of Agricultural Research, New Delhi. 350 p.

ICAR (1971) National germplasm collection of rice. Indian Council of Agricultural Research, Central Rice Research Institute, Cuttack, India. 225 p.

Muralidharan K, Shinde J E, Siddiq E A (1990) Technology for rice production. Directorate of Rice Research, Hyderabad, India. 200 p.

Muralidharan K, Reddy A A, Sandhya P N, Reddy P C, Mangal Sain, Anuradha P N, Reddy A P K (1991) Farmers' constraints in the adoption of recommended rice blast control measures. Oryza 18:229–231.

Muralidharan K, Campbell C L, Pasalu I C, Ramaprasad A S, Rao C S, Siddiq E A (1994) NARP-IPM project report 1994. Directorate of Rice Research, Hyderabad, India. 142 p.

Muralidharan K, Pasalu I C, Rao C S, Ramaprasad A S (1996) NARP-IPM project report 1996. Directorate of Rice Research, Hyderabad, India. 38 p.

Murthy R S, Hirekerur L R, Deshpande S B, Venkatarao B V, Shanakaranarayana H S (1982) Bench mark soils of India: Morphology, characteristics and classification for resource management. All India soil and land use survey, agricultural universities and soil survey organisations of the State Department Agriculture and National Bureau of Soil Survey and Land Use Planning (ICAR), Nagpur, India. 374 p.

Nagarajan S, Muralidharan K (1995) Dynamics of plant diseases. Allied Publishers, Hyderabad, India. 248 p.

NBPGR – National Bureau of Plant Genetic Resources (1995) Annual Report. National Bureau of Plant Genetic Resources (ICAR), New Delhi, India. 159 p.

NBSS – National Bureau of Soil Survey (1995) Annual Report 1994. National Bureau of Soil Survey and Land Use Planning (ICAR), Nagpur, India. 233 p.

Paroda R S (1990) Nationally coordinated programmes: Organisational framework and managerial aspects of the all India coordinate crop improvement projects. Pages 1–32 in Management of Change Workshop, 19–23 February 1990, ISNAR/SACCAR, Zimbabwe.

Paroda R S (1992) All India coordinated crop improvement projects: Organisation and management. In Management of change in all India coordinated crop improvement projects. Proc. Group Dis., ICAR, New Delhi. 216 p.

Pasalu I C, Krishnaiah N V, Bentur J S, Krishnaiah K, Muralidharan K (1996) Occurrence of insect pests in rice and their prediction. In Climate components of rice productivity. ICAR and CAS, Indian Institute of Sciences, Bangalore, India (in press).

Prasad A S R, Prasad G S V, Muralidharan K (1995) Fragile ecosystems: Fragile for rice? Pages 369–393 in Fragile lives in fragile ecosystems. International Rice Research Institute, P.O. Box 933, Manila, Philippines.

Rao Y (1981) The climate of the Indian sub-continent. In Takahashi K, Arrakwa I (Eds.) World survey of climatology. 9. Climates of southern and western Asia. Elsevier Scientific Publishing Co., Oxford, New York. 465 p.

Runnels G D, Shortliffe E H (1987) Advanced computing for medicine. Sci. Am. 257:154–161.

Venkataraman S, Muralidharan K (1988) Sheath rot disease of rice. Pages 1–17 in Crop disease management. Tamil Nadu Agric. University, Coimbatore, India.

Comparison of predictions and observations to assess model performance: a method of empirical validation

P.L. MITCHELL[1] and J.E. SHEEHY[2]

[1]*Department of Animal and Plant Sciences, University of Sheffield, Sheffield S10 2UQ, UK*
[2]*International Rice Research Institute, P.O. Box 933, 1099 Manila, Philippines*

Key words: model, modelling, precision, regression, validation

Abstract
Empirical validation is the comparison of model predictions with observations from the real system, together with an assessment of whether the model is adequate for its purpose. When the data available consist of pairs of predictions and observations covering the range of model operation, they are often presented as a scatter graph, sometimes with a regression line added. There are several reasons why regression is not appropriate for this task. An alternative method described uses the deviations (prediction minus observation) plotted against the observations. From this graph of deviations the bias and precision of the model can be assessed, and how they vary along the range of operation. The two criteria for adequacy of the model are the envelope of acceptable precision and the proportion of points that must lie within it. The envelope of acceptable precision is defined with reference to the purpose of the model, most commonly by comparison with the precision of the observations. The proportion of points in the envelope can be set at 0.95 by analogy with confidence limits. This method is demonstrated on two examples from different models, and it can be extended to other forms of validation data. Although statistical tests are not involved, which avoids the problem of satisfying assumptions, this method is simple, objective and quantitative and can be made rigorous by definition of the criteria in advance.

Introduction

At some point in model development, the predictions of a model have to be compared with observations from the real system and an assessment made of model performance (Amthor and Loomis 1996). This process is generally called empirical validation, to distinguish it from structural or conceptual validation which is concerned with the way that the model represents the real system and the meaning of the parameters. Validation of models is a large subject which has been reviewed from different standpoints by Van Keulen (1976), McCarl (1984) and Beck (1987) and which receives at least some mention in books about modelling (e.g. Burghes and Wood 1980; Jeffers 1982; Gilchrist 1984; Charles-Edwards et al. 1986; Thornley and Johnson 1990; Trenberth 1992).

The concept of validation has been strongly criticized by Oreskes et al. (1994) because it is philosophically impossible to prove that a model is true or correct. Similar doubts have been expressed by Thornley and Johnson (1990) and by Harrison (1990). Modellers have also been inconsistent in the use of the terms validation and verification. For instance, verification has been used by some modellers for either

M.J. Kropff et al. *(eds.): Applications of Systems Approaches at the Field Level, 437–451.*

calibration (Shugart and West 1981) or for structural validation (Jeffers 1982) or for empirical validation (Reckhow et al. 1990).

The distinction between calibration and validation is important (Van Keulen 1976; Gilchrist 1984). Calibration (model tuning or fitting) is the estimation of values for parameters or variables using information from the real system. For validation to be independent of model construction, in particular of calibration, the observations must be a different set of data kept separately or specially collected (Amthor and Loomis 1996). Since modelling is an iterative process, validation can sometimes become calibration when the model is adjusted as a direct result of the validation. The crudest but most effective empirical adjustment is to compute from the validation data an adjustment factor to be applied to the predictions. More subtle adjustment can be made by logical changes to parameter values after inspection of the results of validation. When this is done, the validation step has been transformed into calibration. After this recalibration, especially with an adjustment factor, the model predictions will agree as well as they ever can with the observations on which they are partially based. Further validation is required using a fresh set of observations.

The method we propose for comparing predictions and observations is not a proof of model validity, but emphasizes instead the concept of whether the model is adequate for its purpose; within those terms, it is a method of empirical validation. The demonstration of model adequacy may be required for users of models who are not modellers, or even scientists. Consequently, the ideal method of empirical validation should be simple and understandable with minimal mathematical or statistical knowledge, as well as objective and quantitative.

Background to empirical validation

On methods of empirical validation the references on modelling cited above have little to say. When the validation data consist of pairs of predictions from the model and comparable observations from the real system, they are often plotted on a scatter graph (e.g. Carberry and Abrecht 1991; Clewett et al. 1991; Uehara and Tsuji 1991; Aber and Federer 1992; Warnant et al. 1994). Sometimes a regression line is added (e.g. Nemani and Running 1989; Hammer and Muchow 1991; Keating et al. 1991; Parton et al. 1993; Paruelo and Sala 1995; Woodward et al. 1995), with the implicit or explicit use of the regression statistics as a quantitative measure of how good the model is.

The scatter graph provides an immediate general impression of how closely predictions and observations cluster around the 1:1 line of perfect correspondence which can be drawn on the graph. A model that has reached the stage of empirical validation is certain to have general correspondence of the size of prediction and observation along the range of operation. The scatter graph flatters the model for two reasons. First, only a fraction of the area of the graph is used to display the differences between prediction and observation, so that these differences are made to seem small in the context of the whole graph, most of which is empty. Second, the eye tends to assess the distance between the plotted point and the nearest part of the line. This is similar to the difficulty in truly assessing the vertical gap between two curves that are not horizontal (Cleveland 1994). The true difference between prediction and observation is the horizontal or vertical distance from the point to the 1:1 line. This is 41% greater ($\sqrt{2}$:1) than the shortest distance. A further shortcoming of the scatter graph is that lines to show absolute or relative precision (error) cannot be plotted relative to the 1:1 line. The precision of any point has to be calculated with reference to the position along the observation axis, not with reference to the 1:1 line.

Once the scatter graph is drawn, regression has 'intuitive appeal' (Harrison 1990) as an objective and quantitative method of empirical validation, and it has been promoted by Reckhow et al. (1990), Flavelle (1992) and Mayer et al. (1994). There are several reasons why regression is not appropriate (Harrison 1990; Mitchell 1997): it is a misapplication of regression for which the assumptions are rarely satisfied; null hypothesis tests give ambiguous results; regression lacks sensitivity because distinguishing lines from a random cloud of points is not necessary at this stage in model development; and the fitted line is irrelevant to model performance. It is the differences between prediction and observation that provide information on how far the model fails to simulate the system exactly. We call the differences deviations to distinguish them from residuals in the regression sense. The advantage of computing them as prediction minus observation is that the positive deviations are those where the prediction is larger, and vice versa. The observation is the reference point, taken as the true value from which the prediction differs.

Deviations occur because there can be variation in predictions or observations and often in both. Variation arises because the model is a simplification of the real system: certain factors were omitted or assumed constant and they produce variability in the observations which is not reproduced by the model predictions. Usually there is more than one source contributing to the variability of the deviations. There are several reasons to expect that the deviations will approximate a Normal distribution. They are the several sources of error, the fact that the deviations are differences between two quantities, and the lack of fixed limits. These are the arguments that assist the assumption of Normality for residuals in analysis of variance (Sokal and Rohlf 1981; Mead and Curnow 1983).

A single-figure summary of the deviations is their quadratic mean or root mean square (the square root of the mean of the squared deviations), often called the root mean square error (RMSE). The RMSE is frequently given when predictions are compared with observations but we are not aware of any justification for its use. The calculation of RMSE removes the negative signs and weights the deviations by their squares, thus giving more influence to larger ones. It is independent of the mean but depends on the distribution of the deviations. The RMSE is a poor indicator of the performance of the model, since it gives no indication of bias and cannot be used to compute the precision of the predictions. The mean, standard deviation and confidence limits (assuming that the deviations approximate a Normal distribution) are better quantitative summaries of the deviations, although they are not required in the graphical method used here.

It is inevitable that model predictions are not perfectly precise (deviations all zero).

There may also be bias, i.e. systematic, directional difference of the prediction from the observation. A consistent bias can be corrected temporarily as a final, empirical, stage of calibration until it can be eliminated by further development of the model. (If this is done with validation data, then validation has been superseded by recalibration and further validation is required.) This leads to the central question for empirical validation: 'is the model precise enough to be adequate for its purpose?' The question can be answered by graphical presentation of the deviations and definition of criteria for adequacy with reference to the purpose of the model.

Method for comparing predictions with observations

To save space, the method is given as a checklist of instructions with limited explanation and justification. Further details can be obtained from the authors.

(1) Decide on which of the model outputs is to be validated.

(2) Define the envelope of acceptable precision with reference to the purpose of the model; sketch it on an outline graph on which the deviations will be plotted later.

(The deviations will be plotted against the observations which are the reference values, assumed to truly represent the real system. Compared with the scatter graph, the graph of deviations makes better use of the space and displays the deviations above and below a horizontal reference line of zero deviation. The first criterion of adequacy is an envelope of acceptable precision, that is a region on the graph of deviations in which deviations are considered to be acceptably small. In principle, the envelope of acceptable precision can be any size or shape, made up of absolute or relative precision, or both, and it need not be constant along the x-axis. It can be symmetrical or not about the horizontal line of zero deviation. To be regarded as objective and quantitative, the definition of the envelope should be clearly stated, with arguments to support the values chosen. We can think of four possibilities, in order of preference below.

(a) The model may be designed to replace or supplement observations of the real system, often observations of the type used in empirical validation. It seems unreasonable to expect a model to produce better precision (as 95% confidence limits) than the observations it may replace and it is realistic to allow the predictions somewhat less precision.

(b) The precision of the measurements or estimates of values for parameters or variables in the model can be a reference point.

(c) The envelope of acceptable precision could be defined as 'at least as good as' a previous version of the model or another specified model.

(d) The precision of the model may be specified externally; for example, as a legislated safety standard.)

(3) If a pass/fail test is needed, define the proportion of points inside the envelope for the model to be adequate; use 0.95 unless there is reason to choose a different value.

(The second criterion of adequacy is the proportion of points on the graph of deviations that must lie within the envelope of acceptable precision. If the model is not biased, the deviations will tend to cluster around zero and probably approximate a Normal distribution, as discussed above. It is, therefore, reasonable to borrow the concept of confidence limits and quantify the criterion as 95% of points falling within the envelope. The 95% figure (corresponding to $P=0.05$ in testing hypotheses) is arbitrary but widely accepted as the first threshold at which experimental results are taken to be statistically significant (Sokal and Rohlf 1981; Gilbert 1989). It is consistent to use it in this context as well.)

(4) Collect the observations that will be comparable with the model predictions. Do not examine the observations before this stage. Ensure that these observations are separate from any used to construct the model.

(The greatest rigour is achieved if the criteria of adequacy are defined before the observations for validation are examined, so that accusations of 'moving the goalposts' to make the model adequate can be avoided (Bathurst and O'Connell 1992; Ewen and Parkin 1996). The definitions of the criteria can be regarded as part of model design with respect to the application of the model, and it should perhaps become standard practice to state them when the initial specification of the model is drawn up. The arguments for what proportion of points should lie in the envelope are by analogy with statistics and a default value of 0.95 could find general acceptance. Even if the criteria are not defined in advance, the stated justification in terms of the application of the model provides some objectivity. The basis on which the model is to be regarded as adequate has then been set out clearly, so that it can be debated and revised by others.)

(5) Run the model for the same conditions as the observations (if this has not been done earlier).

(6) Compute the deviations as prediction minus observation and plot them. Are there outlying points for which observation and prediction can be checked for identifiable mistakes?

(Outliers are the points that will frequently cause models not to satisfy the second criterion. They must be retained unless there is clear and objective reason to exclude them – a matter of scientific and personal integrity.)

(7) Draw the envelope of acceptable precision on the graph. What proportion of points lie within the envelope? Is it greater than 95% (or another chosen value)?

(The proportion of points inside the envelope of acceptable precision is a quantitative measure of model performance that can be used to compare models or different versions of the same model or separate sets of validation data. If the envelope turns out to be so wide that all models have a high proportion of points inside, a more exacting envelope can be defined to spread out the models in terms of this measure. Sometimes a qualitative measure of performance as a pass or fail test is required and this is provided by the second criterion of adequacy.)

(8) Assess model performance visually from the graph of deviations. Is the model unbiased and of good precision? Are there trends in bias or precision along the x-axis?

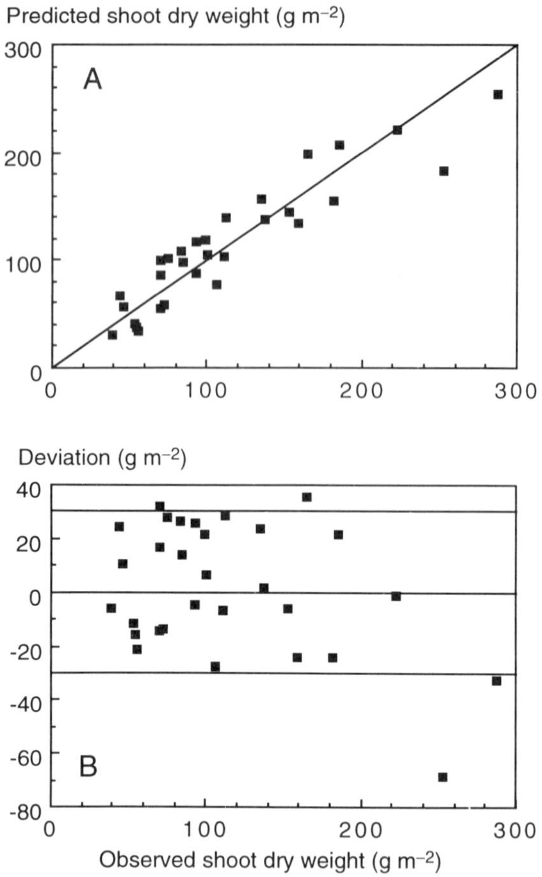

Figure 1. (A) Shoot dry weight of a grass sward predicted by a model compared with observations from harvested plots. The model was driven by the weather records for the harvest locations. The line is the 1:1 line of perfect agreement. (B) Graph of deviations for empirical validation of the grass model. Deviations are prediction minus observation. The envelope of acceptable precision shown as an example is $\pm 30\,g\,m^{-2}$.

(The general pattern of the cloud of points is an indicator of model performance. If the model is not biased, the points will be distributed symmetrically either side of the reference line. If the model has good precision, the points will form a narrow band. Bias and precision may vary along the range of operation of the model and this will be evident from the graph.)

(9) Conclude whether the model is adequate or not.

(10) If not, identify from the graph of deviations whether bias or precision, or both, need improvement and whether this is so for particular ranges along the x-axis. Consider plotting the deviations against other features of the observations or of the model (e.g. driving variables) to gain insight on where the model needs improvement.

Examples

Model of grass growth

This model for the growth of a grass sward (A.C. Terry, pers. commun.), developed from the model of Sheehy et al. (1979, 1980), is driven by daily weather data and simulates growth and regrowth after cutting, to predict dry weight above and below ground. The model was run with the weather records from five locations at which harvests of the grass crop had been made. Cuts were imposed in the model on six harvest dates during the growing season. The variety of harvest dates and locations, well distributed across England, provided a wide range of shoot dry weight, and a scatter graph is given in Figure 1A. The graph of the deviations (Figure 1B) shows more clearly that the deviations are equally scattered above and below the zero line, and there is one outlier where the prediction was nearly 70 g m^{-2} less than the observation. The performance of the model is consistent in the operational range 30–250 g m^{-2}.

As an example, suppose that experience in collecting these and other observations, as sample plots in fields assumed uniform in soil and weather, shows that the precision of the observations (as 95% confidence limits) is ±25 g m^{-2}. The model is not expected to perform as well as this, so a precision of 30 g m^{-2} is defined as the envelope of acceptable precision; this would still be useful in practice. The usual criterion of 95% of points inside the envelope is employed. The envelope, which is in terms of constant absolute precision, is plotted in Figure 1B and it is clear that the model is not adequate because only 26 points (87%) are inside.

If the deviations are plotted against the locations, ranked south to north, no clear pattern can be seen (Figure 2A). However, when the deviations are plotted against harvest date (Figure 2B) it is evident that performance of the model varied: harvest 2 (early June) gave a large spread of deviations, and there was a drift in model bias from a tendency to overestimate shoot dry weight at harvest 3 (early July) and to underestimate it at harvest 6 (late September). Harvest date itself cannot affect model performance but associated with it are the phenological phases of the grass (becoming reproductive from May onwards) and the driving weather variables which vary during the year. This information may be of use in further development of the model to meet the criteria for adequacy that have been defined.

Hemispherical photographs to estimate PAR

Hemispherical (whole-sky) photographs of forest canopies can be used to estimate the amount of photosynthetically active radiation (PAR) received at the ground (Mitchell and Whitmore 1993). The model in this case is not implemented as a single computer program, but as a procedure that includes image analysis of the photographs, equations for solar geometry, a model for the distribution of radiance across the sky and measurements of radiation above the canopy. Hemispherical photographs (hemiphots) were taken in canopy gaps of different size in Bornean rain forest and compared with measurements of PAR made at the same spot in the gap (Whitmore

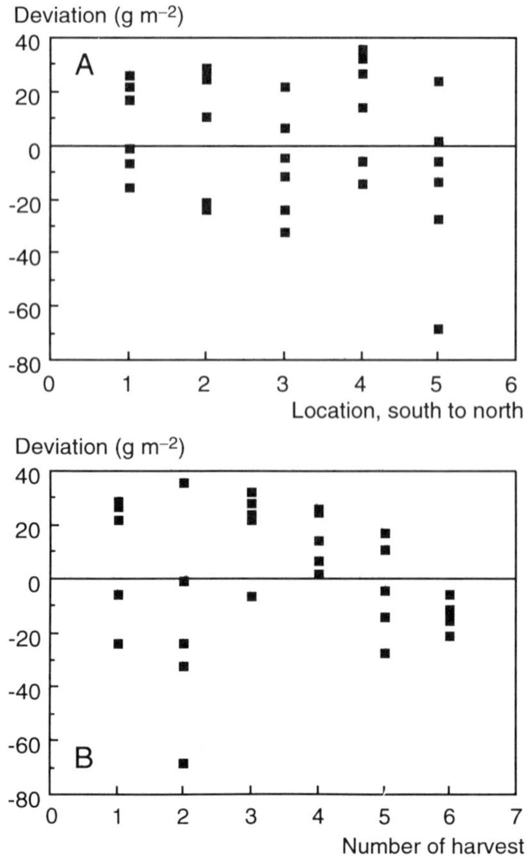

Figure 2. (A) The deviations of the grass model plotted against location from south to north (Devon, Somerset, Herefordshire, Yorkshire, Northumberland). (B) The deviations of the grass model plotted against harvest number. Harvests were at intervals of four weeks from early May to late September.

et al. 1993). The measurements were averaged for the year; there was comparatively little seasonal change, since the location was evergreen forest at 5° N latitude. A scatter graph is shown in Figure 3A. The point close to the 1:1 line at about $35\,\mathrm{E\,m^{-2}\,d^{-1}}$ is for a site in the open; strictly speaking, it should not be included in the validation because its value was used to compute PAR for the other sites. (An Einstein, E, is Avogadro's number of quanta.)

A method of computing PAR from hemiphots would be most useful since it could replace measurements which are difficult to make in the field. To be useful, the model should produce results as unbiased and precise as measurements. A precision of $\pm 10\%$, except below $5\,\mathrm{E\,m^{-2}\,d^{-1}}$ where $\pm 0.5\,\mathrm{E\,m^{-2}\,d^{-1}}$ is applicable, is the best that can be achieved (Mitchell and Whitmore 1993), because field measurements of PAR are extremely variable. This combination of absolute and relative precision constitutes the envelope of acceptable precision, and it is shown in the graph of deviations in

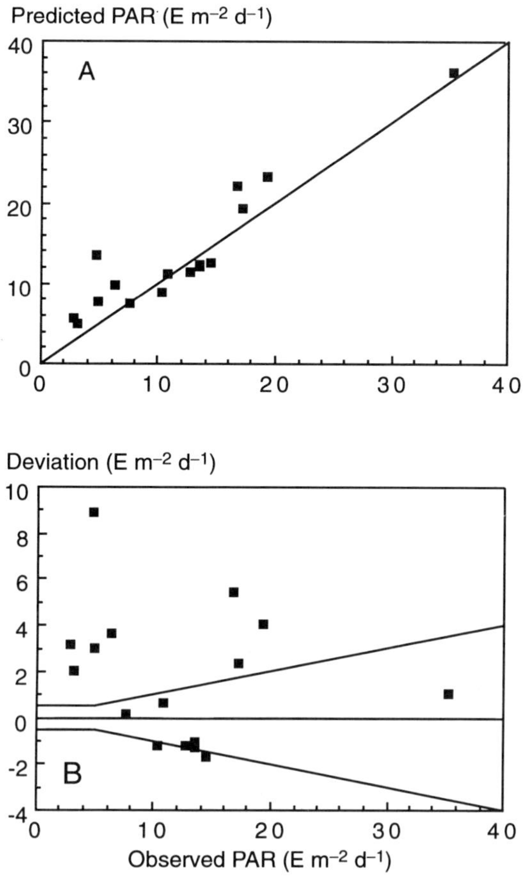

Figure 3. (A) Predicted values of PAR from hemispherical photographs compared with observations made at the same spot in canopy gaps of different sizes in Bornean rain forest. The line is the 1:1 line of perfect agreement. (B) Graph of deviations for empirical validation of hemispherical photographs for predicting PAR under forest canopies. Deviations are prediction minus observation. The envelope of acceptable precision is a combination of absolute and relative precision based on experience of measuring PAR in the field (see text for details).

Figure 3B. The model is far from adequate and further development is needed, particularly to overcome difficulties with the photography and image analysis (Whitmore et al. 1993). Plotting the deviations against the fraction of predicted PAR that was direct from the sun (Figure 4) suggests that the deviations are largest and mostly positive when direct solar radiation contributed more than 55% to the total received beneath the canopy. On the generally shaded forest floor a ray of direct radiation can make a large contribution, since it will occur when the sun is high in the sky. Exact recording on the hemiphot of when a leaf or branch blocks the sun on its track across the sky is essential and this may be the particular shortcoming in the method. Small leaves high up are imperfectly recorded – they tend to have haloes – which probably leads to an overestimate of occasions when the sun is visible from the hemiphot point.

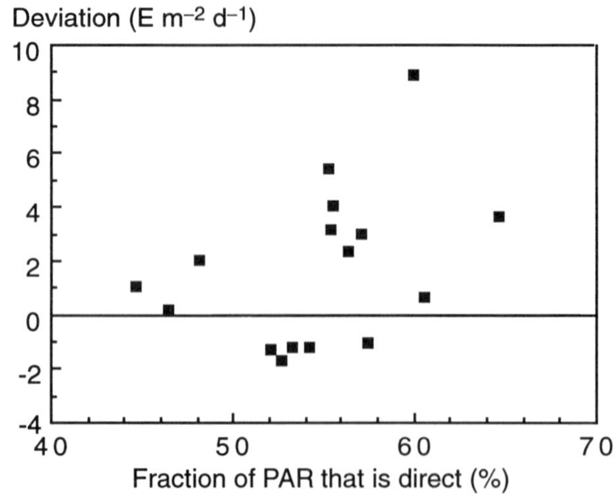

Figure 4. The deviations for empirical validation of hemispherical photographs plotted against the fraction of predicted total PAR that is direct from the sun.

Adaptations of the method

Predictions or observations that are sequences in time or space or are mappable normally show autocorrelation to some degree. Statistical tests that require independence are then not appropriate. The method we propose is easily adapted to these cases, makes no assumption about independence and also provides information about the performance of the model in its operational range of space or time.

The predictions and observations may be time courses, e.g. of growth (Heuvelink 1995), or spatial sequences, e.g. along a river (Ewen and Parkin 1996), where the predictions or observations, or both, are continuous lines. Since either predictions or observations can be points or lines, there are four combinations. The method given in full above can be used when both predictions and observations are points. In other combinations, the graph of deviations is not applicable, but the criteria of the envelope of acceptable precision and the proportion of points within can still be used (cf. Ewen and Parkin 1996). When the prediction is a line continuous along the x-axis (time or space), it can be surrounded by lines above and below denoting the envelope of acceptable precision. Constant relative or absolute precision along the x-axis are the simplest cases, defined as usual with reference to the purpose of the model. However, changes in the definition of the envelope along the x-axis could be justified; for example, if the model was regarded as adequate with a wider or narrower envelope early in the time course – ideally this should be defined in advance. If the observations are points, then the proportion inside the envelope is calculated. If the observations are a line, then the proportion along the x-axis that the observation line lies within the envelope is measured. The fourth combination of point predictions and line observations may be uncommon. The envelope of acceptable precision is defined as usual but implemented as upper and lower limits around each of the point predictions. The proportion of points where the observation line lies within these limits, isolated segments of the envelope, is calculated. In each combination, an adequate model will have at least the defined proportion of line or points within the envelope.

The output of some models (e.g. Prentice et al. 1992; Woodward et al. 1995) is in the form of maps which are compared with maps of observations. For maps of categories, e.g. types of vegetation, the kappa statistic (Monserud and Leemans 1992) has been proposed for empirical validation of this type. There is no absolute standard for the threshold value of kappa and the values used to indicate adequate models are based on experience (Prentice et al. 1992, p. 123), instead of using a statistical test which may not be particularly meaningful. When the mapped predictions and observations are continuous variables, the deviations can be computed and presented as a map (see e.g. Gates et al. (1992, p. 127) the map of deviations in mean temperature for the world from the UK Meteorological Office general circulation model). The envelope of acceptable precision can be defined as usual; it need not be constant, but it can vary with the size of observation or prediction if this can be supported by arguments related to the purpose of the model and knowledge of the precision of the observations. The proportion of mapped locations where the deviation is inside the envelope is calculated and compared with the defined proportion (usually 0.95) for the model to be regarded as adequate. The map of deviations shows how the performance of the model varies geographically and this can be related to the driving variables or parameters of the model through the map location.

Discussion

The purpose of many models is to increase understanding of the system and to emphasize gaps in knowledge, rather than to make precise predictions. Models may be used to examine the relative importance of different driving variables, or to estimate the sensitivity of a system to changes in particular variables. In so far as these models need empirical validation, it can be informal and approximate. For example, a model to demonstrate that nitrogen allocation in a plant could drive carbon partitioning (Sheehy et al. 1996) was validated by showing that field measurements of grass shoot dry weight were consistent with the model output. It was not the purpose of this model to make exact predictions of grassland production. Empirical validation is only a part of validation as a whole and not every model needs it to be employed in full rigour; nevertheless, when it is required for an applied model it should be carried out properly and misuse of statistics should be avoided.

The method proposed here for empirical validation is easy to comprehend through the graphical presentation which concentrates on the deviations. It becomes objective and quantitative when the criteria for adequacy are defined and the reasons for the values chosen are justified with reference to the purpose of the model. The attraction of statistical tests for validation is their objective and quantitative nature (Reckhow et al. 1990; Flavelle 1992; Mayer et al. 1994). But this is an illusion when these properties cannot be related to the purpose of the model or are not attainable because the assumptions of the test are violated. The independence of the individual predictions or observations is an assumption which often may not be met by validation data sets, and this does not matter with the method we propose.

Two examples of empirical validation were chosen which differed in how the envelope of acceptable precision was defined. If a model of grassland is intended to make predictions of production that replace measurements, then empirical validation in this way is necessary although the envelope was defined here merely as an example. The second example showed that hemiphots did not achieve the precision desired and suggested why. Since relative precision converges on zero as the estimate tends to zero, an envelope of acceptable precision based on relative precision always constitutes a tough test for any model that operates with small values. In the second example, a region of absolute precision for low values of PAR was defined in the envelope of acceptable precision (Figure 3B), because for routine field measurements in deep shade a precision of one significant figure in $E\,m^{-2}\,d^{-1}$ was considered realistic (Mitchell and Whitmore 1993). Experience in applying this method of empirical validation may show that few models are adequate when the envelope of acceptable precision is defined as relative precision. A combination of absolute and relative precision similar to that in the second example may prove useful in these cases.

Empirical validation of a mechanistic model offers particular opportunities and problems. In a mechanistic model, the submodels and their outputs have meaning and it may be possible to apply empirical validation to each one, as well as to the overall prediction. For example, a model of crop yield, driven by weather variables, could have submodels for phenological development, photosynthesis and allocation of dry matter to plant parts (see e.g. for rice Horie et al. 1995; Kropff et al. 1995), and the output of each of these submodels could be compared with appropriate observations. This offers the opportunity of identifying areas of weakness within the model and improving the submodel. However, the final prediction of crop yield is liable to carry with it the accumulated uncertainties of the component submodels making empirical validation to the same standard as the submodels more difficult. A full hierarchical error analysis might justify different criteria, as the envelope of acceptable precision of the proportion of points within, for the final prediction. But if the model is intended to be applied – to predict crop yields in place of measurements – then its final output *does* need validation to a suitable standard relevant to the purpose of the model. An empirical model, such as a multiple regression of yield on weather variables for the growing season, may be much easier to validate, because there are no cumulative errors (but will suffer from the usual limitations of empirical modelling in not increasing understanding and in restricted applicability to the range of conditions used in constructing the model).

Application of the method of empirical validation proposed here can produce results that are initially disappointing when the second criterion is set to 95% of points within the envelope of acceptable precision. We believe that this exacting standard is justifiable, if applied models are to be used with confidence, and we note that Amthor and Loomis (1996) regard few crop models as having been adequately validated with independent data. If modelling is to be regarded as rigorous scientific research, and accepted as a reliable technique when applied in the real world, then modellers must evaluate their models realistically.

Acknowledgements

Financial support was from awards to F.I. Woodward through the European Community MEDALUS (Mediterranean Desertification and Land Use) programme and through the Natural Environment Research Council TIGER (Terrestrial Initiative in Global Environmental Research) programme, and from the International Rice Research Institute as funds for P.L.M. to attend the SARP and SAAD2 symposia at IRRI in December 1995. We are grateful for encouragement in this work from F.I. Woodward and participants at the symposia. J.C. Bathurst first set us on the road to objectivity and provided a copy of Ewen and Parkin (1996) before publication.

References

Aber J D, Federer C A (1992) A generalized, lumped-parameter model of photosynthesis, evapotranspiration and net primary production in temperature and boreal forest ecosystems. Oecologia 92:463–474.

Amthor J S, Loomis R S (1996) Integrating knowledge of crop responses to elevated CO_2 and temperature with mechanistic simulation models: Model components and research needs. Pages 317–345 in Koch G W, Mooney H A (Eds.) Carbon dioxide and terrestrial. Academic Press, London, UK.

Bathurst J C, O'Connell P E (1992) Future of distributed modelling: The Système Hydrologique Européen. Hydrological Processes 6:265–277.

Beck M B (1987) Water quality modeling: A review of the analysis of uncertainty. Water Resources Research 23:1393–1442.

Burghes D N, Wood A D (1980) Mathematical models in the social, management and life sciences. Ellis Horwood, Chichester, UK. 287 p.

Carberry P S, Abrecht D G (1991) Tailoring crop models to the semiarid tropics. Pages 157–182 in Muchow R C, Bellamy J A (Eds.) Climate risk in crop production: Models and management for the semiarid tropics and subtropics. CAB International, Wallingford, UK.

Charles-Edwards D A, Doley D, Rimmington G A (1986) Modelling plant growth and development. Academic Press, London, UK. 235 p.

Cleveland W S (1994) The elements of graphic data (revised edition). Hobart Press, Summit, New Jersey, USA. 297 p.

Clewett J F, Howden S M, McKeon G M, Rose C W (1991) Optimising farm dam irrigation in response to climatic risk. Pages 307–328 in Muchow R C, Bellamy J A (Eds.) Climate risk in crop production: Models and management for the semiarid tropics and subtropics. CAB International, Wallingford, UK.

Ewen J, Parkin G (1996) Validation of catchment models for predicting land-use and climate impact. 1 Methodology. J. Hydrology (in press).

Flavelle P (1992) A quantitative measure of model validation and its potential use for regulatory purposes. Advances in Water Resources 15:5–13.

Gates W L, Mitchell J F B, Boer G J, Cubasch U, Meleshko V P (1992) Climate modelling, climate prediction and model validation. Pages 97–134 in Houghton J T, Callander B A, Varney S K (Eds.) Climate change 1992: The supplementary report to the IPCC Scientific Assessment. Report prepared for the Intergovernmental Panel on Climate Change by Working Group I. Cambridge University Press, Cambridge, UK.

Gilbert N (1989) Biometrical Interpretation, 2nd Edition. Oxford University Press, Oxford, UK. 146 p.

Gilchrist W (1984) Statistical modelling. John Wiley and Sons, Chichester, UK. 339 p.

Hammer G L, Muchow R C (1991) Quantifying climatic risk to sorghum in Australia's semiarid tropics and subtropics: Model development and simulation. Pages 205–232 in Muchow R C, Bellamy J A (Eds.) Climate risk in crop production: Models and management for the semiarid tropics and subtropics. CAB International, Wallingford, UK.

Harrison S R (1990) Regression of a model on real-system output: An invalid test of model validity. Agric. Syst. 34:183–190.

Heuvelink E (1995) Dry matter production in a tomato crop: Measurements and simulation. Ann. of Botany 75:369–379.

Horie T, Nakagawa H, Centeno H G S, Kropff M J (1995) The rice crop simulation model SIMRIW and its testing. Pages 51–66 in Matthews R B, Kropff M J, Bachelet D, Van Laar H H (Eds.) Modeling the impact of climate change on rice production in Asia. CAB International, Wallingford; International Rice Research Institute, Manila, Philippines.

Jeffers J N R (1982) Modelling. Chapman and Hall, London, UK. 80 p.

Keating B A, Godwin D C, Watiki J M (1991) Optimising nitrogen inputs in response to climatic risk. Pages 329–358 in Muchow R C, Bellamy J A (Eds.) Climate risk in crop production: Models and management for the semiarid tropics and subtropics. CAB International, Wallingford, UK.

Kropff M J, Matthews R B, Van Laar H H, Ten Berge H F M (1995) The rice model ORYZA1 and its testing. Pages 27–50 in Matthews R B, Kropff M J, Bachelet D, Van Laar H H (Eds.) Modeling the impact of climate change on rice production in Asia. CAB International, Wallingford; International Rice Research Institute, Manila, Philippines.

McCarl B A (1984) Model validation: An overview with some emphasis on risk models. Review of Marketing and Agricultural Economics 52:153–173.

Mayer D G, Stuart M A, Swain A J (1994) Regression of real-world data on model output: An appropriate overall test of validity. Agric. Syst. 45:93–104.

Mead R, Curnow R N (1983) Statistical methods in agriculture and experimental biology. Chapman and Hall, London, UK. 335 p.

Mitchell P L (1997) Misuse of regression for empirical validation of models. Agric. Syst. (in press).

Mitchell P L, Whitmore T C (1993) Use of hemispherical photographs in forest ecology: Calculation of absolute amount of radiation beneath the canopy. Oxford Forestry Institute, Occasional Paper No. 44, Oxford, UK. 39 p.

Monserud R A, Leemans R (1992) Comparing global vegetation maps with the Kappa statistic. Ecological Modelling 62:275–293.

Nemani R R, Running S W (1989) Testing a theoretical climate-soil-leaf area hydrologic equilibrium of forests using satellite data and ecosystem simulation. Agric. For. Meteorol. 44:245–260.

Oreskes N, Shrader-Frechette K, Belitz K (1994) Verification, validation, and confirmation of numerical models in the earth sciences. Science 263:641–646.

Parton W J, Scurlock J M O, Ojima D S, Gilmanov T G, Scholes R J, Schimel D S, Kirchner T, Menaut J-C, Seastedt T, Garcia Moya E, Kamnalrut A, Kinyamario J I (1993) Observations and modelling of biomass and soil organic matter dynamics for the grassland biome worldwide. Global Biogeochemical Cycles 7:785–809.

Paruelo J M, Sala O E (1995) Water losses in the Patagonian steppe: A modelling approach. Ecology 76: 510–520.

Prentice I C, Cramer W, Harrison S P, Leemans R, Monserud R A, Solomon A M (1992) A global biome model based on plant physiology and dominance, soil properties and climate. J. Biogeography 19: 117–134.

Reckhow K H, Clements J T, Dodd R C (1990) Statistical evaluation of mechanistic water-quality models. J. Environm. Engineering 116:250–268.

Sheehy J E, Cobby J M, Ryle G J A (1979) The growth of perennial ryegrass: A model. Ann. of Botany 43: 335–354.

Sheehy J E, Cobby J M, Ryle G J A (1980) The use of a model to investigate the influence of some environmental factors on the growth of perennial ryegrass. Ann. of Botany 46:343–365.

Sheehy J E, Gastal F, Mitchell P L, Durand J-L, Lemaire G, Woodward F I (1996) A nitrogen-led model of grass growth. Ann. of Botany 77:165–177.

Shugart H H, West D C (1981) Long-term dynamics of forest ecosystems. American Scientist 69:647–652.

Sokal R R, Rohlf F J (1981) Biometry: The principles and practice of statistics in biological research, 2nd Edition. Freeman, New York, USA. 859 p.

Thornley J H M, Johnson I R (1990) Plant and crop modelling. Oxford University Press, Oxford, UK. 669 p.

Trenberth K E (Ed.) (1992) Climate system modeling. Cambridge University Press, Cambridge, UK. 788 p.

Uehara G, Tsuji G T (1991) Progress in crop modelling in the IBSNAT Project. Pages 143–156 in Muchow R C, Bellamy J A (Eds.) Climate risk in crop production: Models and management for the semiarid tropics and subtropics. CAB International, Wallingford, UK.

Van Keulen H (1976) Evaluation of models. Pages 23–29 in Arnold G W, De Wit C T (Eds.) Critical evaluation of systems analysis in ecosystems research and management. Pudoc, Wageningen, The Netherlands.

Warnant P, François L, Strivay D, Gérard J-C (1994) CARAIB: A global model of terrestrial biological productivity. Global Biogeochemical Cycles 8:255–270.

Whitmore T C, Brown N D, Swaine M D, Kennedy D, Goodwin-Bailey C I, Gong W-K (1993) Use of hemispherical photographs in forest ecology: Measurement of gap size and radiation totals in a Bornean tropical rain forest. J. of Tropical Ecology 9:131–151.

Woodward F I, Smith T M, Emanuel W R (1995) A global land primary productivity and phytogeography model. Global Biogeochemical Cycles 9:471–490.

Acronyms used in *Applications of Systems Approaches at the Farm and Regional Levels*, and *Applications of Systems Approaches at the Field Level*

AARD	Agency for Agricultural Research and Development
AB-DLO	DLO-Research Institute for Agrobiology and Soil Fertility
AICCIP	All India Coordinated Crop Improvement Project
AIDS	Almost Identical Demand System
AISLUS	All India Soil and Land Use Survey
AMMI	Additive Main effects and Multiplicative Interaction models
APROSC	Agricultural Projects Services Centre
APSRU	Agricultural Production Systems Research Unit
ARMA	AutoRegressive Moving Average
ARO	Advanced Research Organization
ASA	American Society of Agronomy
ASAE	American Society of Agricultural Engineers
AUDPC	Area Under Disease Progress Curve
AZP	Atlantic Zone Programme
BEII	Biocide Environmental Impact Index
BMZ	Federal Ministry for Cooperation
BRGM	Bureau de Recherches Géologiques et Minières
BULOG	National Logistic Board
CAILUP	Computerized Aid to Integrated Land-Use Planning
CASER	Center for Agro-Socio Economic Research
CATIE	Tropical Agronomic Research and Higher Education Center
CCT	Computer Compatible Tapes
CEC	Cation Exchange Capacity
CERES	Crop Estimation through Resource and Environment Synthesis
CGIAR	Consultative Group on International Agricultural Research
CGMS	Crop Growth Monitoring System
CGPRT	Centre for Coarse Grains, Pulses, Roots and Tuber Crops
CIAT	International Center for Tropical Agriculture
CIMMYT	Centro Internacional de Mejoramiento de Maiz y Trigo (International Center for the Improvement of Maize and Wheat)
CIP	Centro Internacional de la Papa (International Potato Center)
CIRAD	Centre de coopération Internationale en Recherche Agronomique pour le Développement (Centre for International Cooperation and Agricultural Development Research)

M.J. Kropff et al. *(eds.): Applications of Systems Approaches at the Field Level, 453–458.*

CIRAD-URPA	Centre de coopération Internationale en Recherche Agronomique pour le Développement, Unite de Recherche 'Prospective et Politique Agricole'
CLIPS	C Language Integrated Production System
CLSU	Central Luzon State University
CMDT	Compagnie Malienne pour le Développement des Textiles
COMAX	CrOp Management eXpert
CPAC	Centro de Pesquisa Agropecuária dos Cerrados
CR	Crop Residues
CRRI	Central Rice Research Institute
CSIRO	Commonwealth Scientific and Industrial Research Organization
Danida	Danish International Development Assistance
DAS	Days After Sowing
DAT	Days After Transplanting
DEM	Digital Elevation Model
DGIS	Directorate General for International Cooperation
DLO	Agricultural Research Department, Ministry of Agriculture, Nature Conservation and Fisheries
DNSI	Direction Nationale de la Statistique et de l'Information
DQC	Dynamic Quality Control
DRR	Directorate of Rice Research
DRSPR	Direction Régionale des Systèmes de Production et de la Gestión des Ressources Naturelles
DSS	Decision Support System
DSSAT	Decision Support System for Agrotechnology Transfer
DTM	Digital Terrain Model
EIA	Environmental Impact Assessment
EMBRAPA	Empresa Brasileira de Pesquisa Agropecuária
EPA	Environmental Protection Agency
EPIC	Erosion Productivity Impact Calculator
ESPGRN	Equipe Systèmes de Production et Gestión des Ressources Naturelles
EU	European Union
EUROSTAT	European Statistical Office
FACE	Free-Air-CO_2-Enrichment
FAO	Food and Agriculture Organization of the United Nations
FCC	False Colour Composites
FHM	Farm-Household Modelling
FPR	Farmer Participatory Research
FSA	Farming Systems Analysis

FSR	Farming Systems Research
FSR/E	Farming Systems Research and Extension
FSS	Farming System Simulator
GATT	General Agreement on Tariffs and Trade
GBPUAT	Govind Ballabh Plant University of Agriculture and Technology
GCM	General Circulation Models
GCREC	Gulf Coast Research and Education Center
GCTE	Global Change and Terrestrial Ecosystems Core Project
GDP	Gross Domestic Product
GE	Grain Equivalents
GFDL	General Fluid Dynamics Laboratory
GIDM	Generic Interactive Dairy Model
GIS	Geographic Information System
GISS	Goddard Institute for Space Studies
GPS	Global Positioning System
GRASS	Geographic Resources Analysis Support System
GTZ	German Agency for Technical Cooperation
G×E	Genotype by Environment
HTIS	HyperText Information System
IARC	International Agricultural Research Center
IARI	Indian Agricultural Research Institute
IBSNAT	International Benchmark Sites Network for Agrotechnology Transfer
ICAR	Indian Council of Agricultural Research
ICARDA	International Center for Agricultural Research in Dry Areas
ICASA	International Consortium for Agricultural Systems Applications
ICLARM	International Center for Living Aquatic Resources Management
ICRAF	International Center for Research in Agroforestry
ICRISAT	International Crops Research Institute for the Semi-Arid Tropics
IDA	Institute for Agricultural Development
IFDC	International Fertilizer Development Center
IFPRI	International Food Policy Research Institute
IGBP	International Geosphere-Biosphere Programme
IIT	Indian Institute of Technology
ILCA	International Livestock Center for Africa
ILEIA	Information Centre for Low External Input and Sustainable Agriculture

IMGLP	Interactive Multiple Goal Linear Programming
INGER	International Network for Genetic Evaluation of Rice
INTA	National Institute for Agricultural Research
IPLU	Integrated Physical Land Units
IRM	Insecticide Resistance Management
IRRI	International Rice Research Institute
IRS	Indian Remote Sensing Satellite
IRU	Integrated Resources Units
ITOPE	Interdisciplinary Tool for Optimizing Productivity and Erosion
IUFRO	International Union of Forest Research Organization
JAU	Jiangxi Agricultural University
JRC	Joint Research Centre
KBS	Knowledge-Based System
KKU	Khon Kaen University
LDC	Less Developed Countries
LE	Land Evaluation
LIS	Livestock Information System
LUST	Land-Use Systems at a defined Technology
MACROS	Modules of an Annual CRop Simulator
MARS	Monitoring Agriculture with Remote Sensing
MCDM	Multiple Criteria Decision Models
MCO	Multi-Criteria Optimization
MDS	Minimum Data Set
MEDALUS	Mediterranean Desertification and Land Use
MET	Multi-Environment Testing
MLNV	Dutch Ministry of Agriculture, Nature Conservation and Fishery
MOLP	Multiple Objective Linear Programming
MP	Mathematical Programming
NARS	National Agricultural Research System
NBD	Nutrient Balance model of a Dairy–crop operation
NDVI	Normalized Difference Vegetation Index
NGO	Non-Governmental Organization
NSADP	North Sinai Agricultural Development Project
NWO-STW	Netherlands Technology Foundation
ORSTOM	Institut Français de Recherche pour le Développement en Coopération

OTC	Open Top Chamber
OTELO	Organisation du Travail Et Langage à Objet
PAR	Photosynthetically Active Radiation
PASW	Plant-Atmosphere-Soil-Water
PCA	Principal Component Analysis
PDF	Probability Density Function
PhilRice	Philippine Rice Research Institute
PI	Panicle Initiation
PPFD	Photosynthetic Photon Flux Density
PRA	Participatory Rapid Appraisal
PRODECER	Cerrados Development Program
PSS	Project Production Soudano Sahélienne
QTL	Quantitative Trait Loci
R,D&E	Research, Development and Extension
RFLP	Restriction Fragment Length Polymorphism
RIDEV	RIce DEVelopment model
RMSE	Root Mean Square Error
RRA	Rapid Rural Appraisal
RSDP	Rice Supply and Demand Project
RUE	Radiation Use Efficiency
SALT	Sustainable Agriculture Land Technology
SANREM CRSP	Sustainable Agriculture and National Resource Management – Collaborative Research Support Program
SARP	Simulation and Systems Analysis for Rice Production
SC-DLO	DLO-Winand Staring Centre
SEBAL	Surface Energy Balance Algorithm for Land
SHIELD	Simulation of Hedgerow Intervention against Erosion and Land Degradation
SIMRIW	Simulation Model for Rice–Weather Relationships
SOI	Survey of India
SPAR	Soil Plant Atmosphere Research
SSG	Spray Strategy Group
SSM	Site Specific Management
SUCROSA	Simple and Universal CROp growth Simulator
SUSENAS	Survey of Social and Economic National
SWAT	Soil and Water Assessment Tool
SWRRB	Simulator for Water Resources in Rural Basins
SYI	Sediment Yield Index
TIGER	Terrestrial Initiative in Global Environmental Research

TLU	Tropical Livestock Units
TNAU	Tamil Nadu Agricultural University
UKMO	United Kingdom Meteorological Office
UN	United Nations
UNDP	United Nations Development Programme
USDA	United States Department of Agriculture
USDA-ARS	United States Department of Agriculture-Agricultural Research Service
USTED	Uso Sostenible de Tierras En el Desarrollo (Sustainable Land Use in Development)
VAMPP	Veterinary Automated Management and Production control Programme
VRSAP	Vietnam River Systems And Plains
WARDA	West Africa Rice Development Association
WAS	West African Sahel
WATMIS	Watershed Management Information System
WAU	Wageningen Agricultural University
WB	World Bank
WCED	World Commission on Environment and Development
WDVI	Weighted Difference Vegetation Index
WEPS	Wind Erosion Prediction System
WRR	Netherlands Scientific Council for Government Policy

Index

M.J. Kropff et al. *(eds.): Applications of Systems Approaches at the Field Level, 459–465.*

Systems Approaches for Sustainable Agricultural Development

1. Th. Alberda, H. van Keulen, N.G. Seligman and C.T. de Wit (eds.): *Food from Dry Lands.* An Integrated Approach to Planning of Agricultural Development. 1992
ISBN 0-7923-1877-3
2. F.W.T. Penning de Vries, P.S. Teng and K. Metselaar (eds.): *Systems Approaches for Agricultural Development.* Proceedings of the International Symposium (Bangkok, Thailand, December 1991). 1993
ISBN 0-7923-1880-3; Pb 0-7923-1881-1
3. P. Goldsworthy and F.W.T. Penning de Vries (eds.): *Opportunities, Use, and Transfer of Systems Research Methods in Agriculture to Developing Countries.* Proceedings of an International Workshop (The Hague, November 1993). 1994
ISBN 0-7923-3205-9
4. J. Bouma, A. Kuyvenhoven, B.A.M. Bouman, J.C. Luyten and H.G. Zandstra (eds.): *Eco-regional Approaches for Sustainable Land Use and Food Production.* Proceedings of a Symposium (The Hague, December 1994). 1995
ISBN 0-7923-3608-9
5. P.S. Teng, M.J. Kroppf, H.F.M. ten Berge, J.B. Dent, F.P. Lansigan and H.H. van Laar (eds.): *Applications of Systems Approaches at the Farm and Regional Levels.* 1996
ISBN 0-7923-4285-2
6. M.J. Kroppf, P.S. Teng, P.K. Aggarwal, J. Bouma, B.A.M. Bouman, J.W. Jones and H.H. van Laar (eds.): *Applications of Systems Approaches at the Field Level.* 1996
ISBN set volume 5 & 6: 0-7923-4287-9; ISBN 0-7923-4286-0

KLUWER ACADEMIC PUBLISHERS – DORDRECHT / BOSTON / LONDON